AF342287

Combinatorial Methods for Chemical and Biological Sensors

Integrated Analytical Systems

Series Editor:
Radislav A. Potyrailo
GE Global Research Center
Niskayuna, NY

Radislav A. Potyrailo • Vladimir M. Mirsky
Editors

Combinatorial Methods for Chemical and Biological Sensors

 Springer

Editors
Radislav A. Potyrailo
General Electric Company
Global Research Center
Niskayuna, NY 12309
USA
potyrailo@crd.ge.com

Vladimir M. Mirsky
Lausitz University of Applied Sciences
Nanobiotechnology
01968 Senftenberg
Germany
vmirsky@fh-lausitz.de

ISBN: 978-0-387-73712-6   e-ISBN: 978-0-387-73713-3
DOI: 10.1007/978-0-387-73713-3

Library of Congress Control Number: 2008940864

Cover illustration: Richard Oudt

Printed on acid-free paper

springer.com

Series Preface

> *In my career I've found that "thinking outside the box" works better if I know what's "inside the box."*
>
> Dave Grusin, composer and jazz musician

> *Different people think in different time frames: scientists think in decades, engineers think in years, and investors think in quarters.*
>
> Stan Williams, Director of Quantum Science Research, Hewlett Packard Laboratories

> *Everything can be made smaller, never mind physics;*
> *Everything can be made more efficient, never mind thermodynamics;*
> *Everything will be more expensive, never mind common sense.*
>
> Tomas Hirschfeld, pioneer of industrial spectroscopy

Integrated Analytical Systems

Series Editor: Dr. Radislav A. Potyrailo, GE Global Research, Niskayuna, NY

The book series *Integrated Analytical Systems* offers the most recent advances in all key aspects of development and applications of modern instrumentation for chemical and biological analysis. The key development aspects include: (i) innovations in sample introduction through micro- and nano-fluidic designs, (ii) new types and methods of fabrication of physical transducers and ion detectors, (iii) materials for sensors that became available due to the breakthroughs in biology, combinatorial materials science and nanotechnology, and (iv) innovative data processing and mining methodologies that provide dramatically reduced rates of false alarms.

A multidisciplinary effort is required to design and build instruments with previously unavailable capabilities for demanding new applications. Instruments with more sensitivity are required today to analyze ultra-trace levels of environmental pollutants, pathogens in water, and low vapor pressure energetic materials in air. Sensor systems with faster response times are desired to monitor transient in-vivo events and bedside patients. More selective instruments are sought to analyze specific proteins in vitro and analyze ambient urban or battlefield air. For these and many other applications, new analytical instrumentation is urgently needed. This book series is intended to be a primary source on both fundamental and practical information of where analytical instrumentation technologies are now and where they are headed in the future.

Looking back over peer-reviewed technical articles from several decades ago, one notices that the overwhelming majority of publications on chemical and biological analysis has been related to chemical and biological sensors and has originated from Departments of Chemistry in universities and Divisions of Life Sciences of governmental laboratories. Since then, the number of disciplines involved in this research field has dramatically increased because of the ever-expanding needs for miniaturization (e.g. for in-vivo cell analysis, embedding into soldier uniforms), lower power consumption (e.g. harvested power), and the ability to operate in complex environments (e.g. whole blood, industrial water, or battle-field air) for more selective, sensitive, and rapid determination of chemical and biological species. Compact analytical systems that have a sensor as one of the system components are becoming more important than individual sensors. Thus, in addition to traditional sensor approaches, a variety of new themes have been introduced to achieve an attractive goal of analyzing chemical and biological species on the micro- and nano-scale.

Foreword

Combinatorial Development
of Sensing Materials:
Where We Are and Where We Are Going

A significant portion of sensor research involves design of new sensing materials. Early sensing schemes were often based on accidental discoveries and labor-intensive rational optimization of materials. In recent years, strategies have changed greatly because rational design often has been complemented by combinatorial approaches. Combinatorial chemistry, initially developed to synthesize large libraries of organic compounds, has experienced a tremendous growth in many respects. In the pharmaceutical industry, thousands or tens of thousands of new compounds are being synthesized and tested daily. It was only a question of time until this technology was transferred to materials research, in particular to the chemistry of organic and inorganic materials for use in sensors. Now, large numbers of different types of such sensing materials can be produced. Once such libraries are available, methods for high-throughput screening of the respective materials have to be found. In fact, development of combinatorial screening tools appears to require a sizable time commitment and effort because each material is likely to be tested for its response to more than just a single analyte. However, as has already been shown by many research groups worldwide, these investments do have significant and fast pay-offs.

This book reflects the exciting developments that have been made in the areas of combinatorial chemistry and high-throughput screening of sensing materials. I have to congratulate the editors for having compiled this book on this extremely timely subject, and the authors for having contributed state-of-the-art chapters to the various sections. This volume will be an invaluable source of information to all scientists and practitioners working in this field.

Regensburg, Germany
Otto S. Wolfbeis

Preface

Sensing materials play a key role in the successful implementation of chemical and biological sensors. The multidimensional nature of the interactions between function and composition, preparation method, and end-use conditions of sensing materials often makes their rational design for real-world applications very challenging. These practical challenges in rational design of sensing materials provide tremendous potential for combinatorial and high-throughput research, which is the applied use of technologies and automation for the rapid synthesis and performance screening of relatively large number of compounds. Ideally, from these experiments, relevant descriptors are determined to understand the details of materials design and to establish quantitative structure–function relationships. This attractive goal has been reached only for a few sensing materials systems. In other cases, empirical screening and simple data mining are being used.

This book is the compilation and critical analysis of work recently performed in research laboratories around the world on the implementation of combinatorial and high-throughput research methodologies for the discovery and optimization of new sensing materials. The book will be of interest to the reader because of its several innovative aspects. First, it provides a detailed description and analysis of strategies for setting up successful processes for screening of sensing materials for both chemical and biological sensors. Second, it summarizes the advances, remaining challenges and suggests the opportunities in combinatorial research in the areas chemical and biosensors based on polymeric, inorganic, biological, and formulated sensing materials. Third, it provides an insight into how to improve the efficiency of the combinatorial screening of sensing materials and generation of new knowledge by combining combinatorial experimentation and advanced data mining techniques.

This book is intended to be a primary source on both fundamental and practical information of where high-throughput analysis technologies are now and where they are headed in the future. It is addressed to the rapidly growing number of active practitioners and those who are interested in starting research in the field of sensing materials for chemical sensors and biosensors, directors of industrial and government research centers, laboratory supervisors and managers, students and lecturers. Combinatorial and high-throughput materials screening approaches analyzed in this book will be also of interest to researchers from many other fields of experimental science working on materials design. The structure of this book offers a basis for a high-throughput instrumentation course at advanced undergraduate or graduate level.

We thank Ken Howell at Springer for his patience during the implementation of this book project and for encouraging us during the various stages of this project. We express our gratitude to the authors for their efforts in preparing their chapters on time and to the referees for reviewing them. R.A.P. thanks the leadership team at GE Global Research for supporting the whole GE Combinatorial Chemistry effort. V.M.M. is grateful to Prof. O.S. Wolfbeis for initiating the development of combinatorial approaches on discovery and optimization of sensing materials in the University of Regensburg. Last, but not least, we thank our families for their patience and enthusiastic support of this project.

Niskayuna, NY   Radislav A. Potyrailo
Senftenberg, Germany   Vladimir M. Mirsky

Contents

Section 4   Biological Receptors

Section 5   Inorganic Gas-Sensing Materials

Section 6   Electrochemical Synthesis of Sensing Materials

Section 7   Optical Sensing Materials

Contributors

Lourdes Basabe-Desmonts
Department of Supramolecular Chemistry and Technology, MESA+ Institute
for Nanotechnology, University of Twente, P.O. Box 217, 7500 AE Enschede,
The Netherlands

Itai Benhar
Department of Molecular Microbiology and Biotechnology, The George S. Wise
Faculty of Life Sciences, Green Building, Room 202, Tel-Aviv University,
Ramat Aviv 69978, Israel
benhar@post.tau.ac.il

Sabine Borgmann
Technische Universität Dortmund, Fachbereich Chemie Biologisch-Chemische
Mikrostrukturtechnik, Otto-Hahn Strasse 6, D-44227 Dortmund, Germany
sabine.borgmann@uni-dortmund.de

Frank V. Bright
Department of Chemistry, The State University of New York, Buffalo,
NY 14260-3000, USA
chefvb@buffalo.edu

Ellen M. Cardone
Department of Chemistry, The State University of New York, Buffalo,
NY 14260-3000, USA
ecardone@buffalo.edu

Young-Tae Chang
Department of Chemistry, and MedChem Program of the Office of Life Sciences,
National University of Singapore, Singapore 117543
and
Laboratory of Bioimaging Probe Development, Singapore Bioimaging
Consortium, Agency for Science, Technology and Research (A*STAR),
Biopolis, Singapore 138667.
chmcyt@nus.edu.sg

Pyeong-Seok Cho
Korea University, Department of Materials Science and Engineering,
Seoul 136-713, Korea
pcaat97@korea.ac.kr

Mercedes Crego-Calama
Department of Supramolecular Chemistry and Technology, MESA+ Institute
for Nanotechnology, University of Twente, P.O. Box 217, 7500 AE Enschede,
m.crego-calama@tnw.utwente.nl

M.E. Díaz-García
Department of Physical and Analytical Chemistry, Faculty of Chemistry,
University of Oviedo, Av. Julián Clavería 8, 33006, Oviedo, Spain
medg@uniovi.es

Masatora Fukuda
Institute of Advanced Energy, Graduate School of Energy Science, Department
of Fundamental Energy Science, Kyoto University Uji, Kyoto 611-0011, Japan

Tetsuya Hasegawa
Institute of Advanced Energy, Graduate School of Energy Science, Department
of Fundamental Energy Science, Kyoto University Uji, Kyoto 611-0011, Japan

Hironori Hayashi
Institute of Advanced Energy, Graduate School of Energy Science, Department
of Fundamental Energy Science, Kyoto University Uji, Kyoto 611-0011, Japan

Tomoya Hirano
Tokyo Medical and Dental University, School of Biomedical Sciences, Tokyo
Medical and Dental University, 2-3-10 Kanda-Surugadai, Chiyoda-ku,
Tokyo 101-0062, Japan
hiraomc@tmd.ac.jp

William G. Holthoff
Joint Expeditionary Forensics Program, Naval Surface Warfare Center Dahlgren,
Asymmetric Operations Technology Branch (Z11), Dahlgren, VA 22448 – 5160,
USA
william.holthoff@navy.mil

Hiroyuki Kagechika
Tokyo Medical and Dental University, School of Biomedical Sciences, Tokyo
Medical and Dental University 2-3-10 Kanda-Surugadai, Chiyoda-ku,
Tokyo 101-0062, Japan
kageomc@tmd.ac.jp

Sun-Jung Kim
Korea University, Department of Materials Science and Engineering,
Seoul 136-713, Korea
jongheun@korea.ac.kr

Francesca Lanza
INFU, University of Dortmund, Otto Hahn Strasse 6, D-44221 Dortmund, Germany

Andrew M. Leach
General Electric Company, Global Research Center, Niskayuna, NY 12309, USA
leach@research.ge.com

Jong-Heun Lee
Korea University, Department of Materials Science and Engineering,
Seoul 136-713, Korea
jongheun@korea.ac.kr

Bo Liedberg
University of Linköping, Division of Molecular Physics, Department of Physics,
Chemistry and Biology, Linköping University, SE-581 83 Linköping, Sweden
bol@ifm.liu.se

G. Pina Luis
Department of Physical and Analytical Chemistry, Faculty of Chemistry,
University of Oviedo, Av. Julián Clavería 8, 33006, Oviedo, Spain

Patrick J. McCloskey
General Electric Company, Global Research Center, Niskayuna, NY 12309, USA
mccloskey@crd.ge.com

Vladimir M. Mirsky
Lausitz University of Applied Sciences, Nanobiotechnology, 01968 Senftenberg,
Germany
vmirsky@fh-lausitz.de

Jun Miyake
Research Institute of Cell Engineering, National Institute of Advanced Industrial
Science and Technology, 1-8-31 Midorigaoka, Ikeda 563-8577 Osaka, Japan
jun-miyake@aist.go.jp

Takashi Morii
Institute of Advanced Energy, Graduate School of Energy Science, Department
of Fundamental Energy Science, Kyoto University Uji, Kyoto 611-0011, Japan
t-morii@iae.kyoto-u.ac.jp

William G. Morris
General Electric Company, Global Research Center, Niskayuna, NY 12309, USA
morris@crd.ge.com

Chikashi Nakamura
Research Institute for Cell Engineering, National Institute of Advanced Industrial
Science and Technology, AIST Tsukuba Central 6, Tsukuba, Ibaraki, Japan
chikashi-nakamura@aist.go.jp

Sergey A. Piletsky
Cranfield Biotechnology Center, Cranfield, University at Silsoe, Bedfordshire,
MK45 4DT, UK
s.piletsky@cranfield.ac.uk

Radislav A. Potyrailo
General Electric Company, Global Research Center, Niskayuna, NY 12309, USA
potyrailo@crd.ge.com

David N. Reinhoudt
Department of Supramolecular Chemistry and Technology, MESA+ Institute for
Nanotechnology, UnivFersity of Twente, P.O. Box 217, 7500 AE Enschede,
The Netherlands

Michael Riepl
Olympus Life Science Research Europa GmbH, Sauerbruchstr. 50,
D-81377 Munich, Germany
michael.riepl@olympus-europa.com

I.A. Rivero-Espejel
Technological Institute of Tijuana, P.O. 1166, Tijuana, 22000, Baja California,
Mexico
iare@tectijuana.mx

Margaret A. Ryan
Jet Propulsion Laboratory, California Institute of Technology, Pasadena,
CA 91109, USA
Margaret.A.Ryan@jpl.nasa.gov

Michael Schäferling
University of Regensburg, Institute of Analytical Chemistry,
Chemo- and Biosensors, D-93040 Regensburg, Germany
michael.schaeferling@chemie.uni-regensburg.de

Eric Schillinger
INFU, University of Dortmund, Otto Hahn Strasse 6, D-44221 Dortmund, Germany

Wolfgang Schuhmann
Ruhr-University Bochum, Analytical Chemistry - Electroanalysis and Sensors,
Universitätsstr. 150, D-44780 Bochum, Germany
wolfgang.schuhmann@rub.de

Börje Sellergren
INFU, University of Dortmund, Otto Hahn Strasse 6, D-44221 Dortmund,
Germany
b.sellergren@infu.uni-dortmund.de

Abhijit V. Shevade
Jet Propulsion Laboratory, California Institute of Technology,
Pasadena, CA 91109, USA
Abhijit.V.Shevade@jpl.nasa.gov

Ellen L. Shughart
U.S. Army Research Laboratory, AMSRD-ARL-SE-EO, Adelphi,
MD 20783 – 1138, USA
ellen.holthoff@us.army.mil

Maike Siemons
RWTH Aachen University, Institute of Inorganic Chemistry,
D-52056 Aachen, Germany
maike.siemons@ac.rwth-aachen.de

Ulrich Simon
RWTH Aachen University, Institute of Inorganic Chemistry,
D-52056 Aachen, Germany
ulrich.simon@ac.rwth-aachen.de

Sreenath Subrahmanyam
Cranfield Biotechnology Center, Cranfield University at Silsoe, Bedfordshire,
MK45 4DT, UK
sri@cranfield.ac.uk

Loraine T. Tan
Department of Chemistry, University at Buffalo, The State University
of New York, Buffalo, NY 14260-3000, USA
lttan@buffalo.edu

Marek Trojanowicz
University of Warsaw, Department of Chemistry, ul. Pasteura 1,
02-093 Warsaw, Poland
trojan@chem.uw.edu.pl

Shenliang Wang
Department of Chemistry, New York University, NY 10003, USA
sw695@nyu.edu

Ronald J. Wroczynski
General Electric Company, Global Research Center, Niskayuna, NY 12309, USA
Wrocznski@crd.ge.com

Section 1
Benefits of Combinatorial Approaches to Sensor Problems

Chapter 1
Introduction to Combinatorial Methods for Chemical and Biological Sensors

Radislav A. Potyrailo and Vladimir M. Mirsky

Abstract Sensing materials play a critical role in advancing selectivity, response speed, and sensitivity of chemical and biological determinations in gases and liquids. The desirable capabilities of sensing materials originate from their numerous functional parameters, which can be tailored to meet specific sensing needs. By increasing the structural and functional complexity of sensing materials, the ability to rationally define the precise requirements that will result in desired materials properties becomes increasingly limited. Combinatorial experimentation methodologies impact all areas of sensing materials research including inorganic, organic, and biological sensing materials.

1 Introduction

A modern sensor system incorporates several key components such as a sample delivery unit, sensing material, transducer, and data processor. Developing sensing materials is a recognized challenge because it requires extensive experimentation not only to achieve the best short-term performance, but also long-term stability, manufacturability, cost, and other practical issues. In a sensor device, a sensing material is applied onto a suitable physical transducer to convert a change in a property of a sensing material into a suitable physical signal. The signal obtained from a single transducer or from an array of transducers is further processed to provide useful information about the identity and concentration of species in the sample.

Compared with sensing based on intrinsic analyte properties (e.g., spectroscopic, dielectric, paramagnetic), sensing that utilizes a responsive sensing material[1–8] dramatically expands the range of detected species, improves sensor performance (for example,

R.A. Potyrailo (✉)
Chemical and Biological Sensing Laboratory, Chemistry Technologies and Material Characterization, General Electric Global Research, Niskayuna, New York, NY 12309, USA
potyrailo@crd.ge.com

V.M. Mirsky
Lausitz University of Applied Sciences,
Department of Nanobiotechnology, 01968 Senftenberg, Germany
vmirsky@fh-lausitz.de

R.A. Potyrailo and V.M. Mirsky (eds.), *Combinatorial Methods for Chemical and Biological Sensors*, DOI: 10.1007/978-0-387-73713-3_1,
© Springer Science + Business Media, LLC 2009

analyte detection limits and analysis time), and is more straightforwardly adaptable for miniaturization.[9–29] These attractive features can be offset by some limitations, for example, insufficient selectivity, poisoning, poor long-term stability, dependence of response rate on analyte concentration and long recovery time.

Sensing materials can be categorized into three general groups that include inorganic, organic, and biological materials. We define inorganic sensing materials as materials that have inorganic signal-generation components (e.g., metals, metal oxides, semiconductor nanocrystals) that may or may not be further incorporated into a matrix. Organic sensing materials comprise indicator dyes, polymer/reagent compositions, conjugated polymers, and molecularly imprinted polymers. Biological sensing materials include receptors such as aptamers, peptides, antibodies, enzymes, etc.

In this chapter, we set a stage with the general principles of combinatorial screening technologies of sensing materials. New parallel synthesis and advanced analytical instruments and data mining tools accelerate discovery and optimization of sensing materials and provide more fundamental knowledge on the material fabrication with tailored initial and the long-term stability properties.

2 Challenges in Rational Design of Sensing Materials

Rational design of sensing materials based on prior knowledge is a very attractive approach because it could avoid time-consuming synthesis and testing of numerous materials candidates.[30–32] However, to be quantitatively successful, rational design[33–38] requires detailed knowledge regarding relation of the intrinsic properties of sensing materials to their performance properties (e.g., affinity to an analyte and interferences, kinetic constants of analyte binding and dissociation, long-term stability, shelf life, resistance to poisoning, best operation temperature, decrease of reagent performance after immobilization, etc.). This knowledge is often obtained from extensive experimental and simulation data. Some examples of successful results of rational design of sensing materials are presented in Table 1.1.

However, with the increase of structural and functional complexity of materials, the ability to rationally define the precise requirements that result in a desired set of properties becomes increasingly limited.[51] Thus, in addition to rational design, a variety of sensing materials ranging from dyes and ionophores, to biopolymers, organic and hybrid polymers, and to nanomaterials have been discovered using detailed experimental observations or simply by chance.[52–60] Such an approach in development of sensing materials reflects a more general situation in materials design that is "still too dependent on serendipity" with only limited capability for rational materials design as recently noted by Eberhart and Clougherty.[61]

Conventionally, the detailed experimentation with sensing materials candidates for their screening and optimization consumes tremendous amount of time and project cost without adding to the "intellectual satisfaction." Fortunately, new synthetic and measurement principles and instrumentation significantly accelerate the development of new materials. The practical challenges in rational sensor material design also provide tremendous prospects for combinatorial materials research.

Table 1.1 Examples of rational design of sensing materials

Group of sensing materials	Rational design	Required knowledge	Ref.
Inorganic	Wavelength of the plasmon excitation in nanoparticles	Physical models for light interaction with appropriate particle structure	39
Inorganic	Gas sensitive materials based on carbon nanotubes	Doping level of impurity atoms in a nanotube	40
Inorganic	Gas response of polycrystalline semiconducting metal oxides	Thermo-mechanical and thermo-chemical effects on percolation-dependent electron transport	41
Inorganic	Hydrophobic coating of gold chemoresistor for rejection of polar interferents	Polarity of analyte and interferents, design of self-assembled monolayers	42
Organic	General spectral features of dyes for sensing applications	Quantum chemical calculations and empirical data	43
Organic	Vapor-sorbing sensing materials	Vapor partition coefficients between air and sensing materials	44
Organic	Polymer matrix components for molecular imprinting	Empirical data on analyte binding to different polymerizable monomers	45
Organic	Electrical wiring of chemosensitive layers by a conducting polymer	Chemosensitive and conductive properties of conjugated polymers	46
Biological	Accuracy of biosensors based on immobilized enzymes	Hydrogen bonding effects between immobilization matrix and enzymes	47
Biological	Application of immobilized DNA as a receptor for enrofloxacines	Data on binding of fluoroqinolones with nucleic acids	48
Biological	Selectivity and sensitivity of immobilized aptamers	Nature and length of covalent linker for immobilization	49
Biological	Biosensors with thermally stable enzyme bioreceptors	Quaternary structure modeling of known tertiary structures of related proteins	50

3 General Principles of Combinatorial Materials Screening

In pharmaceutical and biotechnology industries, combinatorial synthesis and high-throughput analysis methods have found their applications in systematic searching of large parameter spaces for new candidate therapeutic agent molecules.[62] Combinatorial chemistry originated in several laboratories around the world where Frank in Germany,[63] Geysen in Australia,[64] and Houghten in USA[65] developed methods to make more compounds in a shorter period of time.[66]

In materials science, the materials properties depend not only on composition, but also on morphology, microstructure, and other parameters related to the material-preparation conditions and on the end-use environment. As a result of this complexity, a true combinatorial experimentation is rarely performed in materials science with a complete set of materials and process variables explored. Instead, carefully selected subsets of the parameters are often explored in an automated parallel or rapid sequential fashion using high-throughput experimentation (HTE). Nevertheless, the terms "combinatorial chemistry" and "combinatorial materials science" are often applied for all types of automated parallel and rapid sequential materials and process parameters evaluation processes. Thus, an adequate definition of combinatorial and high-throughput materials science is a process that couples the capability for parallel production of large arrays of diverse materials together with different high-throughput measurement techniques for various intrinsic and performance properties followed by the navigation in the collected data for identifying "lead" materials.[67–75]

Individual aspects of accelerated materials development have been known for decades. These include combinatorial and factorial experimental designs,[76] parallel synthesis of materials on a single substrate,[77,78] screening of materials for performance properties,[79] and computer data processing.[80,81] However, it took the innovative scientific vision of Joseph Hanak to suggest in 1970 an integrated materials-development workflow.[82] Its key aspects included (1) complete compositional mapping of a multicomponent system in one experiment, (2) simple rapid nondestructive all-inclusive chemical analysis, (3) testing of properties by a scanning device, and (4) computer data processing. Hanak was truly ahead of his time and "it took 25 years for the world to realize his idea."[68]

In 1995, Xiang, Schultz, and coworkers initiated applications of combinatorial methodologies in materials science with the publication of their paper "A combinatorial approach to materials discovery."[83] Since then, combinatorial materials science has enjoyed much success, rapid progress for over a decade, and tremendous diversification into a wide variety of types of materials. Besides sensing materials, discussed in this book, examples of materials developed using combinatorial and high-throughput screening techniques include superconductor,[83] ferroelectric,[84] magnetoresistive,[85] luminescent,[86] agricultural,[87] structural,[88] hydrogen storage,[89] and organic light-emitting[90] materials; ferromagnetic[91] and thermoelastic[92] shape-memory alloys; heterogeneous,[93] homogeneous,[94] polymerization,[95] and electrochemical[96] catalysts and electrocatalysts for hydrogen evolution[97]; polymers[98]; zeolites[99]; metal alloys[100]; materials for methanol fuel cells,[101] solid oxide fuel cells,[102] and solar cells[103]; automotive,[104] waterborne,[105] vapor-barrier,[106] marine,[107] and fouling-release[108] coatings.

A typical modern combinatorial materials development cycle is outlined in Fig. 1.1. Compared with the initial idea of Hanak,[82] the modern workflow has several new important aspects such as planning of experiments, data mining, and scale up. In combinatorial screening of materials, concepts originally thought as highly automated have been recently refined to have more human input, with only an appropriate level of automation. For the throughput of 50–100 materials formulations per day, it is acceptable to perform certain aspects of the process manually.[109,110] To address numerous materials-specific properties, a variety of high-throughput characterization tools are required. Characterization tools are used for rapid and automated assessment of single or multiple properties of the large number of samples fabricated together as a combinatorial array or "library."[71,111,112]

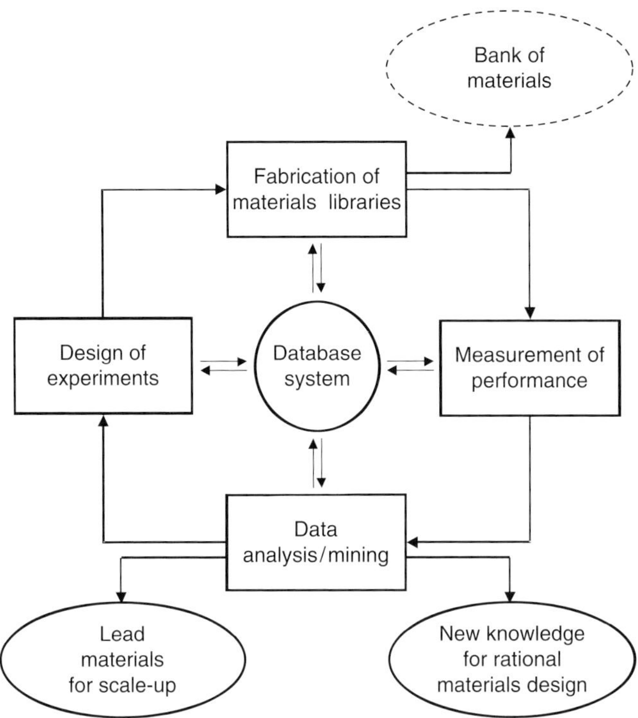

Fig. 1.1 Typical cycle in combinatorial and high-throughput approach for materials development

In addition to the parallel synthesis and high-throughput characterization instrumentation that significantly differ from conventional equipment, the data management approaches are also different from conventional data evaluation.[75] In an ideal combinatorial workflow, one should "analyze in a day what is made in a day"[113] that requires significant computational assistance. In an exemplary combinatorial workflow (Fig. 1.1), design and syntheses protocols for materials libraries are computer assisted, materials synthesis and library preparation are carried out with computer-controlled manipulators, and property screening and materials characterization are also software controlled. Further, materials synthesis data as well as property and characterization data are collected into a materials database. This database contains information on starting components, their descriptors, process conditions, materials testing algorithms, and performance properties of libraries of sensing materials. Data in such a database is not just stored, but also processed with the proper statistical analysis, visualization, modeling, and data-mining tools.

Combinatorial synthesis of materials provides a good possibility for formation of banks of combinatorial materials (Fig. 1.1). Such banks can be used, for example, for further reinvestigation of the materials of interest for some new applications or as reference materials. However, this approach did not find wide applications yet in combinatorial materials science.

4   Opportunities for Sensing Materials

Development of a sensor system with a new sensing material includes several phases such as discovery with initial observations, feasibility experimentation, and laboratory scale detailed evaluation, followed by the transition to the pilot scale and to commercial manufacturing (Fig. 1.2). At the initial stage, performance of the sensing material is matched with the appropriate transducer for the signal generation. The stage of the laboratory scale evaluation is very labor-intensive because it involves a detailed testing of sensor performance. Some of the aspects of this evaluation include optimization of the sensing material composition and morphology, its deposition method, detailed evaluation of response accuracy, stability, precision, selectivity, shelf-life, long-term stability of the response, key noise parameters (e.g., material instability because of temperature, potential poisons), and potential environmental health and safety (EHS) issues with composition and manufacturing of sensing materials. The pilot scale manufacturing focuses on the identification and elimination of manufacturing issues related to the reproducible, high-yield manufacturing of the sensors. During this phase, alpha and beta trials are typically performed. The alpha trials are typically performed by the researchers on an advanced sensor device prototype to identify issues related to sensor operation and functionality. Beta trials are typically performed on the further improved version of the sensor system by the identified group of end-users to seek their feedback on sensor performance, ease of use, failure modes, etc.

It is difficult to predict quantitatively numerous sensor material parameters using rational approaches (Table 1.2). Relevant descriptors must be determined to under-

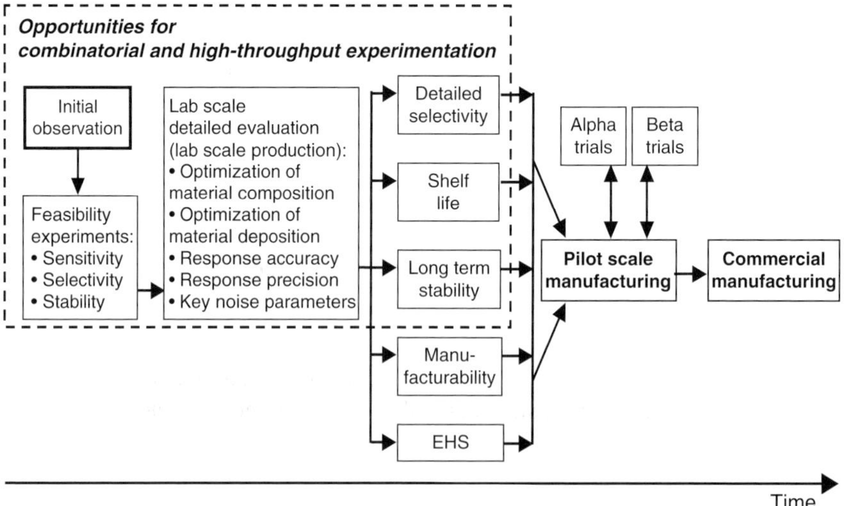

Fig. 1.2 Development phases of new sensing materials in sensors and opportunities for combinatorial and high-throughput experimentation

Table 1.2 Parameters relevant to sensor applications that are difficult or impossible to quantitatively predict and calculate using existing knowledge and rational approaches

Group of sensing materials	Parameter	Ref.
Inorganic	Fundamental effects of volume dopants on base metal oxide materials	114
Inorganic	Hot spots in surface-enhanced Raman scattering	115
Inorganic	Room-temperature gas response of metal oxide nanomaterials	116
Organic	Selectivity of prospective ionophores	3
Organic	Design of polymers for selective complexation with desired metal ions	117
Organic	Collective effects of preparation method of conducting polymers, polymer morphology, properties of the substrate/film interface	24
Organic	Surface ratio of different types of molecules in mixed self-assembled monolayers formed in quasi-equilibrium conditions	118
Biological	Oligonucleotide sequence in aptamers for specific and sensitive target binding	119
Biological	Activity of immobilized bioreceptors	120
Biological	Selection of coupling reagents for immobilization of antibodies	121
Biological	Long-term stability of enzyme-containing polymer films	122
Biological	Affinity properties of antibodies (kinetic association and dissociation constants, binding constant)	123

stand the details of sensor materials design and to establish quantitative structure–function relationships. Although applying combinatorial and high-throughput screening tools can accelerate this process, the number of purely serendipitous combinations of sensor materials components is simply too large to handle even with the "ultra-high throughput screening" in a time and cost-effective manner. It is not feasible to synthesize all possible molecules and apply all possible process conditions to characterize materials function. In this case, a methodology that allows the most promising candidates to be short-listed should be applied. In materials science, focused libraries are often designed, produced, and tested in a high-throughput mode for a subset of a truly "combinatorial" space where the initial subset selection is performed using rational and intuitive approaches.

5 Designs of Combinatorial Libraries of Sensing Materials

Variations of individual parameters of sensing materials result in a modification of a number of response parameters. Thus, the goal of combinatorial and high-throughput development of sensing materials is to determine the structure–function relationships in sensing materials. In combinatorial screening, discrete[76,83] and gradient[77,82,129, 130, 131] arrays (libraries) of sensing materials are employed. A specific type of library layout will depend on the required density of space to be explored, available library-fabrication capabilities, and capabilities of high-throughput characterization techniques.

Discrete or gradient sensor regions are attractive for a variety of different specific applications and can be produced with a variety of tools. Discrete sensor regions are straightforward to produce using a variety of approaches, for example chemical vapor deposition (Fig. 1.3a), pulsed-laser deposition (Fig. 1.3b), dip coating (Fig. 1.3c), spin coating (Fig. 1.3d), electropolymerization (Fig. 1.3e), slurry dispensing (Fig. 1.3f), liquid dispensing (Fig. 1.3g), screen printing (Fig. 1.3h), and some others.

Sensor material optimization can be performed using gradient sensor materials. Spatial gradients in sensing materials can be generated by varying nature and concentration of starting components, processing conditions, thickness, and some others. Gradient sensor regions can be produced using a variety of approaches, for example in situ photopolymerization (Fig. 1.4a), microextrusion (Fig. 1.4b), solvent casting (Fig. 1.4c), self-assembly (Fig. 1.4d), temperature-gradient chemical vapor deposition (Fig. 1.4e), thickness-gradient chemical vapor deposition (Fig. 1.4f), 2-D thickness gradient evaporation of two metals (Fig. 1.4g), gradient surface coverage and gradient particle size (Fig. 1.4h), and some others.

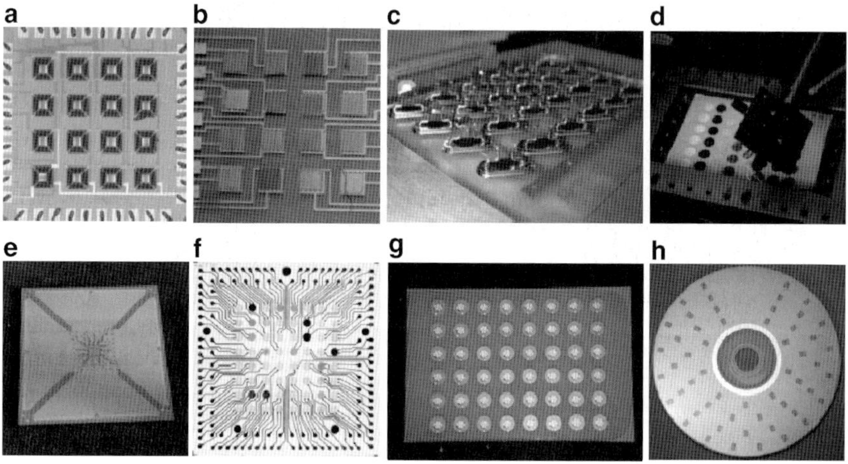

Fig. 1.3 Examples of fabrication methods of discrete sensor arrays (**a**) Chemical vapor deposition: Array of 16 TiO_2 films fabricated at different deposition temperatures. (**b**) Pulsed-laser deposition: Array of 16 compositions of the SnO_2 base material with different dopants. (**c**) Dip coating: A 24-element array of polycarbonate copolymers. (**d**) Spin coating: A 48-element array of polymeric compositions formulated with fluorescent reagents for optical vapor sensing. (**e**) Electropolymerization: A 96-element interdigital electrode array with electropolymerized conductive polymers. (**f**) Robotic slurry dispensing: An array of 64 WO_3 sensing films with different bulk- and surface-dopants. (**g**) Liquid dispensing: A 48-element array of formulated solid polymer electrolyte compositions deposited onto radio-frequency identification sensors. (**h**) Screen printing: A 54-element array of polymeric formulated films for optical ion sensing. (**a**) Reprinted with permission from Taylor and Semancik[132]. (**b**) Reprinted with permission from Aronova et al[133]. Copyright 2003 American Institute of Physics. (**c**) Reprinted with permission from Potyrailo.[134] Copyright 2004 Wiley-VCH Publishers. (**d**) Reprinted with permission from Amis[135]. (**f**) Reprinted with permission from Scheidtmann et al[136]. Copyright 2005 Institute of Physics Publishing. (**h**) Reprinted with permission from Potyrailo et al[137]. Copyright 2007 Optical Society of America

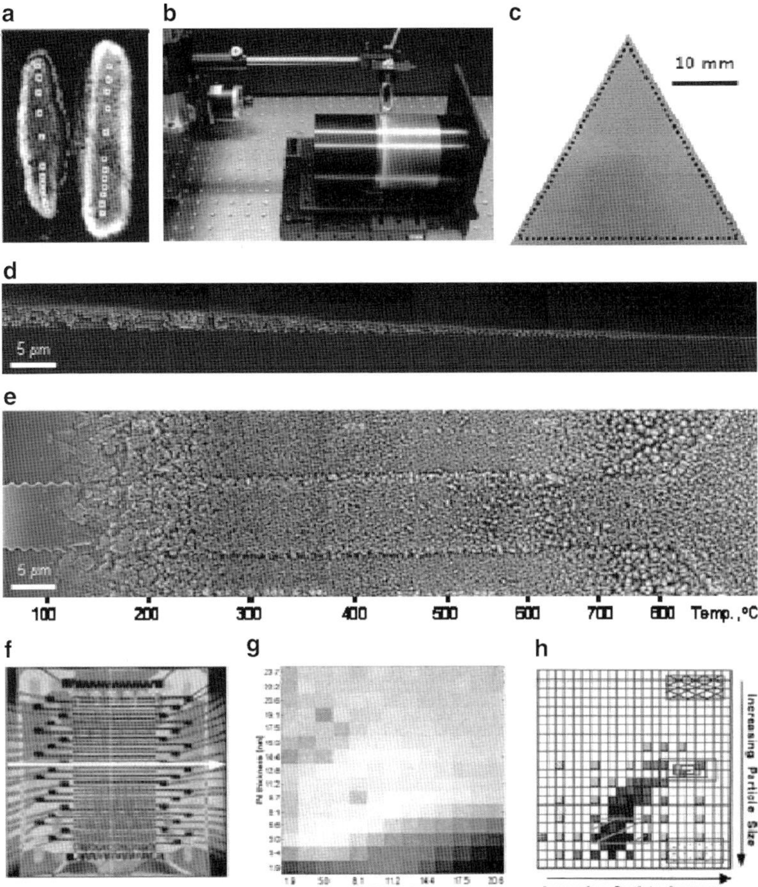

Fig. 1.4 Examples of fabrication methods of gradient sensor arrays: (**a**) In situ photopolymerization: The left stripe is a polymerized gradient array of siloxanes 1 and 2, beginning at the base with siloxane 1 alone and ending at a final monomer ratio of siloxanes 1 and 2. The right stripe is siloxane 1 as control; (**b**) Microextrusion: A coiled 1-D polymer fiber array with a gradient concentration of a temperature-sensitive phosphor along its length; (**c**) Solvent casting: Ternary combination of gradient concentrations of fluorescent reagents in a formulated optical sensor film; (**d**) Self-assembly: A colloidal crystal sensing film of gradient thickness; (**e**) Temperature-gradient chemical vapor deposition. A range of microstructures was produced in a metal oxide film; (**f**) Thickness-gradient chemical vapor deposition. Variable-thickness SiO_2 membrane on top of a metal oxide sensing film in combination with a temperature gradient sensor operation; (**g**) 2-D thickness gradient evaporation. A two metal combination with variable thickness of each metal along two orthogonal directions; (**h**) 2-D immersion: Gradient surface coverage and gradient nanoparticle size along two orthogonal directions in SERS array. (**a**) Reprinted with permission from Dickinson et al[138]. Copyright 1997 American Chemical Society. (**b**) Reprinted with permission from Potyrailo and Wroczynski[139]. Copyright 2005 American Institute of Physics. (**e**) Reprinted with permission from Taylor and Semancik[132]. (**f**) Reprinted with permission from Sysoev et al[140]. Copyright 2004 Molecular Diversity Preservation International. (**g**) Reprinted with permission from Klingvall et al[141]. Copyright 2005 The Institute of Electrical and Electronics Engineers, Inc. (**h**) Reprinted with permission from Baker et al[142]. Copyright 1996 American Chemical Society

Once the gradient or discrete sensing materials array is fabricated, it is exposed to an environment of interest and steady-state or dynamic measurements are acquired. Serial scanning mode of analysis (e.g., optical or impedance spectroscopies) is often performed to provide more detailed information about materials property over parallel analysis (e.g., imaging). When monitoring a dynamic process (e.g., response/recovery time and aging) of sensor materials arranged in an array with a scanning system, the maximum number of elements in sensor library that can be measured with the required temporal resolution can be limited by the data-acquisition ability of the scanning system.[143]

6 Diversity in Needs for Combinatorial Development of Sensing Materials

An appropriate sensing material is based on a successful combination of two somewhat contradicting major material requirements. The first requirement is to have desired material response to the changes in the concentration of species of interest in a sample. It is desirable for the most applications that the response is fast and selective; therefore, the kinetic association constant and the binding constant (the ratio of the kinetic association and kinetic dissociation constants) should be high. The second requirement is to have a reversible sensor response; therefore, the kinetic dissociation constant should be also high. Thus, development of sensing materials needs knowledge on how to fulfill these controversial requirements or to find a suitable compromise.

For diverse applications, specifications on sensing materials are often weighted differently according to an application. High reliability, adequate long-term stability, and resolution top the priority list for industrial sensor users, while often the size and maturity of the technology are the least important factors.[2,11,144] The low false positive rate is very critical for the first responders.[145] In contrast, medical users focus on cost for disposable sensors. Specific requirements for medical in vivo sensors include blood compatibility and minute size.[146] Resistance to gamma radiation during sterilization, the drift-free performance, and cost are the most critical specific requirements for sensors in disposable bioprocess components.[147] The importance of continuous monitoring also differs from application to application. For instance, glucose sensing should be performed 2–4 times a day using home blood glucose biosensors, while blood-gas sensors for use in intensive care should be capable of continuous monitoring with subsecond time resolution.[148,149]

Table 1.3 provides a summary of parameters that should be optimized and controlled for successful development of inorganic, organic, and biological sensing materials. Inorganic sensing materials include catalytic metals for field-effect devices,[141,150,151] metal oxides for conductometric[7,114,132,133,140,152–164] and cataluminescent[165,166] sensors, plasmonic,[142,167–174] and semiconductor nanocrystal[175–177] materials. Organic sensing materials include indicator dyes (free, polymer immobilized, and surface-confined),[178–192] polymeric compositions,[124,125,138,193–195] homo and copolymers,[196–202] conjugated polymers,[203–207] and molecularly-imprinted polymers.[118,208–211] Biological materials include surface and matrix-immobilized bioreceptors.[120,203,212] At present, to find the most appropriate sensing materials, combinatorial methods have become widely accepted.

Table 1.3 Parameters that should be optimized and controlled for successful development of inorganic, organic, and biological sensing materials

Group of sensing materials	Type of sensing material	Optimized and controlled material parameters
Inorganic	Catalytic metals	Surface additives
		Porosity
		Layered structure
		Grain size
		Alloying
		Deposition method
Inorganic	Metal oxide materials	Base single or mixed metal oxides
		Deposition method and conditions of base metal oxide(s)
		Annealing method and conditions
		Dopant(s)
		Doping method and conditions
		Purity of materials
Inorganic	Plasmonic nanostructures and nanoparticles	Substrate type
		Nanoparticle material
		Nanoparticle shape, size, morphology
		Nanoparticles arrangement
		Surface functionality
Inorganic	Plasmonic nanoparticles in polymers	Size of nanoparticle
		Strength of polymer/particle interaction
		Polymer grafting density
		Polymer chain length
Organic	Indicators	Binding constant
		pH-influence
		Redox state
		Selectivity
		Toxicity
		Poisoning agents
Organic	Polymeric compositions	Analyte-responsive reagent
		Polymer matrix
		Analyte-specific ligand
		Plasticizer
		Other agents (stabilizing, phase transfer, etc.)
		Common solvent
Organic	Conjugated polymers	Polymerization conditions
		Types of heterocycles
		Additive(s)
		Side groups
		Dopant
		Oxidation state
		Electrode material
		Thickness
		Morphology

(continued)

Table 1.3 (continued)

Group of sensing materials	Type of sensing material	Optimized and controlled material parameters
Organic	Molecularly imprinted polymers	Functional monomer(s) Template concentration Cross-linker Porogen Monomer(s)/template ratio Physical conditions during polymerization
Biological	Surface-immobilized bioreceptors	Immobilization technique Receptor-surface spacer Receptor-receptor spacer
Biological	Matrix-immobilized bioreceptors	Immobilization technique Receptor density Matrix hydrophilicity Matrix charge Matrix chemical composition Matrix thickness

References

1. Janata, J. *Principles of Chemical Sensors*; Plenum: New York, NY, 1989
2. *Fiber Optic Chemical Sensors and Biosensors*; Wolfbeis, O. S., Ed.; CRC: Boca Raton, FL, 1991
3. Bakker, E.; Bühlmann, P.; Pretsch, E., Carrier-based ion-selective electrodes and bulk optodes. 1. General characteristics, *Chem. Rev.* **1997**, *97*, 3083–3132
4. Potyrailo, R. A.; Hobbs, S. E.; Hieftje, G. M., Optical waveguide sensors in analytical chemistry: Today's instrumentation, applications and future development trends, *Fresenius' J. Anal. Chem.* **1998**, *362*, 349–373
5. Janata, J.; Josowicz, M.; Vanysek, P.; DeVaney, D. M., Chemical sensors, *Anal. Chem.* **1998**, *70*, 179R–208R
6. Wolfbeis, O. S., Fiber-optic chemical sensors and biosensors, *Anal. Chem.* **2006**, *78*, 3859–3874
7. Franke, M. E.; Koplin, T. J.; Simon, U., Metal and metal oxide nanoparticles in chemiresistors: Does the nanoscale matter?, *Small* **2006**, *2*, 36–50
8. Potyrailo, R. A., Polymeric sensor materials: Toward an alliance of combinatorial and rational design tools ?, *Angew. Chem. Int. Ed.* **2006**, *45*, 702–723
9. Bergman, I., Rapid-response atmospheric oxygen monitor based on fluorescence quenching, *Nature* **1968**, *218*, 396
10. Hardy, E. E.; David, D. J.; Kapany, N. S.; Unterleitner, F. C., Coated optical guides for spectrophotometry of chemical reactions, *Nature* **1975**, *257*, 666–667
11. Hirschfeld, T.; Callis, J. B.; Kowalski, B. R., Chemical sensing in process analysis, *Science* **1984**, *226*, 312–318
12. Peterson, J. I.; Vurek, G. G., Fiber-optic sensors for biomedical applications, *Science* **1984**, *224*, 123–127
13. Barnard, S. M.; Walt, D. R., A fibre-optic chemical sensor with discrete sensing sites, *Nature* **1991**, *353*, 338–340
14. Tan, W.; Shi, Z.-Y.; Smith, S.; Birnbaum, D.; Kopelman, R., Submicrometer intracellular chemical optical fiber sensors, *Science* **1992**, *258*, 778–781
15. Charych, D. H.; Nagy, J. O.; Spevak, W.; Bednarski, M. D., Direct colorimetric detection of a receptor–ligand interaction by a polymerized bilayer assembly, *Science* **1993**, *261*, 585–588

16. Dickinson, T. A.; White, J.; Kauer, J. S.; Walt, D. R., A chemical-detecting system based on a cross-reactive optical sensor array, *Nature* **1996**, *382*, 697–700
17. Holtz, J. H.; Asher, S. A., Polymerized colloidal crystal hydrogel films as intelligent chemical sensing materials, *Nature* **1997**, *389*, 829–832
18. Elghanian, R.; Storhoff, J. J.; Mucic, R. C.; Letsinger, R. L.; Mirkin, C. A., Selective colorimetric detection of polynucleotides based on the distance-dependent optical properties of gold nanoparticles, *Science* **1997**, *277*, 1078–1081
19. Lin, V. S.-Y.; Motesharei, K.; Dancil, K.-P. S.; Sailor, M. J.; Ghadiri, M. R., A porous silicon-based optical interferometric biosensor, *Science* **1997**, *278*, 840–843
20. Rakow, N. A.; Suslick, K. S., A colorimetric sensor array for odour visualization, *Nature* **2000**, *406*, 710–713
21. Kong, J.; Franklin, N. R.; Zhou, C.; Chapline, M. G.; Peng, S.; Cho, K.; Dai, H., Nanotube molecular wires as chemical sensors, *Science* **2000**, *287*, 622–625
22. Hagleitner, C.; Hierlemann, A.; Lange, D.; Kummer, A.; Kerness, N.; Brand, O.; Baltes, H., Smart single-chip gas sensor microsystem, *Nature* **2001**, *414*, 293–296
23. Ivanisevic, A.; Yeh, J.-Y.; Mawst, L.; Kuech, T. F.; Ellis, A. B., Light-emitting diodes as chemical sensors, *Nature* **2001**, *409*, 476–476
24. Janata, J.; Josowicz, M., Conducting polymers in electronic chemical sensors, *Nature Mater.* **2002**, *2*, 19–24
25. Li, Y. Y.; Cunin, F.; Link, J. R.; Gao, T.; Betts, R. E.; Reiver, S. H.; Chin, V.; Bhatia, S. N.; Sailor, M. J., Polymer replicas of photonic porous silicon for sensing and drug delivery applications, *Science* **2003**, *299*, 2045–2047
26. Alivisatos, A. P., The use of nanocrystals in biological detection, *Nature Biotechnol.* **2004**, *22*, 47–52
27. Rose, A.; Zhu, Z.; Madigan, C. F.; Swager, T. M.; Bulovic, V., Sensitivity gains in chemosensing by lasing action in organic polymers, *Nature* **2005**, *434*, 876–879
28. Potyrailo, R. A.; Ghiradella, H.; Vertiatchikh, A.; Dovidenko, K.; Cournoyer, J. R.; Olson, E., Morpho butterfly wing scales demonstrate highly selective vapour response, *Nature Photonics* **2007**, *1*, 123–128
29. Armani, A. M.; Kulkarni, R. P.; Fraser, S. E.; Flagan, R. C.; Vahala, K. J., Label-free, single-molecule detection with optical microcavities, *Science* **2007**, *317*, 783–787
30. Njagi, J.; Warner, J.; Andreescu, S., A bioanalytical chemistry experiment for undergraduate students: Biosensors based on metal nanoparticles, *J Chem. Educ.* **2007**, *84*, 1180–1182
31. Shtoyko, T.; Zudans, I.; Seliskar, C. J.; Heineman, W. R.; Richardson, J. N., An attenuated total reflectance sensor for copper: An experiment for analytical or physical chemistry, *J. Chem. Educ.* **2004**, *81*, 1617–1619
32. Honeybourne, C. L., Organic vapor sensors for food quality assessment, *J. Chem. Educ.* **2000**, *77*, 338–344
33. Newnham, R. E., Structure–property relationships in sensors, *Cryst. Rev.* **1988**, *1*, 253–280
34. Akporiaye, D. E., Towards a rational synthesis of large-pore zeolite-type materials?, *Angew. Chem. Int. Ed.* **1998**, *37*, 2456–2457
35. Ulmer II, C. W.; Smith, D. A.; Sumpter, B. G.; Noid, D. I., Computational neural networks and the rational design of polymeric materials: The next generation polycarbonates, *Comput. Theor. Polym. Sci.* **1998**, *8*, 311–321
36. Suman, M.; Freddi, M.; Massera, C.; Ugozzoli, F.; Dalcanale, E., Rational design of cavitand receptors for mass sensors, *J.Am. Chem. Soc.* **2003**, *125*, 12068–12069
37. Lavigne, J. J.; Anslyn, E. V., Sensing a paradigm shift in the field of molecular recognition: From selective to differential receptors, *Angew. Chem. Int. Ed.* **2001**, *40*, 3119–3130
38. Hatchett, D. W.; Josowicz, M., Composites of intrinsically conducting polymers as sensing nanomaterials, *Chem. Rev.* **2008**, *108*, 746–769
39. Schatz, G. C., Using theory and computation to model nanoscale properties, *Proc. Natl. Acad. Sci. USA* **2007**, *104*, 6885–6892
40. Peng, S.; Cho, K., Ab initio study of doped carbon nanotube sensors, *Nano Lett.* **2003**, *3*, 513–517

41. Dmitriev, S.; Lilach, Y.; Button, B.; Moskovits, M.; Kolmakov, A., Nanoengineered chemiresistors: The interplay between electron transport and chemisorption properties of morphologically encoded sno2 nanowires, *Nanotechnology* **2007**, *18*, 055707
42. Mirsky, V. M.; Vasjari, M.; Novotny, I.; Rehacek, V.; Tvarozek, V.; Wolfbeis, O. S., Self-assembled monolayers as selective filters for chemical sensors, *Nanotechnology* **2002**, *13*, 1–4
43. Strohmeier, G. A.; Fabian, W. M. F.; Uray, G., A combined experimental and theoretical approach toward the development of optimized luminescent carbostyrils, *Helvetica Chim. Acta* **2004**, *87*, 215–226
44. Abraham, M. H., Scales of solute hydrogen bonding: Their construction and application to physicochemical and biochemical processes, *Chem. Soc. Rev.* **1993**, *22*, 73–83
45. Salvador, J. P.; Estevez, M. C.; Marco, M. P.; Sanchez-Baeza, F., A new methodology for the rational design of molecularly imprinted polymers, *Anal. Lett.* **2007**, *40*, 1294–1306.
46. Hao, Q.; Wang, X.; Lu, L.; Yang, X.; Mirsky, V. M., Electropolymerized multilayer conducting polymers with response to gaseous hydrogen chloride, *Macromol. Rapid Comm.* **2005**, *26*, 1099–1103
47. Badjic, J. D.; Kostic, N. M., Behavior of organic compounds confined in monoliths of sol-gel silica glass. Effects of guest-host hydrogen bonding on uptake, release, and isomerization of the guest compounds, *J. Mater. Chem.* **2001**, *11*, 408–418
48. Cao, L.; Lin, H.; Mirsky, V. M., DNA-based surface plasmon resonance biosensor for enrofloxacin, *Anal. Chim. Acta* **2007**, *589*, 1–5
49. Potyrailo, R. A.; Conrad, R. C.; Ellington, A. D.; Hieftje, G. M., Adapting selected nucleic acid ligands (aptamers) to biosensors, *Anal. Chem.* **1998**, *70*, 3419–3425
50. Kaneko, H.; Minagawa, H.; Shimada, J., Rational design of thermostable lactate oxidase by analyzing quaternary structure and prevention of deamidation, *Biotechn. Lett.* **2005**, *27*, 1777–1784
51. Schultz, P. G., Commentary on combinatorial chemistry, *Appl. Catal., A* **2003**, *254*, 3–4
52. McKusick, B. C.; Heckert, R. E.; Cairns, T. L.; Coffman, D. D.; Mower, H. F., Cyanocarbon chemistry. VI. Tricyanovinylamines, *J. Am. Chem. Soc.* **1958**, *80*, 2806–2815
53. Bühlmann, P.; Pretsch, E.; Bakker, E., Carrier-based ion-selective electrodes and bulk optodes. 2. Ionophores for potentiometric and optical sensors, *Chem. Rev.* **1998**, *98*, 1593–1687
54. Steinle, E. D.; Amemiya, S.; Bühlmann, P.; Meyerhoff, M. E., Origin of non-nernstian anion response slopes of metalloporphyrin-based liquid/polymer membrane electrodes, *Anal. Chem.* **2000**, *72*, 5766–5773
55. Pedersen, C. J., Cyclic polyethers and their complexes with metal salts, *J. Am. Chem. Soc.* **1967**, *89*, 7017–7036
56. Hu, Y.; Tan, O. K.; Pan, J. S.; Yao, X., A new form of nanosized srtio3 material for near-human-body temperature oxygen sensing applications, *J. Phys. Chem. B* **2004**, *108*, 11214–11218
57. Svetlicic, V.; Schmidt, A. J.; Miller, L. L., Conductometric sensors based on the hypersensitive response of plasticized polyaniline films to organic vapors, *Chem. Mater.* **1998**, *10*, 3305–3307
58. Martin, P. D.; Wilson, T. D.; Wilson, I. D.; Jones, G. R., An unexpected selectivity of a propranolol-derived molecular imprint for tamoxifen, *Analyst* **2001**, *126*, 757–759
59. Potyrailo, R. A.; Sivavec, T. M., Boosting sensitivity of organic vapor detection with silicone block polyimide polymers, *Anal. Chem.* **2004**, *76*, 7023–7027
60. Walt, D. R.; Dickinson, T.; White, J.; Kauer, J.; Johnson, S.; Engelhardt, H.; Sutter, J.; Jurs, P., Optical sensor arrays for odor recognition, *Biosens. Bioelectron.* **1998**, *13*, 697–699
61. Eberhart, M. E.; Clougherty, D. P., Looking for design in materials design, *Nature Mater.* **2004**, *3*, 659–661
62. *A Practical Guide to Combinatorial Chemistry*; Czarnik, A. W.; DeWitt, S. H., Eds.; American Chemical Society: Washington, DC, 1997

63. Frank, R.; Heikens, W.; Heisterberg-Moutsis, G.; Blocker, H., A new general approach for the simultaneous chemical synthesis of large numbers of oligonucleotides: Segmental solid supports, *Nucl. Acid. Res.* **1983**, *11*, 4365–4377

64. Geysen, H. M.; Meloen, R. H.; Barteling, S. J., Use of peptide synthesis to probe viral antigens for epitopes to a resolution of a single amino acid, *Proc. Natl Acad. Sci. USA* **1984**, *81*, 3998–4002

65. Houghten, R. A., General method for the rapid solid-phase synthesis of large numbers of peptides: Specificity of antigen-antibody interaction at the level of individual amino acids, *Proc. Natl. Acad. Sci. USA* **1985**, *82*, 5131–5135

66. Lebl, M., Parallel personal comments on "classical" papers in combinatorial chemistry, *J. Comb. Chem.* **1999**, *1*, 3–24

67. Jandeleit, B.; Schaefer, D. J.; Powers, T. S.; Turner, H. W.; Weinberg, W. H., Combinatorial materials science and catalysis, *Angew. Chem. Int. Ed.* **1999**, *38*, 2494–2532

68. Maier, W.; Kirsten, G.; Orschel, M.; Weiß, P.-A.; Holzwarth, A.; Klein, J., Combinatorial chemistry of materials, polymers, and catalysts, In *Combinatorial Approaches to Materials Development*; Malhotra, R., Ed.; American Chemical Society: Washington, DC, 2002; Vol. 814; 1–21

69. *Combinatorial and Artificial Intelligence Methods in Materials Science*; Takeuchi, I.; Newsam, J. M.; Wille, L. T.; Koinuma, H.; Amis, E. J., Eds.; Materials Research Society: Warrendale, PA, 2002; Vol. 700

70. *Combinatorial Materials Synthesis*; Xiang, X.-D.; Takeuchi, I., Eds.; Marcel Dekker: New York, NY, 2003

71. *High Throughput Analysis: A Tool for Combinatorial Materials Science*; Potyrailo, R. A.; Amis, E. J., Ed.; Kluwer/Plenum: New York, NY, 2003

72. Koinuma, H.; Takeuchi, I., Combinatorial solid state chemistry of inorganic materials, *Nat. Mater.* **2004**, *3*, 429–438

73. *Combinatorial and Artificial Intelligence Methods in Materials Science II*; Potyrailo, R. A.; Karim, A.; Wang, Q.; Chikyow, T., Eds.; Materials Research Society: Warrendale, PA, 2004; Vol. 804

74. *Special Feature on Combinatorial and High-Throughput Materials Research*; Potyrailo, R. A.; Takeuchi, I., Ed.; *Meas. Sci. Technol.*: 2005; Vol. 16, 316

75. *Combinatorial and High-Throughput Discovery and Optimization of Catalysts and Materials*; Potyrailo, R. A.; Maier, W. F., Eds.; CRC: Boca Raton, FL, 2006

76. Birina, G. A.; Boitsov, K. A., Experimental use of combinational and factorial plans for optimizing the compositions of electronic materials, *Zavodskaya Laboratoriya (in Russian)* **1974**, *40*, 855–857

77. Kennedy, K.; Stefansky, T.; Davy, G.; Zackay, V. F.; Parker, E. R., Rapid method for determining ternary-alloy phase diagrams, *J. Appl. Phys.* **1965**, *36*, 3808–3810

78. Hoffmann, R., Not a library, *Angew. Chem. Int. Ed.* **2001**, *40*, 3337–3340

79. Hoogenboom, R.; Meier, M. A. R.; Schubert, U. S., Combinatorial methods, automated synthesis and high-throughput screening in polymer research: Past and present, *Macromol. Rapid Commun.* **2003**, *24*, 15–32

80. Anderson, F. W.; Moser, J. H., Automatic computer program for reduction of routine emission spectrographic data, *Anal. Chem.* **1958**, *30*, 879–881

81. Eash, M. A.; Gohlke, R. S., Mass spectrometric analysis. A small computer program for the analysis of mass spectra, *Anal. Chem.* **1962**, *34*, 713–713

82. Hanak, J. J., The "multiple-sample concept" in materials research: Synthesis, compositional analysis and testing of entire multicomponent systems, *J. Mater. Sci.* **1970**, *5*, 964–971

83. Xiang, X.-D.; Sun, X.; Briceño, G.; Lou, Y.; Wang, K.-A.; Chang, H.; Wallace-Freedman, W. G.; Chen, S.-W.; Schultz, P. G., A combinatorial approach to materials discovery, *Science* **1995**, *268*, 1738–1740

84. Chang, H.; Gao, C.; Takeuchi, I.; Yoo, Y.; Wang, J.; Schultz, P. G.; Xiang, X.-D.; Sharma, R. P.; Downes, M.; Venkatesan, T., Combinatorial synthesis and high throughput evaluation of ferroelectric/dielectric thin-film libraries for microwave applications, *Appl. Phys. Lett.* **1998**, *72*, 2185–2187

85. Briceño, G.; Chang, H.; Sun, X.; Schultz, P. G.; Xiang, X.-D., A class of cobalt oxide magnetoresistance materials discovered with combinatorial synthesis, *Science* **1995**, *270*, 273–275

86. Danielson, E.; Devenney, M.; Giaquinta, D. M.; Golden, J. H.; Haushalter, R. C.; McFarland, E. W.; Poojary, D. M.; Reaves, C. M.; Weinberg, W. H.; Wu, X. D., A rare-earth phosphor containing one-dimensional chains identified through combinatorial methods, *Science* **1998**, *279*, 837–839

87. Wong, D. W.; Robertson, G. H., Combinatorial chemistry and its applications in agriculture and food, *Adv. Exp. Med. Biol.* **1999**, *464*, 91–105

88. Zhao, J.-C., A combinatorial approach for structural materials, *Adv. Eng. Mat.* **2001**, *3*, 143–147

89. Olk, C. H., Combinatorial approach to material synthesis and screening of hydrogen storage alloys, *Meas. Sci. Technol.* **2005**, *16*, 14–20

90. Zou, L.; Savvate'ev, V.; Booher, J.; Kim, C.-H.; Shinar, J., Combinatorial fabrication and studies of intense efficient ultraviolet–violet organic light-emitting device arrays, *Appl. Phys. Lett.* **2001**, *79*, 2282–2284

91. Takeuchi, I.; Famodu, O. O.; Read, J. C.; Aronova, M. A.; Chang, K.-S.; Craciunescu, C.; Lofland, S. E.; Wuttig, M.; Wellstood, F. C.; Knauss, L.; Orozco, A., Identification of novel compositions of ferromagnetic shape-memory alloys using composition spreads, *Nat. Mater.* **2003**, *2*, 180–184

92. Cui, J.; Chu, Y. S.; Famodu, O. O.; Furuya, Y.; Hattrick-Simpers, J.; James, R. D.; Ludwig, A.; Thienhaus, S.; Wuttig, M.; Zhang, Z.; Takeuchi, I., Combinatorial search of thermoelastic shape-memory alloys with extremely small hysteresis width, *Nat. Mater.* **2006**, *5*, 286–290

93. Holzwarth, A.; Schmidt, H.-W.; Maier, W., Detection of catalytic activity in combinatorial libraries of heterogeneous catalysts by IR thermography, *Angew. Chem. Int. Ed.* **1998**, *37*, 2644–2647

94. Cooper, A. C.; McAlexander, L. H.; Lee, D.-H.; Torres, M. T.; Crabtree, R. H., Reactive dyes as a method for rapid screening of homogeneous catalysts, *J. Am. Chem. Soc.* **1998**, *120*, 9971–9972

95. Lemmon, J. P.; Wroczynski, R. J.; Whisenhunt Jr., D. W.; Flanagan, W. P., High throughput strategies for monomer and polymer synthesis and characterization, *Polym. Prepr.* **2001**, *42*, 630–631

96. Reddington, E.; Sapienza, A.; Gurau, B.; Viswanathan, R.; Sarangapani, S.; Smotkin, E. S.; Mallouk, T. E., Combinatorial electrochemistry: A highly parallel, optical screening method for discovery of better electrocatalysts, *Science* **1998**, *280*, 1735–1737

97. Greeley, J.; Jaramillo, T. F.; Bonde, J.; Chorkendorff, I.; Nørskov, J. K., Computational high-throughput screening of electrocatalytic materials for hydrogen evolution, *Nat. Mater.* **2006**, *5*, 909–913

98. Brocchini, S.; James, K.; Tangpasuthadol, V.; Kohn, J., A combinatorial approach for polymer design, *J. Am. Chem. Soc.* **1997**, *119*, 4553–4554

99. Lai, R.; Kang, B. S.; Gavalas, G. R., Parallel synthesis of ZSM-5 zeolite films from clear organic-free solutions, *Angew. Chem., Int. Ed.* **2001**, *40*, 408–411

100. Ramirez, A. G.; Saha, R., Combinatorial studies for determining properties of thin-film gold–cobalt alloys, *Appl. Phys. Lett.* **2004**, *85*, 5215–5217

101. Jiang, R.; Rong, C.; Chu, D., Combinatorial approach toward high-throughput analysis of direct methanol fuel cells, *J. Comb. Chem.* **2005**, *7*, 272–278

102. Lemmon, J. P.; Manivannan, V.; Jordan, T.; Hassib, L.; Siclovan, O.; Othon, M.; Pilliod, M., High throughput screening of materials for solid oxide fuel cells, In *Combinatorial and Artificial Intelligence Methods in Materials Science II. MRS Symposium Proceedings*; Potyrailo, R. A.; Karim, A.; Wang Q.; Chikyow, T., Eds.; Materials Research Society: Warrendale, PA, 2004; Vol. 804; 27–32

103. Hänsel, H.; Zettl, H.; Krausch, G.; Schmitz, C.; Kisselev, R.; Thelakkat, M.; Schmidt, H.-W., Combinatorial study of the long-term stability of organic thin-film solar cells, *Appl. Phys. Lett.* **2002**, *81*, 2106–2108

104. Chisholm, B. J.; Potyrailo, R. A.; Cawse, J. N.; Shaffer, R. E.; Brennan, M. J.; Moison, C.; Whisenhunt, D. W.; Flanagan, W. P.; Olson, D. R.; Akhave, J. R.; Saunders, D. L.; Mehrabi, A.;

Licon, M., The development of combinatorial chemistry methods for coating development I. Overview of the experimental factory, *Prog. Org. Coat.* **2002**, *45*, 313–321

105. Wicks, D. A.; Bach, H., The coming revolution for coatings science: High throughput screening for formulations, *Proceedings of The 29th Int. Waterborne, High-Solids, and Powder Coat. Symp.* **2002**, *29*, 1–24

106. Grunlan, J. C.; Mehrabi, A. R.; Chavira, A. T.; Nugent, A. B.; Saunders, D. L., Method for combinatorial screening of moisture vapor transmission rate, *J. Comb. Chem.* **2003**, *5*, 362–368

107. Stafslien, S. J.; Bahr, J. A.; Feser, J. M.; Weisz, J. C.; Chisholm, B. J.; Ready, T. E.; Boudjouk, P., Combinatorial materials research applied to the development of new surface coatings I: A multiwell plate screening method for the high-throughput assessment of bacterial biofilm retention on surfaces, *J. Comb. Chem.* **2006**, *8*, 156–162

108. Ekin, A.; Webster, D. C., Combinatorial and high-throughput screening of the effect of siloxane composition on the surface properties of crosslinked siloxane-polyurethane coatings, *J. Comb. Chem.* **2007**, *9*, 178–188

109. Potyrailo, R. A.; Chisholm, B. J.; Olson, D. R.; Brennan, M. J.; Molaison, C. A., Development of combinatorial chemistry methods for coatings: High-throughput screening of abrasion resistance of coatings libraries, *Anal. Chem.* **2002**, *74*, 5105–5111

110. Potyrailo, R. A.; Chisholm, B. J.; Morris, W. G.; Cawse, J. N.; Flanagan, W. P.; Hassib, L.; Molaison, C. A.; Ezbiansky, K.; Medford, G.; Reitz, H., Development of combinatorial chemistry methods for coatings: High-throughput adhesion evaluation and scale-up of combinatorial leads, *J. Comb. Chem.* **2003**, *5*, 472–478

111. MacLean, D.; Baldwin, J. J.; Ivanov, V. T.; Kato, Y.; Shaw, A.; Schneider, P.; Gordon, E. M., Glossary of terms used in combinatorial chemistry, *J. Comb. Chem.* **2000**, *2*, 562–578

112. Potyrailo, R. A.; Takeuchi, I., Role of high-throughput characterization tools in combinatorial materials science, *Meas. Sci. Technol.* **2005**, *16*, 1–4

113. Cohan, P. E., Combinatorial materials science applied – mini case studies, lessons and strategies, In *Combi 2002 – the 4th Annual International Symposium on Combinatorial Approaches for New Materials Discovery*; Knowledge Foundation: Arlington, VA, 2002

114. Siemons, M.; Koplin, T. J.; Simon, U., Advances in high throughput screening of gas sensing materials, *Appl. Surf. Sci.* **2007**, *Appl. Surf. Sci.* **2007**, *254*, 669–676

115. Qin, L.; Zou, S.; Xue, C.; Atkinson, A.; Schatz, G. C.; Mirkin, C. A., Designing, fabricating, and imaging raman hot spots, *Proc. Natl. Acad. Sci. USA* **2006**, *103*, 13300–13303

116. Paulose, M.; Varghese, O. K.; Mor, G. K.; Grimes, C. A.; Ong, K. G., Unprecedented ultra-high hydrogen gas sensitivity in undoped titania nanotubes, *Nanotechnology* **2006**, *17*, 398–402

117. Lu, Y.; Liu, J.; Li, J.; Bruesehoff, P. J.; Pavot, C. M.-B.; Brown, A. K., New highly sensitivie and selective catalytic DNA biosensors for metal ions, *Biosens. Bioelectron.* **2003**, *18*, 529–540

118. Hirsch, T.; Kettenberger, H.; Wolfbeis, O. S.; Mirsky, V. M., A simple strategy for preparation of sensor arrays: Molecularly structured monolayers as recognition elements, *Chem. Commun.* **2003**, 432–433

119. Hermann, T.; Patel, D. J., Adaptive recognition by nucleic acid aptamers, *Science* **2000**, *287*, 820–825

120. Cho, E. J.; Tao, Z.; Tang, Y.; Tehan, E. C.; Bright, F. V.; Hicks, W. L., Jr.; Gardella, J. A., Jr.; Hard, R., Tools to rapidly produce and screen biodegradable polymer and sol-gel-derived xerogel formulations, *Appl. Spectrosc.* **2002**, *56*, 1385–1389

121. Mirsky, V. M.; Riepl, M.; Wolfbeis, O. S., Capacitive monitoring of protein immobilization and antigen-antibody reaction on the mono-molecular films of alkylthiols adsorbed on gold electrodes, *Biosens. Bioelectron.* **1997**, *12*, 977–989

122. Rege, K.; Raravikar, N. R.; Kim, D.-Y.; Schadler, L. S.; Ajayan, P. M.; Dordick, J. S., Enzyme-polymer-single walled carbon nanotube composites as biocatalytic films, *Nano Lett.* **2003**, *3*, 829–832

123. Kramer, K.; Hock, B., Antibodies for biosensors, In *Ultrathin Electrochemical Chemo- and Biosensors*; Mirsky, V. M., Ed.; Springer: Berlin, Germany, 2004

124. Matzger, A. J.; Lawrence, C. E.; Grubbs, R. H.; Lewis, N. S., Combinatorial approaches to the synthesis of vapor detector arrays for use in an electronic nose, *J. Comb. Chem.* **2000**, *2*, 301–304

125. Apostolidis, A.; Klimant, I.; Andrzejewski, D.; Wolfbeis, O. S., A combinatorial approach for development of materials for optical sensing of gases, *J. Comb. Chem.* **2004**, *6*, 325–331

126. Deans, R.; Kim, J.; Machacek, M. R.; Swager, T. M., A poly(p-phenyleneethynylene) with a highly emissive aggregated phase, *J. Am. Chem. Soc.* **2000**, *122*, 8565–8566

127. Lavastre, O.; Illitchev, I.; Jegou, G.; Dixneuf, P. H., Discovery of new fluorescent materials from fast synthesis and screening of conjugated polymers, *J. Am. Chem. Soc.* **2002**, *124*, 5278–5279

128. Zhou, Y.; Freitag, M.; Hone, J.; Staii, C.; Johnson, A. T.; Pinto, N. J.; MacDiarmid, A. G., Fabrication and electrical characterization of polyaniline-based nanofibers with diameter below 30 nm, *Appl. Phys. Lett.* **2003**, *83*, 3800–3802

129. Bever, M. B.; Duwez, P. E., Gradients in composite materials, *Mater. Sci. Eng.* **1972**, *10*, 1–8

130. Shen, M.; Bever, M. B., Gradients in polymeric materials, *J. Mater. Sci.* **1972**, *7*, 741–746

131. Pompe, W.; Worch, H.; Epple, M.; Friess, W.; Gelinsky, M.; Greil, P.; Hempel, U.; Scharnweber, D.; Schulte, K., Functionally graded materials for biomedical applications, *Mater. Sci. Eng. A* **2003**, *362*, 40–60

132. Taylor, C. J.; Semancik, S., Use of microhotplate arrays as microdeposition substrates for materials exploration, *Chem. Mater.* **2002**, *14*, 1671–1677

133. Aronova, M. A.; Chang, K. S.; Takeuchi, I.; Jabs, H.; Westerheim, D.; Gonzalez-Martin, A.; Kim, J.; Lewis, B., Combinatorial libraries of semiconductor gas sensors as inorganic electronic noses, *Appl. Phys. Lett.* **2003**, *83*, 1255–1257

134. Potyrailo, R. A., Sensors in combinatorial polymer research, *Macromol. Rapid Comm.* **2004**, *25*, 77–94

135. Amis, E. J., Combinatorial materials science reaching beyond discovery, *Nat. Mater.* **2004**, *3*, 83–85

136. Scheidtmann, J.; Frantzen, A.; Frenzer, G.; Maier, W. F., A combinatorial technique for the search of solid state gas sensor materials, *Meas. Sci. Technol.* **2005**, *16*, 119–127

137. Potyrailo, R. A.; Morris, W. G.; Leach, A. M.; Hassib, L.; Krishnan, K.; Surman, C.; Wroczynski, R.; Boyette, S.; Xiao, C.; Shrikhande, P.; Agree, A.; Cecconie, T., Theory and practice of ubiquitous quantitative chemical analysis using conventional computer optical disk drives, *Appl. Opt.* **2007**, *46*, 7007–7017

138. Dickinson, T. A.; Walt, D. R.; White, J.; Kauer, J. S., Generating sensor diversity through combinatorial polymer synthesis, *Anal. Chem.* **1997**, *69*, 3413–3418

139. Potyrailo, R. A.; Wroczynski, R. J., Spectroscopic and imaging approaches for evaluation of properties of one-dimensional arrays of formulated polymeric materials fabricated in a combinatorial microextruder system, *Rev. Sci. Instrum.* **2005**, *76*, 062222

140. Sysoev, V. V.; Kiselev, I.; Frietsch, M.; Goschnick, J., Temperature gradient effect on gas discrimination power of a metal-oxide thin-film sensor microarray, *Sensors* **2004**, *4*, 37–46

141. Klingvall, R.; Lundström, I.; Löfdahl, M.; Eriksson, M., A combinatorial approach for field-effect gas sensor research and development, *IEEE Sens. J.* **2005**, *5*, 995–1003

142. Baker, B. E.; Kline, N. J.; Treado, P. J.; Natan, M. J., Solution-based assembly of metal surfaces by combinatorial methods, *J. Am. Chem. Soc.* **1996**, *118*, 8721–8722

143. Potyrailo, R. A.; Hassib, L., Analytical instrumentation infrastructure for combinatorial and high-throughput development of formulated discrete and gradient polymeric sensor materials arrays, *Rev. Sci. Instrum.* **2005**, *76*, 062225

144. *Handbook of Chemical and Biological Sensors*; Taylor, R. F.; Schultz, J. S., Eds.; IOP Publishing: Bristol, UK, 1996

145. Carrano, J. C.; Jeys, T.; Cousins, D.; Eversole, J.; Gillespie, J.; Healy, D.; Licata, N.; Loerop, W.; O'Keefe, M.; Samuels, A.; Schultz, J.; Walter, M.; Wong, N.; Billotte, B.; Munley, M.; Reich, E.; Roos, J., Chemical and biological sensor standards study (CBS3), In *Optically Based Biological and Chemical Sensing for Defence*; Carrano, J. C.; Zukauskas, A. Eds.; SPIE – The International Society for Optical Engineering: Bellingham, WA, 2004; Vol. 5617; xi–xiii

146. Meyerhoff, M. E., *In vivo* blood-gas and electrolyte sensors: Progress and challenges, *Trends Anal. Chem.* **1993**, *12*, 257–266

147. Clark, K. J. R.; Furey, J., Suitability of selected single-use process monitoring and control technology, *BioProcess Int.* **2006**, *4(6)*, S16–S20

148. Newman, J. D.; Turner, A. P. F., Home blood glucose biosensors: A commercial perspective, *Biosens. Bioelectron.* **2005**, *20*, 2435–2453

149. Pickup, J. C.; Alcock, S., Clinicians' requirements for chemical sensors for *in vivo* monitoring: A multinational survey, *Biosens. Bioelectron.* **1991**, *6*, 639–646

150. Eriksson, M.; Klingvall, R.; Lundström, I., A combinatorial method for optimization of materials for gas sensitive field-effect devices, In *Combinatorial and High-Throughput Discovery and Optimization of Catalysts and Materials*; Potyrailo, R. A.; Maier, W. F. Eds.; CRC: Boca Raton, FL, 2006; 85–95

151. Lundström, I.; Sundgren, H.; Winquist, F.; Eriksson, M.; Krantz-Rülcker, C.; Lloyd-Spetz, A., Twenty-five years of field effect gas sensor research in Linköping, *Sens. Actuators B* **2007**, *121*, 247–262

152. Goschnick, J.; Koronczi, I.; Frietsch, M.; Kiselev, I., Water pollution recognition with the electronic nose kamina, *Sens. Actuators B* **2005**, *106*, 182–186

153. Mazza, T.; Barborini, E.; Kholmanov, I. N.; Piseri, P.; Bongiorno, G.; Vinati, S.; Milania, P.; Ducati, C.; Cattaneo, D.; Li Bassi, A.; Bottani, C. E.; Taurino, A. M.; Siciliano, P., Libraries of cluster-assembled titania films for chemical sensing, *Appl. Phys. Lett.* **2005**, *87*, 103–108

154. Korotcenkov, G., Gas response control through structural and chemical modification of metal oxide films: State of the art and approaches, *Sens. Actuators B* **2005**, *107*, 209–232

155. Barsan, N.; Koziej, D.; Weimar, U., Metal oxide-based gas sensor research: How to? *Sens. Actuators B* **2007**, *121*, 18–35

156. Semancik, S. Correlation of chemisorption and electronic effects for metal oxide interfaces: Transducing principles for temperature programmed gas microsensors. Final technical report project number: Emsp 65421, grant number: 07-98er62709; US Department of Energy Information Bridge: 2002, pp http://www.osti.gov/bridge/

157. Semancik, S., Temperature-dependent materials research with micromachined array platforms, In *Combinatorial Materials Synthesis*; Xiang, X.-D.; Takeuchi, I., Eds.; Marcel Dekker: New York, NY, 2003; 263–295

158. Simon, U.; Sanders, D.; Jockel, J.; Heppel, C.; Brinz, T., Design strategies for multielectrode arrays applicable for high-throughput impedance spectroscopy on novel gas sensor materials, *J. Comb. Chem.* **2002**, *4*, 511–515

159. Simon, U.; Sanders, D.; Jockel, J.; Brinz, T., Setup for high-throughput impedance screening of gas-sensing materials, *J. Comb. Chem.* **2005**, *7*, 682–687

160. Frantzen, A.; Scheidtmann, J.; Frenzer, G.; Maier, W. F.; Jockel, J.; Brinz, T.; Sanders, D.; Simon, U., High-throughput method for the impedance spectroscopic characterization of resistive gas sensors, *Angew. Chem. Int. Ed.* **2004**, *43*, 752–754

161. Sanders, D.; Simon, U., High-throughput gas sensing screening of surface-doped In_2O_3, *J. Comb. Chem.* **2007**, *9*, 53–61

162. Siemons, M.; Simon, U., Preparation and gas sensing properties of nanocrystalline la-doped $CoTiO_3$, *Sens. Actuators B* **2006**, *120*, 110–118

163. Siemons, M.; Simon, U., Gas sensing properties of volume-doped $CoTiO_3$ synthesized via polyol method, *Sens. Actuators B* **2007**, *Sens. Actuators B* **2007**, *126*, 595–603

164. Siemons, M.; Simon, U., High throughput screening of the propylene and ethanol sensing properties of rare-earth orthoferrites and orthochromites, *Sens. Actuators B* **2007**, *126*, 181–186

165. Nakagawa, M.; Okabayashi, T.; Fujimoto, T.; Utsunomiya, K.; Yamamoto, I.; Wada, T.; Yamashita, Y.; Yamashita, N., A new method for recognizing organic vapor by spectroscopic image on cataluminescence-based gas sensor, *Sens. Actuators B* **1998**, *51*, 159–162

166. Nakagawa, M.; Yamashita, N., Cataluminescence-based gas sensors, *Springer Ser. Chem. Sens. Biosens.* **2005**, *3*, 93–132

167. Kahl, M.; Voges, E.; Kostrewa, S.; Viets, C.; Hill, W., Periodically structured metallic substrates for SERS, *Sens. Actuators B* **1998**, *51*, 285–291

168. Han, M. S.; Lytton-Jean, A. K. R.; Oh, B.-K.; Heo, J.; Mirkin, C. A., Colorimetric screening of DNA-binding molecules with gold nanoparticle probes, *Angew. Chem. Int. Ed.* **2006**, *45*, 1807–1810

169. Dovidenko, K.; Potyrailo, R. A.; Grande, J., Focused ion beam microscope as an analytical tool for nanoscale characterization of gradient-formulated polymeric sensor materials, In *Combinatorial Methods and Informatics in Materials Science. MRS Symposium Proceedings*; Fasolka, M.; Wang, Q.; Potyrailo, R. A.; Chikyow, T.; Schubert, U. S.; Korkin, A., Eds.; Materials Research Society: Warrendale, PA, 2006; Vol. 894; 231–236

170. Bhat, R. R.; Genzer, J., Combinatorial study of nanoparticle dispersion in surface-grafted macromolecular gradients, *Appl. Surf. Sci.* **2005**, *252*, 2549–2554

171. Bhat, R. R.; Tomlinson, M. R.; Wu, T.; Genzer, J., Surface-grafted polymer gradients: Formation, characterization and applications, *Adv. Polym. Sci.* **2006**, *198*, 51–124

172. Bhat, R. R.; Genzer, J., Tuning the number density of nanoparticles by multivariant tailoring of attachment points on flat substrates, *Nanotechnology* **2007**, *18*, 025301

173. Demers, L. M.; Mirkin, C. A., Combinatorial templates generated by dip-pen nanolithography for the formation of two-dimensional particle arrays, *Angew. Chem. Int. Ed.* **2001**, *40*, 3069–3071

174. Ivanisevic, A.; McCumber, K. V.; Mirkin, C. A., Site-directed exchange studies with combinatorial libraries of nanostructures, *J. Am. Chem. Soc.* **2002**, *124*, 11997–12001

175. Potyrailo, R. A.; Leach, A. M., Gas sensor materials based on semiconductor nanocrystal/polymer composite films, In *Proceedings of Transducers'05, the 13th International Conference on Solid-state Sensors, Actuators and Microsystems*, Seoul, Korea, June 5–9, 2005; 1292–1295

176. Potyrailo, R. A.; Leach, A. M., Selective gas nanosensors with multisize cdse nanocrystal/polymer composite films and dynamic pattern recognition, *Appl. Phys. Lett.* **2006**, *88*, 134110

177. Leach, A. M.; Potyrailo, R. A., Gas sensor materials based on semiconductor nanocrystal/polymer composite films, In *Combinatorial Methods and Informatics in Materials Science. MRS Symposium Proceedings*; Wang, Q.; Potyrailo, R. A.; Fasolka, M.; Chikyow, T.; Schubert, U. S.; Korkin, A., Eds.; Materials Research Society: Warrendale, PA, 2006; Vol. 894; 237–243

178. Singh, A.; Yao, Q.; Tong, L.; Still, W. C.; Sames, D., Combinatorial approach to the development of fluorescent sensors for nanomolar aqueous copper, *Tetrahedron Lett.* **2000**, *41*, 9601–9605

179. Szurdoki, F.; Ren, D.; Walt, D. R., A combinatorial approach to discover new chelators for optical metal ion sensing, *Anal. Chem.* **2000**, *72*, 5250–5257

180. Castillo, M.; Rivero, I. A., Combinatorial synthesis of fluorescent trialkylphosphine sulfides as sensor materials for metal ions of environmental concern, *ARKIVOC* **2003**, *11*, 193–202

181. Mello, J. V.; Finney, N. S., Reversing the discovery paradigm: A new approach to the combinatorial discovery of fluorescent chemosensors, *J. Am. Chem. Soc.* **2005**, *127*, 10124–10125

182. Hagihara, M.; Fukuda, M.; Hasegawa, T.; Morii, T., A modular strategy for tailoring fluorescent biosensors from ribonucleopeptide complexes, *J. Am. Chem. Soc.* **2006**, *128*, 12932–12940

183. Wang, S.; Chang, Y.-T., Combinatorial synthesis of benzimidazolium dyes and its diversity directed application toward gtp-selective fluorescent chemosensors, *J. Am. Chem. Soc.* **2006**, *128*, 10380–10381

184. Buryak, A.; Severin, K., Dynamic combinatorial libraries of dye complexes as sensors, *Angew. Chem. Int. Ed.* **2005**, *44*, 7935–7938

185. Buryak, A.; Severin, K., Easy to optimize: Dynamic combinatorial libraries of metal-dye complexes as flexible sensors for tripeptides, *J. Comb. Chem.* **2006**, *8*, 540–543

186. Li, Q.; Lee, J.-S.; Ha, C.; Park, C. B.; Yang, G.; Gan, W. B.; Chang, Y.-T., Solid-phase synthesis of styryl dyes and their application as amyloid sensors, *Angew. Chem. Int. Ed.* **2004**, *46*, 6331–6335

187. Rosania, G. R.; Lee, J. W.; Ding, L.; Yoon, H.-S.; Chang, Y.-T., Combinatorial approach to organelle-targeted fluorescent library based on the styryl scaffold, *J. Am. Chem. Soc.* **2003**, *125*, 1130–1131

188. Shedden, K.; Brumer, J.; Chang, Y. T.; Rosania, G. R., Chemoinformatic analysis of a super-targeted combinatorial library of styryl molecules, *J. Chem. Inf. Comput. Sci.* **2003**, *43*, 2068–2080

189. Basabe-Desmonts, L.; Beld, J.; Zimmerman, R. S.; Hernando, J.; Mela, P.; Garcia Parajo, M. F.; van Hulst, N. F.; van den Berg, A.; Reinhoudt, D. N.; Crego-Calama, M., A simple approach to sensor discovery and fabrication on self-assembled monolayers on glass, *J. Am. Chem. Soc.* **2004**, *126*, 7293–7299

190. Basabe-Desmonts, L.; Zimmerman, R. S.; Reinhoudt, D. N.; Crego-Calama, M., Combinatorial method for surface-confined sensor design and fabrication, *Springer Ser. Chem. Sens. Biosens.* **2005**, *3*, 169–188

191. Basabe-Desmonts, L.; Reinhoudt, D. N.; Crego-Calama, M., Combinatorial fabrication of fluorescent patterns with metal ions using soft lithography, *Adv. Mater.* **2006**, *18*, 1028–1032

192. Basabe-Desmonts, L.; Reinhoudt, D. N.; Crego-Calama, M., Design of fluorescent materials for chemical sensing, *Chem. Soc. Rev.* **2007**, *36*, 993–1017

193. Chojnacki, P.; Werner, T.; Wolfbeis, O. S., Combinatorial approach towards materials for optical ion sensors, *Microchim. Acta* **2004**, *147*, 87–92

194. Potyrailo, R. A., Expanding combinatorial methods from automotive to sensor coatings, *Polym. Mate. Sci. Eng. Polym. Prepr.* **2004**, *90*, 797–798

195. Hassib, L.; Potyrailo, R. A., Combinatorial development of polymer coating formulations for chemical sensor applications, *Polym. Prepr.* **2004**, *45*, 211–212

196. Potyrailo, R. A.; Morris, W. G.; Wroczynski, R. J., Acoustic-wave sensors for high-through-put screening of materials, In *High Throughput Analysis: A Tool for Combinatorial Materials Science*; Potyrailo, R. A.; Amis, E. J., Eds.; Kluwer/Plenum: New York, NY, 2003; ch. 11

197. Potyrailo, R. A.; Morris, W. G.; Wroczynski, R. J., Multifunctional sensor system for high-throughput primary, secondary, and tertiary screening of combinatorially developed materials, *Rev. Sci. Instrum.* **2004**, *75*, 2177–2186

198. Potyrailo, R. A.; McCloskey, P. J.; Ramesh, N.; Surman, C. M. Sensor devices containing co-polymer substrates for analysis of chemical and biological species in water and air; US Patent Application 2005133697: 2005

199. Potyrailo, R. A.; McCloskey, P. J.; Wroczynski, R. J.; Morris, W. G., High-throughput determination of quantitative structure-property relationships using resonant multisensor system: Solvent-resistance of bisphenol a polycarbonate copolymers, *Anal. Chem.* **2006**, *78*, 3090–3096

200. Potyrailo, R. A.; Morris, W. G., Wireless resonant sensor array for high-throughput screening of materials, *Rev. Sci. Instrum.* **2007**, *78*, 072214

201. Wu, X.; Kim, J.; Dordick, J. S., Enzymatically and combinatorially generated array-based polyphenol metal ion sensor, *Biotechnol. Prog.* **2000**, *16*, 513–516

202. Kim, D.-Y.; Wu, X.; Dordick, J. S., Generation of environmentally compatible polymer libraries via combinatorial biocatalysis, In *Biocatalysis in Polymer Science;* American Chemical Society: Washington, DC, 2003; Vol. 840; 34–49

203. Mirsky, V. M.; Kulikov, V., Combinatorial electropolymerization: Concept, equipment and applications, In *High Throughput Analysis: A Tool for Combinatorial Materials Science*; Potyrailo, R.A., Amis, E. J., Eds.; Kluwer/Plenum: New York, NY, 2003; ch 20, pp. 431–446

204. Kulikov, V.; Mirsky, V. M., Equipment for combinatorial electrochemical polymerization and high-throughput investigation of electrical properties of the synthesized polymers, *Meas. Sci. Technol.* **2004**, *15*, 49–54
205. Mirsky, V. M.; Kulikov, V.; Hao, Q.; Wolfbeis, O. S., Multiparameter high throughput characterization of combinatorial chemical microarrays of chemosensitive polymers, *Macromol. Rapid Commun.* **2004**, *25*, 253–258
206. Kulikov, V.; Mirsky, V. M.; Delaney, T. L.; Donoval, D.; Koch, A. W.; Wolfbeis, O. S., High-throughput analysis of bulk and contact conductance of polymer layers on electrodes, *Meas. Sci. Technol.* **2005**, *16*, 95–99
207. Xiang, Y.; LaVan, D., Parallel microfluidic synthesis of conductive biopolymers, *Proc. 2nd IEEE/ASME International Conference on Mechatronic and Embedded Systems and Applications* **2006**, 1–5
208. Mirsky, V. M.; Hirsch, T.; Piletsky, S. A.; Wolfbeis, O. S., A spreader-bar approach to molecular architecture: Formation of stable artificial chemoreceptors, *Angew. Chem. Int. Ed.* **1999**, *38*, 1108–1110
209. Lahav, M.; Katz, E.; Willner, I., Photochemical imprint of molecular recognition sites in two-dimensional monolayers assembled on au electrodes: Effects of the monolayer structures on the binding affinities and association kinetics to the imprinted interfaces, *Langmuir* **2001**, *17*, 7387–7395
210. Prodromidis, M. I.; Hirsch, T.; Mirsky, V. M.; Wolfbeis, O. S., Enantioselective artificial receptors formed by the spreader-bar technique, *Electroanalysis* **2003**, *15*, 1795–1798
211. Tappura, K.; IVikholm-Lundin, I.; Albers, W. M., Lipoate-based imprinted self-assembled molecular thin films for biosensor applications, *Biosens. Bioelectron.* **2007**, *22*, 912–919
212. Cho, E. J.; Tao, Z.; Tehan, E. C.; Bright, F. V., Multianalyte pin-printed biosensor arrays based on protein-doped xerogels, *Anal. Chem.* **2002**, *74*, 6177–6184

Chapter 2
Main Concepts of Chemical and Biological Sensing

Marek Trojanowicz

Abstract Brief historic introduction precedes presentation of main types of transducers used in sensors including electrochemical, optical, mass sensitive, and thermal devices. Review of chemical sensors includes various types of gas sensitive devices, potentiometric and amperometric sensors, and quartz microbalance applications. Mechanisms of biorecognition employed in biosensors are reviewed with the method of immobilization used. Some examples of biomimetic sensors are also presented.

1 Introduction

The technological progress in contemporary world in all branches of human activity is driven by pursuing a goal of better living conditions a well as learning and understanding of surrounding world. In both spheres of this activity, a very significant role is played by analytical chemistry, as a discipline of science and technology indispensable in tracing and control – when necessary – of functioning of living organisms, technological processes, changes in natural environment, and quality of various materials. The progress in analytical instrumentation is related to discoveries of new phenomena and materials, with increasing needs for analytical instrumentation and their functional abilities. This progress is observed in many directions, but certainly one of them is development of sensors for chemical analysis. Table 2.1 shows 20 inventions, arbitrarily selected by author of this text, as milestones in the field of chemical sensors. In the middle of year 2006, one can find in ISI Web of Science, among almost 11 millions of papers in scientific journals abstracted for last ten years, almost 58 thousands papers with keyword "*sensor*," and almost 10,000 papers with keyword "*biosensor.*" This seems to illustrate well the scientific and technological research potential involved in this area.

M. Trojanowicz (✉)
Department of Chemistry, University of Warsaw,
Warsaw, Poland
trojan@chem.uw.edu.pl

R.A. Potyrailo and V.M. Mirsky (eds.), *Combinatorial Methods for Chemical and Biological Sensors*, DOI: 10.1007/978-0-387-73713-3_2,

Table 2.1 Milestones in development of chemical and biochemical sensors

Year	Invented sensor	Reference
1909	Glass pH electrode	Haber and Klemensiewicz[1]
1954	Amperometric oxygen electrode	Clark Jr.[2]
1962	Amperometric enzyme biosensor for glucose	Clark Jr. and Lyons[3]
1964	Piezoelectric sensors	King Jr.[4]
1966	Solid-state fluoride selective electrode	Frant and Ross Jr.[5]
1967	Liquid-state ion-selective electrode with molecular carriers	Stefanac and Simon[6]
1969	Potentiometric enzyme biosensor for urea	Guilbault and Montalvo[7]
1970	Ion-sensitive field effect transistor	Bergveld[8]
1974	Enzyme thermistor	Mosbach and Danielsson[9]
1975	Microbe-based biosensor	Davies[10]
	Immunosensor	Janata[11]
	pCO_2/pO_2 optode	Lübbers and Optiz[12]
	Gas-sensitive semiconductor sensor	Lündstrom, Shivaramn, Svenson, and Lundkvist[13]
1980	Fiber optic pH optode	Peterson, Goldstein, Fitzgerald, and Buckhold[14]
1983	Surface plasmon resonance biosensor	Liedberg, Nylander, and Lündstrom[15]
1986	Tissue-based biosensor	Belli and Rechnitz[16]
1987	Receptor-based biosensor	Taylor, Marenchic, and Cook[17]
	Screen-printed biosensor	Matthews, Bown, Watson, Holaman, Steemson, Hughs, and Scott[18]
1993	DNA biosensor	Millan and Mikkelson[19]
1996	Aptamer biosensor	Drolet, Moon-Dermott, and Roming[20]

The term *"chemical sensor"* in current analytical literature is used with very broad and not always unequivocal meaning. Chemical sensors are defined for instance as *"measurement devices which utilize chemical or biological reactions to detect and quantify a specific analyte or an event"* or *"miniaturized analytical devices which can deliver real-time and on-line information on the presence of specific compounds or ions in complex samples."*

According to definitions of Analytical Division of IUPAC "a chemical sensor is a device that transforms chemical information, ranging from the concentration of a specific sample component to total composition analysis, into an analytically useful signal. The chemical information mentioned above may originate from a chemical reaction of the analyte or from a physical property of the system investigated."[21] In the same work, a classification of chemical sensors is given according to the operating principle of the transducer (Table 2.2).

Chemical sensors can be also classified depending on how they transduce the presence of chemical species into an electrical signal as reactive, physical property, and sorptive sensors.

The chemical sensors array devices (so called "electronic tongues") are defined in IUPAC Technical Report of Analytical Chemistry Division as *"an analytical instrument comprising an array of nonspecific, low selective chemical sensors with*

Table 2.2 The classification of chemical sensors suggested in 1991 by Analytical Chemistry Division of IUPAC[21]

Class of sensors	Operating principle
Optical devices (optodes)	Absorbance
	Reflectance
	Luminescence
	Refractive index
	Optothermal effect
	Light scattering
Electrochemical	Voltammetry (including amperometry)
	Potentiometry
	Chemically sensitized field effect transistor
	Potentiometry with solid electrolytes for gas sensing
Electrical	Metal oxide seminconductivity
	Organic semiconductivity
	Electrolytic conductivity
	Electric permittivity
Mass sensitive	Piezoelectric
	Surface acoustic wave propagation
Magnetic	Changes of paramagnetic gas properties
Thermometric	Heat effects of a specific chemical reaction
Others	Emission of α-, β-, or γ-radiation

high stability and cross-sensitivity to different species in solution, and an appropriate method of PARC (pattern recognition) and/or multivariate calibration for data processing. If properly configurated and trained (calibrated), the electronic tongue is capable of recognizing the qualitative and quantitative composition of multispecies of different natures."[22]

Following another IUPAC recommendations, "biosensor is an analytical device incorporating a biological material or a bio-mimic e.g. tissue, microorganisms organelles, cell receptors, enzymes, antibodies, nucleic acids etc. intimately associated with or integrated with a physicochemical transducers or transducing micro-system using optical, electrochemical, thermometric, piezoelectric or magnetic properties etc". This definition is recommended by Chemistry and the Environment Division.[23] It seems to be worth to recall in this place some limitations indicated in another IUPAC recommendations, published jointly by the Physical Chemistry Division and Analytical Chemistry Division,[24] which were given particularly for electro-chemical biosensors, but they can be extended to all biosensors regardless of transducing principle employed. A biosensor is a self-contained integrated device that should be clearly distinguished from an analytical system, which incorporates additional separation steps, such as high performance liquid chromatography (HPLC), or additional hardware and/or sample processing such as specific reagent introduction, e.g. flow injection analysis (FIA). Thus, on the one hand, a biosensor should be a reagent-less analytical device, although the presence of ambient cosubstrates, such as water for hydrolases or oxygen for oxidoreductases, may be required for the analyte determination.

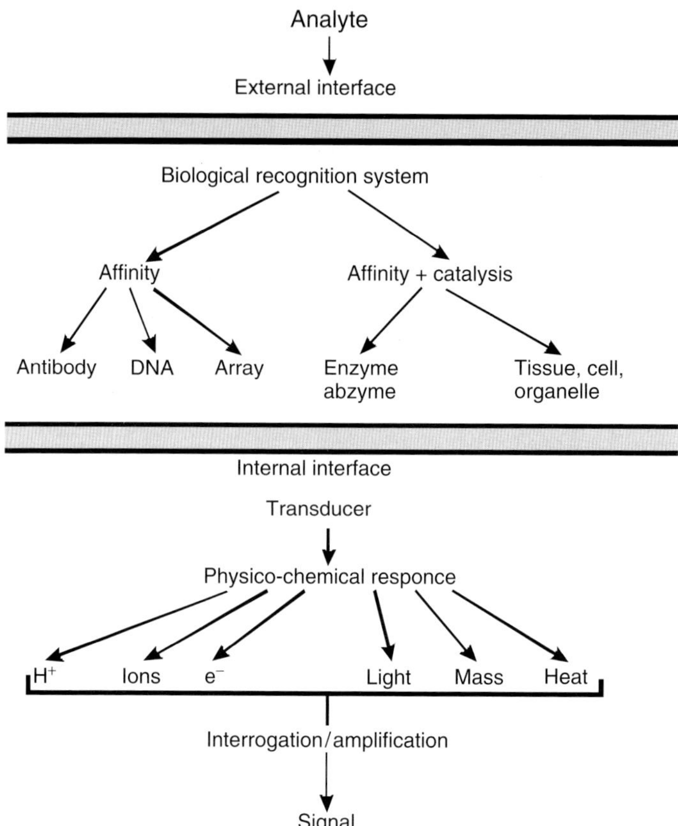

Fig. 2.1 Possible biosensor components and signal transducing[25] – Reproduced by permission of The Royal Society of Chemistry

On the other hand, it may provide as part of some integrated system, some separation or amplification steps achieved by inner or outer membranes or reacting layers (see some simplified schematics in Fig. 2.1).

Chemical sensors are developed for numerous kinds of analytes: gases, vapors and humidity, ions, organic chemicals, and macromolecules. In above cited definitions, there is no practical differentiation from the meaning of term "analytical detector," although, at least at the beginning of development of chemical sensors, their main functional attributes were their use by immersion in the sample solution and easy transfer from one sample to another. These very practical attributes – compared with conventional detectors, where sample has to be introduced – it seems that to some extent can be considered as a driving force in fast progress of this type of devices from indicator strips, through glass pH and ion-selective electrodes, first oxygen and enzyme Clark electrodes to similar in design various electrodes and optodes.

Nowadays, it seems that the borders between meaning and common use of terms chemical sensors, detectors, analytical instrument, or analyzer became very obliterated.

All analytical instruments besides constant improvement, in recent decades are also commonly miniaturized. This is probably also one of the reasons, why miniaturized flow devices (microfluidics) are also named sensors or biosensors, similarly to e.g. microtiter plates, miniaturized mass spectrometers, electrophysiology equipment or chemical labels for imaging. It seems then that terms sensors or biosensors are used in too many meanings (perhaps because it sounds very advanced), and this process seems to be rather irreproducible.

For the task of these considerations, as chemical sensors will be considered portable miniaturized analytical measuring devices, integrating chemical or biochemical recognition element (located on surface or in the material bulk) with signal transducer. The role of recognition element is to create a change of value of some physico-chemical quantity of recognition phase as result of interaction with analyte, with some selectivity and quantitative dependence on analyte concentration. The role of signal transducer is to measure the quantity of change of given parameter caused by analyte, and to convert it into electrical signal produced for appropriate display or read-out.

The recognition event may have a typical chemical nature of interaction of analyte with components of active sensing surface, or it can be combined with electrochemical or photochemical process on the surface with analyte. When in such recognition process, compounds with biological activity (enzymes, receptors, antibodies, DNA, aptamers, tissues or whole microorganisms) are involved, and such sensors are named as biosensors. A large variety of employed phenomena, chemical interactions, biological processes, and electrochemical or photochemical processes is a source of a very intense development for fundamental studies, design, and technology for construction of chemical and biological sensors in last half of a century. In this activity all current achievements of various disciplines of science, material science, electronics and optoelectronics, as well as micromechanics (recently also nanome-chanics) are engaged. Besides, mentioned above, the vast amount of original papers in scientific and technological journals, there are also available several tens of specialized books – some selection is cited below[26–49] and numerous patents. There are also numerous devices commercially available, mostly for clinical and environmental applications – see for example general books on this subject.[50–52] Thanks to the advanced technologies of their mass-production and miniaturization, they become a common tool not only for trained specialists in dedicated laboratories, but also an important driving force for their development in increasing the use of them directly by end-users for monitoring of content of given analytes in various matrices.

The fundamental problems associated with design and applications of chemical sensors are selection of most appropriate chemical mechanism of sensing, immobi-lization of molecular recognition species on the sensing surface, or in the bulk of sensor, using various physico-chemical processes and selection most suitable transducer. The selection of mechanism of sensing depends on kind of analyte, its concentration range expected, and selectivity for future applications. For preparation of recognition layer on the surface of transducer, various methods of coating by adsorption, deposition, evaporation, polymerization, and electrodeposition are used, including formation of highly-organized self-assembled layers or bilayer lipid membranes. Among modern technologies employed for this purpose for mass-production of sensors and biosensors, a screen-printing method should be mentioned. The main groups of transducers employed in sensors and biosensors are electric, electrochemical, electromechanical, optical, thermal, and mass-sensitive ones.

2 Signal Transduction

2.1 Electrochemical Sensors

Earlier, the electrochemical detections were mostly employed in chemical sensors and biosensors, and until now they are most commonly used, especially in commercially available sensors, mainly for clinical and environmental analyses. An intensive development of optical sensors (optodes) in recent 30 years has resulted in numerous designs and commercial products, which are increasingly competitive to electrochemical sensors. A more limited importance, especially as mass production is concerned and applications in routine analyses, have thermal and mass-sensitive sensors and biosensors.

In electrochemical transducers, changes of various electrochemical quantities are utilized, which are associated with electrode processes in the presence of analyte (amperometry, various modes of voltammetry, voltohmmetry, coulometry) or with changes of electrical properties of medium connected with the presence of analyte (potentiometry, conductivity, capacitance, dielectric permittivity, inductance).

Zero-current potentiometry is based on change of ionic distribution in electrode–solution interface caused by the presence of analyte. Potentiometric signals are based on measurements of the difference in potential across the electrodes of electrochemical cells in the absence of appreciable current. In most applications, the electrode potential of one electrode (reference) should be known, constant and insensitive to the composition of measured sample. The response of other electrode in the cell – indicating electrode – depends on the analyte concentration. An ideal indicating electrode responds rapidly and reproducibly to changes in activity of the analyte ion, according to Nernst equation in case of metallic, second kind and redox electrodes, or according to Nikolsky–Eisenman equation for membrane ion-selective electrode. The potential differences between these indicator and reference electrodes are proportional to the logarithm of the ion activity. As no current flows in such measuring system, the potential is usually governed by the thermodynamic properties of the system, and it results from interactions between conducting phases of an electrode and an electrolyte. Changes in the potential of a conducting phase can be affected by altering the charge distribution on or around the phase and coulombic interaction between phases. The most common membrane electrodes are pH glass membrane electrodes, electrodes sensitive to common alkaline metal ions, alkaline earth metal ions, several transition metal ions, ammonium ions, and some common anions (chloride, bromide, iodide, nitrate). The range of practical application of potentiometric membrane sensors is very broad,[26] and also successfully this detection can be used in various biosensors with addition of appropriate biorecognition element to the detection system. Some drawback of this detection is slow response, especially in low concentration level, but its advantage is a wide concentration span of response, unique among all transduction methods employed in chemical and biochemical sensors.

Another common application of potentiometric measurements is detection with filed-effect transistors, invented by the use of palladium layer on gate oxide surface for detection of gaseous hydrogen.[13] In further development of transitor-based sensors, the modification of gates with layers containing ionophores resulted in ion-sensitive field-effect transistors (ISFETs), which are also already commercially available, or by the use of modification with layers containing enzymes or antibodies – in enzyme filed-effect transistors (ENFETs) or immunochemical transistors (IMFETs). In recent years, increasing attention is directed toward the design of enantio selective membrane electrodes with the use of chiral ionophores.[53]

In electrochemical methods based on measurements of Faradic current, where in sensors predominates amperometry at constant potential, detection is based on measurements of current or charge resulting from electron-transfer process, which occurs with participation of analyte or is initiated by presence of analyte. Amperometric measurements are usually conducted in typical voltammetric three electrode system consisting of a working electrode, a reference electrode and a counter electrode that serves to conduct electricity from the source through the solution to a working electrode. The current is determined by a combination of the rate of mass transport of analyte to the edge of the diffusion layer by convection and the rate of transport of analyte from edge of diffusion layer to the electrode surface. Under mass controlled conditions, the current signal has a linear dependence on the bulk concentration of the target analyte, as well as on the area of the working electrode surface.

The working electrodes employed in amperometric sensors have a variety of shapes (e.g., disk, fiber, microband, arrays, interdigitated microbands), and may be of inert metal, such as platinum or gold, various forms of carbon or semiconductors, such as tin or indium oxide. When geometric dimensions of working electrodes are made progressively smaller (e.g., down to micrometer diameter disks), the behavior of electrode departs from that of a large electrode because of changes of the conditions of mass transport from the bulk solution toward the electrode surface. This has several practical implications such as fast steady-signal response, increase of current magnitude because of enhanced mass transport at the electrode boundary, decrease of ohmic drop and increased signal-to-noise ratio.

It is commonly assumed that application of these methods in sensors has started from invention of oxygen Clark electrode,[2] and in biosensors from first glucose biosensor.[3] At present, main sensor application of amperometric and voltammetric detections include, with wide use of oxygen Clark electrode, amperometric sensors based on modification of working electrodes with various materials, and biosensors employing practically all biorecognition species. With the very wide use of the term "sensors," applications of voltammetric detections include also miniaturized screen-printed devices for stripping determinations of, e.g., heavy metal ions.

In recent years, much attention has been focused on the design of arrays of potentiometric or amperometric sensors of various degrees of miniaturization and integration.[54,55] Commonly, although slightly euphemistical, they are described as electronic or bioelectronic noses or tongues, depending on gaseous or liquid sample to be measured. They are constantly improved for multianalyte determinations, including advanced chemometrics for data processing, and in some cases become competitive to high-performance chromatographic methods, which is evidenced by numerous applications,[56,57] and commercially available instrumentation.

The method closely related to voltammetry is voltohmmetry, which is based on the fact that the lateral resistance of a thin metal film depends on the presence or absence of interacting species at its surface. Resistance measurements show a specific selectivity in the electrode potentials, and magnitude of resistance change depends linearly on analyte concentrations in solution from p.p.b. to p.p.m. range.[58] This method is alternative for voltammetry, not only for heavy metals determinations, but also for nonelectroactive species that can be deposited or dissolved under defined potential of resistance.

A large and important group of electrochemical sensors are gas-sensitive metal gate semiconductor devices, invented by Lundström et al., in 1975.[13] Metal oxide semiconductor (MOS) sensors are usually based on the semiconductor silicon, which is oxidized to silicon dioxide film and can be made in two forms as MOS capacitors or MOS transistors. The most popular route toward achieving selectivity of these sensors is via doping. MOS capacitors doped with catalytic metals as the gates have been developed for hydrogen, ammonia, and alcohols, where catalytic metal gate aids in adsorption or dissociation of molecules at a metal–oxide interface. Numerous sensors with various dopands have been also made based on oxides of tin, zinc, titanium, chromium, and tungsten.[59] Besides MOS devices to the group of chemical conductometric sensors, one can include numerous chemiresistors, including ion channel-based sensors, based on blayer lipid membranes, other biomembrane-based sensors employing e.g., self-assembled monolayers and Langmuir–Blodgett films.

Sensors for determination of organic vapors are made using field-effect transistors being an example of organic electronics. Organic thin-film transistors (OTFT) are field-effect devices with organic or polymer thin film semiconductors as channel material. They were shown also to operate in aqueous solutions, and they can be integrated into microfluidics. In the simplest form, OTFT comprises a conductive gate covered by a thin dielectric film that is interfaced to the organic active layer. The active layers of OTFT sensors include thiophene-based polymers and oligomers, naphthalens, phthalocyanines and pentacene, and they were developed for sensing e.g., alcohols, ketones, thiols, esters, and cyclic compounds.

In design of electrochemical sensors (and biosensors) especially helpful is electrochemical impedance spectroscopy (EIS), providing a complete description of an electrochemical system based on impedance measurements over a broad frequency range at various potentials, and determination of all the electrical characteristics of the interface.[60,61] Generally it is based on application of electrical stimulus (known voltage or current) across a resistor through electrodes and observation of response (corresponding current or voltage). Processes occurring include various microscopic processes, such as transport of electrons through electronic conductor and electrode/electrolyte surface, and from charged to uncharged molecular species that results from some oxidation or reduction reaction. Its aim is to determine bound and mobile charged species in the bulk or interfacial regions of solid or liquid material in electrochemical cell.

The electrochemical transducers are widely employed in design of biosensor with various biological recognition species. The main problem to solve is to choose the most convenient biochemical receptor and the way of its application in measurements, the most often being in immobilized form. The way of immobilization determines the optimum use of activity of bioreceptor and stability of measuring system (biosensor). The selectivity of response is determined by the sum of properties of detecting element of transducer, modulation of transport of potential interferents through the layer with immobilized receptor, and by the reactivity of employed process of molecular recognition.

Potentiometric transducing in biosensors with various recognition species is based mostly on the use of membrane ISE for measurement of analyte, which takes part in particular biochemical process. The sensitivity of response of enzymatic biosensors are additionally affected by the kinetics of enzymatic reaction, amount of enzyme used, and buffer capacity. Among the most often used methods of bioreceptor immobilization are cross-linking to ISE membrane, retaining at ISE membrane by dialysis or selectively permeable membrane or physical entrapment in synthetic matrix covering the ISE membrane. The potentiometric ISEs used for these purpose are mostly pH electrodes with glass or plasticized membranes and ammonia electrodes.[62] For the same purpose also ISFETs are used with additional biorecognition layer placed over ion-selective membrane.

The amperometric detection is most often employed in biosensors, especially employing enzymatic biorecognition step. This is mainly due to the presence, in most commonly used enzymes, of redox groups which change their redox state during the biochemical reaction (oxidases and dehydrogenases). By the use of reversible electron mediator (e.g., ferrocene, quinines, or redox active dyes) in measured solution, or coimmobilized with enzyme, an essential decrease of detection potential can be obtained, which often results in elimination of effect of interferences. The most elegant way to connect the redox center of enzyme to electrode surface is via the so-called molecular wiring. The redox center of enzymes can be wired to carbon electrode with e.g., osmium complex containing a redox-containing network. An alternative method for shutting electrons between the enzyme and the electrode is the immobilization of enzyme within a conducting polymer matrix. The amperometric transducing can be also successfully employed in enzyme-label linked immunoassays. The advantageous property of amperometric detection is a possibility of amplification of signal, which can be achieved when analyte can access a cycling process when stoichiometric amount of product is formed and accumulated each time the analyte is turned over in the cycle. The amperometric detection is widely used in the format of screen-printed biosensor for detection of substrates and inhibition of enzymes.

For applications in design of electrochemical biosensors, also various types of capacitance detectors are employed. This can be employed for enzyme sensors to detect substrates, for monitoring cells and cell cultures. A capacitance-based method for specific detection of low levels of pesticides and environmental contaminants has been developed using conducting polymers, which served as chemically sensitive film that transduces an immunoassay into an electrical signal. This can be employed also for DNA-based sensing systems. Change of capacitance can be also monitored for systems with molecularly imprinted polymers. In all these cases, the detection is without label and fast; however, main problems are ambiguity associated with data interpretation and small sensitivity of response.

Impedance transduction can be employed in traditional electrochemical biosensors where electroactive species are generated or destroyed by biological detection, or in direct impedimetric monitoring of bioaffinity reactions without degradative product formation. Numerous applications of this transduction were reported for biosensors with the use of conducting polymers,[61] where sensing elements (receptors, enzymes, antibodies) are entrapped within the conducting polymer, or within a phase adjacent to conducting polymer. Impedance was also used for probing bacteria and cells, which resulted in commercial bactometers for food industry, water control, and pharmaceutical industry.

For design of biosensors, the above mentioned electrochemical impedance spectroscopy (EIS) can be also employed.[63] Immobilization of biomaterials on electrode surfaces alters the capacitance and interfacial electron transfer resistance of the electrodes. The impedance features of the modified electrodes can be translated into equivalent electronic circuits consisting of capacitances and resistances. Kinetics and mechanisms of the electrochemical processes occurring at modified electrode surface could be derived from the analysis of equivalent circuit elements. Different immunosensors that use impedance measurements for the transduction of antibody–antigen (Ab–Ag) complex formation include in-plane impedance measurements between electrodes separated by a nonconductive gap, modified with antigen or Ab molecules. They may also include the amplified detection of Ab–Ag complex formation using biocatalyzed precipitation of insoluble products as amplification route and Faradaic spectroscopy as readout signal. The amplified detection of Ab–Ag complex can be also carried out by the biocatalytic dissolution of polymer film associated with the electrode and Faradaic impedance spectroscopy as transduction method. Impedance spectroscopy offers a very sensitive method to follow time-dependent protein binding to electrode. One of the major problems associated with impedimetric immunosensors is relatively small change of impedance spectra upon direct binding of Ab to Ag-functionalized electrodes. Impedimetric immunosensors very often suffer from nonspecific adsorption molecules.

In a similar way, a DNA biosensors can be designed. Design of DNA biosensors requires immobilization of nucleic acids, acting as the sensing interface, on the electronic transducer and the electronic transduction of hybridization occurring with analyte DNA on the transducer. The analysis of target DNA is based on its hybridization with a complementary nucleic acid on the electronic transducers. Numerous additional steps were introduced in the sensing process to obtain higher amplification of impedimetric signal, including e.g., biocatalytic process. DNA impedimetric biosensor attracts focus to gene analysis, detection of genetic disorders, tissue matching, and forensic applications.

Combination of impedance spectroscopy with other physical methods (SPR, QCM, FTIR spectroscopy) is especially productive in characterization of the interfacial properties of sensing and bioelectronic devices.

2.2 Mass Sensitive Sensors

Mass sensitive sensors, or oscillator sensors, are based on piezoelectricity of crystals (usually suitably cut quartz crystals), where the natural frequency of crystal is determined in part by mass of the crystal, and the resonance frequency varies with any changes in the mass of crystal. When placed in an oscillator circuit, the portion of quartz wafer located between the electrodes vibrates at its fundamental frequency. In bulk acoustic wave (BAW) mode sensor operates in a principle such that very thin films on oscillating piezoelectric crystal causes change of frequency proportional to the mass deposited on the crystal surface. This is the most common application of piezoelectric detection as sensitive microbalance (Quartz Crystal Microbalance – QCM), which is most prominent representative of acoustic sensors with transduction between electrical and mechanical energies.[64] It is widely used both for detection in gaseous phase, and in liquid environment. Several techniques have been used for the application of coating to piezoelectric crystals such as chopping, dipping, spraying, spin coating, and Langmuir–Blodgett technique. Numerous applications have been developed for determinations in gas/vapor and aerosol measurements. The gaseous pollutant is selectively adsorbed by a coating on the surface of the piezoelectric crystal. This can be employed, e.g., for determination of inorganic gases, humidity, hydrocarbons, and other volatile pollutants of ambient air. Biological molecules can be also employed in the coating layer to induce specificity by immobilization methodologies for biocatalysts or immunocomponents. It may be worth to mention that only this technique, and discussed below surface plasmon resonance (SPR), provide labelless method for the direct study of Ab–Ag reactions.

The surface acoustic wave (SAW) piezoelectric devices operate on the base of Rayleigh wave propagation principle at solid thin-film boundaries. Two sets of interdigitated electrodes are deposited on a piezoelectric crystal surface, of which one serves as transmitter and other one as receiver. Radio-wave frequencies applied to the transmitter electrodes produce a synchronous mechanical stress in the crystal, which produces acoustic wave with both longitudinal and vertical shear components. There are generally two types of SAW devices, the delay line and the resonator of which more interest is focused on the device that operates as delay line device, which can be configured to monitor changes in amplitude or velocity. There are numerous examples of application of SAW devices in both gaseous and aqueous phases, with a modified chemically sensitive or biochemically sensitive interface. The velocity of the wave in the sensitive surface layer may be more comparable with that in the piezoelectric material.

Piezoelectric devices can be also used in the thickness shear mode (TSM), which allows the monitoring changes in three-dimensional shape of living cells when they are exposed to chemical, biological or physical stimuli, since the electromechanical analysis of signal offers possibility of extracting a number of relevant parameters in dynamical mechanics. Mechanical models can be transformed into electrical equivalent circuits, which permit a complete description of oscillation in the presence of adsorbent.[65] In monitoring of living cells, it has to be taken into account that with time the state of adhesion to the substrate alters due to changes in the number and type of binding proteins, the strength of adhesion, and alteration of the cytoskeleton.

2.3   Thermal Sensors

The calorimetric detection has considerable potential in design of biosensors, although its important disadvantage is the lack of specificity; hence, all enthalpy changes in measured system contribute to the final result of determination. Numerous enzyme-based biosensors have been designed as calorimetric sensors using thermistors as the means of signal transduction, where electrical resistance is highly temperature dependent. Many enzymes reactions with oxidases or hydrolytic enzymes are associated with significant enthalpy changes; hence, thermistors with immobilized enzymes can detect local temperature changes in the environment near the thermistor surface. In appropriate design of sensing systems, detection is based on temperature measurements with resolution down to 10^{-4}°C.[66] Thermistors are virtually free from drift and fouling, which is important as they do need to be mounted in direct contact with sample. Thermometric enzyme-linked immuno-sorbent assays with immunosorbent immobilized in the microcolumn and the use of enzyme-labeled antigen can be employed for determination of hormones and antibodies. Measurements can be also carried out in organic solvents, and as temperature response depends on the heat capacity the sensitivity of detection can be multiplied, compared with that one in aqueous solutions.

In the second type of thermal sensors, a pyrolytic crystal responds to temperature changes in the same way as piezoelectric crystal responds to stress, i.e., with a change in the spontaneous polarization of the crystal.[67,68] The change in polarization induces a charge on the electrodes, which can be readily measured. This type of sensors is employed for gas sensing with different materials employed as sensing wafers.

2.4   Optical Sensors

In optical sensors and biosensors, detection is based on measurements of change of any optical characteristics of medium that is caused by the presence of analyte (mostly optical absorption, reflection, or luminescence). Although first optic sensors can be traced back to the use of immobilized indicators on cellulosic paper, silica gel or glass slides, since pioneering work by Petersen et al. in 1980,[14] the majority of modern optodes (commonly used name for optical sensors) are based on fiber optics. The optic fibers and planar waveguides are the main components of waveguide-based sensors with cylindrical and planar structures, respectively. However, it has to be admitted that to optical sensors also modern stripe systems with immobilized reagents and instrumental reflective detection, commonly employed in clinical chemistry, should be also included.[68,69]

Cylindrical optode designs using fiber optics can be classified from the construction point of view into several groups, including single or bifuricated fibers with no indicator chemistry for direct, remote spectroscopic measurements, or with immobilized indicator dyes where either reflected or transmitted light can be used for detection.[33] Immobilized, indicating intermediate molecules are needed when analytes to be detected lack of optical measurable properties. They are usually neutral or ionizable dyes, which preferably should have absorption bands between 550 and 650 nm to use low cost radiation sources and work with optical fibers. Commonly, they are immobilized, occluded, or dissolved in supports that are formed by cross-linked polymers, plasticized polymers, or organic and inorganic activated surfaces. In surface optodes, the optically active reagents are immobilized over the surface of optical components (wave-guides, glass, prism) or in a porous matrix, and they are in direct contact with analyte. In bulk optodes reagents are dissolved in immiscible phase that is held up to inert polymeric matrix and serves as sensitive element, and response of sensor depends on change of analyte concentration in the bulk of that phase. Most commonly they are plasticized PVC membranes with incorporated chromoionophore or fluoroionophore.

Various developed wave-guide optochemical sensors employ designs with different area where interaction of light with matter occurs. In extrinsic optodes (conventional design), light temporarily exits the waveguide to interact with analyte or active phase before being directed and guided again. In intrinsic optodes light whole time continues within the waveguide and is modulated to the analyte while it is being guided. The absorption-type sensors include core-based optodes, such as hollow fibers, and direct and coating-based evanescent wave spectroscopic optodes. In the simplest designs, the optical source is light emitting diode or diode laser, while as detectors a PIN photodiodes are used. The development of these sensors is focused on improving ion-selective membranes, as well as on giving ruggedness and portability to the optical system.[70] New area of hollow waveguide applications is the infrared spectroscopy, developed mostly for gas-sensing, but also for detection in aqueous media in the middle IR spectra range using silver halide fibers.

Detection with optodes can be also based on luminescence principles, including emitting or quenching fluorescence of phosphorescence, chemiluminescence, or bioluminescence of conjugates of antibodies or antigens with enzymes. In fluorescent optodes with indicator, the excitation light is introduced into the proximal end of the fiber and travels to the end where it excites immobilized fluorescent indicator. Then some of the isotropically-emitted fluorescence is recaptured by the fiber to a detection system. Many biological molecules show an inherent fluorescence, but generally fluorescent labels are more widely used linked to enzymatic measurements or immunoassays. One of the limitations of sensitive fluorescent indicators can be that of quenching by solution species, which also can be turned to some use in measurements via a quenching agent.

Increasing research activity is focused recently on design of optode arrays. An optical imaging array is a coherent bundle of optical fibers, each possessing its own core and clad. Such array of probes can be used to analyze a sample for several species simultaneously. Patterned array of recognition elements can be also immobilized on the surface of a planar waveguide, and it can be employed, e.g., for detection of hazardous biological agents and environmental pollutants.[71]

Another class of optical sensors is holographic sensors, where combination of rationally designed polymer matrices with incorporated selective recognition elements, together with holographic fringes, provides a relatively inexpensive and generic sensor technology. The holographic element provides not only the analyte-selective polymer matrix, but also the optical interrogation and reporting transducer. Any physical, chemical, or biological mechanism that changes the spacing of the holographic fringes generates observable changes in the wavelength or intensity of the reflection hologram. The holographic sensor (e.g., for trypsin) has been developed from chemically synthesized artificial polymers.[72]

Optical or piezo-resistive read-out detectors can be employed for design of microcantilevers-based nanomechanical biosenors for direct detection of biomolecular interactions.[73] Microcantilevers translate molecular recognition of biomolecules into nanomechanical motion, and high-throughput platforms using arrays of cantilavers can be used for simultaneous measurement and read-out of hundreds of samples.

The surface plasmon resonance (SPR) is another optical method for investigation of the layer structure and thickness, sensing gases, but mostly employed for detection of biochemical interactions.[74,75] They can be sensed both directly or via an optical label. When film of so-called free electron-like metals such as silver or gold attached to a quartz prism is irradiated from the back at a critical angle, energy from light beam will be lost in generating plasmons, which are charge density waves occurring at the interface of metal layer and crystal. This process is affected by the presence of materials at the metal surface and therefore provides a useful method for characterizing binding and recognition events occurring within a thin layer on the metal surface. In commonly used instrumental systems, a convergent light beam together with a photodiode array is used to accurately determine the shift of resonance angle. The applications of SPR reported in the literature include antibody characterization, receptor-ligand interactions, and detection of DNA hybridization.

3   Mechanisms of Chemical Sensing

In chemical sensors, the mechanisms of their response depend directly on interaction of analyte with signal transducer, and main kinds of these interactions were indicated in above discussion of main transducers employed in design of sensors. The strength and selectivity of these interactions are crucial for the sensitivity and selectivity of response of given sensor toward particular analyte. The involvement of bio-chemical molecular recognition step to detection mechanism leads to design of biosensors, which will be separately discussed below. In case of chemical mecha-nisms, the sensitivity and selectivity of response is obtained by the use of various materials in signal-forming element of sensor, by chemical modification of active surface or by including into signal forming mechanism of additional, selective chemi-cal reaction of analyte carried out on/in the sensor. As one can expect, these different procedures are the subject of hundreds of scientific and technological papers in contemporary analytical literature.

As it has been already mentioned, a very large group of chemical sensors are gas sensors of very different design and mechanism of response, and of various and very wide applications. It seems that most widely performed studies and largest number of applications can be assigned to solid-state resistive gas sensors. Although tin oxide is the most popular semiconductor gas sensor materials, other metal oxides, metals, and organic materials have been successfully configurated as conductive gas sensors. Of these, most are sensitive to reducing gases, but some have been developed to sense partial pressure of oxygen and carbon dioxide. Current development efforts are focused on synthesis of materials that have high sensitivity to gases of interest, including e.g. phthalocyanine or conducting polymers. Noble metal additives as Pt, Pd, or Ag are added to metal oxide semiconductor sensors materials to enhance sensor response to a particular gas or class of gases. A number of sensors arrays consisting of an assortment of gas sensors have been developed.[76] Resistive hydrogen sensors are fabricated e.g. by deposition of palladium by vacuum evaporation on aluminum substrate or by the use of thin Pd film as gate electrode on field-effect transistors. Gas sensors are also fabricated by covering the quartz oscillator surfaces with a sputtered noble metals, oxides, or films of other materials selectively but reversibly adsorbing an analyte gas. In the last case, these films can be obtained by chemical polymerization, electrodeposition, or self-assembling. Various materials can be also used for design of cataluminescence-based gas sensors e.g.,[77] where sensing is based on chemiluminescence emitted in a course of catalytic oxidation of particular gaseous analyte. Large research activity in recent years is also focused on possibility of the use of carbon nanotubes (CNT) in gas sensors.[78] The detection in such devices may involve e.g. measuring of changes of resistance of CNT layer, catalytic effect toward chemical oxidation of analyte, changes of electrode impedance with CNTs trapped in interdigitated microelectrode gap, changes of current in Nafion-coated CNT field-effect transistors, piezoelectric detection of volatile analytes, SAW sensing with CNT deposited onto quartz substrates or thermoelectric detection. Other group of gas sensors are gas-sensitive membrane potentiometric electrodes with gas permeable outer membrane and internal ion-selective electrodes.

The mechanism of functioning of potentiometric chemical sensors, of which most commonly are developed and used membrane ion-selective electrodes (ISE), is based on change of distribution of ions on the interface membrane–electrolyte. Thus, there are designed solid-state and plasticized membrane of different chemical composition, which can be tailored to choose and transport ions with a selected charge, sign, or a particular ions among others of the same charge. Interactions of analyte with components of membrane and its transport in the membrane phase decide about magnitude and selectivity of analytical signal. The main direction of development in this field is a search for new materials for solid-state membranes,[79] and new ionophores for liquid or plasticized membranes,[80] including e.g. enantioselective ionophores.[52,81] The limit of detection of ISE can significantly depend on design of electrode and the use of inner solution of appropriate composition.[82] Generally, a key to improvement of detection limit and selectivity is the reduction of transmembrane ion fluxes. This can be achieved e.g. by the reducing of diffusion coefficients in polymeric membrane by increasing their polymer content, by covalent binding the ionophore to the polymer backbone, by decrease of concentration of ion exchanger in the membrane, by increase the thickness of membrane, or by increase of stirring the aqueous phase. Limit of detection of ISEs can be also improved by application of rotating electrode measurements.[83] In recent years, large attention is also focused on the use of conducting polymers in design of ISEs[84] and miniaturization.[85] Commercially available are, also in recent years, ion-selective field-effect transistors with different gate materials.

Faradaic processes of electrode reactions, which are principle mechanism of obtaining analytical signal in amperometric sensors, significantly depend on working electrode material and state of its surface. The common working electrode materials include noble and seminoble metals, solid oxides of various elements and different kinds of carbon materials including carbon nanostructures. They are employed in conventional voltammetric measurements with various modes of electrode polarization, as amperometric chemical sensors, as well as for construction of amperometric biosensors.

Another way of alteration selectivity of response and signal magnitude of amperometric sensors is modification of electrode material by the use of carbon pastes of different compositions or composites of different materials, as well as modification of active surface of sensors. Especially in the last case, numerous procedures of surface modification have been developed.[86] They may induce, e.g., electrocatalytic properties in order to obtain sensitivity for particular analyte, or chemically deposited or electrodeposited layer may enable to bind onto the electrode surface a biomolecule in design of biosensors. Chemical modification of sensor surface can be carried out by formation of polymer layers by direct polymerization on the surface, and by spin or dip coating. Especially advantageous is electropolymerization with well-defined conditions and localization of coatings, and possibilities of the use of monomers with additional functionalities that can be used for further chemical or biochemical alterations.[87] Coating with metallocyanates may provide strong size-charge selectivity toward incorporated counter ions, while e.g. with metalloporphyrins – selective catalytic functioning mimicking various enzymes.

Different other attractive way to modify the active electrode of amperometric sensors, besides deposition of polymers or electropolymerization, is formation of self-assembled structures, e.g., self-assembled monolayers (SAM) on solid supports [88,89] or bilayer lipid membranes (BLM) on various types of support.[90] Both types of theses structures can either induce selectivity of sensor to particular analytes and serve as convenient support for immobilization of other reagents or biomolecules. Ion recognition properties of self-assembled monolayers can be achieved with SAMs containing redox active and inactive receptors.[91]

Numerous same methods of surface coating are used in case of piezoelectric sensors, where retention of analyte on the surface of quartz oscillator, due to different interactions, is a source of analytical signal. Similarly as in case of modification of working surface of voltammetric sensors, these coatings may be employed for formation of next layer(s) with other reagents or biomolecules. Especially attractive for piezoelectric sensors is the use for modification of selectivity of response a molecularly imprinted polymers (MIP) that may provide selective retention of particular analyte or even enantioselectivity.[92] MIPs are formed in a process when functional monomers and cross-linkers are copolymerized in the presence of target analyte. Target acts as molecular template, which form complex with the template. After polymerization, functional groups of monomer are held in their predefined position by highly cross-linked polymeric structure. Template is removed by solvent extraction or chemical cleavage, and remaining binding sites are complementary in size to the analyte. For the use in sensors, electropolymerization can be done on conducting surfaces, but more generally are applicable standard surface coating techniques as spin or spray coating. The other simple way to synthesize a polymer layer on a flat surface is to use sandwich techniques, where imprinting solution is cast between the transducer surface and another flat surface of glass or quartz disk. Preformed polymers in the form of nano or microparticles can be interfaced with the transducer surface by entrapment in layers of different materials, including e.g. electrodeposited conducting polymers. MIPs besides piezoelectric sensors can be employed also on other transducers, including voltammetric or impedometric.[93]

In fiber-optic chemical sensors, the mechanism of inducing selective response and ways of its modulation depends on type of optode. In direct fiber-sensors, no chemistry is involved; hence, optimization of sensing involves geometry of optode, kind of spectroscopy used, or technique of spectroscopic measurements, including those more sophisticated ones, such as evanescent wave spectroscopy or SPR. Much more research is focused on indirect optodes, where color or fluorescence of an immobilized indicator is monitored.[94] In this case, the interplay of indicator, support, and analyte in optical sensor layer are of crucial importance. Sensitivity and selectivity of response depend principally on chemistry involved in changing of selected optical signal. Indicator-based chemical optodes are being developed mostly absorptiometric or fluorimetric, as sensors for gases, vapors, humidity, ions, and selected organic chemicals, such as agrochemicals, pharmaceuticals, or environmental pollutants. Both organic and inorganic polymer materials are used for immobilization of indicator dyes, and they play important but complex role in functioning of optodes. Numerous properties of indicator dyes are taken into account for application of optodes, such as absorption coefficient or emission quantum yields in solution, solvatochromism, selectivity of reaction with analyte, and stability in immobilized form. The indicator dyes are deposited on polymer or incorporated into polymer layer. The largest group of chemical optodes at present are sensors based on the use of fluorescent probes,[95] which have been developed for gases, pH, metal ions, and H_2O_2.[96]

4   Recognition Methods in Biosensing

Including to analytical procedures, a biochemical processes, known from their occurrence in living organisms, is a difficult task, but also a possibility of gaining results that are sometimes impossible to achieve in physico-chemical methods of chemical analysis carried without those processes. Regardless philosophical aspects of this approach, the nature is practically unlimited reservoir of mechanisms and processes, which scientists discover with the progress of science and try to follow for applications in development of various technologies and devices. At current status of development of science and technology, the excellence of numerous natural mechanisms is very often beyond human imagination and technological abilities, and it is tremendously difficult to carry them out of living organism, and to follow them in artificial systems. As a very common example that can be referred, related to design of sensors, is extremely sensitive detecting of feromones by insects or animal and human sight or olfactory ability.

Many biological processes are being exploited in recent decades in chemical analysis and clinical diagnostics, where they allow to perform analytical determinations, which are much more difficult to carry out with other methods, or even impossible at present state of development of analytical instrumentation.[97] One of such applications is design of biosensors. Biological recognition elements can be divided into two types – catalytic and affinity-based.[98] The catalytic ones include enzymes, whole cells, tissues, organelles, and microorganisms. The affinity-based types include antibodies, receptors, nucleic acids, and molecularly imprinted polymers.

The incorporation of a selective step with the use of appropriate biochemical reaction to analytical procedure is a significant field of development of modern analytical methods. With the use of various transducers, a specific interactions can be quantified in the systems of different construction, based on a bulk adsorption of analyte, specific binding at sites in flexible matrices, monolayer or multiplayer interactions, and membrane-bound receptor interactions. In all these mechanisms, the key role is played by strength and selectivity of biorecognition that depend on a multitude of noncovalent interactions between binding partners. The affinity constants of systems employed in design of biosensors span from 10^3 to 10^5 M^{-1} for enzyme–substrate interactions, through 10^7–10^{11} M^{-1} for antibodies–antigens, 10^8–10^{12} M^{-1} for receptors (e.g., with hormones and toxins), and even up to about 10^{15} M^{-1} for avidin–biotin.[99] The high affinity stems mainly from shape complementarity with the help of ionic interactions, hydrogen bondings, van der Waals, and hydrophobic interactions. Because of that, it seems that research widely carried out on bioimprinted MIPs are justified and promising for future applications in sensors.

The success of design of biosensor significantly relies on the way how the biological recognition molecule is immobilized on the sensor surface or in the sensor bulk material. Binding of bioreceptor has been shown often as typical procedure for improving the biomolecule stability, and thus the general biosensor performance. Because of numerous technological, synthetic, and analytical applications of immobilized biomolecules, the scientific literature on this subject is very broad, see for example monographs.[100–104] Developed physical and chemical methods of immobilization of biomolecules include huge variety of supporting materials and methods of attachment of bioreceptors of which selection is essential for obtaining satisfactory activity and half-life of biomolecules, as well as operational stability of biosensor.

There are commonly employed several main types of methods for immobilization of biomolecules employed in design of biosensors, including physical adsorption on the surface of transducer, covalent binding, entrapment and encapsulation, and also cross-linking.

Adsorption methods involve reversible surface interaction between bioreceptor and supporting material, including electrostatic forces, Van der Waals forces, ionic and hydrogen bonds, and hydrophobic bonding. Usually these forces are rather weak, but sufficient for reasonable binding without significant alteration of immobilized species. Most commonly used procedure for preparation of biosensors include conditioning in appropriate chemical and physical conditions the immobilized species at surface of supporting material, followed by washing of unbound species. The most significant drawback of these methods of immobilization is the leakage from support; hence, these methods are mostly employed for preparation of single-use disposable biosensors.

Covalent binding is based on the use of different functional chemical groups of immobilized species, which are not essential for their biological activity, for covalent attachment to chemically activated supports (glass, cellulose or synthetic polymers, etc.). Chemical binding is mostly based on formation of peptide bond, diazo and isourea linkages, and on alkylation reactions. Most important factor in this type of immobilization procedures is to assure inactivation of the active sites of bioreceptor, which is then employed in biosensing mechanism. Several different factors influence the selection of a particular support, but the main factor is hydrophilicity of its surface, which maintains enzyme bioreceptor activity. Before the bioreceptor is added in a coupling reaction, usually functional groups of support material are activated by a specific reagent. For instance, covalent binding of proteins on surfaces activated by means of functional groups or spacers is performed commonly by the use of glutaraldehyde, carbodiimide, silane-compounds, and by formation of self-assembled monolayers. To give more particular examples, cyanogens bromide is often used to activate the hydroxyl groups in polysaccharides, binding protein to support by formation of isourea linkage, whereas corbodiimide group is employed for activation support containing carboxyl groups to bind proteins by formation of peptide bond.

Entrapment methods of immobilization of bioreceptors utilized the lattice structure of particular base material. They include such methods as entrapment behind the membrane, covering the active surface of biosensors, entrapment within a self-assembled monolayers on the biosensor surface, as well as on free-standing or supported bilayer lipid membranes, and also entrapment within a polymeric matrix membranes, or within bulk material of sensor. All these mentioned methods are widely employed in design of biosensors. The essential condition of success of these methods of immobilization is preservation of sufficient mobility of substrate or products of biochemical reaction, involved in sensing mechanism, as matrix may act as a barrier to mass transfer with significant implications for biosensor response. In appropriate conditions, the matrix used for entrapment of bioreceptor may act as barrier against interfering species, leading to enhancement of selectivity of biosensor response. Encapsulation of bioreceptors can be achieved by the enveloping the biological component within a various spatial forms of semipermeable membranes. These methods, however, are of less importance for design of biosensors than entrapment.

Cross-linking immobilization is a support free procedure, based on chemical binding of receptor species to each other to form a large, three-dimensional structure. Binding involves covalent bond formation between bioreceptor species by mans of bi- or multifunctional reagents, involving also often additional species as spacers to minimize the proximity problems affecting activity of bioreceptors. The drawbacks of cross-linking immobilization include of insufficient mechanical properties of cross-linked material and its poor stability.

4.1   Enzymes

Biocatalytic recognition by purified enzymes is the most common mechanism used in design of biosensors. Enzymes are biological catalysts that facilitate conversion of substrate into products by lowering activation energy of the reaction. They are proteins or glycoproteins, and biorecognition properties depend almost entirely on the amino acids of the exposed surface of enzyme molecule.

The first applications of enzymes in bioanalytical chemistry can be dated back to the middle of nineteenth century, and they were also used for design of first biosensors. These enzymes, which have proved particularly useful in development of biosensors, are able to stabilize the transition state between substrate and its products at the active sites. Enzymes are classified regarding their functions, and the classes of enzymes are relevant to different types of biosensors. The increase in reaction rate that occurs in enzyme-catalyzed reactions may range from several up to e.g. 13 orders of magnitude observed for hydrolysis of urea in the presence of urease. Kinetic properties of enzymes are most commonly expressed by Michaelis constant K_M that corresponds to concentration of substrate required to achieve half of the maximum rate of enzyme-catalyzed reaction. When enzyme is saturated, the reaction rate depends only on the turnover number, i.e., number of substrate molecules reacting per second.

In design of biosensors most often used are enzymes from oxidoreductases and hydrolases; however, in very broad literature on this subject, applications of many other enzymes can be found. The analytical characteristics of enzymatic biosensor depend on numerous factors. Besides the origin of enzyme (type of natural material, from which the enzyme was isolated), the most significant is a mode of immobilization. It affects significantly kinetics of enzyme and diffusional limitations of the immobilization matrix. Although there is a general belief that enzymes should be immobilized in a hydrophilic environment, a successful immobilization of enzyme in carbon paste with silicon oil [105] initiated development of numerous biosensors with enzyme immobilized into an organic phase composites.[106] In case of bioconversion of lipophilic compounds, it is desirable to carry out enzymatic reactions in mixtures of water and suitable organic solvent, and numerous enzyme biosensors were successfully used in such measurements. It was shown that catalytic activities of enzymes were maintained in such reaction systems.[107]

For biosensor design enzymes, besides covalent attachment to different supports (discussed below), they can be entrapped both in organic and various inorganic materials.[108] Performance of biosensors can be affected by biocompatibility of the entrapment procedure, permeability of the host structure and hydrophilicity and chemical inertness of the host matrix. Enzyme biosensors be can successfully used for enantiomeric analysis.[81,109] Of great importance is also possibility of alteration of enzyme selectivity using methods of modern genetic engineering, demonstrated e.g. for acetylcholiesterases [110] and laccases.[111]

Although the main analytes that are determined by enzyme biosensor are substrates of enzymes, numerous attempts were reported in the literature about using them for indirect determination of inhibitors.[112,113] Such determination can be also carried out in organic solvents.[114] Many of such inhibition-based methods can be questioned because of numerous fundamental and practical disadvantages.[115] These are, for instance, difficult or more often impossible regeneration of inhibited enzymes, and the fact that different inhibition degree is caused by different inhibitors. One can find example that in some cases, better results are obtained with the use of raw tissues when compared with purified enzyme.[116] It was also pointed out that the most rational use of inhibitory effects is after a separation step for proper removal of species, which can act on the activity of enzyme.

Numerous enzyme-based biosensors have been commercialized in recent three decades, mostly for clinical and environmental applications.[98]

4.2   Cells, Tissues, and Microbes

Another biocatalytic recognition elements employed in biosensors with different transducers are cells and tissues of various origin. When enzymes are isolated from a microbial source or tissues, it makes sense to immobilize the whole cells or tissues treating them in such way that activity of enzyme is retained, while other pathways are deactivated. Although with single purified enzyme the activity of the enzyme decays over time, the kinetics of living cell systems is more complex. In addition to reaction rate of interest, it is also necessary to consider the rates of cell growth and death, as well as kinetics of oxygen transfer and consumption.

In the simplest way reported in numerous papers for biorecognition, such cells or tissues are selected where the occurence of particular enzyme is known. Plant tissues can be grinded and placed at the transducer surface using dialytic membrane, e.g. spinach leaves for amperometric biosensor for catechol.[117] Numerous sensors with potentiometric detection have been reported with slices of animal tissues attached to the transducer surface in the similar way. Cells can be also entrapped in different materials, e.g. algal cells were successfully entrapped into a alginate gel or pyrrole-alginate matrix,[118] and numerous papers were published on entrapment of cells and tissues in carbon pastes.[106] This class of biocatalytic materials maintains the enzyme of interest in its natural environment, which results in a considerable stabilization of desired enzymatic activity. Additionally, tissue materials have been shown to provide sufficiently high activities for the construction of certain biosensors, where isolated enzymes have failed. The selectivity of biosensors based on whole tissue cells must be considered in detail owing to the large number of suspected biocatalytic activities within these cells.

More generally, cells are able to respond in sensitive way to changes in their environment by changing cell metabolism and cell morphology. For sensing the specific signals from outside, cell uses specific receptor proteins transmitting and amplifying signals to nucleus to change gene activities and to further functional units to elicit immediate response. If cells are successfully interfaced to appropriate microsensor, these changes can be monitored in real time and noninvasively. This can be achieved by growing cells directly on the surface of silicon or glass-based sensor chips. Chips may integrate different types of sensing structures with potentiometric, amperoemtric, or impedimetric principles of response. Living cells can be also used for detection with various optical readout systems (optical microscopes, fiber optics, CDD cameras) to detect absorptive or luminescent signals from cells and tissues. Depending on the transducers used, a various cellular parameters can be measured, for instance extracellular acidification with ISFETs,[119] or light-activated potentiometric sensors, oxygen consumption with amperometric or optical sensors changes in cellular morphology with electrical impedance sensors.

In spite of large research activity in the use of these biorecognition elements, it is still mostly a proof-of-principle or demonstration phase, and not close to extensive or commercial use outside of academia.[120] The exception can be for instance biochemical oxygen demand (BOD) sensors employing omnivorous yeast,[121] and produced commercially.

The catalytic biorecognition takes place also when bacteria are used in design of biosensors with various transducers. In such case, mechanism of detection is based on production o metabolites measured by appropriate transducers, but also on assimilation of organic compounds by microorganisms, or change of respiration activity. Many early studies on design of microbial biosensors have been focused on biotechnological applications (determination of sugars, simple carboxylic acids, ethanol), mostly by immobilization of suitable bacteria on amperometric oxygen electrode. Analytes were assimilated by immobilized microorganisms with consumption of oxygen. Soma bacteria are able to produce hydrogen, e.g. from formic acid, hence as transducer in such a case a fuel-cell can be employed.[122] Nitrifying bacteria, which utilize ammonia as energy source, can be used to obtain ammonia biosensor. Several different bacterial biosensors have been reported with the use of luminescent bacteria, which can degrade refractory organic compounds, and were employed in BOD biosensors.[121] Microbial sensors have been designed also for some other environmental applications e.g. for cyanide with bacteria that aerobically biodegrade cyanide,[123] or for trichloroethylene using also a selectively degrading bacteria.[124]

Similarly to modification of enzymes, for application in biosensors also, luminescence-based microbial biosensors were reported with the use of genetically engineered bacteria for such important environmental applications as determination of polycyclic aromatic hydrocarbons,[125] or for genotoxicity monitoring.[126]

4.3 *Immuno-Systems*

The principle of immunochemical determinations employed in analysis for almost half a century is specific interaction of antibody (Ab)–antigen (Ag). Antibodies are one of main classes of proteins. They are glycoproteins of a specific Y-shape of molecules that posses two identical binding sites per molecule. They are produced in living organisms (excluding plants) via the immuno response, as response to immunogen. Antigens are usually macromolecular species of a molecular weight >1 kDa, which are capable of binding selectively to antibodies. Smaller molecules, called in this case haptens, may induce also immuno response, after earlier formation of a conjugate with carrier protein. The sequence of amino acids at the top of two identical arms of Y-shape structure (light polypeptide chains) determines the specific antigen-binding properties.

The selectivity of Ab–Ag (or hapten–Ab) interactions is analogous to the selectivity of substrate–enzyme interaction. In contrary to enzymes, however, antibodies do not function as catalysts, but reversibly bind antigens into stable complex. In the immuno system response to antigen, the lymphocytes produce many different Ab molecules, all directed to different parts of antigen (e.g., proteins), and they are described as polyclonal antibodies. Monoclonal antibodies with strictly defined specificity and affinity can be also produced in complex procedures and they are much more expensive, but they offer a very important advantages compared with polyclonal Abs. They are produced with relatively little batch-to-batch variations, and their selectivity allows to minimize cross-reactions. The immunochemical interactions between binding partners are considered as primary ones, which involve specific recognitions of antigens determinant with binding site of antibody, and secondary ones, which occurs as result of antigen multivallency and leads to agglutination or precipitation of a polymeric Ab–Ag network. Because of large stability of immunocomplexes their dissociation requires rather extreme conditions, e.g. high or low pH vales, high ionic strength, presence of chaotropic ions, or denaturants.

Number of antibodies that can be produced is practically unlimited, hence wide development of immunoassays not only for determination of macromolecules, which induce immunoreactions, but also for small molecule analytes inducing such reaction after binding into suitable conjugates. Besides immunoassays carried out in various formats and immunoaffinity liquid chromatography, since many years very broad studies are carried out on design of integrated immunosensors.[127–129] In principle their construction should simplify the immunochemical determination by elimination various steps necessary in conventional immunoassays based on the use of labels to monitor the immunochemical reaction. In immunochemical process, the association rate is determined by diffusion rates of Ab and Ag, while dissociation rate is governed by the strength of Ab–Ag complex. The immobilization of Ab in immunosensors affects those reaction rates. Nonlabeled immunosensors are usually designed in such a way that immunocomplex is formed on the surface of transducer, which provides corresponding physical changes (electrical, optical, piezoelectrical, etc.). Similarly to format of immunoassays, also labeled immunosensors are designed with immobilized Ab, where concentration of analyte to be monitored is correlated with the amount of labeled antibody (added to measured solution). In case of both type immunosensors, they can be designed for determination of antibody by immobilization of appropriate antigen. For simplification of measuring procedure and elimination of immunocomplex dissociation step, very often immunosensors are considered as a single-use devices. Then, if changes of transducer signal provides insufficient limit of detection, indirect systems are being developed which relay on the use of enzyme- or fluorescence-tagged reagents. The unique approach is the use of Ab with catalytic properties, which not only binds Ag but also chemically transforms the target molecule to produce easy detectable product.[130]

In recent years, an evident progress is observed in design of biosensors not only for clinical diagnostics – the main area of early applications of immunoassays – but also for the needs of environmental analysis e.g., determination of residue pesticides,[129] or detection of biological threat agents.[130] In early stage of development of immunosensors, they were designed mostly with electrochemical transducers. They were e.g. amperometric sensors with enzyme labels on antigen or antigen labeled with electroactive species, or potentiometric transducers with various membrane electrodes.[131] As it is evident from reviews cited above, nowadays immunosensors are produced with numerous other transducers such as piezoelectric, capacitive, conductimetric, optical in various modes of measurements, and SPR.

The main problems that require more research and optimization of immunosensor design are nonspecific adsorption of coexisting substances, further involvement of genetic engineering in production of novel antibodies, and search for the same purpose of smaller proteins with antibody properties (e.g., anticalins).

4.4 Receptors

The other type of biomolecules researched since the middle of 1980s as elements of molecular biorecognition are receptors (also called chemoreceptors) [132] from a natural sources. Their role in organisms is strictly related to molecular mechanisms of human and animal senses. They are special type of proteins of protein complexes with carbohydrates and/or lipids, which interact specifically with biological and chemical substance (known as messengers, transmitters, or ligands). Ligands can be low molecular weight compounds (neurotransmitters, amino acids, hormones), and larger molecules (peptide hormones, peptide toxins, proteins), or even total life forms such as viruses and bacteria. In all cases, the binding of ligand to its specific receptor in living organism results in activation or blockade of a series of biochemical reactions, leading to some physiological event in receptor-bearing organism. Therefore, biological receptors are critical mediators of both intra and intercellular communication. For instance, in olfactory sensing from more than a thousand genes, each olfactory neuron is supposed to express only one receptor subtype.[133]

Unlike antibodies, the receptors are selective but rather nonspecific and react with classes of substances bearing a common structural component able to bind to the receptor binding site(s). So, similarly to antibodies, with some different properties, they can be immobilized on the surface of transducer to form receptor-based biosensor. The main advantage of such sensors is their unique capability to be used to screen samples for the class of analytes that bind to the receptor.

Receptors can be isolated and immobilized on the transducer surface, but for the purpose a complete organelle can be employed, e.g., animal or insect antennule, or nerve fibers,[132] where different sensitivity was demonstrated for various stimuli (amino acids and closely-related analogs). For comparison, the biosensor made with isolated olfactory receptor proteins, immobilized on the surface of piezoelectric crystal, exhibited sensitivity to different volatile compounds.[134]

Numerous applications have been reported in the literature on the use of cell membrane receptors which require the use of ligand bilayer or phospholipid vesicles to maintain functionality, but other immobilization methods were also successfully used. The biosensing of e.g. acetylcholine and cholinergies has been reported with different transducers (ISFETs, interdigitated electrodes with measurements of capacity changes and optical fiber optode with fluorescence detection).

The role of bioreceptor can be also played by viruses covalently attached to the electrode surface, which was shown for resistance detection of antibody and prostate-specific membrane antigen.[135] Taking into account up to 10^{20} unique species, phage-displayed libraries provide a vast pool of candidate receptors to practically any analyte, including small molecules, proteins and nucleic acids. Piezoelectric virus sensor has been developed e.g. for detection of distinctive antigens of the human cytomegalovirus.[136]

4.5 Nucleic Acids: Genosensors

A very broad research activity in recent years is focused on investigation of possibilities of applications of nucleic acids for biosensing, especially DNA. As DNAs in organisms function as carrier of genetic information, for biosensors employing DNA as molecular recognition element a name "genosensors" appears recently in analytical literature.

Biosensing applications of DNA are being investigated in two directions. One of them is biosensing of DNA hybridization to determine sequence of DNA for genetic testing, detection of various types of biological agents, and early diagnosis of numerous human diseases. The second field is application as alternative kind to antibodies and receptors, as the affinity-type biorecognition element for small molecular weight molecules. Results of these studies have been a subject of several recent reviews.[137–141]

In case of DNA hybridization detector, usually a single-stranded DNA probe (15–20 nucleotides long) is immobilized on transducer surface to form a recognition element. When target DNA (analyte) contains a sequence that exactly matches that of the DNA probe, a hybrid duplex is formed at the transducer surface through noncovalent interactions, described as base pairing, which is a highly specific process. This hybrid formation is then translated into analytical signal by means of appropriate transducer. Especially electrochemical transducers are suitable for direct detection of specific DNA sequences, although numerous other transducers for this detection are being developed. In the first-generation of electrochemical DNA biosensor, the detection was based on measuring changes in the peak current of redox marker, which binds to the target-probe duplex. It takes place when duplex is exposed to solution of indicator (oligonucleotide with redox marker). In the second generation electrochemical DNA biosensors, a direct detection of hybridization is employed. This can be based on the use of conducting polymers where base-pair recognition can be associated with switching the electronic properties of conducting polymer, or on detection of hybridization based on interfacial properties (impedance, capacitance, conductivity across bilayer lipid membrane). For this purpose also intrinsic electroactivity of DNA can be exploited or photocurrent produced on irradiation of a probe-modified surface, which is suppressed upon hybridization. Numerous advantages of the use of nanostructures for detection of DNA sequence with different transducers have been recently reviewed.[142] Described applications included the use of metal and oxide nanoparticles, Si nanowires, ZnS and CdSe quantum dots, and carbon nanotubes. The most impressive limits of detection (1 fM) for nucleic assay were reported with piezoelectric detection employing Au nanoparticles,[143] and fluorescence with fluorescent dye-doped silica nanoparticles.[144]

Among numerous reported applications of genosensors for DNA hybridization as few examples, one can refer to a disposable DNA sensor for detection of hepatitis B virus genome DNA,[145] biosensor systems for homeland security using DNA microarrays,[146] and DNA electrochemical biosensor with conducting polymer film and nanocomposite as matrices for detection of HIV DNA sequences.[147]

Other applications of DNA biosensors are based on utilization of interactions of DNA with various compounds that was also discussed in the cited reviews.[138,140,141] The main types of these interactions involve generally nonspecific electrostatic binding along the exterior of DNA helix, groove binding, where analyte interacts with the edges of base pairs, and intercalation of planar aromatic ring systems between base pairs, which can be accompanied with unwinding or bending of DNA structure. In electrochemical DNA biosensors, such interactions can be analytically utilized mainly as result of change of oxidation currents for guanidine and adenine. For the same purpose also changes of oxidation or reduction currents of analyte, caused by interaction with DNA, may be also employed, as well as shifts of formal potential of analyte caused by intercalation of nucleic acid-binding molecules into DNA helix. Examples of analytical applications of such interactions can be shown, e.g., determination of antitumor agent Mitomycin C based on changes of guanine peak as a result of interaction with ssDNA,[148] or determination of anthracycline antibiotic and antitumor drug daunamycin with dsDNA based on decrease of voltammetric current of drug due to its intercalation.[149] In another example of application, the antioxiant sensor based on sDNA modified electrode was developed, where damages occurring by dsDNA oxidation by photogenerated radicals are detected by adding Methylene Blue as an intercallant probe.[150]

Analytical determinations based on DNA recognition events can be carried out with various transducers – e.g., see review on optical transducers,[151] and they can be treated as biosensing methods complementary to immunochemical methods and those utilizing molecular and whole cell receptors.

5   Biomimetic Sensors

Development of artificial synthetic systems mimicking functioning of biological systems and processes, which exist in the nature always, was and still is a tremendous challenge for science, and it can be considered as potentially significant step toward understanding of the mystery of life.[152] Studies on design of analytical sensors in recent decades are also carried out toward development of synthetic materials, which might exhibit properties mimicking the biological recognition materials. As they are synthetically produced, sensor based on their use cannot be strictly considered as biosensor, and it seems that more appropriate term for them might be biomimetic sensors.

Current chemical literature contains numerous examples of attempts to synthesize compounds corresponding to biocatalysts, affinity receptors, or natural nucleic acids, together with examples of their analytical applications.

The synthetic compound that are considered as mimicking biocatalytic properties of enzymes often originate from well-developed field of chemical modifications of surface of working electrodes in voltammetry. For numerous modifiers, the electro-catalytic properties have been documented, and for the use in amperometric sensors they can play the same role as modification of electrodes with immobilized enzymes. For instance, one can find a polymeric system based on electropolymerization of metalloporphyrin-bearing pyrrole in acetonitrile as biomimetic devices, which present a remarkable similarity with the active site of a variety of vitally important enzyme systems and has been applied for sensing of dissolved oxygen.[153] Similarly, iron-tetraaminophthalocyanine was reported for hydrazine oxidation,[154] and glassy carbon electrode modified with a copper dipyridyl ion-exchanged Nafion film as biomimetic catalytic amperometric sensor for phenols.[155] Numerous other systems have been demonstrated mimicking the activity of enzymes, e.g., uricase,[156] glutathione peroxidase,[157] or peroxidase.[158]

The role of affinity-based biorecognition materials, such as antibodies or receptors, can be played by above-mentioned molecular imprinted polymers (MIP).[159–162] As antibody combining site mimics, MIPs show binding affinities and cross-reactivity profiles comparable to their biological counterparts and have been employed e.g. in medical diagnostic assays.[163] Besides most commonly used covalent and noncovalent molecular imprinting, also bio-imprinting and alternative molecular imprinting have been developed.[161] In the bio-imprinting technique, the existing recognition site in an enzyme might be modified by the presence of a print molecule. In alternative imprinting, the molecular recognition sites are formed at the same time as the molecular imprinting materials are prepared from the polymer solution. The molecular recognition sites can be introduced into various polymeric materials such as oligopeptide derivatives, natural and synthetic polymers. MIPs can be employed for a wide range of different transducers.[160] For instance a capacitance sensor for phenylalanine has been reported with MIP deposited on a gold electrode,[164] while QCM sensor was developed e.g. with S-propanolol-imprinted polymer, which was able to discriminate enantiomers of propranolol with selectivity coefficient of 5.[165]

MIPs compared with antibodies have several advantages, such as no requirement for laboratory animals for their preparation, then they are robust and cost of their production is low. There are also some drawbacks compared with antibodies, such as difficult to obtain site heterogeneity and polymer–template–solvent compatibility, as well as difficulty in preparation of MIPs for biological macromolecules.

As selective molecular recognition element in biomimetic sensors instead of antibodies, also aptamers can be used, which are synthetic nucleic acid ligands.[166] Because they are small molecules for instance of DNA, which were chemically synthesized, the additional chemistries can be tacked on, usually without a loss in function. Chemical groups can be attached to the ends of aptamers to increase their life spans, target them to particular locations, or help immobilize them to a surface. Because aptamers are smaller molecules than antibodies, a denser receptor layer can be generated on the surface that may lead to a better limit of detection in sensing. In the process of synthesis their binding affinity, specificity, and stability can be manipulated. Comparing with antibodies a better homogeneity can be achieved via chemical synthesis, and they are more resistant to denaturation.

To demonstrate a way of the use of aptamers in design of biomimetic sensors, two examples will be cited from the recent literature. The piezoelectric sensor for protein IgE has been developed with the use of commercially available anti-IgE aptamer oligonucleotide.[167] The obtained sensor shows specificity and sensitivity equivalent to these of immunosensor, but for aptamer-based sensor a less decrease of sensitivity after consecutive cycles of analyte binding and regeneration, as well as relative heat resistance and stability over several weeks was shown. A more complex mechanism of sensing was employed in adenosine aptamer-based sensor.[168] Detection was based on enzymatic activity measurements by fluorescence polarization with the use of aptameric enzyme subunit, which was a DNA aptamer composed of enzyme-inhibiting aptamer and adenosine-binding aptamer.

6   Closing Remarks

Numerous investigations of the market of chemical and biological sensors, especially in USA, were published recently. In-Star/MDR (http://www.instat.com) reported an annual growth rate of 11.5% for the market of biological and chemical sensors from

2002 to 2007. Biosensors are well established in the diabetes market for use in blood glucose testing, and have been successfully coupled with microfluidics technology for use in point-of-care diagnostics. In addition, there has been a consistently strong focus on the use of biological and chemical sensors for industrial process control (especially for food quality and safety), for environmental monitoring, and recently also in the war against terrorism.

In the report of Global Information Inc. (http://www.the-infoshop.com) a 7.3% annual growth is predicted for 2009 for chemical and biological sensors market. This market growth is argumented by the population aging and by increased number of chronic diseases. Terrorist attacks in recent years resulted in a rapid increase of funding of the development of chemical sensors. New chemical sensor systems targeted at biological pathogens, explosives, and toxic chemicals will benefit from strong demand by public facilities and military.

Among different types of chemical sensors, the fastest growth in 2008–2013 is predicted by Freedonia Group (http://www.freedonia.ecnext.com) for optical sensors and biosensors; this growth will be driven by the development of quick diagnostic tests besides glucose sensors. Optical sensors will benefit from falling prices, which will allow them to compete with other sensors in key markets such as gas detection. Demand on electrochemical sensors will benefit from increasing use in the motor vehicle market, although maturity in established industry safety markets will pull down overall gains.

Chemical sensor market can be considered as a part of a wider market of artificial sensing. The BBC Research (http://www.bccresearch.com) forecasts an annual growth rate 4.6% for commercial and medical sensing applications between 2007 and 2012 in a global market overview.

In author's opinion, from the fundamental point of view, especially important directions of further progress will be a wider application of artificial chemical nanostructures, micro- and nanomechanics and further progress in the area of application of biomimetic mechanisms and design of biomimetic devices.

References

1. Haber, F.; Klemensiewicz, Z., Über elektrische Phasengrenzekräfte, *Z. Phys. Chem.* **1909**, 67, 385–431
2. Clark Jr., L. C., Monitor and control of blood tissue O_2 tensions, *Trans. Am. Soc. Artif. Intern. Organs* **1956**, 2, 41–48
3. Clark Jr., L. C.; Lyons, C., Electrode systems for continuous monitoring in cardiovascular surgery, *Ann. N.Y. Acad. Sci.* **1962**, 102, 29–45
4. King Jr., W. H., Piezoelectric sorption detector, *Anal. Chem.* **1964**, 36, 1735–1739
5. Frant, M. S.; Ross Jr., J. W., Electrode for sensing fluoride ion activity in solution, *Science* **1966**, 154, 1553–1555
6. Stefanac, Z.; Simon, W., Ion specific electrochemical behavior of macrotetrolides in membranes, *Microchem. J.* **1967**, 12, 125–132
7. Guilbault, G. G.; Montalvo, J., A urea specific enzyme electrode, *J. Am. Chem. Soc.* **1969**, 91, 2164–2165

8. Bergveld, P., Development of an ion-sensitive solid-state device for neurophysiological measurements, *IEEE Trans. Biomed. Eng.* BME-19, **1970**, 17, 70–71

9. Mosbach, K.; Danielsson, B., Enzyme thermistor, *Biochim. Biophys. Acta* **1974**, 364, 140–145

10. Davies, C., Ethanol oxidation by an Acetobacter xylinium microbial electrode, *Ann. Microbiol.* **1975**, 126, 175–86

11. Janata, J., An immunoelectrode, *J. Am. Chem. Soc.* **1975**, 97, 2914–2916

12. Lübbers, D. W.; Optiz, N., Die pCO_2/pO_2-Optrode: Eine neue pCO_2-bzw. pO_2-Messonde zur Messung des pCO_2 oder pO_2 von Gasen und Flüssigkeiten, *Z. Naturf. C.* **1975**, 30, 532–533

13. Lündstrom, I.; Shivaramn, S.; Svenson, C.; Lundkvist, L., A hydrogen-sensitive MOS field-effect transistor, *Appl. Phys. Lett.* **1975**, 26, 55–57

14. Peterson, J. I.; Goldstein, S. R.; Fitzgerald, R. V.; Buckhold, D. K., Fiber optic pH probe for physiological use, *Anal. Chem.* **1980**, 52, 864–869

15. Liedberg, B.; Nylander, C.; Lündstrom I., Surface-plasmon resonance for gas-detection and biosensing, *Sens. Actuators* **1983**, 4, 299–304

16. Belli, S. L.; Rechnitz, G. A., Prototype potentiometric biosensor using intact chemoreceptor structures, *Anal. Lett.* **1986**, 19, 403–416

17. Taylor, R. F.; Marenchic, I. G.; Cook, E. J., Receptor-based biosensors, US Patent 5,001,048: **1987**

18. Matthews, D. R.; Bown, E.; Watson, A.; Holaman, R. R.; Steemson, J.; Hughs, S.; Scott, D., Pen-sized digital 30-second blood-glucose meter, *Lancet* **1987**, 8536, 778–779

19. Millan, K.; Mikkelson, S. R., Sequence-selective biosensor for DNA based on electroactive hybridization indicators, *Anal. Chem.* **1993**, 65, 2317–2323

20. Drolet, D. W.; Moon-Dermott, L.; Roming, T. S., An enzyme-linked oligonucleotide assay, *Nat. Biotechnol.* **1996**, 14, 1021–1025

21. Hulanicki, A.; Glab, S.; Ingman, F., Chemical sensors definitions and classifications, *Pure Appl. Chem.* **1991**, 63, 1247–1250

22. Vlasov, Y..; Legin, A.; Rudnitskaya, A.; Di Natale, S.; D'Amico, A., Nonspecific sensor arrays ("electronic tongue") for chemical analysis of liquids, *Pure Appl. Chem.* **2005**, 77, 1965–1983

23. Stephenson, G. R.; Ferris, I. G.; Holland, P. T.; Nordberg, M., Glossary of terms relating to pesticides, *Pure Appl. Chem.* **2006**, 78, 2075–2154

24. Thevenot, D. R.; Toth, K.; Durst, R. A.; Wilson, G. S., Electrochemical biosensors: Recommended definitions and classification, *Pure Appl. Chem.* **1999**, 71, 2333–2348

25. Vadgama, P.; Crump, P.W., Biosensors: Recent trends. A review, *Analyst* **1992**, 117, 1657–1670

26. Ion-Selective electrodes in analytical chemistry; Freiser, H. Ed.; Plenum: New York, NY, 1978

27. Biosensors: Fundamentals and applications; Turner, A. P. R.; Karube, I.; Wilson, G., Eds.; Oxford University Press: Oxford, 1987

28. Chemical sensors; Edmonds T. E., Ed.; Chapman and Hall: New York, NY, 1988

29. Janata, J. Principles of chemical sensors; Plenum: New York, NY, 1989

30. Biosensors – A practical approach; Cass A. E. G. Ed.; IRL, Oxford University Press: Oxford, 1990

31. Hall, E. A. H. Biosensors; Open University Press: Buckingham, 1990

32. Wise D.; Wingard, L. B. Biosensors with fiberoptics; Humana: Clifton NJ, 1991

33. Fiber optic chemical sensors and biosensors; Wolfbeis, O. S. Ed.; CRC: Boca Raton, FL, 1991

34. Hauptman, P.; Pownall, T. Sensors principles and applications; Prentice-Hall: New York, NY, 1993

35. Gardner, J. W. Microsensors: Principles and applications; Wiley: Chichester, 1994

36. Optical fiber sensor technology; Gatan, K. T. V.; Grattan, K. T. Eds.; Chapman and Hall: London, 1996

37. Handbook of chemical and biological sensors; Taylor, R. F.; Schults, J. S. Eds.; Institute of Physics Publishing: Bristol, 1996

38. Ballantine Jr., S. D.; White, R. M.; Martin, S. J.; Ricco, A. J.; Zellers, E. T.; Frye, G. C.; Wohltjen, H. Acoustic wave sensors: Theory, design and physico-chemical applications (Application of modern acoustics), Academic: London, 1996
39. Boisde, G.; Harmer, A. Chemical and Biochemical Sensing with Optical fibres and waveguides, Artech House: Boston, MA, 1996
40. Brignell, J.E.; White, N.M. Intelligent sensor systems; Institute of Physics Publishing: Bristol, 1996
41. Di Natale, C.; D'Amico, A. Eds.; Sensors and microsystems; World Scientific: Singapore, 1996
42. Fraden, J. Handbook of modern sensors; Springer: Berlin, 1997
43. Cattrall, R. W. Chemical sensors; Oxford University Press: Oxford, 1997
44. Diamond, D. Principles of chemical and biological sensors; Wiley: New York, 1998
45. Solomon, S. Sensors handbook; McGraw-Hill: New York, 1998
46. Frontiers in chemical sensors, Vol. 1, Optical sensors; Narayanaswamy, R.; Wolfbeis, O. S. Eds.; Springer: Berlin, 2004
47. Frontiers in chemical sensors, Vol. 2. Ultra-thin electrochemical chemo- and biosensors; Mirsky, V. Ed.; Springer: Berlin, 2006
48. Frontiers in chemical sensors, Vol. 3, Novel principle and techniques; Orellana, G.; Moreno-Bondi, M.C. Eds.; Springer: Berlin, 2004
49. Frontiers in chemical sensors, Vol. 4, Surface plasmon resonance; Homola, J. Ed.; Springer: Berlin, 2005
50. Campbell, M. Sensor systems for environmental monitoring: Sensor technologies; Kluwer: Dordrecht, 1996
51. Commercial biosensors. Application to clinical, bioprocess and environmental samples; Ramsey, G. Ed.; Wiley: New York, NY, 1998
52. Novel approaches in biosensors and rapid diagnostic assays; Liron, Z.; Bormberg, A.; Fisher, M. Eds.; Kluwer: New York, NY, 2001
53. Trojanowicz, M.; Wcisło, M., Electrochemical and piezoelectric enantioselective sensors and biosensors, *Anal. Lett.* **2005**, 38, 523–547
54. Cho, E. J.; Bright, F. V., Pin-printed chemical sensor arrays for simultaneous multianalyte quantification, *Anal. Chem.* **2002**, 74, 1462–1466
55. Held, M.; Schuhmann, W.; Jahreis, K.; Schmidt, H-L., Microbial biosensor array with transport mutants of *Escherichia coli* K12 for the simultaneous determination of mono- and disaccharides, *Biosens. Bioelectron.* **2002**, 17, 1089–1094
56. Krantz-Rulcker, C.; Stenberg, M.; Winquist, F.; Lundstrom, L. Electronic tongues for environmental monitoring based on sensor arrays and pattern recognition: a review, *Anal. Chim. Acta* **2001**, 426, 217–225
57. Deisingh, A. K.; Stone, D. C.; Thompson, M., Applications of electronic noses and tongues in food analysis, *Int. J. Food Sci. Technol.* **2004**, 39, 587–604
58. Schöning, M. J.; Hüllenkremer, B.; Glück, O.; Lüth, H.; Emons, E., Voltohmmetry – a novel sensing principle for heavy metal determination in aqueous solutions, *Sens. Actuators B* **2001**, 76, 275–280
59. Eranna, G.; Joshi, B. C.; Runthala, D. P.; Gupta, R. P., Oxide materials for development of integrated gas sensors – A comprehensive review, *Crit. Rev. Solid State Mat. Sci.* **2004**, 29, 111–188
60. Park, S. N.; Yoo, J. S., Electrochemical impedance spectroscopy for better electrochemical measurements, *Anal. Chem.* **2003**, 75, 455A–462A
61. Lillie, G.; Payne, P.; Vadgama, P., Electrochemical impedance spectroscopy as a platform for reagentless bioaffinity sensing, *Sens. Actuat. B.* **2001**, 78, 249–256
62. Koncki, R.; Hulanicki, A.; Glab, S., Biochemical modifications of membrane ion-selective sensors, *Trends Anal. Chem.* **1997**, 16, 528–536
63. Ho, W. O.; Krause, S.; McNeil, C. J.; Pritchard, J. A.; Armstrong, R. D.; Athey, D.; Rawson, K., Electrochemical sensor for measurement of urea and creatinine in serum based on ac impedance measurement of enzyme-catalyzed polymer, *Anal. Chem.* **1999**, 71, 1940–1946

64. O'Sullivan, C. K.; Guilabault, G. G., Commercial quartz crystal microbalances – theory and applications, *Biosens. Bioelectron.* **1999**, 14, 663–670
65. Yang, M.; Thompson, M., Multiple chemical information from the thickness shear mode acoustic wave sensor in the liquid phase, *Anal. Chem.* **1993**, 65, 1158–1168
66. Danielsson, B., Calorimetric biosensors, *J. Biotechnol.* **1990**, 15, 187–200
67. De Frutos, J.; Jimenez, B., Pure and calcium-modified lead tiotanate ceramics for pyroelectric sensors, *Sens. Actuators A.* **1992**, 32, 393–395
68. Zipp, A., Development of dry reagent chemistry for the clinical laboratory, *J. Autom. Chem.* **1981**, 3, 71–74
69. Walter, B., Dry reagent chemistries in clinical analysis, *Anal. Chem.* **1983**, 55, 498A–506A
70. Wolfbeis, O.S., Fiber-optic chemical sensors and biosensors, *Anal. Chem.* **2006**, 78, 3859–3873
71. Golden, J. P.; Saaski, E. W.; ShriverLake, L. C.; Anderson, G. P.; Ligler, F. S., Portable multichannel fiber optic biosensor for field detection, *Opt. Eng.* **1997**, 36, 1008–1013
72. Mayes, A. G.; Blyth, J.; Millington, R. B.; Lowe, C. R., A holographic sensor based on a rationally designed synthetic polymer, *J. Mol. Recognit.* **1998**, 11, 168–174
73. Carrascosa, L. G.; Moreno, M.; Alvarez, M.; Lechiga, L. M., Nanomechanical biosensors: a new sensing tool, *Trends Anal. Chem.* **2006**, 25, 196–206
74. Liedberg, B.; Nylander, C.; Lundström, I., Biosensing with surface plasmon resonance – how it all started, *Biosens. Bioelectron.* **1995**, 10, i–ix
75. Homola, J.; Yee, S. S.; Gauglitz, G., Surface plasmon resonance sensors: review, *Sens. Actuators B* 1999, 54, 3–15
76. Gardner, J.W.; Pike, A.; de Rooij, N.F.; Koudelka-Hep, M.; Clerc, P.A.; Hierlemann, A.; Göpel, W., Integrated array sensor for detecting organic solvents, *Sens. Actuators B* 1995, 26–27, 135–139
77. Okabayashi, T.; Fujimoto, T.; Yamamoto, I.; Utsunomiya, K.; Wada, T.; Yamashita, Y.; Yamashita, N.; Nakagawa, M., High sensitive hydrocarbon gas sensor utilizing cataluminescence of γ-Al_2O_3 activated with Dy3+, *Sens. Actuators B* **2000**, 64, 54–58
78. Trojanowicz, M., Analytical applications of carbon nanotubes, *Trends Anal. Chem.* **2006**, 25, 480–489
79. De Marco, R.; Jiang, Z. T.; Becker, T.; Clarke, G.; Murgatroyd, G.; Prince, K., Response mechanisms and new approaches with solid-state ion-selective electrodes: A powerful multi-technique materials characterization approach, *Electroanalysis* **2006**, 18, 1273–1281
80. Buhlmann, P.; Pretsch, E.; Bakker, E., Carrier-based ion-selective electrodes and bulk optodes. 2. Ionophores for potentiometric and optical sensors, *Chem. Rev.* **1998**, 98, 1593–1687
81. Horvath, V.; Takacs, T.; Horvai, G.; Huszthy, P.; Bradshaw, J. S.; Izatt, R. M., Enantiomer-selectivity of ion-selective electrodes based on a chiral crown-ether ionophore, *Anal. Lett.* **1997**, 30, 1591–1609
82. Ceresa, A.; Sokalski, T.; Pretsch, E., Influence of key parameters on the lower detection limit and response function of solvent polymeric membrane ion-selective electrodes, *J. Electroanal. Chem.* **2001**, 501, 70–76
83. Ye, Q.; Meyerhoff, M. E., Rotating electrode potentiometry: lowering the detection limits of nonequilibrium polyion-sensitive membrane electrodes, *Anal. Chem.* **2001**, 73, 332–336
84. Maksymiuk, K., Chemical reactivity of polypyrrole and its relevance to polypyrrole based electrochemical sensors, *Electroanalysis* **2006**, 18, 1537–1551
85. Morf, W. E.; De Rooij, N. F., Micro-adaptation of chemical sensor materials, *Sens. Actuators A* **1995**, 51, 89–95
86. Zen, J. M.; Kumar, A. S.; Tsai, D. M., Recent updates of chemically modified electrodes in analytical chemistry, *Electroanalysis*, **2003**, 15, 1073–1087
87. Trojanowicz, M., Application of conducting polymers in chemical analysis. A review on recent advances, *Microchim. Acta* **2003**, 143, 75–91
88. Mirsky, V. M.; Vasjari, M.; Novotny, I.; Rehacek, V.; Tvarozek, V.; Wolfbeis, O. S., Self-assembled monolayers as selective filters for chemical sensors, *Nanotechnology* **2002**, 13, 175–178

89. Love, J. C.; Estroff, L. A.; Kriebel, J. K.; Nuzzo, R. G.; Whitesides, G. M., Self-assembled monolayers of thiolates on metals as a form of nanotechnology, *Chem. Rev.* **2005**, 105, 1103–1169

90. Trojanowicz, M., Miniaturized biochemical sensing devices based on planar bilayer lipid membranes, *Fresenius J. Anal. Chem.* **2001**, 371, 246–260

91. Zhang, S.; Cardona, C. C.; Echegoyen, L., Ion recognition properties of self-assembled monolayers (SAMs), *Chem. Comm.* **2006**, 4461–4473

92. Kriz, D.; Ramstrom, O.; Mosbach, K., Molecular imprinting – new possibilities for sensor technology, *Anal. Chem.* **1997**, 69, 345A–349A

93. Panasyuk-Delaney, T.; Mirsky, V. M.; Ulbricht, M.; Wolfbeis, O. S., Impedometric herbicide chemosensors based on molecularly imprinted polymers, *Anal. Chim. Acta* **2001**, 435, 157–162

94. Wolfbeis, O. S., Fiber-optic chemical sensors and biosensors, *Anal. Chem.* **2006**, 78, 3859–3874

95. Wolfbeis, O. S., Materials for fluorescence-based optical chemical sensors, *J. Mater. Chem.* **2005**, 15, 2657–2669

96. Dürkop, A.; Wolfbeis, O. S., Nonenzymatic direct assay of hydrogen peroxide at natural pH using the Eu_3Tc fluorescent probe, *J. Fluoresc.* **2005**, 15, 755–761

97. Mikkelsen, S. R., Corton, E. Bioanalytical chemistry; Wiley-Interscience: Huboken, NJ, 2004

98. Nakamura, H.; Karube, I., Current research activity in biosensors, *Anal. Bioanal. Chem.* 2003, 377, 446–468

99. Wilchek, M.; Hofstetter, H.; Hofstetter, O., The application of biorecognition, In Novel approaches in biosensors and rapid diagnostic assays; Liron, E. Ed.; Kluwer/Plenum: New York, 2000

100. Taylor, R. F. Protein immobilization: fundamentals and applications; Marcel Dekker: New York, 1991

101. Uses of immobilized biological compounds; Guilbault, G.G.; Mascini, M., Eds.; Kluwer, Amsterdam, 1993

102. Bickerstaff, G.F. Immobilization of enzymes and cells; Humana: Totowa, NJ 1997

103. Cass, I.T.; Light, F.S. Immobilized biomolecules in analysis: a practical approach; Oxford University Press: Oxford, 1998

104. Immobilization of enzymes and cells; Guisan, J.M., Ed.; Humana: Totowa, NJ 2006

105. Matuszewski, W.; Trojanowicz, M., Graphite paste based enzymatic glucose electrode for flow injection analysis, *Analyst* **1988**, 113, 735–738

106. Gorton, L., Carbon paste electrodes modified with enzymes, tissues, and cells, *Electroanalysis* **1995**, 7, 23–45

107. Campanella, L.; Favero, G.; Sammartino, M. P.; Tomassetti, M., Further development of catalase, tyrosinase and glucose oxidase based organic phase enzyme electrode response as a function of organic solvent properties, *Talanta* **1998**, 46, 595–606

108. Cosnier, S.; Mousty, C.; Gondran, C.; Lepellec, A., Entrapment of enzyme within organic and inorganic materials for biosensor applications: Comparative study, *Mat. Sci. Eng. C* **2006**, 26, 442–447

109. Schügerl, K.; Ulber, R.; Scheper, T., Developments of biosensors for enantiomeric analysis, *Trends Anal. Chem.* **1996**, 15, 56–62

110. Schultze, H.; Vorlova, S.; Villatte, F.; Bachmann, T. T.; Schmid, R. D., Design of acetylo-cholinesterases for biosensor applications, *Biosens. Bioelectron.* **2003**, 18, 201–209

111. Kulys, J.; Vidziunaite, R., Amperometric biosensors based on recombinant laccases for phenols determination, *Biosens. Bioelectron.* **2003**, 18, 319–325

112. Krawczyński vel Krawczyk, T., Analytical applications of inhibition of enzymatic reactions, *Chem. Anal. (Warsaw)* **1998**, 43, 135–158

113. Amine, A.; Mohammadi, H.; Bourais, I.; Palleschi, G., Enzyme inhibition-based biosensors for food safety and environmental monitoring, *Biosens. Bioelectron.* **2006**, 21, 1405–1423

114. Wang, J.; Dempsey, E.; Eremenko, A.; Smyth, M.R., Organic-phase biosensing of enzyme inhibitors, *Anal. Chim. Acta* **1993**, 279, 203–208

115. Luque de Castro, M. D.; Herrera, M. C., Enzyme inhibition-based biosensors and biosensing systems: questionable analytical devices, *Biosens. Bioelectron.* **2003**, 18, 279–294
116. Mazzei, F.; Botre, F.; Botre, C., Acid phosphatase/glucose oxidase-based biosensors for the determination of pesticides, *Anal. Chim. Acta* **1996**, 336, 67–75
117. Uchiyama, S.; Tamata, M.; Tofuku, Y.; Suzuki, S., A catechol electrode based on spinach leaves, *Anal. Chim. Acta* **1988**, 208, 287–290
118. Ionescu, R. E.; Abu-Rabeh, K.; Cosnier, S.; Durrieu, C.; Chovelon, J. M.; Marks, R. S., Amperometric algal Chlorella vulgaris cell biosensors based on alginate and polypyrrole-alginate gels, *Electroanalysis*, **2006**, 18, 1041–1046
119. Baumann, W. H.; Lehmann, M.; Schwinde, A.; Ehret, R.; Brischwein, M.; Wolf, W., Microelectronic sensor system for microphysiological application on living cells, *Sens. Actuators B* **1999**, 55, 77–89
120. Harms, H.; Wells, M. C.; van der Meer, J. R., Whole-cell living biosensors – are they ready for environmental application? *Appl. Microb. Biotechnol.* **2006**, 70, 273–280
121. Nomura, Y.; Chee, G.J.; Karube, I., Biosensor technology for determination of BOD, *Field Anal. Chem. Technol.* **1998**, 2, 333–340
122. Matusnaga, T.; Karube, I.; Suzuki, S., A specific microbial sensor for formic acid, *Appl. Microbiol. Biotechnol.* **1980**, 10, 235–245
123. Lee, J.I.; Karube, I., A novel microbial sensor for the determination of cyanide, *Anal. Chim. Acta* **1995**, 313, 69–74
124. Han, T. S.; Kim, Y. C.; Sasaki, S.; Yano, K.; Ikebukuro, K.; Kitahara, A.; Nagamine, T.; Karube, I., Microbial sensor for trichloroethylene determination, *Anal. Chim. Acta* **2001**, 431, 225–230
125. Ripp, S.; Nivens, D. E.; Werner, C.; Sayler, G. S., Bioluminescent most-probable-number monitoring of a genetically engineered bacterium during a long-term contained field release, *Appl. Microbiol. Biotechnol.* **2000**, 53, 736–741
126. Billinton, N.; Baker, M. G.; Michel, C. E.; Knight, A. W.; Heyer, W. D.; Goddard, N. J.; Fielden, P. R.; Walmsley, R. M., Development of a green fluorescent protein reporter for a yeast genotoxicity biosensor, *Biosens. Bioelectron.* **1998**, 13, 831–838
127. Killard, A. J.; Deasy, B.; O'Kennedy, R. and Smyth, M. R., Antibodies: production, functions and applications in biosensors, *Trends Anal. Chem.* **1995**, 14, 257–266
128. Marco, M-P.; Gee, S.; Hammock, B. D., Immunochemical techniques for environmental analysis. I. Immunosensors, *Trends Anal. Chem.* **1995**, 14, 341–350
129. Suri, C. R.; Raje, M.; Varshney, G. C., Immunosensors for pesticide analysis: Antibody production and sensor development, *Crit. Rev. Biotechnol.* **2002**, 22, 15–32
130. Iqbal, S. S.; Mayo, M. W.; Bruno, J. G.; Bronk, B. V.; Batt, C. A.; Chambers, J. P., A review of molecular recognition technologies for detection of biological threat agents, *Biosens. Bioelectron.* 2000, 15, 549–578
131. Blackburn, G. F., Talley, D. B., Booth, P. M., Durfor, C. N., Martin, M. T., Napper, A. D. and Rees, A. R. (1990) Potentiometric biosensor employing catalytic antibodies as the molecular recognition element, *Anal. Chem.* **1990**, 62, 2211–2218
132. Belli, S. L.; Rechnitz, G. A., Biosensors based on native chemoreceptors, Fresenius Z. Anal Chem. 1988, 331, 439–447; Subrahmanyam, S.; Piletsky, S. A.; Turner, A. P. F., Application of natural receptors in sensors and assays, *Anal. Chem.* **2002**, 74, 3942–3951
133. Breer, H., Olfactory receptors: molecular basis for recognition and discrimination of odors, *Anal. Bioanal. Chem.* **2003**, 377, 427–433
134. Wu, T.Z., A piezoelectric biosensor as an olfactory receptor for odour detection: electronic nose, *Biosens. Bioelectron.* **1999**, 14, 9–18
135. Yang, L. C.; Tam, P. Y.; Murray, B. J.; McIntire, T. M.; Overstreet, C. M.; Weiss, G. A.; Penner, R. M., Virus electrodes for universal biodetection, *Anal. Chem.* **2006**, 78, 3265–3270
136. Susmel, S.; O'Sullivan, C. K.; Guilbault, G. G., Human cytomegalovirus detection by a quartz crystal microbalance immunosensor, *Enzyme Microb. Technol.* **2000**, 27, 631–645
137. Wang, J., Towards genoelectronics: Electrochemical biosensing of DNA hybridization, *Chem. Eur. J.* **1999**, 5, 1681–1685
138. Palecek, E.; Fojta, M., DNA hybridization and damage, *Anal. Chem.* **2001**, 73, 75A–83A

139. Gooding, J.J., Electrochemical DNA hybridization biosensors, *Electroanalysis* **2002**, 14, 1149–1156
140. Erdem, A.; Ozsoz, M., Electrochemical DNA biosensors based on DNA-drug interactions, *Electroanalysis* **2002**, 14, 965–974
141. Pividori, M. I.; Merkoci, A.; Alegret, A., Electrochemical genosensor design: immobilization of oligonucleotides onto transducer surfaces and detection methods, *Biosens. Bioelectron.* **2000**, 15, 291–303
142. Rosi, N. L.; Mirkin, C. A., Nanostructures in biodiagnostics, *Chem. Rev.* **2005**, 105, 1547–1562
143. Weizmann, Y.; Patolsky, F.; Willner, I., Amplified detection of DNA and analysis of single-base mismatches by the catalyzed deposition of gold on Au-nanoparticles, *Analyst* **2001**, 126, 1502–1504
144. Zhao, X. J.; Tapec-Dytioco, R.; Tan, W. H., Ultrasensitive DNA detection using highly fluorescent bioconjugated nanoparticles, *J. Am. Chem. Soc.* **2003**, 125, 11474–11475
145. Hashimoto, K.; Ito, K.; Ishimori, Y., Microfabricated disposable DNA sensor for detection of hepatitis B virus DNA, *Sens. Actuators B* **1998**, 46, 220–225
146. Bruckner-Lea, C. J., Biosensor systems for homeland security, *Electrochem. Soc. Interface* **2004**, Summer, 36–42
147. Fu, Y.; Yuan, R.; Chai, Y.; Zhou, L.; Zhang, Y., Coupling of a reagentless electrochemical DNA biosensor with conducting polymer film and nanocomposite as matrices for the detection of the HIV DNA sequence, *Anal. Lett.* **2006**, 39, 467–482
148. Marin, D.; Perez, P.; Teijeiro, C.; Palecek, E., Interactions of surface-confined DNA with acid-activated mitomycin C, *Biophysical Chem.* **1998**, 75, 87–95
149. Wang, J.; Ozsoz, M.; Cai, X.; Rivas, G.; Shiraishi, H.; Grant, D. H.; Chicharro, M.; Fernandes, J.; Palecek, E., Interactions of antitumor drug daunomycin with DNA in solution at the surface, *Bioelectrochem. Bioenerg.* **1998**, 45, 33–40
150. Liu, J.; Roussel, C.; Lagger, G.; Tacchini, P.; Girault, H. H., Antioxidant sensors based on DNA-modified electrodes, *Anal. Chem.* **2005**, 77, 7687–7694
151. Yang, M.; McGovern, M. E.; Thompson, M., Genosensor technology and the detection of interfacial nucleic acid chemistry, *Anal. Chim. Acta* **1997**, 346, 259–275
152. Fiammengo, R.; Crego-Calama, M.; Reinhoudt, D. N., Synthetic self-assembled models with biomimetic functions, *Curr. Opinion Chem. Biol.* **2001**, 5, 660–673
153. Cosnier, S.; Gondran, C., Wessel, R.; Montforts, F.-P.; Wedel, M., Poly(pyrrole-metallodeuteroporphyrin)electrodes: towards electrochemical biomimetic devices, *J. Electroanal. Chem.* **2000**, 488, 83–91
154. Ardiles, P.; Trollund, E.; Isaacs, M.; Armijo, F.; Canales, J. C.; Aguirre, M. J.; Canales, M. J., Electrocatalytic oxidation of hydrazine at polymeric iron-tetraminophthalocyanine modified electrodes, *J. Mol. Catal.* **2001**, 165, 169–175
155. Sotomayor, M. D. P. T.; Tanaka, A. A.; Kubota, L. T., Development of an amperometric sensor for phenol compounds using a Nafion (R) membrane doped with copper dipyridyl complex as a biomimetic catalyst, *J. Electroanal. Chem.* **2002**, 536, 71–81
156. Mifune, M.; Odo, J.; Iwado, A.; Saito, Y.; Motohashi, N.; Chikuma, M.; Tanaka, H., Uricase-like catalytic activity of ion-exchange resins modified with metalloporphyrins, *Talanta* **1991**, 38, 779–783
157. Iwado, A.; Mifune, M.; Harada (nee Noguchi), R.; Mukuno, T.; Motohashi, N.; Saito, Y., Glutathione peroxidase-like catalytic activities of ion-exchange resins modified with metalloporphyrins, *Anal. Sci.* **1998**, 14, 515–518
158. Iwado, A.; Mifune, M.; Hazawa, T.; Mukuno, T.; Oda, J.; Motohashi, N.; Saito, Y., Peroxidase-like activity of ion exchange resin modified with metal-porphine in fluorescent flow injection analysis, *Anal. Sci.* **1999**, 15, 841–846
159. Yano, K.; Karube, I., Molecularly imprinted polymers for biosensor applications, *Trends Anal. Chem.* **1999**, 18, 199–204
160. Haupt, K.; Mosbach, K., Molecularly imprinted polymers and their use in biomimetic sensors, *Chem. Rev.* 2000, 100, 2495–2504

161. Yoshikawa, M., Molecularly imprinted polymeric membranes, *Bioseparation* **2002**, 10, 277–286

162. Nicholls, I. A.; Rosengren, J. P., Molecular imprinting of surfaces, *Bioseparation* **2002**, 10, 301–305

163. Vlatakis, G.; Andersson, L. I.; Müller, R.; Mosbach, K., Drug assay using antibody mimics made by molecular imprinting, *Nature* **1993**, 361, 645–647

164. Panasyuk, T. L.; Mirsky, V. M.; Piletsky, S. A.; Wolfbeis, O. S., Electropolymerized molecularly imprinted polymers as receptor layers in a capacitive chemical sensors, *Anal. Chem.* **1999**, 71, 4609–4613

165. Haupt, K.; Noworyta, K.; Kutner, W., Imprinted polymer-based enantioselective acoustic sensor using a quartz crystal microbalance, *Anal. Commun.* **1999**, 36, 391–393

166. Mukhopadhyay, R., Aptamers are ready for the spotlight, *Anal. Chem.* **2005**, 77, 114A–118A

167. Liss, M.; Petersen, B.; Wolf, H.; Priohaska, E., An aptamer-based quartz crystal protein biosensor, *Anal. Chem.* **2002**, 74, 4488–4495

168. Yoshida, W.; Sode, K.; Ikebukuro, K., Aptameric enzyme subunit for biosensing based on enzymatic activity measurement, *Anal. Chem.* **2006**, 78, 3296–3303

Section 2
Self-Assembled Monolayers and Nanoparticles

Chapter 3
Self-Assembled Monolayers with Molecular Gradients

Michael Schäferling, Michael Riepl, and Bo Liedberg

Abstract In recent years, biosensors and sensor arrays have developed into very important analytical tools, which found applications in many fields such as pharmaceutical (high-throughput) screening, medical diagnosis, or industrial process control. One of the major challenges for material research is the preparation of appropriate sensor surfaces, providing an interface with a high sensitivity and selectivity toward a given analyte. This chapter discusses some straightforward and flexible approaches to study structure and/or composition–function relationships and response characteristics of polymeric and molecular sensor materials. The controlled continuous deposition of self-assembled monolayers (SAMs), e.g. of substituted thiols or silanes, paves the way for the generation of molecular gradients on solid surfaces. These are useful for the preparation of interfaces with spatially controlled chemical composition and/or physical properties. These tools can help to improve the selectivity and specificity of surfaces for biosensors and biochips. They can also be utilized for the study of fundamental protein adsorption and exchange phenomena.

1 Introduction

Combinatorial and high-throughput techniques are essential tools for the synthesis and screening of pharmaceutical agents, as well as for the rational design of novel sensor and catalytically active materials. Polymers or hydrogels constitute an important and highly flexible class of sensor materials that are compatible with combinatorial approaches.[1–3] An attractive methodology "combichem" for the development of functionalized polymers was described by Menger et al.[4] In this pioneering work, multiple functional groups were randomly attached to a polyamine backbone via amide linkages, and a huge number of such polymers, with various

M. Schäferling (✉)
Institute of Analytical Chemistry, Chemo- and Biosensors,
University of Regensburg, 93040 Regensburg, Germany
michael.schaeferling@chemie.uni-regensburg.de

R.A. Potyrailo and V.M. Mirsky (eds.), *Combinatorial Methods for Chemical and Biological Sensors*, DOI: 10.1007/978-0-387-73713-3_3,
© Springer Science + Business Media, LLC 2009

functionalities and relative compositions, were screened for their catalytic activity toward phosphate ester hydrolysis. Bio- and chemical sensors relying on combinatorial methodologies also have been developed, including hybrid materials with unique molecular recognition properties,[5] ion-complexing (co)polymers,[6] vapor-sorbing (co)polymers,[7,8] gas-sensitive conducting polymers,[9] and molecular imprinted polymers,[10] just to mention a few.

Other widely used strategies for the preparation of sensor materials involve self-assembled monolayers (SAMs) on solid supports.[11] The main advantage of SAMs is their straight-forward and inexpensive preparation. SAMs are well defined and robust structures, and several surface analytical techniques are available for thorough structural and functional characterization.[12,13] Monolayer architectures of terminal functionalized alkanethiols on gold or organosilanes on glass have found wide-spread applications in biosensor and microarray technology.[14,15] SAMs on glass have been also used for the fabrication of fluorescent chemosensors for determination of metal ions.[16,17] Self assembly techniques for the development of sensor surfaces were complemented with a combinatorial approach for the fabrication of *mixed* monolayers consisting of metal complexing molecules as receptors and commercially available fluorophores as reporters.[18] The preparation was carried out in a three-step procedure. First, a basic monolayer of an amino-terminated alkylsilane was formed on a glass substrate. Then, the amino-reactive fluorophore and the metal-binding group were coupled subsequently to the monolayer by covalent binding. As a result, the sensing properties of the layer are determined by the nature of the receptor group, the properties of the fluorophore, and the structure of the underlying monolayer. In this particular case, a library of sensitive fluorescent SAMs comprising different combinations of fluorophores and receptors were randomly distributed on the surface.

The present review discusses a rapid and flexible approach to optimize and improve the understanding of structure and/or composition–function relationships and response characteristics of polymeric and molecular sensor materials. The topics covered herein concern primarily the so-called molecular gradients,[19] which enable spatially controlled (continuous) deposition of molecular building blocks, e.g substituted thiols, on a solid surface. The molecular gradient approaches should be regarded as an extension of mixed SAMs as it offers the generation of a gradually changing chemical composition on a single substrate surface. Besides being an excellent tool to optimize the selectivity and specificity of biosensor surfaces and biochips, they also have found numerous applications for studies of fundamental protein adsorption and exchange phenomena, as outlined in the next sections.

2   Self-Assembled Monolayers with Molecular Gradients

Gradually functionalized materials play an important role in highly diversified fields of science and technology. Gradients of attracting molecules have been used to guide the growth of axons, demonstrating that gradual modified culture media and membranes are useful tools in neuronal research. Moreover, the use of pH gradients in polyacrylamide gel electrophoresis is inevitable for efficient separation of proteins. Starting from a short outline of applications of gradients in gels and polymers, this chapter focuses on SAMs with molecular gradients.

2.1 General Aspects of Gradually Modified Materials and Surfaces

Typical techniques for creating molecular gradients in materials such as agarose, nitro-cellulose, or cell culture media are based on diffusion or surface tension. Such substrates have been applied in neuronal research[20–22] or for the analysis of chemotactic responses.[23] Isoelectronic focusing (IEF) is an essential technique in two-dimensional polyacrylamide gel electrophoresis (2-D PAGE), which is based on pH gradients. Immobilized pH gradients (IPGs) are created by means of a set of well-defined buffering acrylamide derivative monomers ("immobilines") copolymerized with the acrylamide matrix.[24,25] The use of IPGs improved PAGE in terms of stability of the applied pH gradients and reproducibility compared with the conventional carrier ampholyte (CA) method. Following the CA strategy, electric field-induced gradients of charged fluorescent probes have been generated in bilayer membranes,[26] showing the applicability of electrical techniques in this area.

The controlled formation of gradients is a well-established technique in polymer science. This includes copolymer structures as well as grafting density gradients,[27–29] which can be produced by well-elaborated (co)polymerization techniques. Gradients of chemical groups on polymer surfaces can be achieved by radio frequency plasma discharge (RFPD). Applying RFPD, the surface can be reacted with a variety of gases such as ammonia or sulfur dioxide. The principle is shown in Fig. 3.1.[30] Different analytical topics were targeted by gradient modified polymers, e.g., the study of "mushroom-to-brush" phase transitions;[31] insertion of elasticity,[32] hydrophobicity,[30] or wettability[33] gradients, the control of protein or cell adsorption on

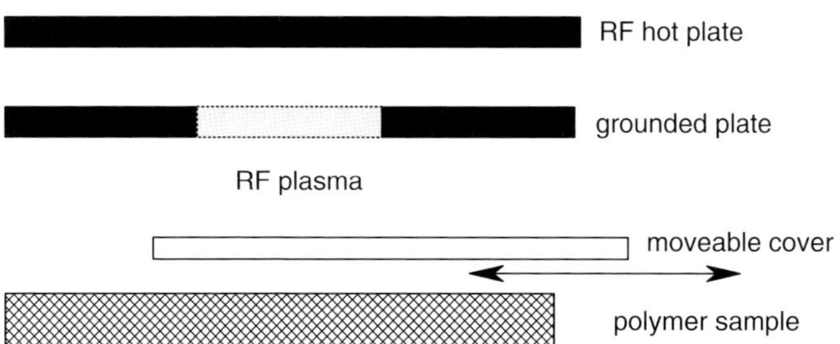

Fig. 3.1 Apparatus used for producing a gradient with radio frequency plasma discharge (RFPD). The RF plasma appears between the RF hot plate and the grounded plate. The plasma is exposed to the polymer surface at various lengths of time as regulated by the movable cover[30, 55]

surfaces;[34] and advanced materials for polyacrylamide gel electrophoresis,[35] optical fibers or multifocal lenses.[36] The following sections will highlight the application of self-assembled monolayers with molecular gradients.

2.2 Silane Monolayers on Glass or Silicon Substrates

Silanized glass or silicon wafers are major substrates in biosensors and protein/ DNA microarray technology. Trichloroalkylsilanes or trialkoxyalkylsilanes form SAMs *in situ* as a 2-D polysiloxane network on hydroxylated surfaces (SiOH). SAMs based on ω-substituted alkyl silanes are used for the immobilization of bio-molecules, either by electrostatic adsorption (amino-terminated monolayers) or cova-lently by amide bond formation (aldehyde, epoxy, or active ester-terminated monolayers). A few commercially available silane reagents for the derivatization of glass or silicone chips are shown in Fig. 3.2. A great variety of silanization protocols pertaining to reaction time, temperature, and solvent have been described in litera-ture.[12,13,37–40] Besides a thorough cleaning of the surface, the amount of water in the organic solvent has to be controlled for the formation of high-quality SAMs.

Mixed monolayers provide a useful tool for surface engineering at the molecular level. The proper spacing of chemical functionalities is a fundamental condition for an effective coupling of biomolecules,[38,39] or to control surface free energy for sub-sequent atomic layer deposition.[41] Coadsorption of alkyltrichlorosilanes with different

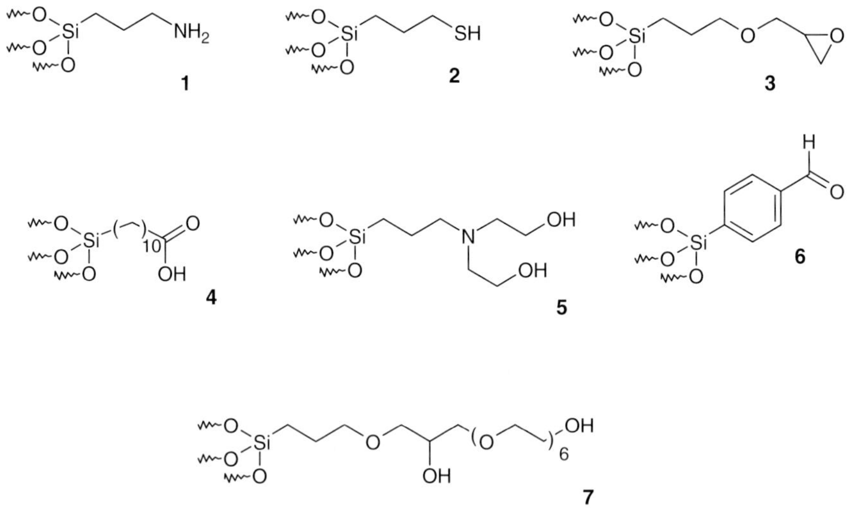

Fig. 3.2 Commonly used alkylsilane reagents for functionalization of glass surfaces. (1) *APTES* aminopropyltriethoxysilane, (2) *MPTS* 3-mercaptopropyl-triethoxysilane, (3) *GPTS* glycidoxy-propyltrimethoxysilan, (4) *TETU* triethoxy-silane undecanoic acid, (5) *HE-APTS* bis(hydroxyethyl) aminopropyltriethoxy-silane, (6) 4-trimethoxysilylbenzaldehyde, (7) *GPTS/HEG* glycidoxypropyltri-methoxysilane-hexaethylene glycol

terminal functionalities is a straightforward route to control the density of reactive groups on a surface. In case of mixed monolayers of methyl-terminated alkyltrichlorosilanes with [ω-(vinylalkyl)]- or [ω–(naphthylalkyl)]-tri-chlorosilane, the composition of the layer was found to be equal to the composition of the loading solution.[42–44] In the following years, numerous examples of binary monolayers have been prepared by competitive chemisorption,[45] some of which appear as statistically mixed SAMs, whereas others phase separate into single component islands of varying size. Phase separation in binary monolayer mixtures can be achieved by applying two very different alkylsilane amphiphiles. Separation can be driven by strong cohesive interaction of one component and exclusive hydrogen bonding interactions of the second. This approach generates isolated nano-islands of chemical functionalities.[46] Nanoscaled structures of SAMs with different components can be also formed by soft lithography-based contact printing methods.[47] Chemical and lithographic patterning methods of alkylsilane-SAMs have been reviewed by Onclin et al.[40]

The idea of forming molecular gradients was developed by Elwing et al.[48] They presented in their pioneering work a diffusion technique for generating wettability gradients on oxidized silicon surfaces. The native silicon dioxide layer bears silanol groups at the interface with which the chlorosilane can react to form a covalent bond. A hydrophobic gradient is produced by letting dichlorodimethylsilane (DDS) diffuse between two organic solvents forming a density gradient. During incubation of the silicone substrate, the two organic solvents diffuse into each other, and DDS, initially dissolved only in the lower, high density organic solvent, binds covalently to the free surface silanol groups (Fig. 3.3). In this manner, a ~5 mm long gradient of hydrophobic methyl groups is developed on the surface. In the ensuing years, these substrates were used in numerous studies to explore absorption and exchange processes of proteins at the liquid/solid interface by means of ellipsometry[48–51] or total internal reflection fluorescence (TIRF).[52]

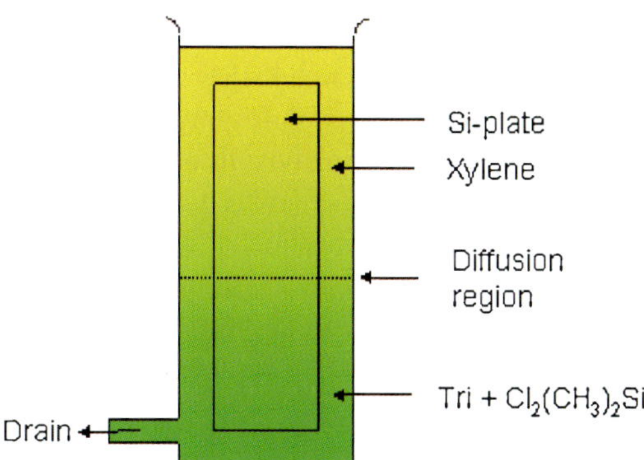

Fig. 3.3 Diffusion method used to prepare surface gradients of methyl groups according to Elwing et al.[48,54] Silicon or glass substrates are placed in a container filled with xylene. Trichloroethylene (Tri) with 0.1% (v/v) DDS is bedded under the xylene phase because of its higher density. The methylsilane diffuses together with Tri into the xylene phase and binds simultaneously to the SiOH groups on the silicon surface. The experiment is interrupted by removal of the solutions through the drain

A related technique for making DDS gradients on quartz slides is the density gradient method. Here, a stable density gradient is formed in a cuvette before the substrate is incubated.[53] The different preparation and characterization methods have been reviewed by Elwing and Gölander.[54] Later, also gradients of long-chain alkylsilanes were obtained by the diffusion technique.[55] A gradient of amino-propyldimethylsilane on silica has been used to couple aldehyde-terminated poly(ethylene glycol) (PEG). The resulting surface density gradient of PEG serves as platform for protein adsorption studies.[56] Hydrophobicity gradients on silicon wafers can also be generated by vapor deposition techniques, e.g. by exposing the surface to a diffusing front of a vapor of decyltrichlorosilane.[57]

The diffusion-controlled vapor deposition method paved the way for the fabrication of two-component gradient SAMs. A gradient of the first component is prepared by allowing the evaporated trichlorosilane compound to diffuse over a silicone wafer. This density profile is formed by covalent binding of the silanes onto the surface. Specifically, a terminal-functionalized trichlorosilyl-component is mixed with paraffin oil, placed in an open container that is positioned close to an edge of a silicon wafer, and the mixture is heated. As the silane evaporates, it diffuses in the vapor phase, generating a concentration gradient along the silicon substrate. The molecular gradient can be tuned by varying the diffusion time and the flux of the silane molecules. The latter can be conveniently adjusted by varying the ratio of silane and paraffine oil and the temperature of the source. In a second step, the functionalized chips are immersed in an organic solution of the second trichlorosilane component, forming a binary gradient. By means of this method, anchored polymers with a gradual variation in grafting densities on solid substrates were fabricated.

The technique for generating such structures comprises the formation of a molecular gradient of a polymerization initiator on the solid substrate and subsequent polymerization from the substrate-bound initiators.[58–60] For this purpose, one of the alkylsilane components is terminated by the initiator moiety. Gradient polymer brushes can be engineered if the binary monolayer is terminated by two different polymerization initiators (Fig. 3.4).[61]

Subsequent gradual modification of silane SAMs can be achieved by exposure to a gradient of UV-ozone radiation. This UV-ozonolysis (UVO) induces a gradual oxidation of the terminal methyl groups to carboxylic acid groups.[62] Such architectures are discussed as reference substrates for scanning probe microscopy[63] and are part of the project "Gradient Reference Specimens for Advanced Scanned Probe Microscopy (SPM)" by the American National Institute of Standards and Technology (NIST).[64] For these purposes, a gradient micropattern is engineered. In a first step, a SAM forming a strip micropattern is deposited by a combination of vapor deposition and soft lihography techniques. In the second step, UVO gradually oxidizes the end-groups of the SAM-strips in one direction (Fig. 3.5). These so-called μp specimens are useful for gauging tip quality and calibrating image contrast in SPM.[65]

Finally, silanized glass surfaces can be gradually modified with biopolymers by combining electrophoretical methods with soft lithography.[66] In this case, a micrometer-sized polydimethylsiloxane (PDMS) stamp is dipped into a soaking solution that is subjected to an electric field. Thereby, the electrophoretically induced gradient in the soaking

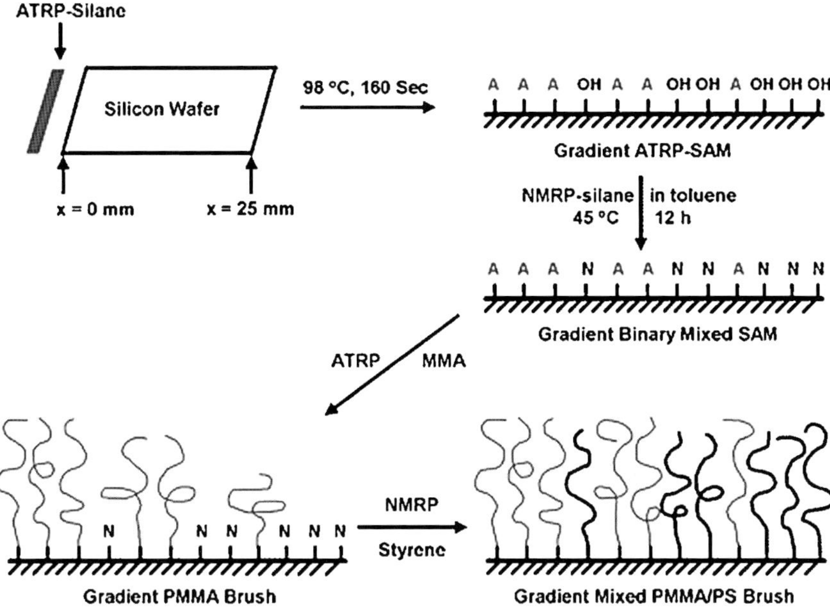

Fig. 3.4 Synthesis of gradient-mixed PMMA/PS brushes on silicon wafers via a gradient mixed SAM of 11′-trichlorosilylundecyl-2-bromo-2-methylpropionate (ATRP) and 1-(3′-oxa-1′-phenyl-14′-trichlorosilyltetradecyloxy)-2,2,6,6-tetra-methyl-piperidine (NMRP), forming a mixed polymerization initiator layer (reproduction with permission from ACS Publications)[61]

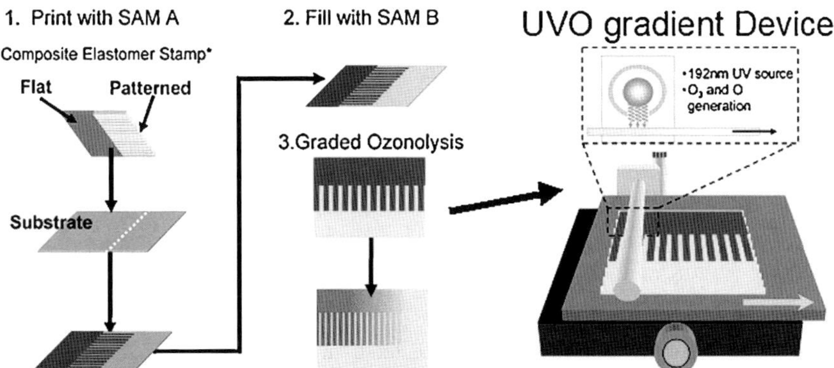

Fig. 3.5 Fabrication of μp specimens for SPM calibration. The first step (1) comprises soft-lithography of appropriate SAM molecules onto a planar substrate. A composite stamp, which has both flat and corrugated areas, allows printing of the comb-patterned strip with the adjacent solid calibration field. Next, a graded UV-ozonolysis (UVO) systematically modifies the chemistry of the patterned SAM (and calibration field) along one direction (3). For example, methyl-terminated alkyl chain SAMs (hydrophobic) can be gradually converted into carboxylic acid terminated (hydrophilic) chains. Subsequent "filling" with a hydrophilic SAM completes the "matrix" of the specimen (2)[64]

solution is transferred to the stamp. This biomolecular profile can be stamped on aminopropylsilane-modified glass substrates by means of microcontact printing (μ-CP).[67–69] This approach was evaluated by printing gradients of fluorescent-labeled poly-lysine.

2.3 *Alkanethiol Monolayers on Gold Surfaces*

Homogeneous SAMs of ω-substituted alkylthiols and disulfides on metal surfaces such as Au, Ag, Cu, or Pd are well-defined molecular systems, which have been studied for more than 25 years.[11–13,70–74] They form spontaneously organized, densely packed, and very stable monolayers. Therefore, alkanethiol SAMs have been applied in diverse scientific and technological fields such as molecular electronics, micro- and nanoscale pattering, biomaterials and (bio)chemical sensing. Preparation of alkanethiol monolayers is very straightforward. Many standard molecules (e.g., 16-mercaptohexadecanoic acid (MHA) or octadecanethiol (ODT) are commercially available. Prior to monolayer preparation, gold surfaces should be cleaned thouroghly with piranha solution (3:1 H_2SO_4/H_2O_2) or ammonia solution (5:1:1 $H_2O/H_2O_2/NH_4OH$; 60°C) to remove organic contaminants. Formation of the SAMs is simply achieved by incubating the gold-coated slides into alkanethiol solutions (in a concentration range from $10\,\mu mol\ L^{-1}$ to $1\,mmol\ L^{-1}$ in EtOH) for several hours, preferably over night.

This procedure works for single thiol solutions as well as for mixed thiol compositions. The latter results in mixed self-assembled monolayers with tailored surface properties.[75–77] It is worth to note that the composition of the SAM is not necessarily identical to the composition in the loading solution. The self-assembly of alkanethiols on gold offers an attractive alternative for the formation of molecular gradients on the nanometer to centimeter length scale. Liedberg and cowokers[19,78] developed a cross-diffusion strategy where the diffusion of two different alkanethiols from opposite sides of a polysaccharide matrix could be used to generate monolayer gradients on gold surfaces (Fig. 3.6). Ethanolic solutions of each of the two thiols are simultaneously injected into two glass filters at opposite ends of the gold substrate. The presence of the polysaccharide gel controls the diffusion, and the thiol attachment to the surface is slow enough to enable the formation of a gradient of macroscopic dimensions, e.g. with a length of a few millimeter up to ~20mm. These gradients offer an enormous flexibility in terms of chemical properties. As a first approach, gradients with CH_3 and OH-tail groups were assembled to form surfaces with controlled wettability properties. Later, protein adsorption was studied on so-called conformational gradients generated from oligo(ethylene glycol)-terminated (OEG) thiols with varying chain length. Proteins adsorb readily on alkanethiols with short OEG segments, e.g. on a diethyleneglycol (EG2)-terminated monolayer with all-*trans* conformation, whereas molecules bearing longer OEG-segments, such as EG6, which adopts the helical conformation, were resistant to protein adsorption (Fig. 3.7, example of an EG2–EG6 gradient). These data correlate excellently with results of mixed monolayers where each composition of EG2 and EG6 was prepared individually.[79,80]

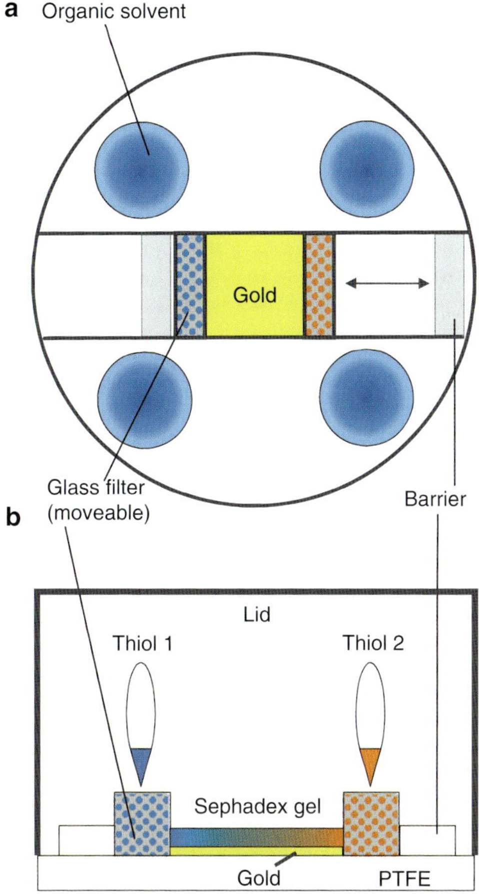

Fig. 3.6 The *top* (**a**) and the *side* (**b**) views of the cross-diffusion geometry used for the preparation of the molecular gradients. The glass filters are used as a preconcentration zone for the thiols. The geometry of the cell, including Teflon barriers, allows only diffusion in one direction. Several reservoirs for the organic solvent (swelling fluid) were added to guarantee a permanent wetting of the gel. The cell was covered with a lid to avoid evaporation

Another application of cross-diffusion gradients is the surface-based separation of proteins. COOH and NH_2-terminated thiols can be used to fabricate isoelectric point (pI) gradients. Exposed to proteins with largely differing isoelectric points (e.g., pepsin with a pI about 1 and lysozyme with a pI about 11), a separation corresponding to the negatively charged COO^- and positively charged NH_3^+ side was observed. The results were studied by contact angle, ellipsometry, FT-IR, AFM, and XPS.[78]

Geissler and colleagues[81] demonstrated the formation of nano-scale gradients of MHA and ODT by edge-spreading lithography. Silica beads with a size of 1.6 μm formed a 2-D array on a gold surface. A mixture of the thiols in ethanol was inked with a PDMS stamp. The thiol molecules diffused with different speed via the silica

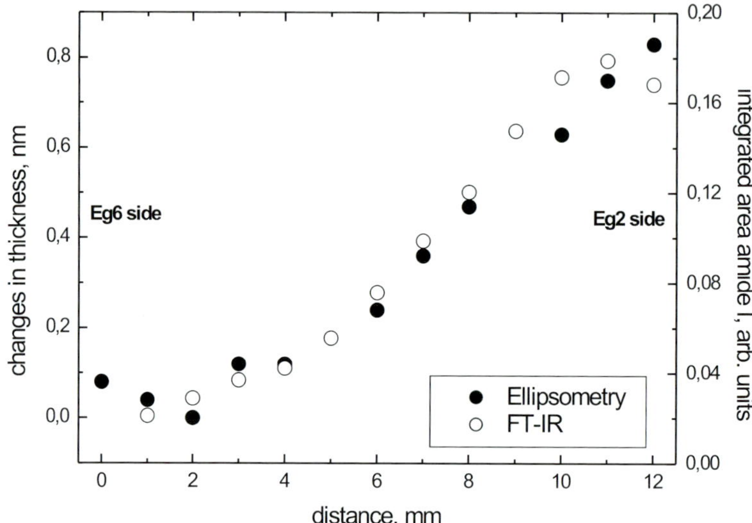

Fig. 3.7 Adsorption of fibrinogen on an EG2/EG6 gradient at pH 7.4. The incremental changes in ellipsometric thickness (*filled circles*) and the integrated area of the amide I peak obtained by FTIR-spectroscopy (*open circles*) are displayed

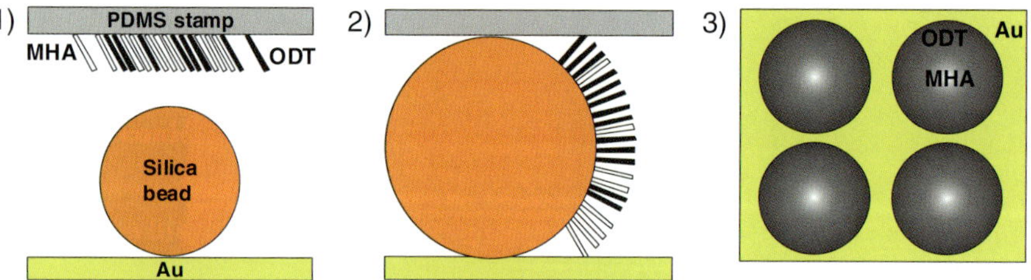

Fig. 3.8 The gradient approach of Geissler et al.[55] A PDMS stamp is used to ink a mixture of MHA (*white*) and ODT (*black*) onto a 2D array of silica beads immobilized on a gold substrate (1). Molecules are moving from the stamp to the gold surface with different speed (2). By reaching the gold, the molecules assemble into molecular rings whose composition creates a gradient from the inside to the outside (3)

beads on the gold surface. The width of the gradient was regulated by the composition of the mixture (Fig. 3.8). Metallic nanoparticles with a size of 13 nm and a significant higher affinity to MHA were used to contrast the gradients.

Another gradient method was presented by Mougin et al.[82] In this case, a cystamine sublayer was used for the formation of a Gaussian-shaped PEG-gradient. The terminal amino-group of the cystamine was linked with chemically activated (via NHS and EDC) PEG molecules with long chain length (about 100 ethylene units) (Fig. 3.9). The gradient had a width of several millimeters from center to edge. The quality of biomoelcular repulsion was demonstrated by incubation of the gradually-modified surface into cell solutions.

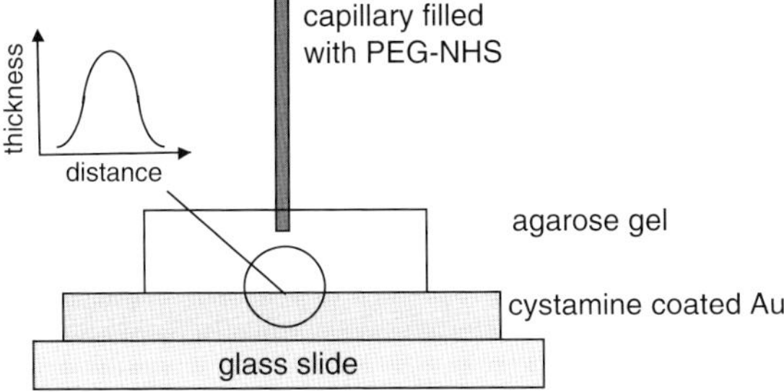

Fig. 3.9 The gel diffusion method[82] for the construction of PEG surface gradients. The width of the gradient (center to edge) is in millimeter scale

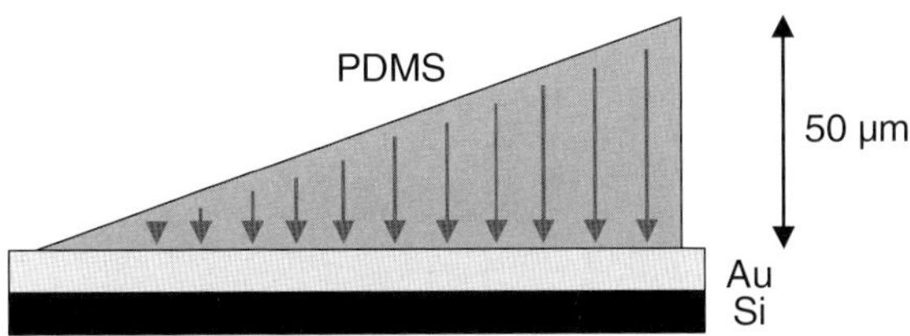

Fig. 3.10 Molecular printing of gradients. Thiol diffuses from the PDMS stamp to the gold surface and creates a monolayer gradient. The length of the *arrows* correlates with the amount of soaked up thiol

Two alternative devices for creating two-component gradients of alkanethiols on gold have been developed by Spencer and coworkers. In their first approach, a gold sample was mounted on a linear-motion driven device to control the adsorption time of a very dilute solution (about $5\,\mu M$) of dodecanethiol on the surface. The speed of the device – and therefore the immersion time – was adjusted between $2.5\,\mu m\ s^{-1}$ and $2.5\,mm\ s^{-1}$. The width of the gradient was up to $35\,mm$.[83–85] The resulting monolayers were disordered compared with perfectly assembled monolayers. To achieve complete coverage free of defects, the sample was incubated in an ethanolic solution of a complementary hydrophilic alkanethiol (11-mercapto-undecanoic acid or 1-undecanol) for several hours. The second method used μ-CP[67–69] applying PDMS stamps of variable thickness to form linear or radial gradients with a width of several millimeter.[86] First, hexadecanethiol diffuses from an ink pad to the stamp. It was shown that a higher amount of thiol can penetrate in the thicker part of a wedge-shaped stamp than in the thinner one. The height difference of the stamp from one side to the other was $50\,\mu m$. The thiol is transferred to a gold surface by means of contact printing (Fig. 3.10). The surface is only covered partially with a complete monolayer on the thinner edge. Nonoccupied holes in the SAM were filled by immersion of the surface in solutions of a perfluorinated thiol. The shape of the gradient is limited by the fabrication of the PDMS stamp. In both cases, linear motion drive and μ-CP, replacement of the first disordered thiol plays a critical role.

The quality of the gradient was observed by contact angle measurements, infrared spectroscopy, XPS, ellipsometry, and lateral force microscopy.

Herbert et al.[87] created micropatterning gradients by photolithograhic activation. A peptide was coupled photochemically to a monolayer consisting of an EG6-alkanethiol. The amount of peptide on the surface had a linear relationship to the light exposure time. The terminal Arg-Gly-Asp sequences of the peptides have a high binding affinity to cells, while the EG6-dominated regions are repellant. Cell adhesion on the density gradient of RGD-containing peptides was studied by optical microscopy and radiolabeling techniques.

The group of Wang and Bohn[88] prepared gradients of thiol SAMs by formation of a homogeneous monolayer and subsequent tailored desorption of molecules by impressing different electrochemical potentials. Varying potentials from −0.4 to −1.4 V were applied for several minutes to prepare the gradient with different width, using a three electrode set-up. A second thiol species was exposed to fill the holes in the monolayer structure. For example, a punched monolayer of 11-amino-1-undecanethiol was refilled by PEG-thiol molecules. The epidermal growth factor (EGF) protein was coupled to the remaining aminothiol units via EDC/NHS activation. Quantification of protein immobilization was performed by MALDI-MS, SPR, or XPS. Gradients of RGD-terminated peptides on gold were fabricated by a similar electrochemical process.[89] Such substrates can be used to probe interactions with extracellular matrix proteins. Plummer and Bohn[90] described the formation of a gradient based on amino-terminated alkanethiols with a width of 1 mm. A potential between −200 and −1,000 mV was applied to the modified gold surface for 5–300 s. Electrolysis was performed in a 0.5 M KOH solution with a Pt counter electrode and an Ag/AgCl reference electrode. The remaining SAM molecules were covalently linked with fluorescently doped polystyrene nanospheres. The gradient formation can be investigated by means of fluorescence microscopy or SPR. It was found that the tagged gradient yielded spatial intensity profiles of a sigmoidal shape.

3   Conclusion and Outlook

Polymeric and molecular/chemical gradients have been extensively used as model surfaces in numerous studies to improve the understanding of fundamental adsorption and wetting phenomena. The pioneering works of Pitt[33] and Elwing[48] on wettability gradients for studies of protein adsorption and exchange reactions have undoubtedly encouraged scientists from diverse areas of science and technology to employ the gradient concept in their own field of specialization. This includes, for example, the development of radial hydrophobicity gradients as coating materials in heat exchangers,[91] hydrophobicity gradients for ice nucleation and phase transition studies,[64] conformational gradients in biomaterials,[78,92] biotin gradients as a platform for the development of biosensing surfaces,[78] fluorescent chemosensors for metal ion detection,[16,17] and RGD peptide gradients for cell interaction studies.[89] Apparently, the vast majority of publications in the open literature concern biological applications of molecular/chemical gradients, because it is easy to create linear composition gradients that can be used to address issues relevant for specific as well as unspecific interactions of biomolecules with surfaces.

The use of gradients for identification of critical surface parameters is not only a rapid technique, but also economically advantageous, because it eliminates tedious preparation and characterization work with multiple "single composition" samples. The gradient approach is, therefore, expected to stimulate scientists and engineers from diverse fields to try the concept for screening applications. Moreover, a broad range of surface analytical and imaging techniques with sufficient sensitivity and lateral resolution are available for the characterization of the composition, conformation, and distribution of molecular species on surfaces. Complementary real time imaging techniques are well suited for a subsequent biological evaluation of binding phenomena on the gradient surface. The methodologies used to create linear gradients can be extended to include more complex geometries including, for example, radial[92] and 2D gradients. The latter approach was recently demonstrated by Riepl et al.,[93] who created a 2D-gradient containing a COO^-/NH_3^+ gradient in one direction and a conformational oligo(ethylene glycol)-gradient in an orthogonal direction. We foresee the use of molecular gradients in confined geometrics (tubes, channels, etc.) for the development of smart devices and microfluidics for Lab-on-Chip applications. It is also conceivable that the gradient concept can be developed into a generic platform for the guidance, manipulation, and interconnection/fusion of cells and cellular components on nonnatural surfaces

References

1. Potyrailo, R. A., Polymeric sensor materials: Toward an alliance of combinatorial and rational design tools?, *Angew. Chem. Int. Ed.* **2006**, 45, 702–723
2. Meier M. A. R.; Schubert, U. S., Combinatorial polymer research and high-throughput experimentation: powerful tools for the discovery and evaluation of new materials, *J. Mater. Chem.* **2004**, 14, 3289–3299
3. Kohn, J. A., New approaches to biomaterials design, *Nat. Mater.* **2004**, 3, 745–747
4. Menger, F. M.; Eliseev, A. V.; Migulin, V. A., Phosphatase catalysis developed via combinatorial organic chemistry, *J. Org. Chem.* **1995**, 60, 6666–6667
5. Diaz-Garcia, M. E.; Pina-Luis, G.; Rivero, I., Combinatorial solid-phase organic synthesis for developing materials with molecular recognition properties, *Trends Anal. Chem.* **2006**, 25, 112–121
6. Wu, X.; Kim, J.; Dordick, J. S., *Biotechnol. Prog.* **2000,** 10, 513–516
7. Portyrailo, R. A.; May, R. J.; Sivavec, T. M., *Sens. Lett.* 2, **2004**, 31–36
8. Hierlemann, A.; Zellers, E. T.; Ricco, A. J., *Anal. Chem.* **2001**, 73, 3458–3466
9. Mirsky, V. M.; Kulikov, V.; Hao, Q.; Wolfbeis, O. S., Multiparameter high throughput characterization of combinatorial chemical microarrays of chemosensitive polymers, *Macromol. Rapid Comm.* **2004**, 25, 253–258
10. Batra, D.; Shea, K. J., *Curr. Opin. Chem. Biol.* **2003**, 7, 434
11. Liedberg, B.; Cooper, J., Bioanalytical applications of self-assembled monolayers, In Immobilized Biomolecules in Analysis: A practical approach; Cass, T.; Liegler, F. S., Eds.; PAS series, Oxford University Press, Oxford 1998

12. Ulman A., An Introduction to Ultrathin Organic Films, Academic, San Diego, CA 1991
13. Ulman A, Formation and structure of self-assembled monolayers, *Chem. Rev.* **1996**, 96, 1533–1554
14. Schäferling, M.; Kambhampati, D., Protein microarray surface chemistry and coupling schemes, In Protein Microarray Technology; Kambhampati D., Ed.; Wiley, Weinheim 2004
15. Schäferling, M.; Schiller, S.; Paul, H.; Kruschina, M.; Pavlickova P.; Meerkamp, M.; Giammasi, C.; Kambhampati, D., Application of self-assembly techniques in the design of biocompatible protein microarray surfaces, *Electrophoresis*, **2002**, 23, 3097–3105
16. Crego-Calama, M.; Reinhoudt, D. N., New materials for metal ion sensing by self-assembled monolayers on glass, *Adv. Mater.* **2001**, 13, 1171–1174
17. Basabe-Desmonts, L.; Beld, J.; Zimmerman, R. S.; Hernando, J.; Mela, P.; Garcia Parajo, M. F.; van Hulst, N. F.; van den Berg, A.; Reinhoudt, D. N., A simple approach to sensor discovery and fabrication on self-assembled monolayers on glass, *J. Am. Chem. Soc.* **2004**, 126, 7293–7299
18. Basabe-Desmonts, L.; Zimmerman, R. S.; Reinhoudt, D. N.; Crego-Calama, M., Combinatorial method for surface-confined sensor design and fabrication, In Springer Series on Chemical Sensors and Biosensors Vol. 3: Frontiers in Chemical Sensors; Orellana, G.; Moreno-Bondi, M.C., (Eds.); Springer, Berlin 2005
19. Liedberg, B.; Tengvall, P., Molecular gradients of ω-substituted alkanethiols on gold: preparation and characterization, *Langmuir* **1995**, 11, 3821–3827
20. McKenna, M. P.; Raper, J. A., Growth cone behavior on gradients of substratum bound laminin, *Develop. Biol.* **1988**, 130, 232–236
21. Halfter, W., The behavior of optic axons on substrate gradients of retinal basal lamina proteins and merosin, *J. Neurosci.* **1996**, 16, 4389–4401
22. Bailly, M.; Yan, L.; Whitesides, G. M.; Condeelis, J. S.; Segall, J. E., Regulation of protrusion shape and adhesion to the substratum during chemotactic responses of mammalian carcinoma cells, *Exp. Cell Res.* **1998**, 241, 285–299
23. Baier, H.; Bonhoeffer F., Axon guidance by gradients of a target-derived component, *Science* **1992**, 255, 472–475
24. Bjellqvist, B.; Ek, K.; Righetti, P. G.; Gianazza, E.; Görg, A.; Westermeier, R.; Postel, W., Isoelectric focusing in immobilized pH gradients: Principle, methodology and some applications, *J. Biochem. Biophys. Methods* **1982**, 6, 317–339
25. Görg, A.; Postel, W.; Günther, S., The current state of two-dimensional electrophoresis with immobilized pH gradients, *Electrophoresis* **1988**, 9, 531–546
26. Groves, J. T.; Boxer, S. G., Electric field-induced concentration gradients in planar supported bilayers, *Biophys. J.* **1995**, 69, 1972–1975
27. Matyjaszewski, K.; Ziegler, M. J.; Arehart, S. V.; Greszta, D.; Pakula, T., Gradient copolymers by atom transfer radical copolymerization, *J. Phys. Org. Chem.* **2000**, 13, 775–786
28. Bhat, R. R.; Tomlinson, M. R.; Wu, T.; Genzer, J., Surface-grafted polymer gradients: formation, characterization, and applications, *Adv. Polym. Sci.* **2006**, 198, 51–124
29. Shen, M.; Bever, M. B., Gradients in polymeric materials, *J. Mater. Sci.* **1972**, 7, 741–746
30. Gölander, C. G.; Pitt, W. G., Characterization of hydrophobicity gradients prepared by means of radio frequency plasma discharge, *Biomaterials* **1990**, 11, 32–35
31. Wu, T.; Efimenko, K.; Vicek, P.; Subr, V.; Genzer, J., Formation and properties of anchored polymers with a gradual variation of grafting densities on flat substrates, *Macromolecules* **2003**, 36, 2448–2453
32. Askadskii, A. A., Development and properties of gradient polymeric materials, *Russian Polym. News* **1999**, 4, 34–37
33. Pitt, W. G. Fabrication of a continous wettability gradient by radio frequency plasma discharge, *J. Colloid Interface Sci.* **1989**, 133, 223–237
34. Bhat, R. R.; Tomlinson, M. R.; Genzer, J., Orthogonal surface-grafted polymer gradients: A versatile combinatorial platform, *J. Polym. Sci. B* **2005**, 43, 3384–3394
35. Gersten, D. M.; Bijwaard, K. E., Polyacrylamide gel electrophoresis in vertical, inverse and double-crossing gradients of soluble polymers, *Electrophoresis* **1992**, 13, 282–286
36. Kryszewski, M., Gradient polymers and copolymers, *Polym. Adv. Technol.* **1998**, 9, 244–259

37. Le Grange, J. D.; Markham, J. L.; Kurjian, C. R., Effects of surface hydration on the deposition of silane monolayers on silica, *Langmuir* **1993**, 9, 1749–1753

38. Zammatteo, N.; Jeanmart, L.; Hamels, S.; Courtois, S.; Louette, P.; Hevesi, L.; Remacle, J., Comparison between different strategies of covalent attachment of DNA to glass surfaces to build DNA microarrays, *Anal. Biochem.* **2000**, 280, 143–150

39. Sullivan, T. P.; Huck, W. T., Reactions on monolayers: Organic synthesis in two dimensions, *Eur. J. Org. Chem.* **2003**, 17–29

40. Onclin, S.; Ravoo, B. J.; Reinhoudt, D., Engineering silicon oxide surfaces using self-assembled monolayers, *Angew. Chem. Int. Ed.* **2005**, 44, 6282–6304

41. Lee, J. P.; Jang, Y. J.; Sung, M. M., Aomic layer deposition of TiO2 thin films on mixed self-assembled monolayers studied as a function of surface free energy, *Adv. Funct. Mater.* **2003**, 13, 873–876

42. Wasserman, S. R.; Tao, Y. T.; Whitesides, G. M., Structure and reactivity of alkylsiloxane monolayers formed by reaction of alkyltrichlorosilanes on silicon substrates, *Langmuir* **1989**, 4, 1074–1087

43. Mathauer, K.; Frank, C. W., Naphthalene chromophore tethered in the constrained environment of a self-assembled monolayer, *Langmuir* **1993**, 9, 3002–3008

44. Mathauer, K.; Frank, C. W., Binary self-assembled monolayers as prepared by successive adsorption of alkyltrichlorosilanes, *Langmuir* **1993**, 9, 3446–3451

45. Fadeev, A. Y.; McCarthy, T. J., Binary monolayer mixtures: modification of nanopores in silicon-supported tris(trimethylsiloxy)silyl monolayers, *Langmuir* **1999**, 15, 7238–7243

46. Fan, F.; Maldarelli, C.; Couzis, A., Fabrication of surfaces with nanoislands of chemical functionality by the phase separation of self-assembling monolayers on silicon, *Langmuir* **2003**, 19, 3254–3265

47. Finnie, K. R.; Nuzzo, R. G., The phase behavior of multicomponent self-assembled monolayers directs the nanoscale texturing of Si(100) by wet etching, *Langmuir* **2001**, 17, 1250–1254

48. Elwing, H.; Welin, S.; Askendal A.; Nilsson U.; Lundström, I., A wettability gradient method for studies of macromolecular interactions at the liquid/solid interface, *J. Colloid Interface Sci.* **1987**, 119, 203–210

49. Elwing, H.; Askendal, A.; Lundström, I., Competition between adsorbed fibrinogen and high-molecular weight kiniogen on solid surfaces incubated in human plasma (the Vroman effect): influence of solid surface wettability, *J. Biomed. Mat. Res.* **1987**, 21, 1023–1028

50. Elwing, H.; Askendal A.; Lundström I., Desorption of fibrinogen and gamma-globulin from solid surfaces induced by a nonionic detergent, *J. Colloid Interface Sci.* **1989**, 128, 296–300

51. Welin-Klintström, S.; Askendal, A.; Elwing, H., Surfactant and protein interactions on wettability gradient surfaces, *J. Colloid Interface Sci.* **1993**, 158, 188–194

52. Gölander, C.-G.; Lin, Y.-S.; Hlady, V.; Andrade, J. D., Wetting and plasma-protein adsorption studies using surfaces with a hydrophobicity gradient, *Colloid Surf.* **1990**, 49, 289–302

53. Gölander, C.-G.; Caldwell, K.; Lin, Y.-S., A new technique to prepare gradient surfaces using density gradient solutions, *Colloid Surf.* **1989**, 42, 165–172

54. Elwing, H.; Gölander, C.-G., Protein and detergent interaction phenomena on solid surfaces with gradients in chemical composition, *Adv. Colloid Interface Sci.* **1990**, 32, 317–339

55. Lin, Y.-S.; Hlady, V., Human serum albumin adsorption onto octadecyl-dimethylsilyl-silica gradient surface, *Colloids Surf. B: Biointerfaces* **1994**, 2, 481–491

56. Lin, Y.-S.; Hlady, V.; Gölander, C.-G., The surface density gradient of grafted poly(ethylene glycol): preparation, characterization and protein adsorption, *Colloids Surf. B: Biointerfaces* **1994**, 3, 49–62

57. Chaudhury, M. K.; Whitesides, G. M., How to make water run uphill, *Science* **1992**, 256, 1539–1541

58. Wu, T.; Efimenko, K.; Genzer, J., Combinatorial study of the mushroom-to-brush crossover in surface anchored polyacrylamide, *J. Am. Chem. Soc.* **2002**, 124, 9394–9395

59. Wu, T.; Efimenko, K.; Vlcek P., Subr, V., Genzer, J. Formation and properties of anchored polymers with a gradual variation of grafting densities on flat substrates, *Macromolecules* **2003**, 36, 2448–2453

60. Wu, T.; Gong, P.; Szleifer, I.; Vlcek, P.; Subr V.; Grenzer, J., Behavior of surface-anchored poly(acrylic acid) brushes with grafting density gradients on solid substrates: 1. experiment, *Macromolecules* **2007**, 40, 8756–8764

61. Zhoa, B., A combinatorial approach to study solvent-induced self-assembly of mixed poly (methyl methacrylate)/polystyrene brushes on planar silica substrates: effect of relative grafting density, *Langmuir* **2004**, 20, 11748–11755

62. Roberson, S. V.; Fahey, A. J.; Sehgal, A.; Karim, A., Multifunctional ToF-SIMS: combinatorial mapping of gradient energy substrates, *Appl. Surf. Sci.* **2002**, 200, 150–164

63. Fasolka, M. J.; Julthongpiput, D.; Briggman, K. A., Gradient reference surfaces for scanning probe microscopy, *PMSE* **2004**, 90, 721

64. http://www.polymers.msel.nist.gov/researcharea/combi/Gradient_Reference_Specimens_Advanced_Scanned_Probe_Microscopy.cfm

65. Julthongpiput, D.; Fasolka, M. J.; Zhang, W.; Nguyen, T.; Amis, E. J., Gradient chemical micropatterns: a reference substrate for surface nanometrology, *Nano Lett.* **2005**, 5, 1535–1540

66. Venkateswar, R. A.; Branch, D. W.; Wheeler, B. C., An electrophoretic method for microstamping biomolecule gradients, *Biomed. Microdevices* **2000**, 2, 255–264

67. Kumar, A.; Whitesides, G. M., Features of gold having micrometer to centimeter dimensions can be formed through a combination of stamping with an elastomeric stamp and an alkanethiol "ink" followed by chemical etching, *Appl. Phys. Lett.* **1993**, 63, 2002–2004

68. Wilbur, J.L.; Kumar, A.; Kim E.; Whitesides, G. M., Microfabrication by microcontact printing of self-assembled monolayers, *Adv. Mater.* **1994**, 6, 600–604

69. Libioulle, L.; Bietsch, A.; Schmid, H.; Michel, B.; Delamarche, E., Contact-inking stamps for microcontact printing of alkanethiols on gold, *Langmuir* **1999**, 15, 300–304

70. Nuzzo, R. G.; Allara, D. L., Adsorption of bifunctional organic disulfides on gold surfaces, *J. Am. Chem. Soc.* **1983**, 105, 4481–4483

71. Porter, M. D.; Bright T. B.; Allara, D. L.; Chidsey, C. E. D., Spontaneously organized molecular assemblies. 4. Structural characterization of *n*-alkyl thiol monolayers on gold by optical ellipsometry, infrared spectroscopy, and electrochemistry, *J. Am. Chem. Soc.* **1987**, 109, 3559–3568

72. Bain, C. D.; Whitesides G. M., Correlations between wettability and structure in monolayers of alkanethiols adsorbed on gold, *J. Am. Chem. Soc.* **1988**, 110, 3665–3666

73. Kumar, A.; Biebuyck H. A.; Whitesides, G. M., Patterning self-assembled monolayers: applications in material science, *Langmuir* **1994**, 10, 1498–1511

74. Schreiber, F., Structure and growth of self-assembling monolayers, *Prog. Surf. Sci.* **2000**, 65, 151–256

75. Roberts, C.; Chen, C. S.; Mrksich, M.; Martichonok V., Ingber, D. E.; Whitesides, G. M., Using mixed self-assembled monolayers presenting RGD and $(EG)_3OH$ groups to characterize long-term attachment of bovine capillary endothelial cells to surfaces, *J. Am. Chem. Soc.* **1998**, 120, 6548–6555

76. Knoll, W.; Liley, M.; Piscevic, D.; Spinke, J.; Tarlov, M. J., Supramolecular architectures for the functionalization of solid surfaces, *Adv. Biophys.* **1997**, 34, 231–251

77. Riepl, M.; Enander, K.; Liedberg, B.; Schäferling, M.; Kruschina, M.; Ortigao, F., Functionalized surfaces of mixed alkanethiols on gold as a platform for oligonucleotide microarrays, *Langmuir* **2002**, 18, 7016–7023

78. Riepl, M.; Östblom, M.; Lundström, I.; Svensson, S. C. T.; van der Gron, A. W. D.; Schäferling, M.; Liedberg, B., Molecular gradients: an efficient approach for optimizing the surface properties of biomaterials and biochips, *Langmuir* **2005**, 21, 1042–1050

79. Valiokas, R.; Svedhem, S.; Svensson, S. C. T.; Liedberg, B., Self-assembled monolayers of oligo(ethylene glycol)-terminated and amide group containing alkanethiolates on gold, *Langmuir* **1999**, 15, 3390–3394

80. Benesch, J.; Svedhem, S.; Svensson, S. C. T.; Valiokas, R.; Liedberg, B.; Tengvall, P., Protein adsorption to oligo(ethylene glycol) self-assembled monolayers: experiments with fibrinogen, heparinized plasma, and serum, *J. Biomat. Sci.* **2001**, 12, 581–597

81. Geissler, M.; Chalsani, P.; Cameron, N. S.; Veres, T., Patterning of chemical gradients with submicrometer resolution using edge-spreading lithography, *Small* **2006**, 2, 760–765

82. Mougin, K.; Ham, A. S.; Lawrance, M. B.; Fernandez, E. J.; Hiller, A. C., Construction of a tethered poly(ethylene glycol) surface gradient for studies of cell adhesion kinetics, *Langmuir* **2005**, 21, 4809–4812

83. Morgenthaler, S. M.; Lee, S.; Zürcher, S.; Spencer, N. D., A simple, reproducible approach to the preparation of surface-chemical gradients, *Langmuir* **2003**, 19, 10459–10462

84. Morgenthaler, S. M.; Lee, S.; Spencer, N. D., Submicrometer structure of surface-chemical gradients prepared by a two-step immersion method, *Langmuir* **2006**, 22, 2706–2711

85. Venkataraman, N. V.; Zürcher, S.; Spencer, N. D., Order and composition of methyl-carboxyl and methyl-hydroxyl surface-chemical gradients, *Langmuir* **2006**, 22, 4184–4189

86. Kraus, T.; Stutz, R.; Balmer, T. E.; Schmid, H.; Malaquin, L.; Spencer, N. D.; Wolf, H., Printing chemical gradients, *Langmuir* **2005**, 21, 7796–7804

87. Herbert, C. B.; McLernon, T. L.; Hypolite, C. L.; Adams, D. N.; Pikus, L.; Huang, C. C.; Fields, G. B.; Letourneau, P. C.; Distefano, M. D.; Hu, W.-S., Micropatterning gradients and controlling surface density of photoactivatable biomolecules on self-assembled monolayers of oligo(ethylene glycol) alkanethiolates, *Chem. Biol.* **1997**, 4, 731–737

88. Wang, Q.; Bohn, P.W., Surface composition gradients of immobilized cell signaling molecules. Epidermal growth factor on gold, *Thin Solid Films* **2006**, 513, 338–346

89. Wang, Q.; Jakubowski, J. A.; Sweedler, J. V.; Bohn, P. W., Quantitative submonolayer spatial mapping of Arg-Gly-Asp-containing peptide organo-mercaptan gradients on gold with matrix-assisted laser desorption/ionization mass spectrometry, *Anal. Chem.* **2004**, 76, 1–8

90. Plummer, S. T.; Bohn, P.W., Spatial dispersion in electrochemically generated surface composition gradients visualized with covalently bound fluorescent nanospheres, *Langmuir*, **2002**, 18, 4142–4149

91. Daniel, S.; Chaudhury, M. K.; Chen, J. C., Fast drop movements resulting from the phase change on a gradient surface, *Science* **2001**, 291, 633–636

92. Lestelius, M.; Engquist, I.; Tengvall, P.; Chaudhury, M. K.; Liedberg, B., Order/disorder gradients of *n*-alkanethiols on gold, *Colloids Surf., B: Biointerfaces* **1990**, 15, 57–70

93. Riepl, M.; Lundström, I.; Liedberg, B., New methods for the preparation of (bio)sensor surfaces: Molecular gradients and mixed monolayers containing oligo(ethylene glycols), Proceedings 2. Deutsches Biosensor-Symposium, Tübingen, Germany, 2001

Chapter 4
Combinatorial Libraries of Fluorescent Monolayers on Glass

Lourdes Basabe-Desmonts, David N. Reinhoudt,
and Mercedes Crego-Calama

Abstract Fluorescent self-assembled monolayers (SAMs) on glass surfaces are discussed as new sensing materials for metal ions and inorganic anions. The sensing SAMs are created by sequential deposition of two building blocks, a fluorophore and a ligand molecule onto an amino terminated SAM on glass slides. A large number of different systems can be fabricated by combinatorial techniques and parallel synthesis. A collection of sensing SAMs constitute a cross-reactive sensor array, with which analytes can be identified by differential sensing using the collective response of the SAMs array, instead of the individual response of a single SAM. Arrays of fluorescent SAMs have been produced both in microtiter plate and in multichannel microfluidic chip formats. Additionally, the glass substrates coated with fluorescent SAMs have been used as substrates for chemical patterning.

1 Introduction

Chemical sensing is a critical aspect of life. Our ability to sense the environment is essential for vision, reproduction, olfaction, and auditory or tactile stimulation. In analogy with "natural sensing," chemical sensing is essential for the evolution of science. It has been an active area involving scientists from many disciplines since sensing is the tool to discover and quantify analytes in any chemical or biological environment. Despite many efforts, there is still a large number of interesting analytes that cannot be easily detected. Thus, new probes for rapid and low-cost testing methods must be designed for application in medical diagnostics, industrial manufacturing, and national security. Among the different types of chemical sensors, optical sensors and biosensors are expected to grow the fastest. Fluorescence appears as one of the

M. Crego-Calama (✉)
Department of Supramolecular Chemistry and Technology, MESA⁺ Institute for
Nanotechnology, University of Twente, P.O. Box 217
7500 AE Enschede, The Netherlands
m.crego-calama@tnw.utwente.nl

R.A. Potyrailo and V.M. Mirsky (eds.), *Combinatorial Methods for Chemical and Biological Sensors*, DOI: 10.1007/978-0-387-73713-3_4,

most powerful transduction mechanisms,[1] because of its high sensitivity and the number of different analytical parameters that can change in the presence of the target analytes.[2] Additionally, and in contrast with electrochemical methods, light can travel without a physical wave-guide, facilitating the technical requirements.[3] In the last two decades, a large number of new fluorescent sensing probes have been synthesized. Until now the general trend in the production of sensing materials has been the immobilization of those sensing probes in polymeric matrices. A new emerging tendency in chemical sensing is the production of functional materials with intrinsic sensing properties.[4] This has simplified the sensor device implementation and allowed the production of nanosensors.[5] The production of new fluorescent functional materials able to report continuously and reversibly chemical recognition events plays an important role in the development of chemical sensors since the sensor performance depends very much on the properties of the sensory material. A number of fluorescent materials from polymers, sol–gels, mesoporous materials, surfactant aggregates, quantum dots, and glass or gold surfaces are being currently developed for chemical sensing.[4]

Our ability to predict the structural requirements for a perfect fluorescent probe for each analyte is limited, and a trial and error approach is still widely used for chemical sensor design.[6] Therefore, combinatorial approaches to the discovery of both binding and fluorescence building blocks would be powerful if effective library schemes could be invented. The combinatorial concept is based on the relative ease of production of a large number of potential compounds or devices. It is clearly different from the "classical" rational design and individual production of specific targets.[7] It allows the production of a large number of targets, which can be tested to determine successful hits. Linked to a proper screening methodology and data processing, it allows for facile search and optimization of a target lead structure, e.g., drug discovery, catalysis, bimolecular interaction studies, or sensitive probe discovery.[8] Many different types of combinatorial methods have been already employed to obtain new sensitive optical probes. Combinatorial approaches can also be used for the fabrication of cross-reactive sensor arrays. These cross-reactive sensor arrays, (artificial noses[9] and tongues[10]) are a collection of sensor, (or combinatorial library) created such that specificity is distributed across the array's entire reactivity pattern rather than contained in a single recognition element. The response of the array of sensors to the analyte presence is a pattern of responses along the array. Such patterns can be then incorporated in an artificial neural network for recognition of mixtures of analytes.[11,12]

Thus combinatorial methods and the fabrication of sensor arrays, either to select the best system or to enhance the performance of nonselective systems by the fabrication of cross-reactive sensor arrays, are paving the way toward efficient sensors.[4]

Surface confined chemical sensing offers many advantages over physical entrapping methods of fluorophores in polymers or sol–gels; it avoids, for example, leaking problems, offering long-term stability. Self-assembled monolayers (SAMs) adsorbed on gold surfaces or covalently bound to silicon oxide surfaces (glass, silicon, or quartz) are suitable interfaces for sensing.[13] They produce fast responses since all the receptors are exposed to the surface–liquid interface. SAMs are synthetically flexible so that they can be tailored to be chemically independent, and they are cheap, durable, and easy to immobilize on the transducer surface.[14] SAMs can be easily and inexpensively manipulated to yield families of materials that provide independent chemical responses in the presence of target analytes.[13] In spite of the fact that different functionalisation of SAMs seems to be a convenient method for fabrication of fluorescent chemosensors, the realization of such sensors is very scarce. SAMs on gold or other metallic surfaces have been extensively applied to chemo- and biosensing by electrochemical methods. But SAMs-based fluorescent sensors development encountered difficulties due to an efficient fluorescent quenching by the metal surfaces.[15,16] Only few reports have been published on the detection of fluorescence from self-assembled monolayer on gold.[16–18]

Because glass does not display the problems of gold, i.e., it is transparent to light, it has been frequently used for fluorescent bioassays for biological studies (protein, DNA microchips, etc.)[19,20] and to prove energy transfer by assembly of donor and acceptor chromophores as mixed monolayers.[21,22] The first effective sensing systems using covalently bonded dyes to glass were used for pH sensing.[23,24] Almost 20 years ago Wolfbeis et al.[25] reported the covalent immobilization of fluorescent acridinium and quinolinium indicators on a glass surface to create the first optical sensor for halides and pseudohalides. The sensors are able to indicate the concentration of halides in solution by virtue of the decrease in fluorescence intensity due to the quenching process. Another example of specific sensing probes covalently bound to glass surfaces was reported by Xavier et al.[26] They developed a molecular oxygen sensor in nonaqueous media by covalently attaching luminescent Ru(II) complexes via sulfonamide bonds to amino-derivatized porous glass. In this very interesting work, the authors outline the influence of the immobilization procedure used for optical sensing in terms of sensitivity and stability. In contrast to physical techniques such as dissolution, adsorption, and entrapment in a porous network, covalent immobilization of the luminescent indicator has been probed to increase the long-term stability of the sensitive system. Porous glass materials provide robust nonswelling rigid supports that can be easily modified with a number of chemical reactions. The resulting material displays strong emission above 600 nm, which is effectively quenched by oxygen in both organic solvents and aqueous media, with a detection limit of 6.2 μM.

Recent developments in the chemistry of SAMs on glass have opened a new possibility for fluorescent chemosensor design. The parts of a fluorescent sensor can be covalently attached to glass, silica, and quartz in one or more synthetic steps relying on the ability of trialkoxysilanes or halogensilanes to react with the hydroxylated surfaces of the substrates and to form SAMs.[27] The first examples of fluorescent sensing on glass using SAMs were reported by Reinhoudt et al.[28] They monitored the concentration of an aqueous β-cyclodextrin solution in the milimolar range using a dansyl modified amino-terminated SAM on glass.[28] SAMs of dansyl adsorbates were prepared attaching a 3-amino-propyltriethoxysilane (APTES) monolayer to a glass plate and converting this layer into the desired dansyl SAM by reaction with dansyl chloride. The selective binding of β-cyclodextrin to the dansyl moieties produces an enhancement of the fluorescence intensity of the monolayer accompanied by a shift of the fluorescence maximum from 510 to 480 nm.

Soon after another paper from this group reported the covalent attachment of a selective fluorescent calix-based receptor to a SAM for the detection of Na$^+$ in methanol down to 3.6 μM.[29] This was the first example of the detection of metal ions by fluorescence using a monolayer of a selective receptor. This novel work offered an alternative to physical immobilization of fluoroionophores in membranes. They probed that the fluoroionophore on the surface functions independently and that the confinement in the monolayer does not affect the complexation behavior. Similar work has been published recently by Wasielewski et al.[30] They attached two identical fluorophores to the upper ring of a calix[4]arene, while the lower ring was functionalized to be attached either directly to a glass surface or to an amino terminated monolayer. Nevertheless, they have not reported the use of this fluorescent monolayer for sensing purposes. Supramolecular interactions such as π–π have also been exploited to make sensitive glass surfaces. Raymo et al. have reported the functionalisation of glass surfaces (quartz, glass slides, and silica particles) with 2,7-diazapyrene derivatives for the detection of catecholamine neurotransmitters as dopamine.[31] The association of the 2,7-diazapyrenium acceptors with dopamine donors at the solid–liquid interface produces a fluorescent quenching. The layers responded to submillimolar concentrations of dopamine, and they showed selectivity for dopamine in presence ascorbic acid, which is the main interference in conventional dopamine detection protocols. Differently of what occurs in polymers-based sensors, the response time of SAMs is normally faster since all the recognition sites are directly exposed to the liquid interface. However, their sensitivity and dynamic range are restricted by the limited number of receptors inherent to a planar surface. Probably in the future, new strategies as functionalisation of the monolayers with dendrimers will be used to increase the number of recognition sites.[21] Recently, the immobilization of a pH sensitive fluorophore (Oregon Green 514) inside of a glass microchannel has yielded the first monolayer-functionalized microfluidic devices for optical sensing of pH.[32]

Finally combinatorial methods for the fabrication of chemical sensors have been proven to enhance significantly the performance of differential sensors. In this chapter, we show the use of a combinatorial parallel fabrication of fluorescent SAMs.

2 Fluorescent Monolayers on Glass

Fluorescent SAMs on glass have been developed by Crego-Calama et al. as new sensor materials. In 1999, the phenomenon of proximal but spatially separable receptor–fluorophore communication was recorded in solution by Tonellato et al.[33] and in 2001 Crego et al. demonstrated for the first time that disconnection between fluorophore and receptor can be applied to the preparation of stable sensitive fluorescent materials for metal ion sensing.[34] They used SAMs on glass substrates as a 2D scaffold to impart sufficient molecular orientation to separately deposit various binding functionalities (rather than the entire receptor molecule) and the fluorophore on the surface to achieve analyte selectivity. This approach avoids the synthesis and

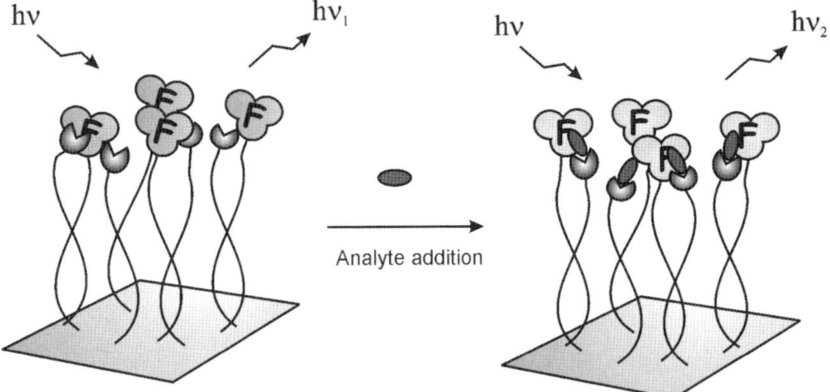

Fig. 4.1 Fluorescent sensitive monolayer (SAM) on a glass surface. The sensitive fluorescent monolayer comprises a monolayer modified with fluorophores (F) and the binding molecules. In the presence of an analyte, the fluorescence emission of the SAM changes due to interaction of the analyte with the layer[36] – Reproduced with permission of The Royal Society of Chemistry

optimization of highly specific ionophores, which is a difficult and laborious task. By sequential deposition of a fluorescent probe and nonspecific complexing functionalities, sensitive monolayers are produced. The randomly distributed fluorophores and functionalities generate a surface with a large number of sensing pockets (Fig. 4.1). The approach is a parallel combinatorial fabrication of sensing SAMs. Libraries composed of SAMs with different complexing and sensing properties can be easily fabricated using different combinations of fluorophores and complexing functionalities.

Fluorescent monolayers libraries have been used for the sensing of metal ions, and inorganic anions in organic solvents (Sect. 2.3.1)[34,35] and in water (Sect. 2.3.2).[36] The selectivity of these systems is not large but their performance is enhanced by the realization of cross-reactive sensor arrays. Fluorescent SAMs can also be formed on the walls of glass microchannels,[35] resulting in the fabrication of sensing microchannels. Detection of analytes can be performed by simple flow of the problem mixture through the functionalized channels (Sect. 3.2). A different approach for the fabrication of fluorescent monolayers implies microcontact printing (µCP), a soft lithography technique that permits to easily make controlled-size features down to 100 nm (Sect. 2.1).[35] Both the coating of glass microchannels and the use of lithography techniques permits the miniaturization of the systems to the micro and nanosize range; what makes this approach adequate to the fabrication of high density cross-reactive sensor arrays.

Beside the modular nature of this sensing system based mix monolayers on glass, which allows the use of a combinatorial methodology for the fabrication of sensing monolayers libraries, these 2D sensing materials offer a major novelty, and they can be used for the fabrication of combinatorial metal ion and luminescent patterns.[4] Using patterning techniques as µCP and dip pen nanolithography (DPN), different metal ions can be deposited on discrete areas of the sensing surfaces

(Sects. 4.1 and 4.2). The fluorescence of the substrate changes upon complexation of the metal ions, thus a fluorescent pattern is produced at the same time than the metal ion pattern on the substrate. A combinatorial approach for the fabrication of metal ions and luminescent patterns on 2D materials can be envisioned, where libraries of patterns can be produced from different combinations of metal ions and individual substrates of a monolayer library.

2.1 Synthesis of Combinatorial Libraries of Fluorescent Monolayers

The synthesis of the fluorescent SAMs is depicted in Fig. 4.2 A three-step procedure is followed. First, an amino terminated monolayer is formed by silylation of the quartz slides and the silicon wafers with (*N*-[3-(trimethoxysilyl)propyl]ethylenediamine) (TPEDA). Subsequently, this layer is converted into a fluorescent-SAM by reaction with an amino reactive fluorescent probe. Then the residual amino groups (the steric hindrance renders impossible the reaction of every surface amino group with a fluorophore molecule) are reacted with a small molecule to form a binding group such as amide, urea, thiourea, or sulfonamide, yielding the sensitive fluorescent SAM.

Two different methods can be used for the fabrication of these sensing monolayers. These methods differ in the technique used for the covalent attachment of the fluorescent probes (Fig. 4.2a ii) while the first and the third fabrication steps (silanation (Fig. 4.2a i) and binding molecule attachment (Fig. 4.2a iii) are identical for both methods. The first method is a solution-based procedure, and the second method is a μCP-based procedure. In the solution-based procedure, all the steps are carried out by sequential dipping of the glass substrate into the different solutions containing the appropriate reactive compound (silane, fluorophore, and binding molecule). In the case of the μCP approach, the covalent attachment of the fluorophore is carried out by soft lithography. A poly(dimethylsiloxane) stamp (PDMS) is submersed in the ink solution (fluorophore) and brought into contact with the amino-reactive surface for few seconds. This process results in the covalent attachment of the fluorophore to the glass slide only in the contact areas between the amino terminated substrate and the PDMS stamp. The fluorescently patterned glass slide is then immersed in a solution of an amino reactive molecule for the attachement of the binding groups onto the surface at the sites of the unreacted surface amino groups (Fig. 4.3).

Using the solution-based procedure, fully covered substrates are obtained, while by the μCP-based procedure discrete areas of coverage are patterned on the substrate (Fig. 4.3), which is convenient in terms of microarray fabrication because different fluorophore probes can be placed on specific, discrete regions of the same glass substrate. Such an array could then be exposed to a guest solution and subsequently the surface fluorescence emission scanned by a confocal microscope for the simultaneous acquisition of the optical data from the individually addressable areas.[37–40]

Using any of the both synthetic approaches, solution or μCP-based approach, combinatorial libraries of fluorescent SAMs, can be obtained by the sequential

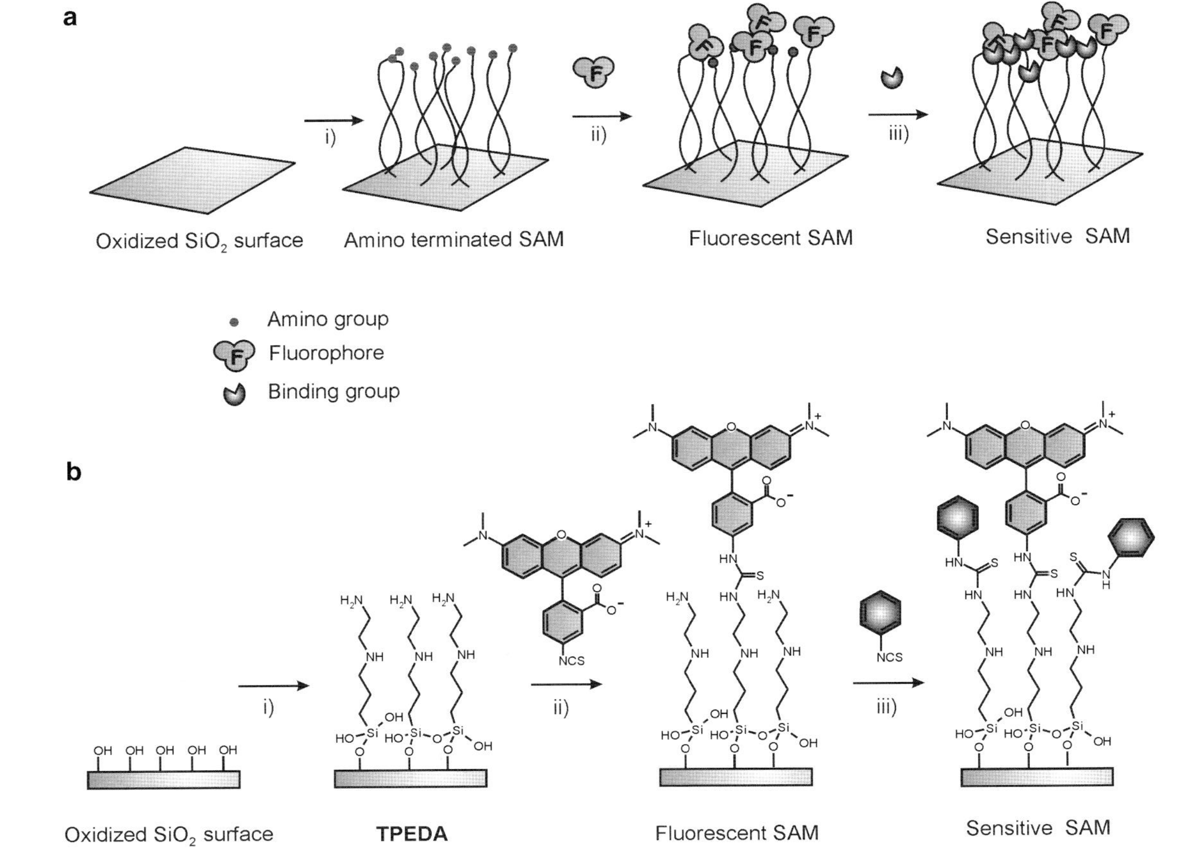

Fig. 4.2 General schematic cartoon (**a**), and chemical structures (**b**) of the fabrication procedure of a fluorescent sensitive monolayer on glass (i) Silanation of the glass slide with N-[3-(trimethoxysilyl)propyl]ethylenediamine to form the amino terminated monolayer TPEDA, (ii) reaction with an activated fluorophore, and (iii) covalent attachment of a binding molecule

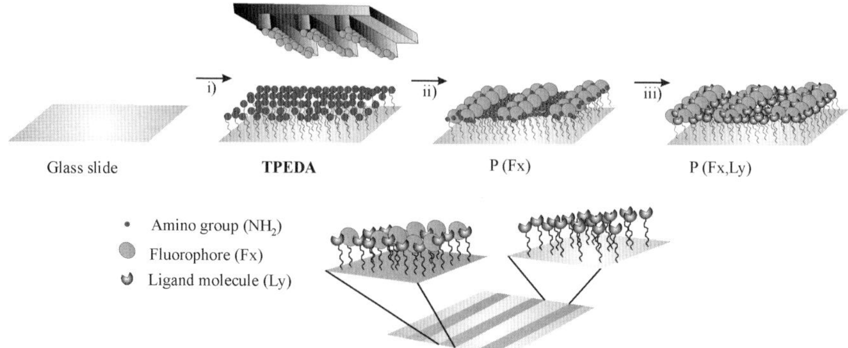

Fig. 4.3 Cartoon of the fabrication process of fluorescent patterned glass substrates P (F$_x$, L$_y$). (i) Functionalisation of a glass slide with *N*-[3-(trimethoxysilyl)propyl]ethylenediamine, (toluene, rt, 3.5 h) to form an the amino-terminated SAM TPEDA; (ii) μCP of the fluorophore solution; (iii) Reaction of the patterned fluorescent SAM with amino reactive binding molecules

funtionalization of amino terminated monolayers with different combinations of fluorophores, Lissamine, rhodamine B sulfonyl chloride (L), 5-(and-6)-carboxytetramethylrhodamine, succinimidyl ester (5(6)-TAMRA, SE) (TM) or tetramethylrhodamine-5-(and-6)- isothiocyanate (5(6)-TRITC) (T), and binding molecules, isocyanates, thioisocyanates, or acid chlorides.

In Fig. 4.4 the preparation of two different monolayers libraries, Library 1 and Library 2, is depicted.

2.2 Characterization of Fluorescent Monolayers on Glass

Fluorescent SAMs are characterized by contact angle goniometry, ellipsometry, and fluorescence spectroscopy. To assure the introduction of the binding groups, X-ray photoelectron spectroscopy (XPS) measurements can also be performed. By contact angle measurements,[41] the wettability properties of the surfaces, which depend on the hydrophobicity or hydrophilicity of the monolayer components as well as the layer packing, can be analyzed. By ellipsometry,[42] the estimation of thickness of the thin organic film is possible. Fluorescence spectroscopy[43] confirms the presence of the fluorescent molecules by detection of the characteristic spectral emission peaks of the fluorophores. XPS[44] provided elemental analysis data on the composition of the monolayer.

As an example, herein we explain the characterization of the combinatorial Libraries 1 and 2, mentioned above, which were synthesized using the solution-based procedure. Contact angle and ellipsometry data of the monolayer Libraries 1 and 2 are summarized in Fig. 4.1. Amino terminated SAMs formed from *N*-[3-(trimethoxysilyl)propyl]ethylenediamine (TPEDA) have an advancing contact angle (θ_a) of 60°, and a hysteresis of 25°.[34] Hysteresis ($\theta_a - \theta_r$) values range from

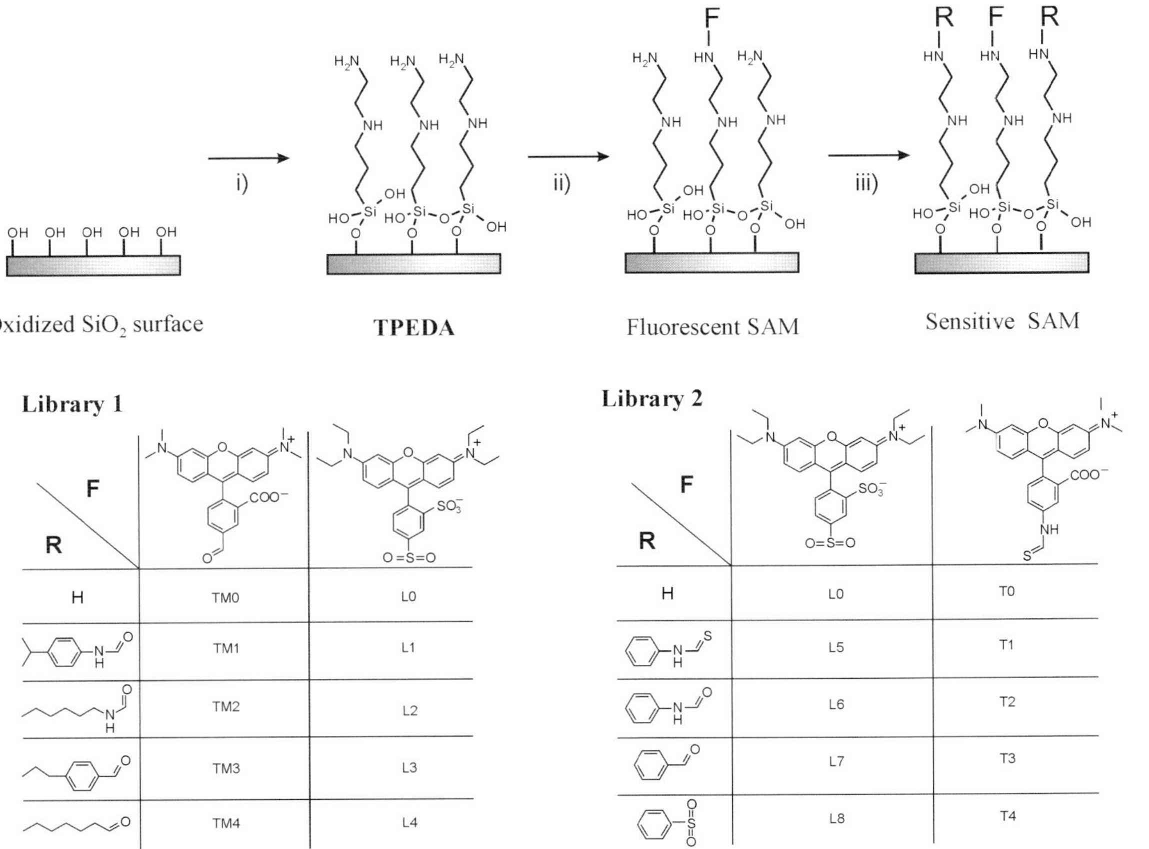

Fig. 4.4 Synthetic scheme for the preparation of Library 1 (composed of layers TM0–TM4 and L0–L4) and Library 2 (composed of layers L0, L5–L8, and T0–T4) of fluorescent SAMs on glass and silicon surfaces. (i) N-[3-(trimethoxysilyl)propyl]ethylenediamine, toluene, room temperature (rt), 3.5 h; (ii) amino reactive fluorophore Lissamine, rhodamine B sulfonyl chloride, 5-(and-6)-carboxytetramethylrhodamine, succinimidyl ester (5(6)-TAMRA, SE) (TM) or tetramethylrhodamine-5-(and-6)- isothiocyanate (5(6)-TRITC) (T), acetonitrile, rt, 4 h; (iii) isocyanates, thioisocyanates or acid chlorides used as binding molecules, chloroform, rt, 16 h

Table 4.1 Advancing and receding water contact angles ((θ_a) and (θ_r)),[a] and ellipsometric thicknesses (nm) of TPEDA SAMs of Library 1 and Library 2

SAM	θ_a (°)	θ_r (°)	Ell. thickness (nm)
TPEDA	60	35	1.35 ± 0.30
TM0	69	32	1.33 ± 0.04
TM1	83	53	1.83 ± 0.20
TM2	55	35	1.13 ± 0.28
TM3	82	58	1.48 ± 0.31
TM4	89	75	1.28 ± 0.30
L0	63	35	1.32 ± 0.15
L1	62	34	1.43 ± 0.09
L2	45	44	1.62 ± 0.19
L3	76	50	1.27 ± 0.11
L4	76	58	2.11 ± 0.23
L5	40	17	[b]
L6	72	30	[b]
L7	72	40	1.89 ± 0.14
L8	65	30	2.40 ± 0.43
T0	57	22	1.39 ± 0.1
T1	65	25	1.35 ± 0.16
T2	78	40	1.56 ± 0.15
T3	72	45	1.42 ± 0.18
T4	74	40	1.76 ± 0.19

TPEDA N-[3-(trimethoxysilyl)propyl]ethylenediamine
[a]Deviation of the contact angle values are bellow 30 in all the cases except TM2 ($\theta_a = 55 \pm 6.6$; $\theta_r = 35 \pm 7$) and L2 ($\theta_r = 44 \pm 13$)
[b]Ellipsometry thickness was not measured

approximately 25 to 40° indicating high disorder of the monolayers.[41] Changes in the values of the individual advancing/receding contact angles as well as changes in the hysteresis reflect a variation in the surface hydrophobicity, and therefore variation in the surface functionalisation.

Ellipsometry measurements were performed on silicon wafers. The experimental monolayer thicknesses are in good agreement with the thickness modeled with CPK models (software WebLab Viewer v2.01). The calculated values of the layer thickness considering the extended adsorbate lengths are 1.16 nm for TPEDA SAM, and 2.24 nm for L0, TM0, and T0 SAMs assuming that the molecules of the monolayer are in a perpendicular orientation with respect to the surface. Variations in thicknesses from the calculated values can be due to two factors: (a) the molecules are not perpendicular to the surface but tilted, and (b) the fluorophore and/or complexing functionalities are lying flat on the amino-surface (TPEDA). The introduction of the complexing molecules on the layer does not influence the thickness of the layers because they are smaller molecules than the previously attached fluorophores.

Fluorescent spectroscopy confirms the introduction of the fluorophores; maximum emission peaks were found at λ of 588 nm for L0, 585 nm for TM0, and 575 nm for

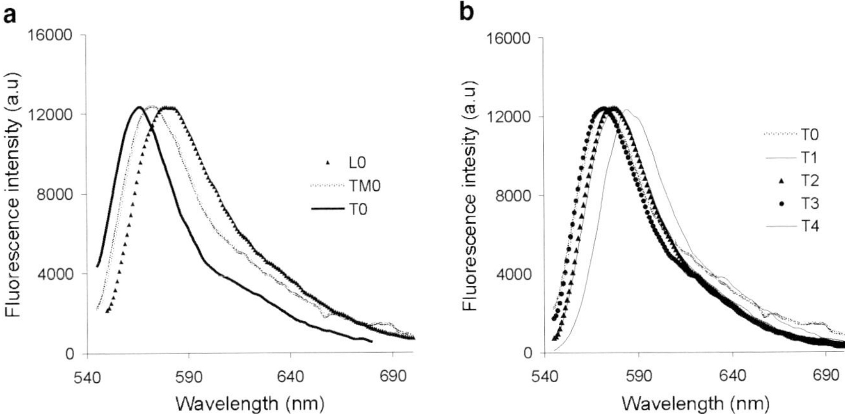

Fig. 4.5 Fluorescence emission spectra of (**a**) amino-terminated L0, TM0 and T0 SAMs (λ_{ex} = 545 nm for L0 and TM0 and 535 nm for T0) and (**b**) T1–T4, SAMs functionalized with different binding molecule (amino-terminated SAM T0 also included) (λ_{ex} = 535 nm). All the measurements are done in acetonitrile

T0 in the fluorescence spectra of the functionalized quartz slides (Fig. 4.5a). The maximum emission may be shifted by a few nanometers when the binding molecules are introduced in the fluorescent monolayers. As an example, the fluorescence spectra of layers T0–T4 are depicted in Fig. 4.5.

The surface elemental composition of the TPEDA monolayer and the other monolayers was evaluated by XPS analysis. The XPS analysis of the TPEDA monolayer indicated a carbon to nitrogen ratio (C1s/N1s) of 69.4%: 30.6% in close agreement with the calculated 71.4%:28.6%. To study the fluorophore attachment to the TPEDA SAM, XPS analysis of the monolayer functionalized with lissamine was performed (L0, Fig. 4.4). In this case, the two sulfur atoms of lissamine can be used as XPS sensitive label. The XPS spectrum of L0 showed a sulfur concentration (S2s) of 3.2% in the elemental composition, indicating successful monolayer functionalisation with the fluorophore. The surface coverage of the samples after their functionalisation with the complexing groups was investigated by marking the binding molecule with an XPS sensitive label.[35]

2.3 *Chemical Sensing by Fluorescent Monolayers on Glass*

The metal ion and inorganic anion sensing properties of the fluorescent monolayers was studied, in organic solvents first, and later the study was expanded to the more important analysis of aqueous solutions.

2.3.1 Cation and Anion Sensing in Organic Solvents

Library 1 (Fig. 4.4) was used for the sensing of Cu^{2+}, Co^{2+}, Ca^{2+}, and Pb^{2+} metal ions. This library is composed of ten different monolayers (TM0–TM4 and L0–L4), and it is the result of the binary combination (one fluorophore–one binding group) of two fluorophores, Lissamine (Lissamine™ rhodamine B sulfonyl chloride *mixed isomers*) (L) and TAMRA ((5(6)-TAMRA, SE) *mixed isomers*) (TM), TRITC (tetramethylrhodamine-5-(and-6)-isothiocyanate (5(6)-TRITC) *mixed isomers*) (T), and five different functional groups (amino, 4-isopropylphenylurea, hexylurea, 4-propylbenzylamide, and hexylamide). Fluorescence measurements were performed with 10^{-4} M acetonitrile solutions of Cu^{2+}, Co^{2+}, Pb^{2+}, and Ca^{2+} as perchlorate salts. In each case, the layer responded to the analytes within seconds. Upon washing with 0.1N HCl, the layers were recycled and reproducible results were obtained. The different combinations of binding group and fluorophore resulted in a range of different fluorescent responses (Fig. 4.6).

Comparing the responses within one fluorophore series, it can be seen that changes in the nature of the binding group significantly affect the binding profiles. For example, both of the fluorophore TAMRA layers, the ureido-substituted TM1 and the amido-substituted TM3, showed an increase in fluorescence upon addition of Ca^{2+} and Pb^{2+}, with a larger increase for the TM1 than for the TM3 layer. More impressive is the comparison of ureido-substituted layers TM1 and TM2. TM1 shows increased fluorescence for both Pb^{2+} (65%) and Ca^{2+} (90%), while TM2, which has a hexyl group instead of a p-isopropylphenyl group, shows very little change in fluorescence in the presence of these cations. The differences between binding groups substituted with an aryl vs. an alkyl group are remarkable considering that these substituents, in principle, should not directly coordinate either Pb^{2+} or

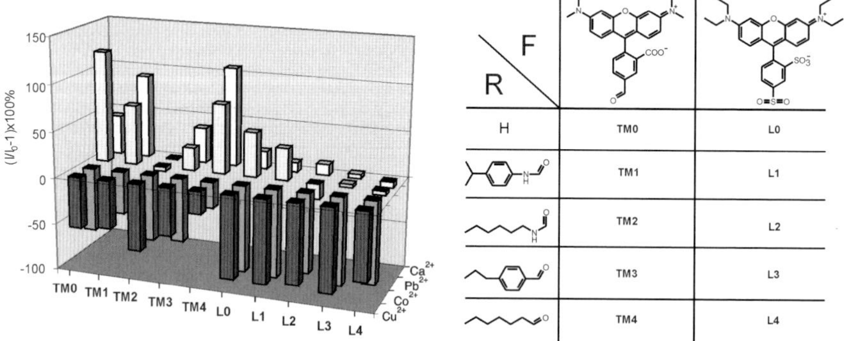

Fig. 4.6 Relative fluorescence intensity of monolayers TM0–TM4 and L0–L4 in the presence of 10^{-4} M solutions of Pb^{2+}, Ca^{2+}, Co^{2+}, and Cu^{2+} as perchlorate salts in acetonitrile. These data have been normalized; in the absence of metal cations the maximum fluorescence emission of each layer is set to 0. Positive values correspond to an enhancement in the fluorescence emission intensity of the layer while negative values represent a quenching of the fluorescence emission intensity of the layer.[35] Reprinted with permission from[35]. Copyright 2004 American Chemical Society

Ca^{2+}, and it is a typical example of the power of the combinatorial method. Because of the ease of generating different sensing system pairs, it requires a minimal effort to synthesize layers with these small variations, which would usually be disregarded in solution state receptor synthesis.

When comparing responses between the two fluorophores, there are marked differences between the hexylamido TAMRA system TM4, which shows fluorescence intensity increases for both Ca^{2+} (107%) and Pb^{2+} (75%), and the corresponding hexylamido lissamine system L4, which exhibits virtually no response for these cations. Overall, all systems exhibited a marked fluorescence intensity decrease for Co^{2+} and Cu^{2+}, although more quenching was seen in the lissamine layers than in the corresponding TAMRA layers. Additionally, TM0 is the best for the sensing of Pb^{2+} (121%) vs. the other ions due to the large response differences (Ca^{2+} 43%, Co^{2+} −68%, and Cu^{2+} −55%). A detection limit study was performed at concentrations from 10^{-10} to 10^{-3} M. Responses below 10^{-7} M were negligible.

The mechanism for the perturbation of the fluorescence properties of the monolayers by the metal ions is not well understood yet. The type of ligating functionality and its distribution across the layer, together with possible steric constraints or additional surface interactions, such as monolayer packing, van der Waals forces, and cation–π, and π–π interactions, may determine the properties of the layers, and therefore the response toward different metal ions. Cation-controlled photoinduced processes, such as photoinduced electron transfer and charge transfer, may be responsible for the fluorescence perturbation.[45]

Similar sensing properties can be observed in fluorescent monolayers fabricated by μCP. Several experiments have demonstrated good correlation of the sensing properties of fluorescent monolayers fabricated by both methods, solution-based and μCP.[35] Figure 4.7 depicts an example of a glass slide patterned with the monolayer TM1 fabricated by μCP as explained before. The patterned fluorescent monolayer displays a fluorescent enhancement upon addition of Ca^{2+} ions.

Anion sensing development in terms of both sensitivity and selectivity lags behind that of cations for the entire range of sensing systems, from solution state to

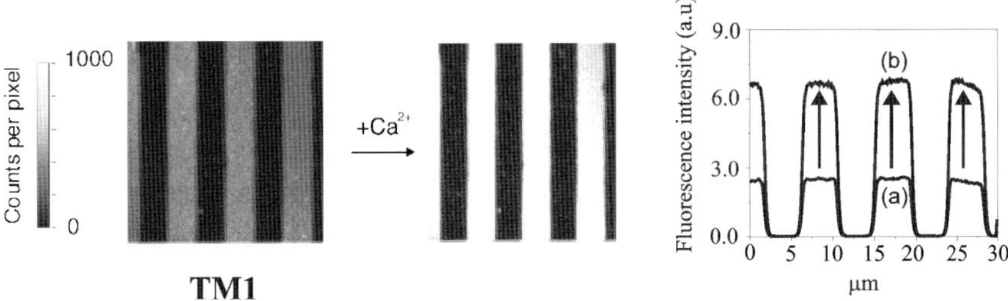

Fig. 4.7 Fluorescence confocal microscopy images of the systems TM1 in contact with acetonitrile and in contact with Ca^{2+} (10^{-4} M, acetonitrile) (*left*). Profiles of cross sections of the images of TM1 before (**a**) and after (**b**) addition of Ca^{2+} (right).[35] Reprinted with permission from[35]. Copyright 2004 American Chemical Society

CHEMFETs and ion selective electrodes.[46] The reasons for this include such factors as their larger size-to-charge ratio, large heat of hydration, and geometrical constraints.[47] However, many analytes of interest are anionic, and there is a great interest in pushing forward the field of anion sensing. To study the possibility of anion sensing by fluorescent monolayers on glass Library 2 (Fig. 4.4) was made. Amino terminated SAMs on quartz slides were functionalized sequentially, in the manner previously described for Library 1, with one fluorophore (TRITC or lissamine) and one anion-binding functional group (i.e., amino, amide, sulfonamide, urea, or thiourea, Fig. 4.4). The fluorescence response of the layers to 10^{-4} M acetonitrile solutions of tetrabutylammonium salts of HSO_4^-, NO_3^-, $H_2PO_4^-$, and CH_3COO^- (AcO^-) anions was recorded (Fig. 4.8). As in the case of cation sensing, each layer revealed a different response pattern for the different anions.

As a general trend, the TRITC functionalized layers T0–T4 exhibited a larger response to all the anions than the corresponding lissamine layers L0, L5–L8. This difference is likely due to the linker functionality. TRITC forms a thiourea bond when reacted with the layer, while lissamine forms a sulfonamide bond.

Important for sensing purposes is the response of the TRITC functionalized layers T1–T3. The fluorescence response of these layers increased between 24 and 87% in the presence of HSO_4^-, while the same layers showed decreased fluorescence intensity in the presence of $H_2PO_4^-$ between 35 and 56%. These varied responses (especially the increases in fluorescence) across the library will help to decrease the chance of false positives for the individual analytes. Detection limit studies showed that systems T0–T4, L0, and L5 already exhibit sensitivity towards both AcO^- and HSO_4^- at anion concentrations of 10^{-5} M.

These findings confirmed the ability of fluorescent monolayers to be used for anion sensing.

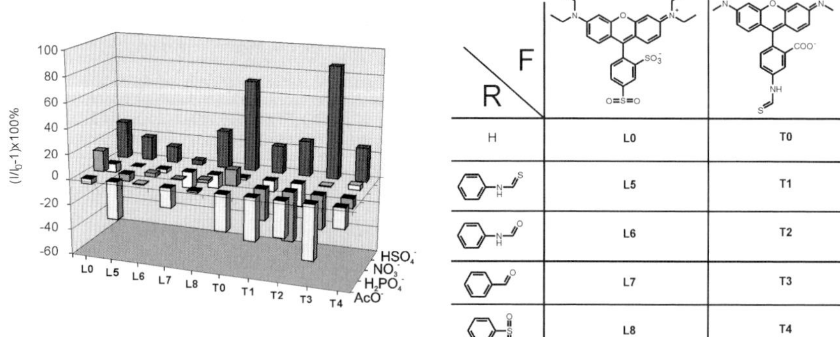

Fig. 4.8 Relative fluorescence intensity of the monolayers L0, L5–L8, and T0–T4 in the presence of 10^{-4} M solutions of HSO_4^-, NO_3^-, $H_2PO_4^-$ and AcO^- as tetrabutylammonium salts in acetonitrile. These data have been normalized; in the absence of the analytes the maximum fluorescence emission of each layer is set to 0. Positive values correspond to an enhancement in the fluorescence emission intensity of the layer, while negative values represent a quenching of the fluorescence emission intensity of the layer.[35] Reprinted with permission 35. Copyright 2004 American Chemical Society

2.3.2 Cation and Anion Sensing in Water

Because the most interesting analytes are present in aqueous environment, sensing in water is very important, especially for medical and environmental applications. Thus, the extension of the approach to metal ion sensing in water was the next step. The synthesis of new fluorescent monolayers stable in water was performed, and their sensing abilities toward cations and anions in aqueous environment were investigated.

Our original SAM sensing systems used TPEDA (*N*-[3-(trimethoxysilyl)propyl]-ethylenediamine) to form an amino-terminated monolayer on a glass surface[34] to which subsequently the fluorophore and complexing functionality were attached. However, it is known that direct attachment of amino-terminated silanes onto glass does not result in a highly ordered monolayer,[48] and it is likely that the lack of a well-ordered hydrophobic layer allows water to penetrate the layer and to hydrolyze the Si–O bonds, which will eventually destroy the layer. Because of the instability of these layers, a stepwise chemical synthesis of a long-chain amino-terminated alkylsilane monolayer of 1-amino-11-silylundecane (2) was performed (Fig. 4..9). This two-step synthesis of the amino-terminated monolayer yields a well-defined structure of the siloxane network increasing in general the stability of the monolayer in water.[49]

Subsequently, a small library of monolayers based on the functionalisation of amino-terminated SAM (**2**, Fig. 4.9) was made to test the response of different monolayers to metal cations in water. Two binding molecules (benzenesulfonyl chloride and *p*-isopropylphenyl isocyanate) and three fluorophores (dansyl chloride, 7-dimethylaminocoumarin-4-acetic acid succinimidyl ester), and TAMRA were sequentially reacted with the amino-terminated SAM **2** (Fig. 4.9) in different combinations, resulting in a library of nine sensing surfaces with different complexing functionality-fluorophore pairs (Fig. 4.9). Each sensing layer has one complexing functionality known to bind to cations (i.e., amino (A), aryl-urea (U), aryl-sulfonamide (S)) and one fluorophore (i.e., short excitation wavelength dansyl (D) or coumarin (C), or long excitation wavelength TAMRA (TM)).

After formation of the stable monolayers, their cation sensing properties were studied. The chloride salts of Hg^{2+}, Ca^{2+}, Co^{2+}, and Cu^{2+} were used as analytes. Each of the layers of the sensing library was placed in a spectrofluorometer cuvette filled with 0.1 M aqueous solution of HEPES buffer (pH 7.0) and the fluorescence spectrum was measured. A solution of the corresponding cation was added so that the concentration of the analyte in the cuvette was 10^{-4} M, and the fluorescence spectrum was measured again. Two typical examples of the layer fluorescence emission spectra in the presence of analytes are shown in Fig. 4.10.

Figure 4.11 shows the fluorescence response of all the layers to 10^{-4} M solutions of Hg^{2+}, Cu^{2+}, Co^{2+} and Ca^{2+} as chloride salts in water (pH 7.0, 0.1 M HEPES).

Confirming the results shown in acetonitrile, the nature of the functionalities of the fluorophore and the other ligating functionalities seems to influence the sensing ability of the layer resulting in a range of responses to different analytes. Yet whether one component will be predominant in determining the fluorescence response or whether they will work together to form a unique sensing system is unpredictable. There are a few overall trends in the library responses when we compare the amino, sulfonamide,

     L. Basabe-Desmonts et al.

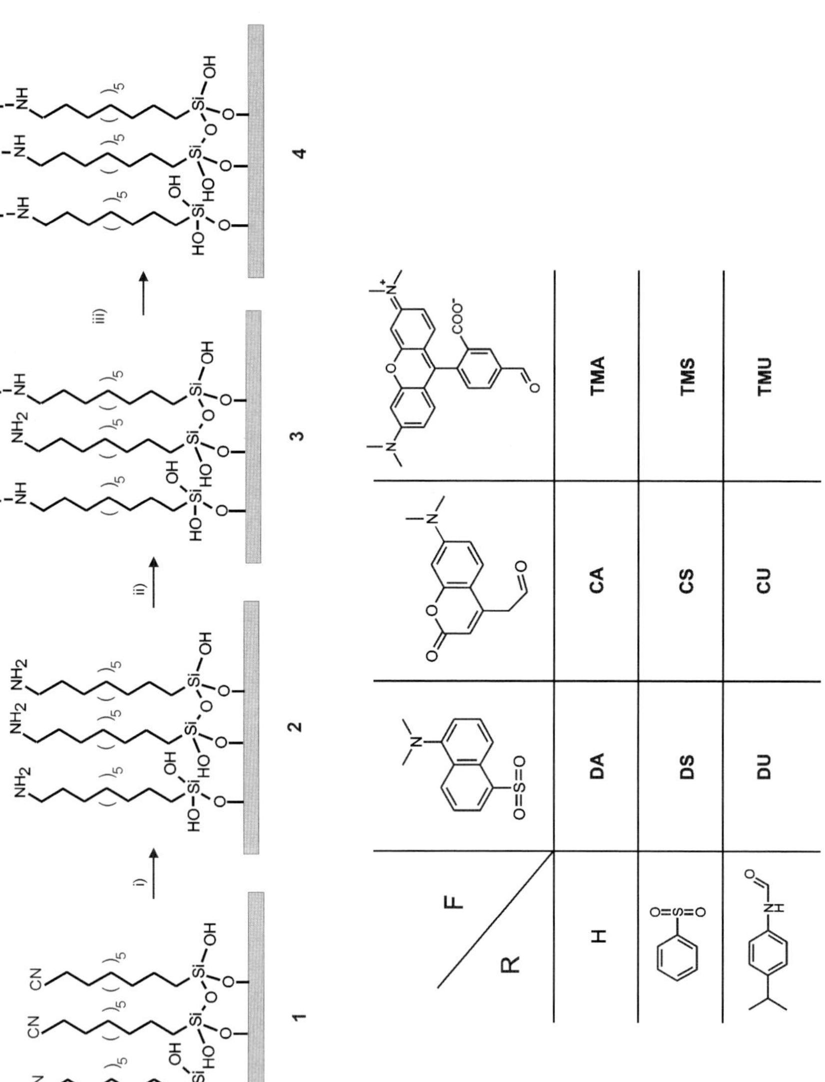

Fig. 4.9 Synthesis scheme for the preparation of the fluorescent SAMs (DA, CA, TMA, DS, CS, TMS, DU, CU, and TMU) on glass and silicon surfaces. (i) Red Al, toluene, 45°C; (ii) dansyl chloride, 7-dimethylaminocoumarin-4-acetic acid succinimidyl ester, or 5-(and-6)-carboxytetramethylrhodamine, succinimidyl ester, acetonitrile, rt; (iii) benzenesulfonyl chloride or *p*-isopropylphenyl isocyanate, acetonitrile, rt. 36 Reproduced with permission of The Royal Society of Chemistry

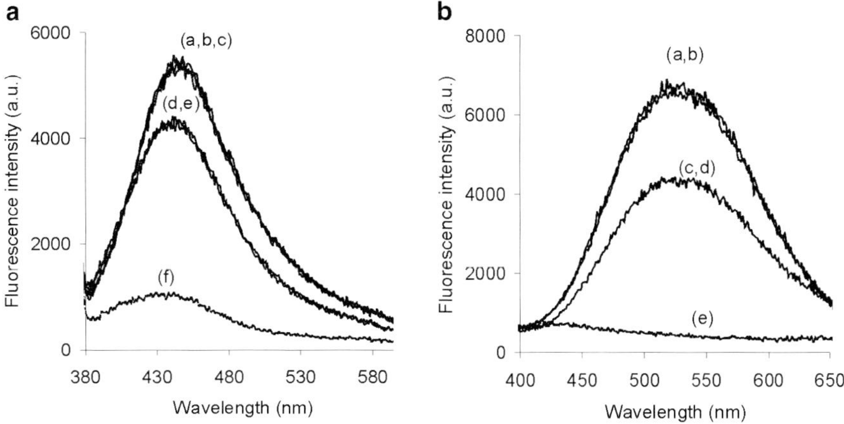

Fig. 4.10 *Left*: Spectra of CA layer in 0.1 M HEPES solution (**a**), after 3 min. (**b**), after 5 min. (**c**), in 10^{-4} M aqueous solution of HgCl (**d**), 3 min. later (**e**), and spectrum of the residual solvent after removal of the layer from the fluorescence cuvette (**f**). *Right*: Spectra of DA layer in 0.1 M HEPES solution (**a**), after 5 min. (**b**), in 10^{-4} M aqueous solution of $CuCl_2$ (**c**), 3 min. later (**d**), and spectrum of the residual solvent after removal of the layer from the fluorescence cuvette (**e**). Units of the *y*-axis are counts per second (cps). 36 Reproduced with permission of The Royal Society of Chemistry

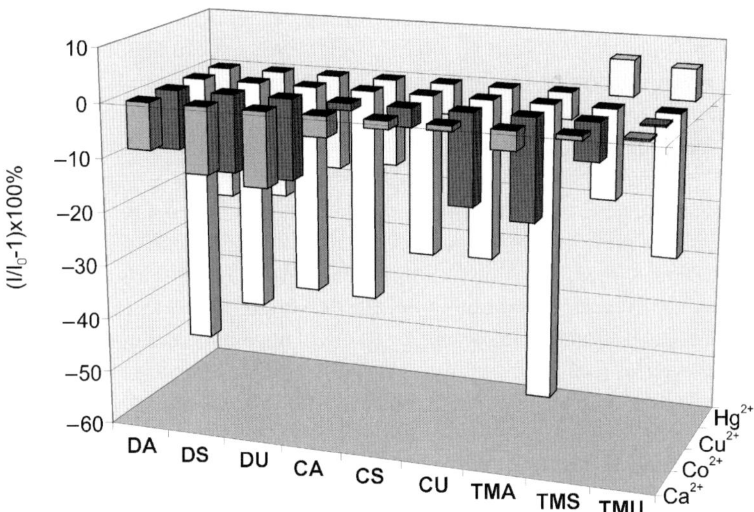

Fig. 4.11 Relative fluorescence intensity of DA, DS, DU, CA, CS, CU, TMA, and TMU. SAMs in the presence of 10^{-4} M solutions of Hg^{2+}, Cu^{2+}, Co^{2+}, and Ca^{2+} as chloride salts in water (pH 7.0, 0.1 M HEPES). These data have been normalized, in the absence of metal cations the maximum fluorescence emission of each layer is set to 0. Positive values correspond to enhancement in the fluorescence emission intensity of the layer, while negative values represent quenching of the fluorescence emission intensity of the layer

and urea functionalities and monitor the effect of the fluorophores on the response. For example, the amino layers (DA, CA, TMA) consistently give a high response for Cu^{2+} for all three fluorophores, whereas not all of the sulphonamide (DS, CS, TMS) and urea (DU, CU, TMU) layers respond as strongly. And when the influence of the different fluorophores (dansyl, coumarin, and TAMRA) on the monolayer response is compared, the dansyl monolayers (DA, DS, DU) exhibit more fluorescence quenching than the monolayers with the other fluorophores.

The unpredictability of what components will constitute a successful sensing layer underlines the power of utilizing a 2D combinatorial parallel approach to the discovery of successful sensing systems in aqueous media. The library response toward metal cations can be used to search for either a unique response (individual "hit") or a whole "fingerprint" of responses. Here, the "fingerprint" is the collection of the individual responses of each sensing layer to one cation. Rapid inspection of the library "fingerprint" (Fig. 4.11) provides a unique response for each cation.

3 Combinatorial Monolayer Array Fabrication

The intrinsic characteristics of this sensing system, i.e., flat and transparent surfaces, make it a good candidate for the fabrication of cross-reactive fluorescent sensor arrays.[50] The fabrication of cross-reactive sensor arrays is an emerging sensing approach[51] that is inspired by the mammalian natural way of sensing, where the olfactory system, with a set of nonspecific receptors, generates a response pattern that is perceived by the brain as a particular odor. We have studied the applicability of combinatorial libraries of fluorescent monolayers for the generation of such cross-reactive sensor arrays. First, arrays of fluorescent monolayers have been made on small glass microtiter plates. Previous works have shown the applicability of plastic microtiter plates to the fabrication of cross-reactive sensor arrays.[52] But the use of microfabication techniques makes possible the manufacture of glass microtiter plates with the dimensions of a standard glass slide, which allows the high throughput screening of the microtiterplate with a fluorescent scanner.

A step forward from the fabrication of fluorescent monolayers arrays in micotiterplate is the integration of such monlayer arrays in microfluidic chips. Microfluidics appears a powerful tool for miniaturization of analytical assays.[53] The glass surface of a microchannel behaves similarly to macro scale glass surfaces, and thus can be chemically activated and modified with SAMs.[54] The integration of (functional) SAMs into microchannels can be very advantageous for the fabrication of sensors because it combines the advantages of monolayer chemistry with those of the microfluidics devices[13] Microfluidics chips are being extensively exploited for antibody, nucleic acid, and protein analysis.[55] Although potentially very useful as cross sensor arrays of nonspecific sensing probes for the analysis of small molecules, microfluidics chips have not been exploited for differential sensing.[37,50,52,56] A series of individually addressable microchannels each containing a different sensing probe would provide a cross-reactive differential array integrated in a chip.[57]

3.1 Combinatorial Monolayer Array in a Microtiter Plate Format for Metal Ion Sensing

A sensing array containing 21 different fluorescent monolayers was prepared in a $25 \times 75\,mm^2$ 140 wells microtiter-plate (Fig. 4.12). It was found that the exposure of the monolayer array to several metal ions generates a characteristic pattern for each analyte, a "finger print" response, which was imaged by confocal microscopy and fluorescence laser scanning.[58] The power and utility of this new fluorescent array fabrication and detection methodology is demonstrated for the analysis of Cu^{2+}, Co^{2+}, Pb^{2+}, Ca^{2+}, and Zn^{2+}.

The general procedure for the fabrication of the sensing monolayers on the glass surface of the wells involves the functionalisation of the MTP surface with the amino terminated SAM *N*-[3-(trimethoxysilyl)propyl]ethylenediamine (TPEDA) and its sequential modification with a different fluorophore–ligand molecule pair in each well (Fig. 4.13). This allows the direct covalent attachment of the array building blocks amino-reactive fluorescent probes (F) ($\lambda_{ex} = 500–550\,nm$, $\lambda_{em} > 550\,nm$) and ligands (R) to the MTP wells.

A library of SAMs is fabricated in the microtiterplate by coating of each well with a different SAM. The synthesis of SAMs on the glass surface of the microtiterplate wells is performed as described above, in Sect. 2.1. First, the whole microtiter plate is coated with the amino terminated monolayer (TPEDA, Fig. 4.13) and subsequently the proper solutions are pipetted on different columns and rows of the MTP resulting in an array of 21 different sensing monolayers (TM0–TM6, T0–T6, L0–L6).

To facilitate the data quantification, the MTP was divided in two parts (Fig. 4.14). The first part was used as a reference, and the second part, the analysis library, was incubated with the corresponding analyte solutions.

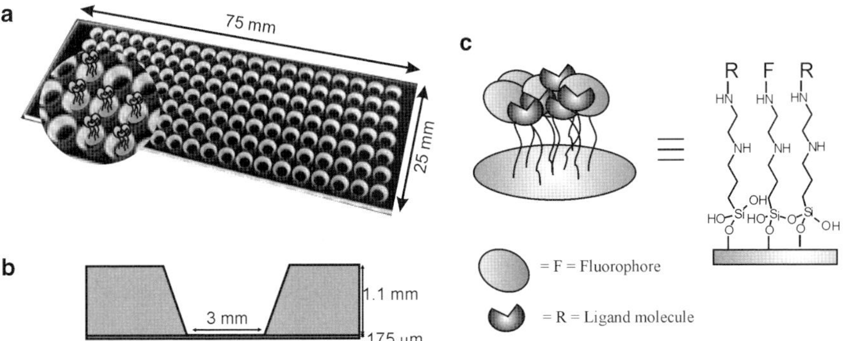

Fig. 4.12 (**a**) Picture of the 140 well glass microtiter plate (MTP) with a schematic representation (enlargement) of the wells substitution). Schematic representation of (**b**) one microtiter plate well with the corresponding dimensions and (**c**) the self-assembled monolayers formed in each well of the MTP 58. Reproduced with permission

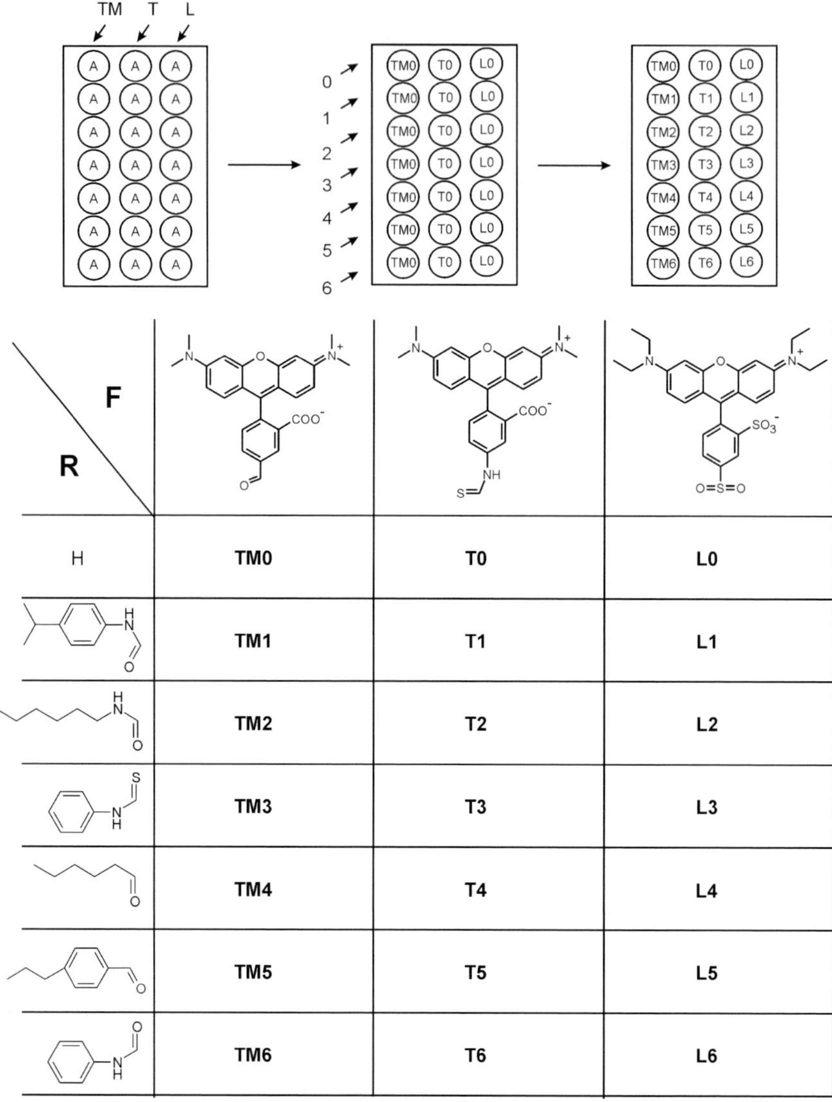

Fig. 4.13 *Top*: Schematic representation of the monolayer library synthesis (TM0–TM6, T0–T6, L0–L6) in a microtiter plate. Three different acetonitrile solutions containing each a different fluorophore, TAMRA (TM), TRITC (T), or Lissamine (L) are pipetted in the wells of three consecutive columns of a MTP coated with TPEDA monolayer (A in this figure). Subsequently, six solutions (1–6) containing each a different ligand are pipetted in consecutive rows. No ligand solution is added on the *top row* (wells TM0, L0, T0). *Bottom*: Chemical composition of the library of fluorescent SAMs (TM0–TM6, L0–L6, T0–T6) prepared on the MTP glass surface 58. Reproduced with permission

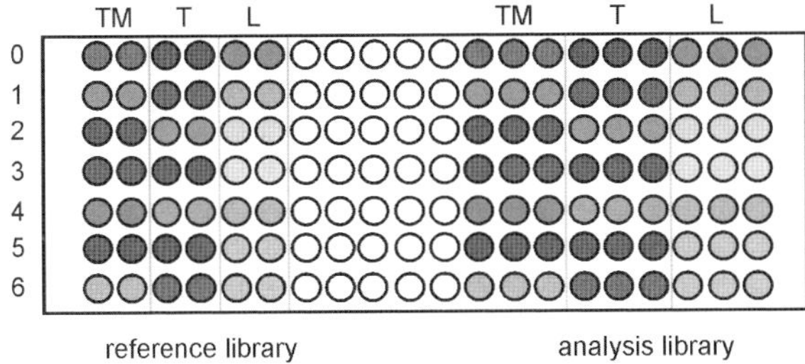

Fig. 4.14 (**a**) Fluorescence microscope images of two neighboring wells functionalized with L0 SAM (*left*) and TPEDA SAM (*right*), after rinsing the MTP with propylamine 58. Reproduced with permission

In total, 105 wells were functionalized with 21 different sensing systems. To demonstrate the generality of this sensing scheme, the metal ions were chosen to represent a wide range of ionic species. The response of the above described sensor array to Ca^{2+}, Cu^{2+}, Co^{2+}, Zn^{2+}, or Pb^{2+} was studied. Depending on the interaction between metal ion and SAM, the fluorescence emission can be enhanced or quenched.[35] Qualitative and quantitative analysis of the microtiter plate response in the presence of the analytes are both feasible. The fluorescence intensity of each well was measured by LSCM, and one colored image was generated for each well. By setting threshold values luminescence intensity of the array can be scale up and down, to get high or low fluorescence intensity, into a visible pattern by eye. Different color within a gray scale is assigned to different fluorescence intensity values, and with a particular graphical software, the gray scale pictures are transformed into colored photos with a red to yellow scale. The ratio of fluorescence intensities of the reference and the analysis library was calculated providing a quantitative measure of the analyte influence in each well. The interaction of each metal ion with the different sensing wells results in an individual fluorescent signal for each well. The combination of the 21 individual responses generates a characteristic pattern signature (fingerprint) for each analyte. Figure 4.15 shows the fluorescence intensity of each MTP well after incubation with the metal ion solutions. It can be clearly seen that each analyte in the array produces a unique fluorescent pattern. In this way, each analyte is easily distinguished from the others. Every monolayer was made in triple in consecutive wells in the analysis library of the MTP. One image of each well is obtained after the incubation with the analytes. But only one image (out of three) for each of the 21 systems in the array is selected for the final picture.

As mentioned before, the MTP used to form the monolayer array had the dimensions of a standard microscope slide ($75 \times 25\,mm^2$), suitable for measurements in commercial fluorescence scanners. Using a fluorescent scanner, the whole array can be read in a shorter time with high image quality, and data reproducibility for accurate and reliable measurements. The simplicity of the analysis technology should enhance the performance of our sensing scheme, allowing fast and accurate measurements. To prove whether the sensing SAMs array can be measured using a standard fluorescence array scanner, sensing and recycling experiments were

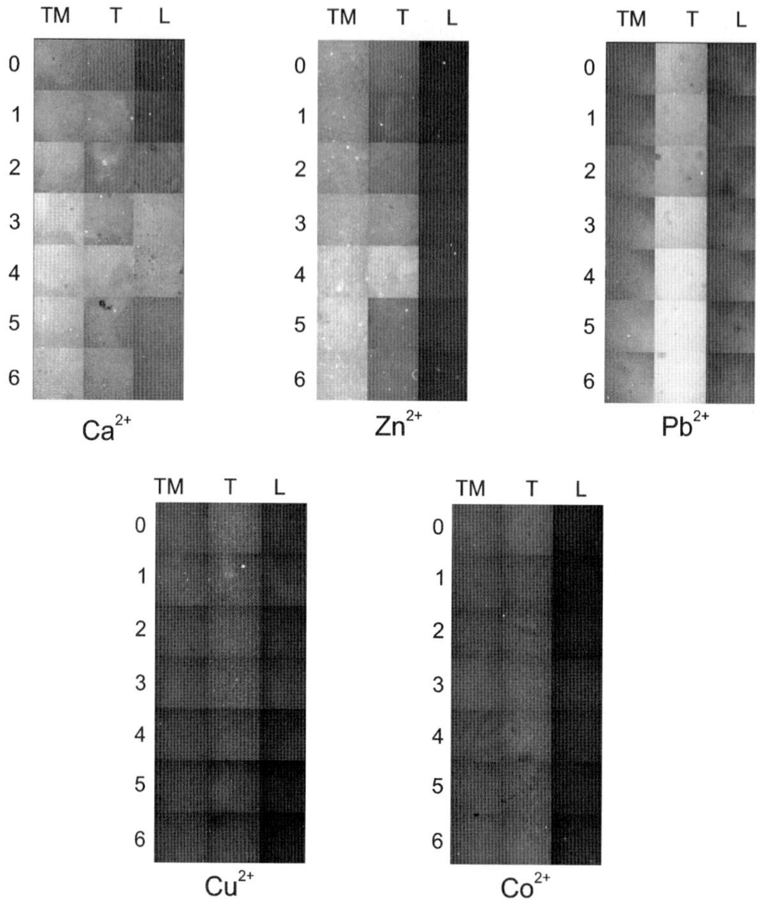

Fig. 4.15 Fluorescence microscopy images of the MTP wells after each plate was incubated with Ca^{2+}, Cu^{2+}, Co^{2+}, Zn^{2+}, and Pb^{2+} (10^{-4} M, acetonitrile). Each *square* in every image represents an individual signal, and the collection of the individual signals constitutes the response pattern of the array, (for well composition see Fig. 4.13) 58. Reproduced with permission

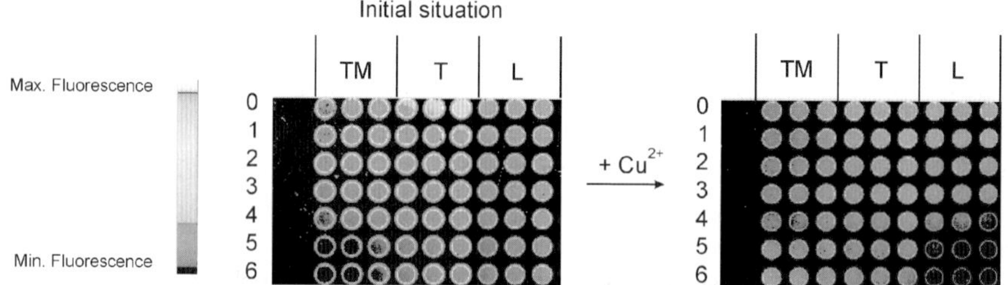

Fig. 4.16 Fluorescence scanner images of a functionalized microtiter plate with a library of 21 SAMs configured into an array (for library composition see Fig. 4.13), before and after exposure to a 10^{-4} M acetonitrile solution of $Cu(ClO_4)_2$, respectively 58. Reproduced with permission

performed. Figure 4.16 shows the pictures of the plate functionalized with the SAM array before and after Cu^{2+} addition. The scanning of the microtiter plate takes less than 40 s. The fluorescence intensity patterns of the functionalized MTP after exposure to the analyte provided a signature characteristic for the analyte comparable with that obtained previously using the confocal microscope. The fluorescence laser scanner allows imaging a sensing SAM array in a short time what constitutes an extra advantage in the screening methodologies for sensor arrays discussed in this chapter.

Recycling of this MTP confined sensing array was also studied, and it was found that the complexation of metal ions at the surface is reversible, and the surface can be easily recycled for further analysis upon rinsing the surface with a metal chelate agent.

The results herein demonstrated the successful application of parallel combinatorial methods to generate libraries of sensing SAMs covalently immobilized in the wells of a glass microtiter-plate. The fluorescence pattern after exposure of the array to different metal ion solutions allows identification of Cu^{2+}, Co^{2+}, Ca^{2+}, Zn^{2+} and Pb^{2+} at 10^{-4} M concentration by laser confocal microscopy and fluorescence laser scanner. The collection of the unselective response of the monolayers in the presence of the cations generates a characteristic fluorescent pattern, a "fingerprint" of each analyte in the array.

3.2   Combinatorial Monolayer Array in a Microfluidic Chip

SAMs on microchannel walls have been studied for surface properties in microreactors,[59] to control surface wetting,[60] to create zones for specific immobilization of proteins and biomolecules,[61] and to conduct catalytic reactions.[62] And a pH sensing monolayer confined to a glass microchannel has been reported by our group.[32]

Recently, the fabrication of a sensor array in a chip by covalent attachment of fluorescent monolayers on the walls of glass microchannels has been demonstrated. The monolayer array (confined in a multichannel chip) is prepared by parallel synthesis of monolayers functionalized with different fluorophore–ligand pairs, resulting in a sensing chip, which is able to generate a fingerprint of the network with a single fluorescence "snapshot" for different analytes.[63]

This method combines the advantages of microfluidic real-time analysis with the cost-effectiveness of microanalytical devices. In contrast with the monolayer array confined to a microtiter-plate, it prevents problems such as solvent evaporation during the monolayer formation. Furthermore, it reduces dramatically the occupied space and the amount of reagents, enhances the automation of the fabrication process, and allows for continuous monitoring of analyte solutions. Regeneration of the channel activity for the sequential testing of multiple analytes is also possible.

To automate the process of SAM synthesis in the microchannels for the generation of sensing arrays, a multichannel chip suitable for continuous flow of reagents was designed. A microchannel chip ($4 \times 2.2\,cm^2$) with five parallel channels confined in an area of 200 µm was fabricated. Each channel can be individually addressed and

                                                       L. Basabe-Desmonts et al.

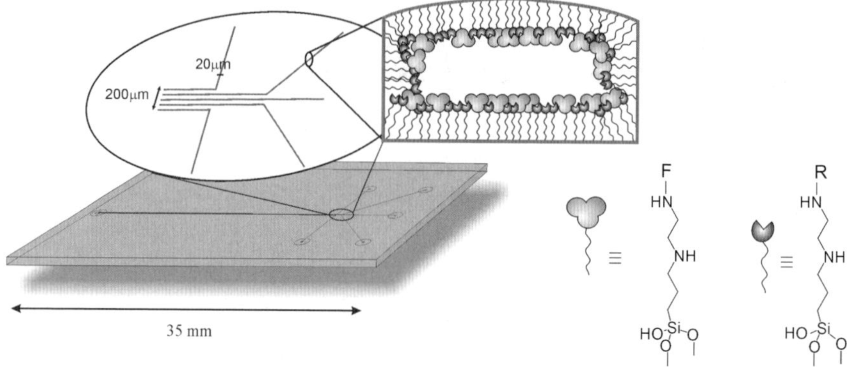

Fig. 4.17 (**a**) Schematic representation of the five-channel chip and the channel walls functionalized with a fluorescent SAMs 63

functionalized with chemically different fluorescent sensing monolayers to produce a sensing array chip (Fig. 4.17). The five channels integrated in the chip are 35-mm long, 20-μm wide, and 2-μm deep. The thickness of the glass bottom plate was 175 μm to allow fluorescent microscopy imaging of the bottom surface of the channel even with oil inmersion objectives for aquisition of higher resolution images if needed. They share a common outlet and each of them has an independent inlet.

Figure 4.17 depicts the protocol for the parallel synthesis of five sensing SAMs in the multichannel chip. After the activation with piranha and the formation of the amino-terminated TPEDA SAM, the unique functionalisation of each monolayer can be achieved by introducing consecutively different fluorophores (F_x, $x = 1$–5) and ligands (L_y, $y = 1$–5) in each channel. As a result an array of five fluorescent SAMs (F_x,L_y) confined to a single chip can be achieved. The result is a chip with an array of five fluorescent SAMs.

To flush the reagents and the solvents through the channels, the chip is mounted in a holder and connected to the fluid feeders by means of fused silica fibres. The holder is designed for fitting fused silica fibers into the inlet and outlet chip reservoirs using commercially available Nanoport™ assemblies. This system allows flow control in each channel, avoiding cross contamination between different channels. Additionally, the holder is designed to fit on an inverted fluorescence microscope to image the channels.

After the reaction, the channels were rinsed and the fluorescence microscopy images of the chip were acquired. Lissamine, TAMRA, and TRITC fluorophores are suitable for excitation with green light while the FITC and OG, with maximum absorption at 494 nm 511 nm, respectively, are suitable for blue light excitation. For visualization of the channels, two fluorescent images were recorded; first green excitation light ($510 \leq \lambda \leq 550$ nm) was used and red emission light ($\lambda > 590$ nm) was detected, followed by a second imaging using blue excitation light ($450 \leq \lambda \leq 480$ nm) and green emission ($\lambda > 515$ nm). After acquisition an overlay image was generated with digital imaging software (Fig. 4.18).

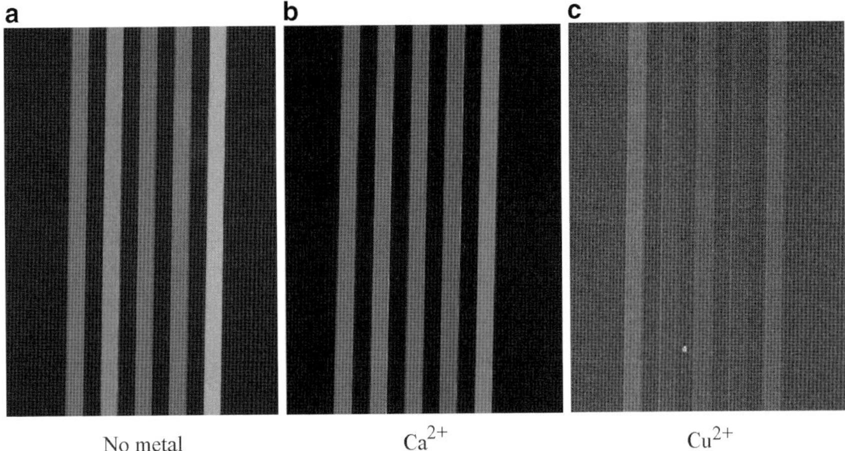

Fig. 4.18 Overlay image of the multichannel chip functionalized with an array of five different fluorescent monolayers filled with acetonitrile (**a**), Ca^{2+} (perchlorate salt, 10^{-4} M, acetonitrile) (**b**), and Cu^{2+} (perchlorate salt, 10^{-4} M, acetonitrile) (**c**) 68

Subsequently, the sensing properties of the monolayer array for Ca^{2+} and Cu^{2+} were evaluated (Fig. 4.18). Two fluorescence images were obtained each time and then combined, one using blue light and the other using green light for excitation of the fluorophores. The channels were filled with acetonitrile (Fig. 4.18a) and Ca^{2+} (perchlorate salt, 10^{-4} M, acetonitrile) (Fig. 4.18b). The fluorescence intensity of each SAM-coated channel varied, which created a fluorescent pattern in the presence of Ca^{2+}. Afterwards a flow of Cu^{2+} (perchlorate salt, 10^{-4} M acetonitrile) was injected during 5 min at a flow rate of 0.04 μL min^{-1} and different levels of fluorescent quenching in the channels were observed (Fig. 4.18c). These results are in agreement with those previously showed. The collective response of the sensing chip results in a different finger print for each metal ion.

In conclusion, the advantages of microfluidic devices, parallel synthesis, and combinatorial approaches can be merged to integrate a fluorescent chemical sensor array in a microfluidic chip. Fluorescent microchannel array can be produced by parallel synthesis of fluorescent monolayers covalent attached to the walls of glass microchannels.

4 Combinatorial Fabrication of Luminescent and Metal Ion Patterns on Glass

Pattern fabrication is an important issue in many fields ranging from microelectronics to biological microarray production and nanotechnology.[64] Soft and probe lithography techniques such as microcontact printing (μCP) and dip-pen nanolithography (DPN) are frequently used to pattern surfaces.[65] Conventional μCP is an efficient and low-cost method for patterning with feature sizes between ca. $0.5\,\mu m$ and few millimeters. μCP uses conformal contact of a patterned elastomer with a surface to transfer a chemical ink from the elastomer to localized areas of the surface due to the physical properties of the rubber-elastic polymer stamp.[65,65–68,68–75]

Dip-pen nanolithography is a high resolution patterning technique that enables the creation of patterns from the sub 100 nm to many micrometers length scale.[76] This technique uses an ink-coated AFM (atomic force microscopy) tip as a nanopen. The ink molecules are transported from the tip to a substrate, normally by capillary forces when the tip is in contact with the surface of the substrate.[77] The driving force for such transport is chemisorption of the ink to the underlying substrate due to a chemical[78] or electrochemical force.[79]

Even though soft-lithography techniques (μCP) have been applied to combinatorial methods, the scope of these studies has been limited to the immobilization of arrays of proteins[80] or nanocrystals[81] on surfaces, to optimize the patterning of molecular organic semiconductors by organic vapor jet printing,[82] and to optimize the properties of polymers anchored to surfaces.[83] DPN can be used to generate a large number of different patterns from the same or different chemical inks, which can be screened under identical conditions. Combinatorial approaches using DPN have been used to fabricate particle arrays,[84] to generate libraries of patterns for studying ink exchanges between the ink and nanostructures on the substrate[85] and to deposited monolayer-based resists with micrometer to sub 100 nm dimensions on different substrates for the generation of solid-state features.[86] All these cases are restricted to a unique, printable substrate, and they are not systematically expanded to the fabrication of different substrates and to the use of different ligand–substrate combinations. In contrast to these specificity-based patterning methods, an unprecedented combinatorial approach for the generation of libraries of luminescent patterns on glass has been developed in our labs. The complexing properties of the fluorescent SAM-coated substrates have been exploited to create and easily visualize metal ion patterns on glass substrates (Fig. 4.19). Patterning of fully covered fluorescent surfaces with metal ions (Ca^{2+}, Co^{2+}, Pb^{2+}, Cu^{2+}) has been carried out by soft and probe lithography techniques. Microcontact printing (μCP) and DPN of metal ion salts onto the fluorescent SAMs on glass resulted on the successful transfer of the metal ions to the surface. Fluorescent metal ion patterns are achieved by deposition of the metal ion salts on discrete areas of glass substrates fully covered with fluorescent SAMs (SAM (F_x,L_y)). First, a library of nonpatterned fluorescent self-assembled monolayer SAMs (F_x,L_y) ($x,y = 1,2,3,\ldots, n$; Fig. 4.19) is generated, as explained before, by sequential deposition of fluorophores F_x and ligand molecules L_y onto an amino-terminated monolayer on glass using solution synthesis. Then fluorescence patterns on the glass substrate are achieved by delivering different metal ions M_z ($z = 1, 2, 3,\ldots, n$) onto the layer by μCP (Sect. 4.1) or DPN (Sect. 4.2). The fluorescence modulation is produced in the contact areas between the stamp (μCP) and the AFM tip (DPN) and the fluorescent SAM. A library of fluorescent patterns is obtained where each pattern $P(F_x,L_y,M_z)$ ($x,y,z = 1,2,3,\ldots,n$), is the result of the combination of three building blocks, i.e., a fluorophore, a ligand molecule and a metal ion.

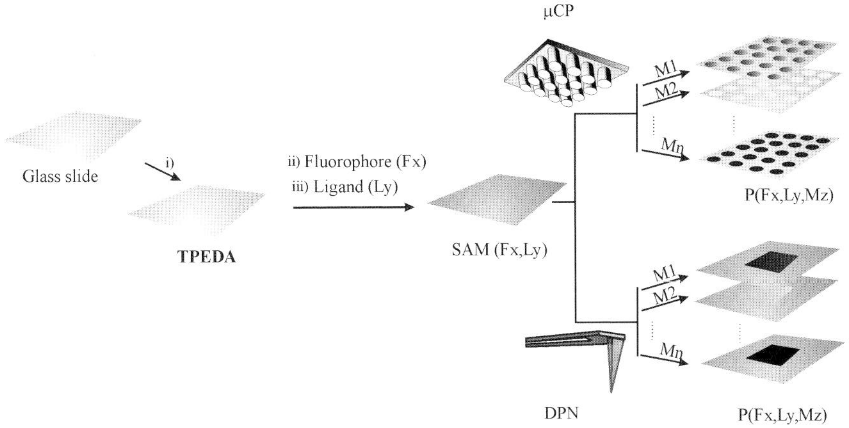

Fig. 4.19 The fabrication of metal ion patterns on glass by μCP and DPN: (i) Functionalisation of a glass slide with the amino-terminated TPEDA SAM; (ii) functionalisation of the TPEDA SAM with a fluorophore (F_x) and, (iii) with a ligand molecule (L_y) in solution to produce a fully covered nonpatterned fluorescent SAM ($F_x L_y$). Subsequently, μCP or DPN are used to transfer metal ions (M_z) from a PDMS stamp or an AFM tip onto the substrates thus creating metal ion patterns. Modulation of the fluorescence of the substrate is observed where the metal ions have been deposited.[87] Copyright Wiley-VCH Verlag GmbH & Co. KGaA. Reproduced with permission

The basic concept of these approaches is the fast generation and screening of substrates that store fluorescent and/or metal ion patterns. Besides the simplicity in the pattern generation, the approaches presented here offer the advantage of easy high throughput analysis. Visualization of the generated patterns is simply analyzed by fluorescence microscopy.

4.1 Fabrication of Metal Ion Patterns on Glass by Microcontact Printing

For the fabrication of metal ion patterns on glass by μCP, after the fabrication of the fluorescent monolayers, a PDMS stamp was used to deliver metal ions by μCP to the specific areas where the PDMS stamp was brought into contact with the functionalized substrate. Metal ions Cu^{2+}, Co^{2+}, Ca^{2+}, and Pb^{2+} were transferred by μCP to a glass slide coated with the TM4 SAM (see Fig. 4.4 for SAM composition).[87] After the μCP process, the layers were imaged with a fluorescence microscope. The images showed that the fluorescence emission intensity of the glass substrate in the areas where the metal ion was printed had changed (Fig. 4.20)

When comparing the patterns produced by the printing of different metal ions, it can be seen that Pb^{2+} and Ca^{2+} induced a fluorescence intensity enhancement of the native monolayer, creating a pattern with brighter dots while Co^{2+} and Cu^{2+} quenched the initial fluorescence intensity, resulting in a pattern with darker dots (Fig. 4.21a). Ca^{2+} produced higher fluorescence intensity enhancement than Pb^{2+}.

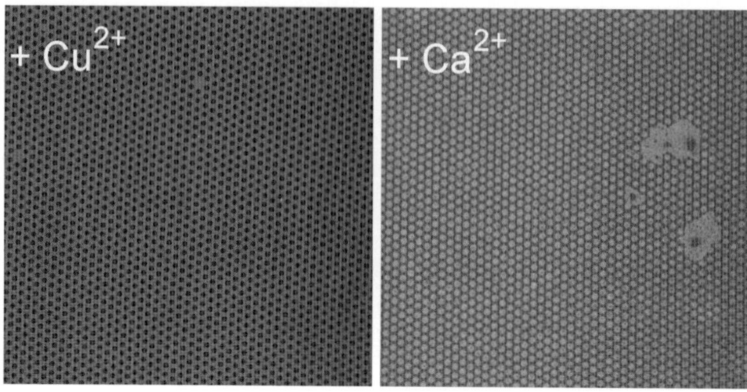

Fig. 4.20 $650 \times 650\,\mu m^2$ fluorescence microscopy image of the TM4 fluorescent monolayer on glass in which Cu^{2+} and Ca^{2+} (Cu $(ClO_4)_2$ and Ca $(ClO_4)_2$, 10^{-3} M, acetonitrile) were printed with PDMS stamps with an array of $10\,\mu m$ dots. This image clearly shows that the fluorescence of the glass slide was quenched and enhanced where the Cu^{2+} and the Ca^{2+} ions, respectively, were printed, i.e. the dotted pattern 87. Copyright Wiley-VCH Verlag GmbH & Co. KGaA. Reproduced with permission

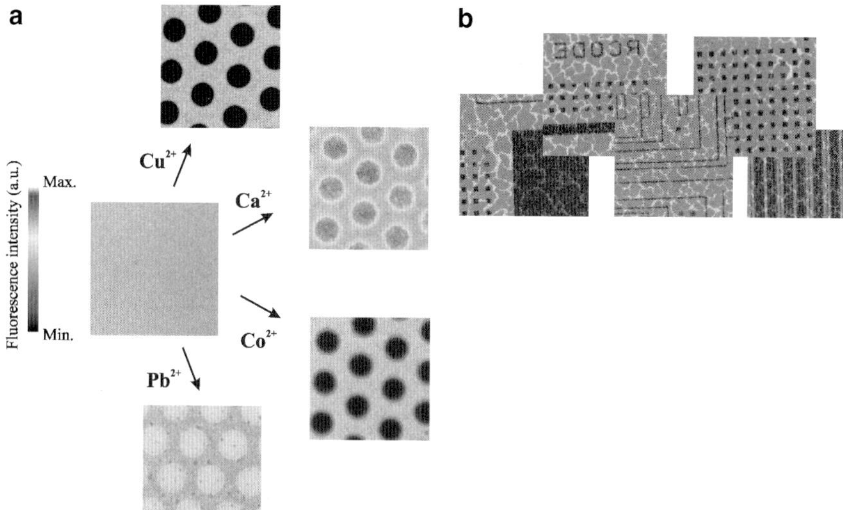

Fig. 4.21 (**a**) Fluorescence microscopy images of the TM4 fluorescent monolayer on glass in which 10^{-3} M acetonitrile solutions of different metal ions were printed with a PDMS stamp with $10\,\mu m$ diameter dot features separated by a period of $5\,\mu m$ in an array of dots. The image without dots corresponds to the initial fluorescence image of the TM4 SAM before printing the metal ions. The *color bar* on the *left side* represents the fluorescence emission intensity scale of the images; (**b**) Different sections of a fluorescence confocal microscopy image of a glass slide coated with L0 SAM in which Cu^{2+} (10^{-3} M Cu $(ClO_4)_2$ in water) was printed with a PDMS stamp with 2–100 μm circuit features. The *dark features* correspond to the areas where the fluorescence is quenched upon deposition of Cu^{2+} 87. Copyright Wiley-VCH Verlag GmbH & Co. KGaA. Reproduced with permission

The quenching effect produced by Cu^{2+} deposition was slightly higher than the quenching by the Co^{2+} ions. Figure 4.21b shows another example. A circuit shape pattern of Cu^{2+} was successfully transfered onto a L0 SAM on glass. The created metal ion pattern with features ranging from 2 to $100\,\mu m$ is well defined and directly visualized by fluorescence microscopy because of the fluorescence quenching of the layer where Cu^{2+} ions have been deposited.

4.2 Fabrication of Metal Ion Patterns on Glass by Dip Pen Nanolithography

DPN has some advantages over microcontact printing. Direct creation of arrays by different surface modification with different inks is possible with direct-write DPN, while simple μCP process is normally a single ink process.[88] Because of the flexibility and relatively high throughput of DPN, it allows the preparation of a large number of different patterns on one or more different substrates in a combinatorial fashion. DPN can be carried out as a serial process, and parallel-probe cantilevers have been fabricated by IBM[89] to be used for high-density data storage. Cantilever arrays with 10,000 cantilevers have been fabricated by Liu[90], and "fountain pens" have been developed as a nanodispensers[91] by integration of microfluidics systems to control the inking of individual cantilevers in a parallel probe array. These features make DPN a powerful tool for the development of the combinatorial generation of metal ion (and) fluorescent patterns.

DPN has been used to transfer metal ions from an AFM tip to fluorescent SAMs (F_x, L_y)-coated glass slides. In this way, metal ion patterns can be created on glass, nonconductive substrates. Additionally, the transfer of metal ions to discrete areas of the fluorescent glass slides results in the modulation of the fluorescence intensity in such area of the glass slide.[92] Here we show two examples, the transfer of Cu^{2+} and Ca^{2+} to a TM4 SAM and Ca^{2+} to a T3 (see Fig. 4.4 for SAM composition) SAM using DPN (Fig. 4.22). To perform the experiments, AFM tips were immersed in a solution of $Cu(ClO_4)_2$ or $Ca(ClO_4)_2$ (10^{-2} M, acetonitrile or ethanol) then dried and used to scan square areas of $20 \times 20\,\mu m^2$ on a glass substrate functionalized with the fluorescent SAM. The resulting patterns can be imaged by laser scanning confocal microscopy (LSCM) and by AFM. LSCM images of the surfaces written with Cu^{2+} showed a $20 \times 20\,\mu m^2$ pattern where the initial fluorescence of the TM4 was quenched (Fig. 4.22a). In a second experiment, Ca^{2+} (10^{-2} M, acetonitrile) was used as the ink and TM4 as the fluorescent substrate. A bright fluorescent square feature was observed in the LSCM image of the AFM-written area (Fig. 4.22b). The fluorescence intensity of both substrates has been modulated in the area where the metal ion has been transferred.

These results demonstrated the transfer of metal ions to nonconductive functionalized glass substrates by DPN. In some cases, the metal ion pattern is only visible by fluorescence microscopy and not by AFM, most presumably because a lower metal ion concentration have been deposited on the surface than in those cases in which the pattern is visible also in the friction AFM image. Thus, a strong advantage of the system

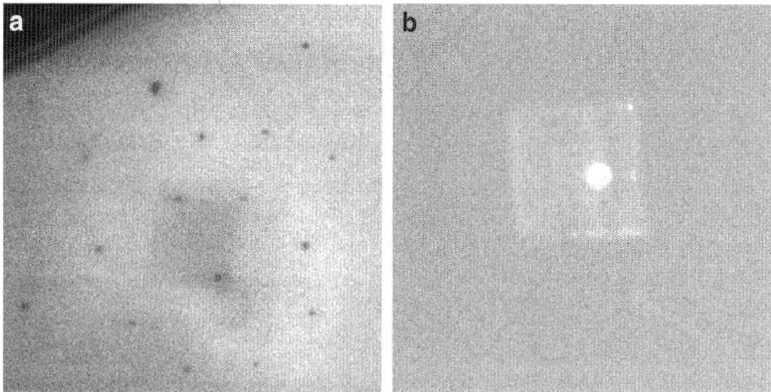

Fig. 4.22 Friction and height AFM $40 \times 40\,\mu\mathrm{m}$ images and confocal microscopy (LSCM) images with different sizes of the fluorescent monolayers TM4 (images **a–f** in which $20 \times 20\,\mu\mathrm{m}^2$ metal ion patterns were written by DPN. TM4 SAMs were patterned with Cu^{2+} (Cu $(ClO_4)_2$, 10^{-2} M, acetonitrile) (images **a–c**), or with Ca^{2+} (Ca $(ClO_4)_2$, 10^{-2} M, acetonitrile) (images **d–f**)

is that these new fluorescent substrates allow the pattern visualization even in cases where AFM does not reveal any feature. These fluorescent monolayers constitute a material more sensitive than AFM for the visualization of metal ion patterns.

5 Conclusions and Outlook

Fluorescent SAMs on glass surfaces have been studied as a new material for chemical sensing. The sensing SAMs are created by sequential deposition of two building blocks, a fluorophore and a ligand molecule, onto an amino terminated SAM on glass slides. This produces a flat glass surface provided with a large number of binding pockets. Perturbation of the fluorophore upon binding of the analyte to the surface creates a measurable signal making the sensing of analytes possible. In this way, a large number of different systems are fabricated by combinatorial techniques and parallel synthesis. Combinatorial libraries of fluorescent SAMs can be easily fabricated and they find applications in high throughput techniques for sensor discovery or development and in the fabrication of cross reactive sensory arrays.

Fluorescent SAMs constitute a new sensing material, which allows a label free sensing approach for several analytes. Their performance has been proven for metal ion and inorganic anions sensing in both organic solvents and aqueous solution. The sensing SAMs displayed good sensitivity for a number of ions with detection limits of 10^{-6} M. These sensing SAMs are also used as cross-reactive sensor arrays in which the analyte is identified by differential sensing using the collective response of a series of different SAMs to the analyte instead of the individual response of a single SAM. Arrays of fluorescent SAMs have been produced both in microtiterplate and in multichannel microfluidic chip formats. The scope of the method is such that many possibilities can be envisioned between combinations of fluorophores and any other complexing groups, reactive molecule, biomolecules, or specific receptors with a compatible functionality for attachment to the surface. In addition, fluorescent SAMs on glass have been used as substrates for chemical patterning. Different metal ion patterns have been created onto fluorescent SAMs coated glass slides. Because of the sensing properties of the substrates, modulation of fluorescence occurs in the localized areas where the metal ions are deposited. Therefore, the chemical pattern is easily revealed by a luminescent pattern, which is visualized by simple fluorescent microscopy.

The examples of small combinatorial libraries presented in this chapter were all fabricated manually, but automation of the process could be implemented using for example the already developed techniques for microarray fabrication, microspotter robots for fabrication, and fluorescence scanner for analysis. We hope that this chapter constitutes a clear explanation of the fabrication of combinatorial libraries of fluorescent self assembled monolayers and their applications.

Acknowledgments This work was supported by the Royal Netherlands Academy of Arts and Sciences (KNAW), and the MESA$^+$ Institute for Nanotechnology.

References

1. *Encyclopedia of Sensors, 10-Volumen Set*; Grimes, C. A.; Dickey, E. C.; Pishko, M. V.; The Pennsylvania State University: University Park,USA, 2005
2. *Topics in Fluorescence Spectroscopy*; Lakowicz, J. R.; Kluwer: New York, NY, 2002; Vol. 1
3. Lakowicz, J. R. *Probe Design and Chemical Sensing;* Lakowicz, J. R., Eds; Topics in Fluorescence Spectroscopy. Kluwer: New York, NY, 2002; Vol. 4
4. Basabe-Desmonts, L.; Reinhoudt, D. N.; Crego-Calama, M. Design of Fluorescent Materials for Chemical Sensing. *Chem. Soc. Rev.* **2007**, 36, 993–1017
5. Medintz, I. L.; Uyeda, H. T.; Goldman, E. R.; Mattoussi, H. Quantum Dot Bioconjugates for Imaging, Labelling and Sensing. *Nat. Mater.* **2005**, 4, 435–446
6. Bell, J. W.; Hext, N. M. Supramolecular Optical Chemosensors for Organic Analytes. *Chem. Soc. Rev.* **2004**, 33, 589–598
7. Lavigne, J. J.; Anslyn, E. V. Sensing a Paradigm Shift in the Field of Molecular Recognition: From Selective to Differential Receptors. *Angew. Chem. Int. Ed.* **2001**, 40, 3119–3130
8. Gauglitz, G. Optical Detection Methods for Combinatorial Libraries. *Curr. Opin. Chem. Biol.* **2000**, 4, 351–355
9. Dickinson, T. A.; White, J.; Kauer, J. S.; Walt, D. R. Current Trends in 'artificial-Nose' Technology. *Trends Biotechnol.* **1998**, 16, 250–258
10. Goodey, A.; Lavigne, J. J.; Savoy, S. M.; Rodríguez, M. D.; Curey, T.; Tsao, A.; Simmons, G.; Wright, J.; Yoo, S. J.; Sohn, Y.; Anslyn, E. V.; Shear, J. B.; Neikirk, D. P.; McDevitt, J. T. Development of Multianalyte Sensor Arrays Composed of Chemically Derivatized Polymeric Microspheres Localized in Micromachined Cavities. *J. Am. Chem. Soc.* **2001**, 123, 2559–2570
11. C. M., B. *Neural Networks for Pattern Recognition*; Oxford University Press: New York, NY, 2004
12. Lyons, W. B.; Lewis, E. Neural Networks and Pattern Recognition Techniques Applied to Optical Fibre Sensors. *Trans. Inst. Measurm. Control* **2000**, 22, 385–404

13. Crooks, R. M.; Ricco, A. J. New Organic Materials Suitable for Use in Chemical Sensor Arrays. *Acc. Chem. Res.* **1998**, 31, 219–227

14. Chechik, V.; Crooks, R. M.; Stirling, C. J. M. Reactions and Reactivity in Self-Assembled Monolayers. *Adv. Mater.* **2000**, 12, 1161–1171

15. Dulkeith, E.; Morteani, A. C.; Niedereichholz, T.; Klar, T. A.; Feldmann, J.; Levi, S. A.; Van Veggel, F. C. J. M.; Reinhoudt, D. N.; Möller, M.; Gittins, D. I. Fluorescence Quenching of Dye Molecules Near Gold Nanoparticles: Radiative and Nonradiative Effects. *Phys. Rev. Lett.* **2002**, 89, 203002

16. Imahori, H.; Norieda, H.; Nishimura, Y.; Yamazaki, I.; Higuchi, K.; Kato, N.; Motohiro, T.; Yamada, H.; Tamaki, K.; Arimura, M.; Sakata, Y. Chain Length Effect on the Structure and Photoelectrochemical Properties of Self-Assembled Monolayers of Porphyrins on Gold Electrodes. *J. Phys. Chem. B* **2000**, 104, 1253–1260

17. Motesharei, K.; Myles, D. C. Molecular Recognition in Membrane Mimics – a Fluorescence Probe. *J. Am. Chem. Soc.* **1994**, 116, 7413–7414

18. Sun, X. Y.; Liu, B.; Jiang, Y. B. An Extremely Sensitive Monoboronic Acid Based Fluorescent Sensor for Glucose. *Anal. Chim. Acta* **2004**, 515, 285–290

19. Panicker, R. C.; Huang, X.; Yao, S. Q. Recent Advances in Peptide-Based Microarray Technologies. *Comb. Chem. High Throughput Screening* **2004**, 7, 547–556

20. Walsh, D. P.; Chang, Y. T. Recent Advances in Small Molecule Microarrays: Applications and Technology. *Comb. Chem. High Throughput Screening* **2004**, 7, 557–564

21. Adronov, A.; Robello, D. R.; Frechet, J. M. J. Light Harvesting and Energy Transfer Within Coumarin-Labeled Polymers. *J. Polym. Sci., Part A: Polym. Chem.* **2001**, 39, 1366–1373

22. Chrisstoffels, L. A. J.; Adronov, A.; Frechet, J. M. J. Surface-Confined Light Harvesting, Energy Transfer, and Amplification of Fluorescence Emission in Chromophore-Labeled Self-Assembled Monolayers. *Angew. Chem. Int. Ed.* **2000**, 39, 2163–2167

23. Saari, L. A.; Seitz, W. R., pH sensor based on immobilized fluores ceinamine. *Anal. Chem.* **1982**, 54, 821

24. Harper, B. G., Reusable glass-bound pH indicators. *Anal. Chem.* **1975**, 47, 348

25. Urbano, E.; Offenbacher, H.; Wolfbeis, O. S. Optical Sensor for Continuous Determination of Halides. *Anal. Chem. Abstract* **1984**, 56, 427–429

26. Xavier, M. P.; García-Fresnadillo, D.; Moreno-Bondi, M. C.; Orellana, G. Oxygen Sensing in Nonaqueous Media Using Porous Glass With Covalently Bound Luminescent Ru(II) Complexes. *Anal. Chem.* **1998**, 70, 5184–5189

27. Sullivan, T. P.; Huck, W. T. S. Reactions on Monolayers: Organic Synthesis in Two Dimensions. *Eur. J. Org. Chem.* **2003**, 17–29

28. Flink, S.; Van Veggel, F. C. J. M.; Reinhoudt, D. N. A Self-Assembled Monolayer of a Fluorescent Guest for the Screening of Host Molecules. *Chem. Commun.* **1999**, 2229–2230

29. Van der Veen, N. J.; Flink, S.; Deij, M. A.; Egberink, R. J. M.; Van Veggel, F. C. J. M.; Reinhoudt, D. N. Monolayer of a Na⁺-Selective Fluoroionophore on Glass: Connecting the Fields of Monolayers and Optical Detection of Metal Ions. *J. Am. Chem. Soc.* **2000**, 122, 6112–6113

30. Van der Boom, T.; Evmenenko, G.; Dutta, P.; Wasielewski, M. R. Self-assembly of Photofunctional Siloxane-based Calix[4]arene on Oxide Surfaces. *Chem. Mater.* **2005**, 15, 4068–4074

31. Cejas, M. A.; Raymo, F. M. Fluorescent Diazapyrenium Films and Their Response to Dopamine. *Langmuir* **2005**, 21, 5795–5802

32. Mela, P.; Onclin, S.; Goedbloed, M. H.; Levi, S.; García-Parajó, M. F.; Van Hulst, N. F.; Ravoo, B. J.; Reinhoudt, D. N.; Van den Berg, A. Monolayer-Functionalized Microfluidics Devices for Optical Sensing of Acidity. *Lab Chip* **2005**, 5, 163–170

33. Grandini, P.; Mancin, F.; Tecilla, P.; Scrimin, P.; Tonellato, U. Exploiting the Self-Assembly Strategy for the Design of Selective Cu–Ii Ion Chemosensors. *Angew. Chem., Int. Ed.* **1999**, 38, 3061–3064

34. Crego-Calama, M.; Reinhoudt, D. N. New Materials for Metal Ion Sensing by Self-Assembled Monolayers on Glass. *Adv. Mater.* **2001**, 13, 1171–1174

35. Basabe-Desmonts, L.; Beld, J.; Zimmerman, R. S.; Hernando, J.; Mela, P.; Parajó, M. F. G.; Van Hulst, N. F.; Van den Berg, A.; Reinhoudt, D. N.; Crego-Calama, M. A Simple Approach

to Sensor Discovery and Fabrication on Self-Assembled Monolayers on Glass. *J. Am. Chem. Soc.* **2004**, 126, 7293–7299

36. Zimmerman, R. S.; Basabe-Desmonts, L.; Van der Baan, F.; Reinhoudt, D. N.; Crego-Calama, M. A combinatorial approach to surface-confined cation sensors in water. *J. Mater. Chem.* **2005**, 15, 2772–2777

37. Albert, K. J.; Lewis, N. S.; Schauer, C. L.; Sotzing, G. A.; Stitzel, S. E.; Vaid, T. P.; Walt, D. R. Cross-Reactive Chemical Sensor Arrays. *Chem. Rev.* **2000**, 100, 2595–2626

38. Lavigne, J. J.; Savoy, S.; Clevenger, M. B.; Ritchie, J. E.; Mcdoniel, B.; Yoo, S. J.; Anslyn, E. V.; Mcdevitt, J. T.; Shear, J. B.; Neikirk, D. Solution-Based Analysis of Multiple Analytes by a Sensor Array: Toward the Development of an "Electronic Tongue". *J. Am. Chem. Soc.* **1998**, 120, 6429–6430

39. Pirrung, M. C. Spatially Addressable Combinatorial Libraries. *Chem. Rev.* **1997**, 97, 473–488

40. Cremer, P. S.; Yang, T. L. Creating Spatially Addressed Arrays of Planar Supported Fluid Phospholipid Membranes. *J. Am. Chem. Soc.* **1999**, 121, 8130–8131

41. Adamson, A. W. *Physical Chemistry of Surfaces*, Wiley: Chicester **1993**

42. Azzam, R. M. A.; Bashara, N. M. *Ellipsometry and Polarized Light*, North-Holland: New York **1987**

43. Mathauer, K.; Frank, C. W. Naphthalene Chromophore Tethered in the Constrained Environment of a Self-Assembled Monolayer. *Langmuir* **1993**, 9, 3002–3008

44. Wagner, C. D.; Riggs, W. M.; Davis, L. E.; Moulder, J. F. *Handbook of X-Ray Photoelectron Spectroscopy*, Muilenberg, G.E.; Ed., Perkin-Elmer Corporation: Eden Prairie, Minesota **1979**

45. Valeur, B.; Leray, I. Design Principles of Fluorescent Molecular Sensors for Cation Recognition. *Coord. Chem. Rev.* **2000**, 205, 3–40

46. Antonisse, M. M. G.; Snellink-Ruel, B. H. M.; Lugtenberg, R. J. W.; Engbersen, J. F. J.; Van den Berg, A.; Reinhoudt, D. N. Membrane Characterization of Anion-Selective Chemfets by Impedance Spectroscopy. *Anal. Chem.* **2000**, 72, 343–348

47. Beer, P. D.; Gale, P. A. Anion Recognition and Sensing: the State of the Art and Future Perspectives. *Angew. Chem. Int. Ed.* **2001**, 40, 487–516

48. Bierbaum, K.; Kinzler, M.; Woll, C.; Grunze, M.; Hahner, G.; Heid, S.; Effenberger, F. A. Near-Edge X-Ray-Absorption Fine Structure Spectroscopy and X-Ray Photoelectron-Spectroscopy Study of the Film Properties of Self-Assembled Monolayers of Organosilanes on Oxidized Si(100). *Langmuir* **1995**, 11, 512–518

49. Onclin, S.; Mulder, A.; Huskens, J.; Ravoo, B. J.; Reinhoudt, D. N. Molecular Printboards: Monolayers of Beta-Cyclodextrins on Silicon Oxide Surfaces. *Langmuir* **2004**, 20, 5460–5466

50. Rakow, N. A.; Suslick, K. S. A Colorimetric Sensor Array for Odour Visualization. *Nature* **2000**, 406, 710–713

51. Lundstrom, I. Artificial Noses – Picture the Smell. *Nature* **2000**, 406, 682–683

52. Mayr, T.; Igel, C.; Liebsch, G.; Klimant, I.; Wolfbeis, O. S. Cross-Reactive Metal Ion Sensor Array in a Micro Titer Plate Format. *Anal. Chem.* **2003**, 75, 4389–4396

53. Andersson, H.; Van den Berg, A. Microfluidic Devices for Cellomics: a Review. *Sens. Actuators, B* **2003**, 92, 315–325

54. Munro, N. J.; Huhmer, A. F. R.; Landers, J. P. Robust Polymeric Microchannel Coatings for Microchip-Based Analysis of Neat PCR Products. *Anal. Chem.* **2001**, 73, 1784–1794

55. Auroux, P. A.; Koc, Y.; Demello, A.; Manz, A.; Day, P. J. R. Miniaturised Nucleic Acid Analysis. *Lab Chip* **2004**, 4, 534–546

56. Schauer, C. L.; Steemers, F. J.; Walt, D. R. A Cross-Reactive, Class-Selective Enzymatic Array Assay. *J. Am. Chem. Soc.* **2001**, 123, 9443–9444

57. Wolfbeis, O. S. Materials for Fluorescence-Based Optical Chemical Sensors. *J. Mater. Chem.* **2005**, 15, 2657–2669

58. Basabe-Desmonts, L.; van der Baan, F.; Zimmerman, R. S.; Reinhoudt, D. N.; Crego-Calama, M. Cross-Reactive Sensor Array for Metal Ion Sensing Based on Fluorescent SAMs. *Sensors* **2007**, 7, 1731–1746

59. Brivio, M.; Oosterbroek, R. E.; Verboom, W.; Goedbloed, M. H.; Van den Berg, A.; Reinhoudt, D. N. Surface Effects in the Esterification of 9-Pyrenebutyric Acid Within a Glass Micro Reactor. *Chem. Commun.* **2003**, 1924–1925

60. Zhao, B.; Moore, J. S.; Beebe, D. J. Surface-Directed Liquid Flow Inside Microchannels. *Science* **2001**, 291, 1023–1026

61. Smith, E. A.; Thomas, W. D.; Kiessling, L. L.; Corn, R. M. Surface Plasmon Resonance Imaging Studies of Protein–Carbohydrate Interactions. *J. Am. Chem. Soc.* **2003**, 125, 6140–6148

62. Kobayashi, J.; Mori, Y.; Okamoto, K.; Akiyama, R.; Ueno, M.; Kitamori, T.; Kobayashi, S. A Microfluidic Device for Conducting Gas-Liquid-Solid Hydrogenation Reactions. *Science* **2004**, 304, 1305–1308

63. Basabe-Desmonts, L.; Benito-Lopez, F.; Gardeniers, H. J. G. E.; Duwel, R.; van den Berg, A.; Reinhoudt, D. N.; Crego-Calama, M. Fluorescent Sensor Array in a Microfluidic Chip. *Anal. Bioanal. Chem.* **2008**, 390, 307–315

64. Xia, Y. N.; Rogers, J. A.; Paul, K. E.; Whitesides, G. M. Unconventional Methods for Fabricating and Patterning Nanostructures. *Chem. Rev.* **1999**, 99, 1823–1848

65. Geissler, M.; Xia, Y. N. Patterning: Principles and Some New Developments. *Adv. Mater.* **2004**, 16, 1249–1269

66. Delamarche, E.; Donzel, C.; Kamounah, F. S.; Wolf, H.; Geissler, M.; Stutz, R.; Schmidt-Winkel, P.; Michel, B.; Mathieu, H. J.; Schaumburg, K. Microcontact Printing Using Poly(Dimethylsiloxane) Stamps Hydrophilized by Poly(Ethylene Oxide) Silanes. *Langmuir* **2003**, 19, 8749–8758

67. Zheng, H. P.; Rubner, M. F.; Hammond, P. T. Particle Assembly on Patterned "Plus/Minus" Polyelectrolyte Surfaces Via Polymer-on-Polymer Stamping. *Langmuir* **2002**, 18, 4505–4510

68. Auletta, T.; Dordi, B.; Mulder, A.; Sartori, A.; Onclin, S.; Bruinink, C. M.; Peter, M.; Nijhuis, C. A.; Beijleveld, H.; Schonherr, H.; Vancso, G. J.; Casnati, A.; Ungaro, R.; Ravoo, B. J.; Huskens, J.; Reinhoudt, D. N. Writing Patterns of Molecules on Molecular Printboards. *Angew. Chem. Int. Ed.* **2004**, 43, 369–373

69. Yang, K. L.; Cadwell, K.; Abbott, N. L. Contact Printing of Metal Ions Onto Carboxylate-Terminated Self-Assembled Monolayers. *Adv. Mater.* **2003**, 15, 1819–1823

70. Lahiri, J.; Ostuni, E.; Whitesides, G. M. Patterning Ligands on Reactive SAMs by Microcontact Printing. *Langmuir* **1999**, 15, 2055–2060

71. Bernard, A.; Renault, J. P.; Michel, B.; Bosshard, H. R.; Delamarche, E. Microcontact Printing of Proteins. *Adv. Mater.* **2000**, 12, 1067–1070

72. Renault, J. P.; Bernard, A.; Juncker, D.; Michel, B.; Bosshard, H. R.; Delamarche, E. Fabricating Microarrays of Functional Proteins Using Affinity Contact Printing. *Angew. Chem. Int. Ed.* **2002**, 41, 2320–2323

73. Mahalingam, V.; Onclin, S.; Peter, M.; Ravoo, B. J.; Huskens, J.; Reinhoudt, D. N. Directed Self-Assembly of Functionalized Silica Nanoparticles on Molecular Printboards Through Multivalent Supramolecular Interactions. *Langmuir* **2004**, 20, 11756–11762

74. Mulder, A.; Onclin, S.; Peter, M.; Hoogenboom, J. P.; Beijleveld, H.; Ter Maat, J.; García-Parajň, M. F.; Ravoo, B. J.; Huskens, J.; Van Hulst, N. F.; Reinhoudt, D. N. Molecular Printboards on Silicon Oxide: Lithographic Patterning of Cyclodextrin Monolayers with Multivalent, Fluorescent Guest Molecules. *Small* **2005**, 1, 242–253

75. Onclin, S.; Huskens, J.; Ravoo B. J.; Reinhoudt, D. N. Molecular Boxes on a Molecular Printboard: Encapsulation of Anionic Dyes in Immobilized Dendrimers. *Small* **2005**, 852–857

76. Ginger, D. S.; Zhang, H.; Mirkin, C. A. The Evolution of Dip-Pen Nanolithography. *Angew. Chem. Int. Ed.* **2004**, 43, 30–45

77. Tseng, A. A.; Notargiacomo, A.; Chen, T. P. Nanofabrication by Scanning Probe Microscope Lithography: a Review. *J. Vac. Sci. Technol. B* **2005**, 23, 877–894

78. Lim, J. H.; Mirkin, C. A. Electrostatically Driven Dip-Pen Nanolithography of Conducting Polymers. *Adv. Mater.* **2002**, 14, 1474–1477

79. Li, Y.; Maynor, B. W.; Liu, J. Electrochemical AFM "Dip-Pen" Nanolithography. *J. Am. Chem. Soc.* **2001**, 123, 2105–2106

80. Stoll, D.; Templin, M. F.; Schrenk, M.; Traub, P. C.; Vohringer, C. F.; Joos, T. O. Protein Microarray Technology. *Frontiers Biosci.* **2002**, 7, C13–C32

81. Vossmeyer, T.; Jia, S.; Delonno, E.; Diehl, M. R.; Kim, S. H.; Peng, X.; Alivisatos, A. P.; Heath, J. R. Combinatorial Approaches Toward Patterning Nanocrystals. *J. Appl. Phys.* **1998**, 84, 3664–3670

82. Shtein, M.; Peumans, P.; Benziger, J. B.; Forrest, S. R. Direct Mask-Free Patterning of Molecular Organic Semiconductors Using Organic Vapor Jet Printing. *J. Appl. Phys.* **2004**, 96, 4500–4507

83. Wu, T.; Tomlinson, M.; Efimenko, K.; Genzer, J. A Combinatorial Approach to Surface Anchored Polymers. *J. Mater. Sci.* **2003**, 38, 4471–4477

84. Demers, L. M.; Mirkin, C. A. Combinatorial Templates Generated by Dip-Pen Nanolithography for the Formation of Two-Dimensional Particle Arrays. *Angew. Chem., Int. Ed.* **2001**, 40, 3069–3071

85. Ivanisevic, A.; Mccumber, K. V.; Mirkin, C. A. Site-Directed Exchange Studies with Combinatorial Libraries of Nanostructures. *J. Am. Chem. Soc.* **2002**, 124, 11997–12001

86. Weinberger, D. A.; Hong, S. G.; Mirkin, C. A.; Wessels, B. W.; Higgins, T. B. Combinatorial Generation and Analysis of Nanometer- and Micrometer-Scale Silicon Features Via "Dip-Pen" Nanolithography and Wet Chemical Etching. *Adv. Mater.* **2000**, 12, 1600–1603

87. Basabe-Desmonts, L.; Reinhoudt, D. N.; Crego-Calama, M. Combinatorial Fabrication of Fluorescent Patterns with Metal Ions Using Soft Lithography. *Adv. Mater.* **2006**, 18, 1028–1032

88. Multiple inking is possible using microfluidic networks over the PDMS stamps.Papra, A.; Bernard, A.; Juncker, D.; Larsen, N. B.; Michel, B.; Delamarche, E. Microfluidic Networks Made of Poly(Dimethylsiloxane), Si, and Au Coated With Polyethylene Glycol for Patterning Proteins Onto Surfaces. *Langmuir* **2001**, 17, 4090–4095

89. Vettiger, P.; Despont, M.; Drechsler, U.; Durig, U.; Haberle, W.; Lutwyche, M. I.; Rothuizen, H. E.; Stutz, R.; Widmer, R.; Binnig, G. K. The "Millipede" - More Than One Thousand Tips for Future AFM Data Storage. *IBM J. Res. Dev.* **2000**, 44, 323–340

90. Zhang, M.; Bullen, D.; Chung, S. W.; Hong, S.; Ryu, K. S.; Fan, Z. F.; Mirkin, C. A.; Liu, C. A Mems Nanoplotter With High-Density Parallel Dip-Pen Manolithography Probe Arrays. *Nanotechnology* **2002**, 13, 212–217

91. Deladi, S.; Tas, N. R.; Berenschot, J. W.; Krijnen, G. J. M.; De Boer, M. J.; De Boer, J. H.; Peter, M.; Elwenspoek, M. C. Micromachined Fountain Pen for Atomic Force Microscope-Based Nanopatterning. *Appl. Phys. Lett.* **2004**, 85, 5361–5363

92. Submitted for publication. Basabe-Desmonts, L.; Wu, C.; van der Werf, K. O.; Peter, M.; Bennik, M.; Otto, C.; Velders, A. H.; Reinhoudt, D. N.; Subramaniam, V.; Crego-Calama, M. Fabrication and Visualization of Metal Ion Patterns on Glass by Dip Pen Nanolithography

Chapter 5
High-Throughput Screening of Vapor Selectivity of Multisize CdSe Nanocrystal/ Polymer Composite Films

Radislav A. Potyrailo and Andrew M. Leach

Abstract We have achieved selective gas sensing based on different size semiconductor nanocrystals incorporated into rationally selected polymer matrices. From the high-throughput screening experiments, we have found that when CdSe nanocrystals of different size (2.8 and 5.6 nm diameter) were incorporated into different types of polymer films, the photoluminescence (PL) response patterns upon laser excitation at 407-nm and exposure to polar and nonpolar solvent vapors were dependent on the nature of polymer. We analyzed the spectral PL response from both sizes of CdSe nanocrystals using multivariate analysis tools. Results of this multivariate analysis demonstrate that a single film with different size CdSe nanocrystals serves as a selective sensor. The stability of PL response to vapors was evaluated upon 16 h of continuous exposure to laser excitation.

1 Introduction

Formulated polymeric sensor films are finding their applications when there is a need to expand the range of types of species detectable by sensors or to improve sensor selectivity.[1] Depending on the analyte-reagent response mechanism, sensors with formulated polymeric films are built based on transduction principles that involve radiant, electrical, mechanical, and thermal energy.[2–6] Optical sensors based on organic dyes formulated into polymeric materials are among the most popular sensors because this approach allows to tailor the response sensitivity and selectivity by a proper reagent–polymer combination.[1,7,8]

It was shown that polymeric matrices have their sensing response stability over several years.[9–11] However, in fluorescent chemical sensors, photobleaching of matrix-immobilized organic dyes significantly hinders the applicability of such

R.A. Potyrailo (✉)
Chemical and Biological Sensing Laboratory, Chemistry Technologies and Material Characterization, General Electric Global Research, Niskayuna, New York, NY 12309, USA
potyrailo@crd.ge.com

R.A. Potyrailo and V.M. Mirsky (eds.), *Combinatorial Methods for Chemical and Biological Sensors*, DOI: 10.1007/978-0-387-73713-3_5,
© Springer Science + Business Media, LLC 2009

sensing films. To reduce these problems, several approaches can be used, e.g., self-recovery after photobleaching[12] and adaptive exposures.[13]

An attractive alternative to organic fluorophores in formulated polymeric sensor films for chemical detection is to use inorganic luminescent materials. Semiconductor nanocrystals have a dramatically improved photostability and are attractive as luminescent labels.[14,15] However, some nanocrystals also exhibit photoluminescence (PL) that is sensitive to the local environment. For example, it was observed that PL of CdSe nanocrystals incorporated into polymer thin films responded reversibly to different gases.[16] Because in sensing applications nanomaterials bring previously unavailable capabilities[17–19] and unexpected results,[20–23] we explored the environmental sensitivity of semiconductor nanocrystals upon their incorporation in different rationally selected polymeric matrices.[24–26]

From our high-throughput screening of different polymeric matrices for CdSe nanocrystals for gas sensing applications, we have found that when CdSe nanocrystals of different size (2.8 and 5.6 nm diameter) passivated with tri-n-octylphosphine oxide (TOPO) using known methods[27,28] were incorporated into a certain polymer film and photoactivated, each size of CdSe nanocrystals unexpectedly demonstrated its own PL response pattern under a 407-nm laser excitation upon exposure to polar and nonpolar vapors. When this composite response was processed using multivariate analysis, a single film with different size CdSe nanocrystals served as a selective sensor. The concept for selective chemical sensing based on the multisize nanocrystals incorporated into a polymer film is illustrated in Fig. 5.1. In this chapter, we provide the details of multivariate analysis of the spectral PL response from both sizes of CdSe nanocrystals and demonstrate that a single film with different size CdSe nanocrystals serves as a selective sensor. To efficiently test the effects of different polymer matrices on the sensor performance of CdSe nanocrystals, we applied spectroscopic high-throughput analysis tools that we have developed for screening of polymerization catalysts,[29] automotive coatings,[30,31] performance additives,[32,33] and colorimetric sensor materials.[34] Our new sensor materials based on polymer-embedded semiconductor nanocrystal reagents of different size promise to overcome photobleaching and short shelf-life limitations of traditional organic reagent-based sensor materials.

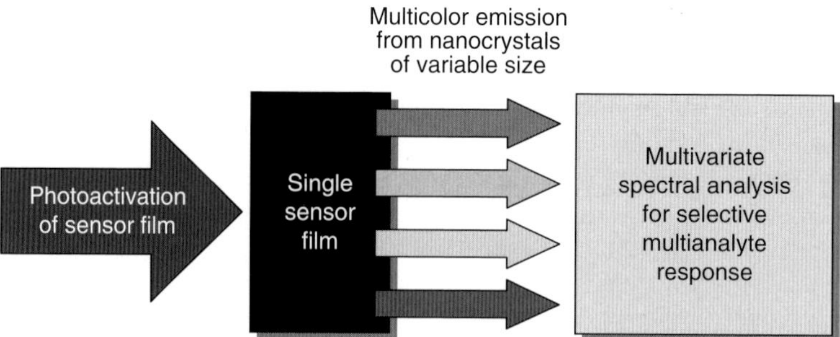

Fig. 5.1 Concept for selective gas sensing based on multisize CdSe nanocrystal/polymer composite films and dynamic pattern recognition. Reprinted with permission from Potyrailo and Leach[25]; copyright 2006 American Institute of Physics

2 Materials Characterization

CdSe nanocrystal solutions in toluene were obtained from Evident Technologies (Troy, NY). Concentrations of 2.8 and 5.6-nm CdSe nanocrystals were 2.5 mg/mL. Sensor films were spin-cast from 5% vol. solutions of nanocrystals in polymer/toluene. Polymers tested as matrices for the CdSe nanocrystals are listed in Table 5.1. Film thickness measured using an interference microscope was found to be under 1 μm.

For high-throughput screening of photoluminescence response, multiple films with different polymers were arranged as an one-dimensional array. Automatic measurements of PL spectra of the sensor films array were performed using a modular automatic scanning system assembled in house (Fig. 5.2). The system included a 407-nm portable GaN laser (2 mW output power), an in-line optical filter holder, a portable spectrometer, and an X–Y translation stage. Laser light was focused into one of the arms of a bifurcated fiber-optic reflection probe. The common arm of the probe illuminated an element of the library at a small (~10°) angle relative to the normal to the surface. The same portable laser was used for photoactivation and PL excitation of the sensor films. Illumination conditions were selected to have minimal contributions from light directly reflected from the sensor coating back into the probe. The second arm of the probe was coupled to the spectrometer. Data acquisition was achieved with a computer program that provided an adequate control of the data acquisition parameters and a real-time communication with the translation stage.

The fabricated sensor films were exposed at room temperature to methanol and toluene vapors at the vapor pressure of 46 and 11 Torr, respectively. This corresponded to concentrations of 61,000 ppm and 14,000 ppm of methanol and toluene, respectively.

Table 5.1 Polymer matrices for incorporation of different size CdSe nanocrystals

Polymer no.	Polymer type	Rationale for selection as sensor matrix
1	Poly(trimethyl-silyl) propyne	Largest known solubility of oxygen, candidate for efficient oxidation of CdSe nanocrystals
2	Poly(methyl methacrylate)	Matrix for solvatochromic dyes
3	Silicone block polyimide[35]	Polymer for sorbing of organic vapors
4	Polycapro-lactone	Matrix for solvatochromic dyes
5	Polycarbonate	Polymer with high Tg for sorbing of organic vapors
6	poly(iso-butylene)	Polymer with low Tg for sorbing of organic vapors
7	Poly(dimethyl-aminoethyl) methacrylate	Polymer for surface passivation of semiconductor nanocrystals
8	Polyvinyl-pyrrolidone	Polymer for sorbing of polar vapors
9	Styrene-Butadiene ABA block copolymer	Polymer for sorbing of nonpolar vapors

Reprinted with permission from Leach and Potyrailo[26]; copyright 2006 Materials Research Society

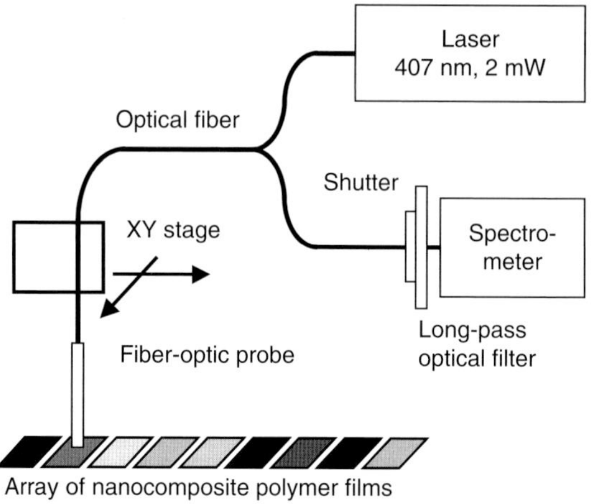

Fig. 5.2 Modular automatic scanning system for high-throughput evaluation of PL spectra of the sensor films array. Reprinted with permission from Leach and Potyrailo[26]; copyright 2006 Materials Research Society

Exposures were done automatically with multiple replicates. Processing of collected spectra was performed using KaleidaGraph (Synergy Software, Reading, PA) and PLS_Toolbox Software operated with Matlab (The Mathworks Inc., Natick, MA). Principal components analysis (PCA) was done upon autoscaling collected spectra in the range from 450 to 700 nm without baseline correction.

3 Spectral Properties of Photoactivated Sensing Films

As expected, solutions containing mixtures of nanocrystal sizes exhibited spectral profiles corresponding to the sum of the spectra of the individual sizes (Fig. 5.3). The initial PL peak positions of 2.8 and 5.6-nm CdSe nanocrystals in toluene solution were at 541 and 636 nm, respectively. We incorporated the CdSe nanocrystal mixtures into polymer matrices rationally selected based on polymer-sensing properties (Table 5.1). CdSe nanocrystals exhibited spectral shifts upon their immobilization in polymer films and photoactivation with the 407-nm diode laser. Photoactivation is an important aspect of the performance of the CdSe and other semiconductor crystals as chemical sensor materials.[16] Different spectral shifts of steady-state PL emission and different emission intensity were observed upon incorporation of the 2.8 and 5.6-nm CdSe nanocrystals in polymer films as illustrated in Fig. 5.4. Polymer 2 (PMMA) was selected for the detailed evaluation because it has been used previously as a matrix for incorporation of CdSe nanocrystals.[16,36]

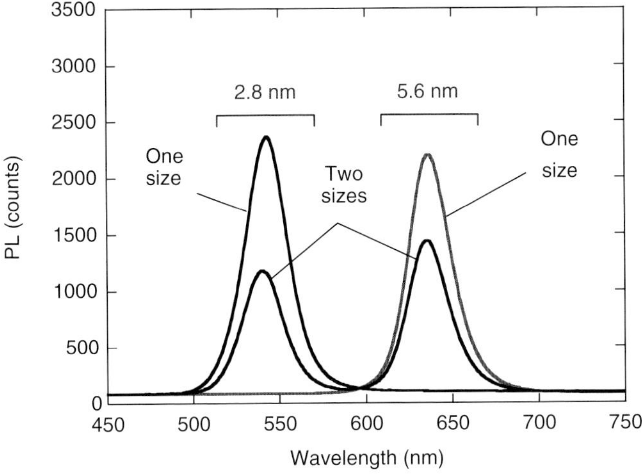

Fig. 5.3 PL spectra of individual size CdSe nanocrystals and their mixture in toluene. Reprinted with permission from Leach and Potyrailo[26]; copyright 2006 Materials Research Society

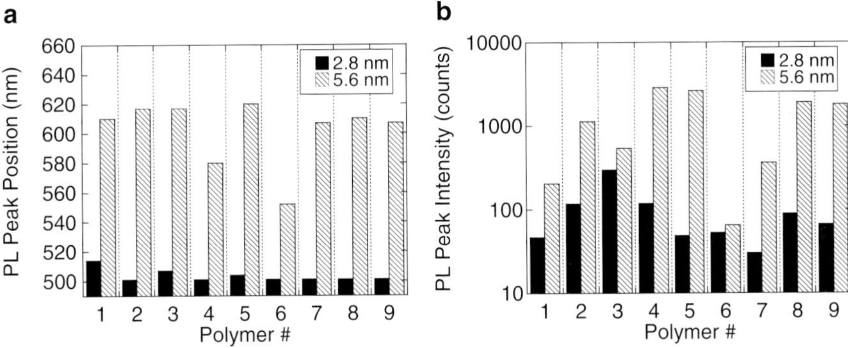

Fig. 5.4 Positions (**a**) and intensities (**b**) of PL peaks of CdSe nanocrystals of 2.8 and 5.6-nm in polymer films 1–9 upon photoactivation. Initial PL peak positions of 2.8 and 5.6-nm CdSe nanocrystals in toluene solution are at 541 and 636 nm, respectively

Figure 5.5 illustrates an example of PL spectra, where 2.8 and 5.6 nm diameter nanocrystals were incorporated into a PMMA film. The 2.8 nm nanocrystals had an initial emission maximum in toluene at 541 nm, which was slightly red-shifted to 543 nm upon an addition of a polymer solution. A strong blue shift to 519 nm was observed in a dry film containing 2.8 nm nanocrystals and PMMA, with an additional blue shift to 504 nm upon an extended photoactivation. As shown in Fig. 5.6, the peak shift of 5.6-nm nanocrystals in PMMA was less dramatic upon photoactivation, but mirrored the behavior of the 2.8-nm nanocrystals. The peak positions for

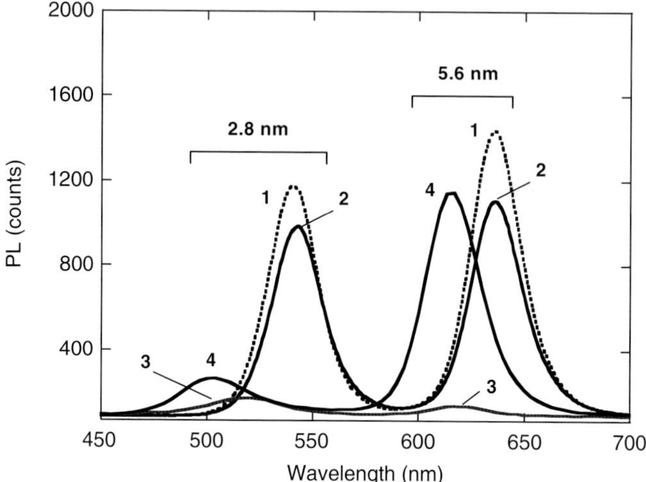

Fig. 5.5 PL spectra of a mixture of 2.8 and 5.6-nm CdSe nanocrystals in different environments: (1) Toluene; (2) Toluene + PMMA; (3) PMMA film before photoactivation; (4) PMMA film after photoactivation. Reprinted with permission from Potyrailo and Leach[25]; copyright 2006 American Institute of Physics

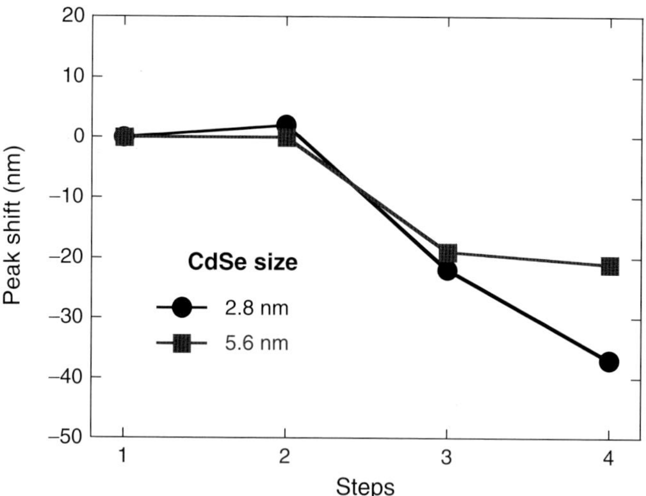

Fig. 5.6 Spectral shifts of 2.8 and 5.6-nm CdSe nanocrystals with different steps of film fabrication and use: (1) Solution in toluene; (2) Solution in toluene + PMMA; (3) PMMA film before activation; and (4) PMMA film after activation. Negative spectral shift is to lower wavelengths of PL emission. Reprinted with permission from Leach and Potyrailo[26]; copyright 2006 Materials Research Society

the 5.6 nm nanocrystals were 636, 636, 617, and 615 nm in toluene, toluene and PMMA, dry initial film, and a photoactivated film, respectively. Photoactivation has been shown to blue shift nanocrystal emission. The observed shifts can be attributed to the difference in nanoparticle size and the photo-oxidation of the nanocrystal surface [37,38]. A ~20-nm greater blue shift of the final peak position of the 2.8-nm nanocrystals over the 5.6-nm nanocrystals demonstrates larger magnitude oxidation of the nanocrystals of the smaller size.

4 Sensing Response Patterns

Results of exposure of all prepared sensor films to methanol and toluene vapors are presented in Fig. 5.7. These data indicate a striking diversity in response of the two-size CdSe nanocrystals to these polar and nonpolar vapors. Results presented in this chapter detail the performance of the sensor film based on PMMA, represented as polymer #2.

When a single wavelength was selected at the peaks of PL spectra of each size of the nanocrystals in PMMA sensor film, two dynamic responses to these vapors were obtained. Figure 5.8 shows baseline-corrected response patterns of gas-dependent PL of the two-size CdSe nanocrystals/PMMA sensor film. Emission of the 2.8-nm nanocrystals was measured at 511 nm and emission of the 5.6-nm nanocrystals was measured at 617 nm. The noise in PL measurements was ~ 0.6 and ~1.5 PL counts per integration period (500 ms) at 511 and 617 nm, respectively. We believe that the difference in the response patterns of the nanocrystals is related to the combined effects of the dielectric medium surrounding the nanocrystals, their size, and surface oxidation state.

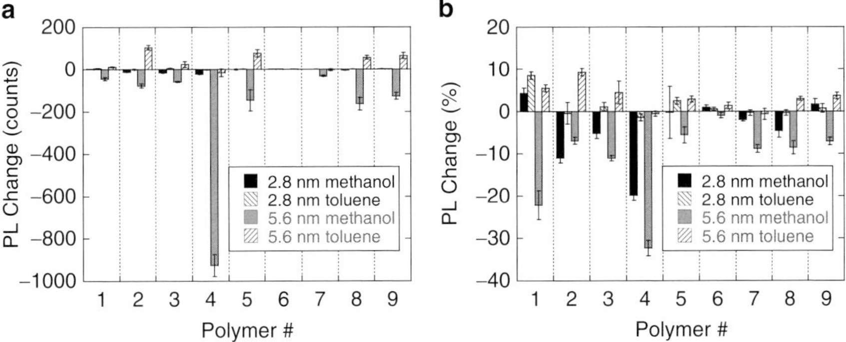

Fig. 5.7 Diversity in steady-state response of the two-size CdSe nanocrystals embedded in polymers 1–9 to polar (methanol) and nonpolar (toluene) vapors: absolute (**a**) and relative (**b**) PL response values. Error bars are 1 SD from replicate ($n = 3$) measurements

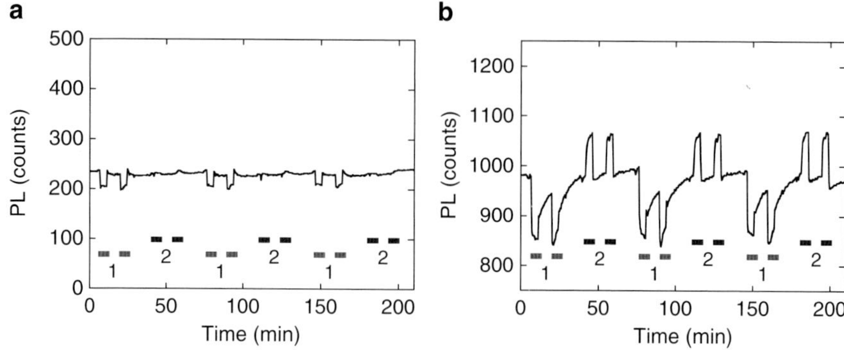

Fig. 5.8 Gas-dependent PL of the two-size CdSe nanocrystals sensor film: (**a**) Emission of 2.8-nm nanocrystals at 511 nm and (**b**) Emission of 5.6-nm nanocrystals at 617 nm. Numbers 1 and 2 are replicate exposures of sensor film to methanol and toluene, respectively. Reprinted with permission from Leach and Potyrailo[26]; copyright 2006 Materials Research Society

The sensitivity of these sensors was defined as a signal change upon exposure to the known concentrations of vapors. Sensitivity of the 2.8-nm CdSe nanocrystals was ~0.8 PL counts/Torr of methanol with almost no detectable sensitivity to toluene. The sensitivity of the 5.6-nm CdSe nanocrystals was 2.9 PL counts/Torr of methanol and 8.8 PL counts/Torr of toluene. Although this environmental sensitivity was compatible with earlier reported sensors based on polished or etched bulk CdSe semiconductor crystals[39,40] and polymer-nanocrystals composites,[16] the sensor reported here had a more selective response to polar and nonpolar vapors due to the multiwavelength PL from different-size nanocrystals incorporated into the polymer film. The response and recovery kinetics of PL from the 2.8-nm nanocrystals in PMMA upon exposure to methanol were very fast (<0.5 min). However, 5.6-nm nanocrystals in the same sensor film exhibited a much longer response and recovery times upon interactions with methanol, 4 and 20 min, respectively. The 5.6-nm nanocrystals had 4-min response and 0.5-min recovery times upon interactions with toluene.

Computational studies of the effects of dielectric media on CdSe nanocrystal electronic properties show that the absolute value of the dipole moment of the nanocrystal increases with elevated dielectric constant of the surrounding environment.[41] The magnitude of the change in dipole moment increases with particle size. The elevated diameter and surface area of larger particles means that even small changes in the surface charge of the nanocrystals will result in significant changes in the particles dipole moment. Both sizes of nanocrystals in our sensor films exhibited a decrease in emission upon exposure to methanol ($\varepsilon = 33$), which causes a large increase in the local dielectric constant relative to the PMMA medium ($\varepsilon = 2.7$–3.0). Toluene ($\varepsilon = 2.2$) has a slightly lower dielectric constant than PMMA resulted in an increase in the emission of the larger nanocrystals alone. The different size nanocrystals also can have different coverage with the capping TOPO ligand[42] and when immobilized into a polymer matrix[28] can contribute to the variable response pattern.

5   Multivariate Spectral Analysis

The capability for discriminating vapors with a single sensor film with incorporated CdSe nanocrystals of different size was evaluated from individual PL responses of nanocrystals of different size using a principal components analysis (PCA) technique.[43] PCA is a multivariate data analysis tool that projects the data set onto a subspace of lower dimensionality with removed collinearity. PCA achieves this objective by explaining the variance of the data matrix X in terms of the weighted sums of the original variables with no significant loss of information. These weighted sums of the original variables are called principal components (PCs). Upon applying the PCA, the data matrix X is expressed as a linear combination of orthogonal vectors along the directions of the principal components:[44]

$$X = t_1 \, p^\mathrm{T}_1 + t_2 \, p^\mathrm{T}_2 + \ldots + t_\mathrm{A} \, p^\mathrm{T}_\mathrm{K} + E \tag{1}$$

where t_i and p_i are, respectively, the score and loading vectors, K is the number of principal components, E is a residual matrix that represents random error, and T is the transpose of the matrix.

For spectral analysis using PCA, we had 482 samples (spectra) × 75 variables (wavelengths over the range from 450 to 700 nm) collected during the repetitive ($n = 3$) exposures of the sensor film to methanol and toluene vapors without baseline drift correction over 210 min (Fig. 5.9). These 482 PL spectra and 75 dynamic signatures at different wavelengths of two-size CdSe nanocrystals in PMMA film to replicate exposures to methanol and toluene vapors are plotted in Fig. 5.9.

The variance captured by the built PCA model for each of PCs is summarized in Table 5.2. We evaluated dynamic behavior of each of the PCs upon exposure of the

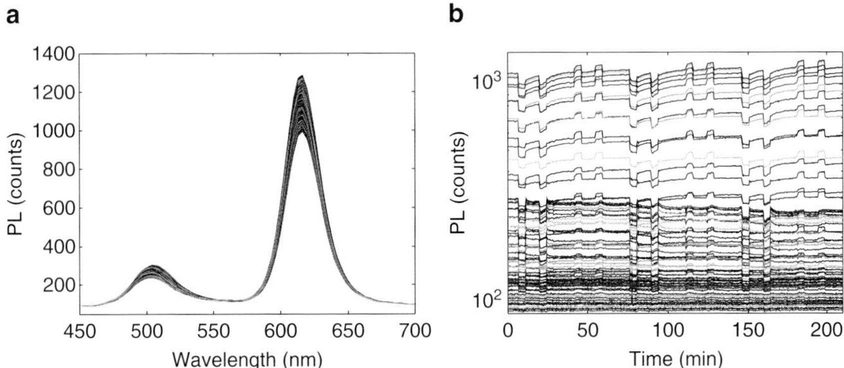

Fig. 5.9   Response of two-size CdSe nanocrystals in PMMA film to methanol and toluene vapors: (a) PL spectra (total number = 482). (b) Dynamic signatures at different wavelengths (total number = 75). Reprinted with permission from Leach and Potyrailo[26]; copyright 2006 Materials Research Society

Table 5.2 Percent variance captured by PCA model

Principal component number	% variance captured by this principal component	% variance captured total
1	53.96	53.96
2	27.86	81.82
3	5.09	86.91
4	3.41	90.32
5	1.45	91.77

Reprinted with permission fromLeach and Potyrailo[26]; copyright 2006 Materials Research Society

two-size CdSe nanocrystals in PMMA sensor film to methanol and toluene (Fig. 5.10). The largest variance (captured by the first principal component of the PCA model) exhibited the strongest response to both vapors. The second PC captured the drift contributions in the response along with the vapor responses. The third PC also captured vapor response, however, without the drift contributions. All principal components from #1 to #5 demonstrated a progressive increase in the noise contributions, in agreement with the conventional PCA interpretations.

The relationship between the collected data was described by plotting scores of relevant principal components of the PCA model vs. each other as shown in Fig. 5.11. These plots demonstrated that responses of the sensor film to different vapors were well-separated in the PCA space. By plotting PC#1 vs. PC#2, some drift effects were pronounced (Fig. 5.11a). However, the drift effect was removed by plotting PC#1 vs. PC#3 (Fig. 5.11b). The remaining response scatter in regions 1 and 2 of Fig. 5.11b was due to nonequilibrated sensor responses during the kinetic experiments.

Analysis of diversity in response to polar and nonpolar vapors of all screened polymers was performed using PCA analysis followed by cluster analysis. The scores plots of the first three PCs (Fig. 5.12) illustrate the diversity of performance of all sensing polymers. The larger the distance between polymer data points, the larger the difference in the response pattern between the respective CdSe/polymer nanocomposites. To quantify this diversity, cluster analysis was further performed where the distances based on principal component scores were adjusted to unit variance.[45] This distance measure, known as Mahalanobis distance, accounts for the different amount of variation in different directions. An example of such difference is shown in Fig. 5.12a, b, where the distance between polymer 4 and other polymers is much larger on the plot of PC#1 vs. PC#2 when compared with the plot of PC#2 vs. PC#3.

Results of cluster analysis are demonstrated in the dendrogram in Fig. 5.13. The dendrogram was constructed using Mahalanobis distance on three PCs. From this dendrogram, it is clear that polymers 6 and 7 are the most similar in their vapor response with studied CdSe nanocrystals as demonstrated by a very small distance to K-nearest neighbor between them. Polymer 4 is the most different from the rest of polymers as indicated by the largest diversity distance to K-nearest neighbor. Such data mining tools provide a means to quantitatively evaluate polymer matrices. When coupled with quantitative structure–property relationships simulation tools

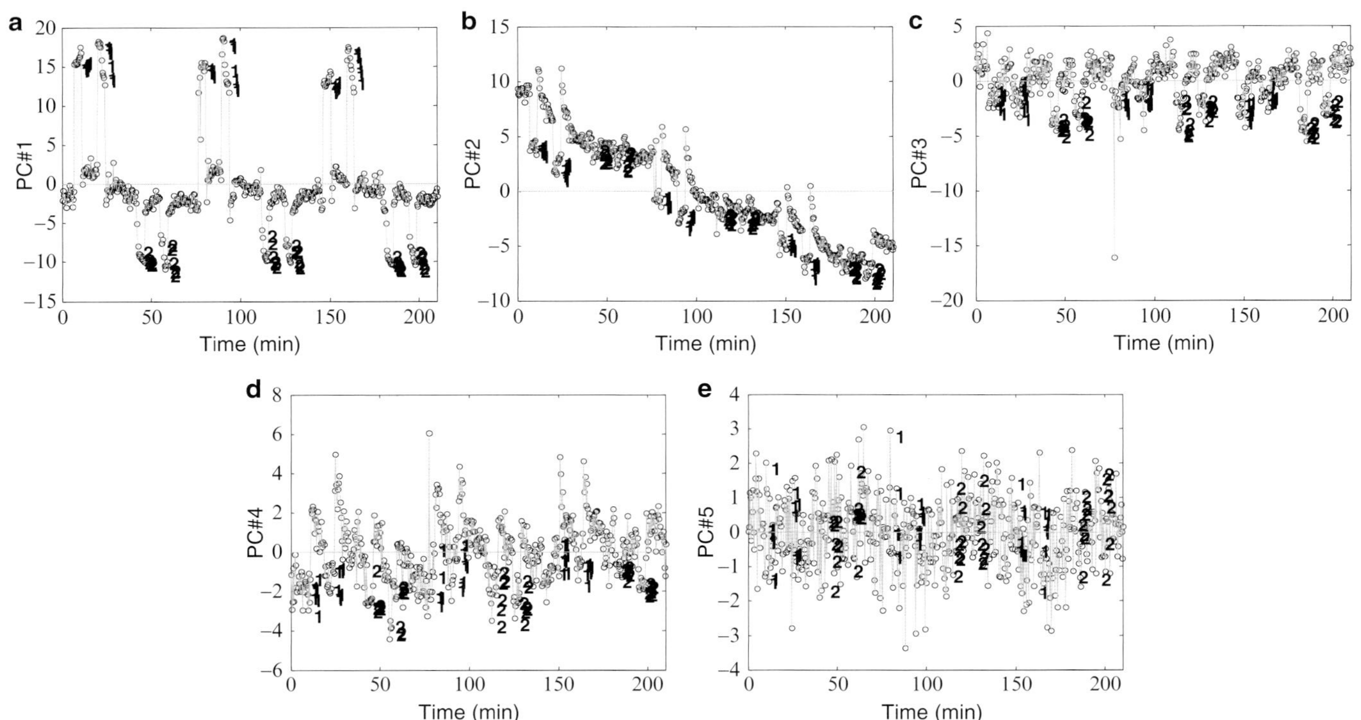

Fig. 5.10 Dynamic behavior of the first five principal components (**a**)–(**e**) in response to replicate exposures to (1) methanol and (2) toluene. Unlabeled data points are when the film was exposed to a blank (dry air). Reprinted with permission from Leach and Potyrailo[26]; copyright 2006 Materials Research Society

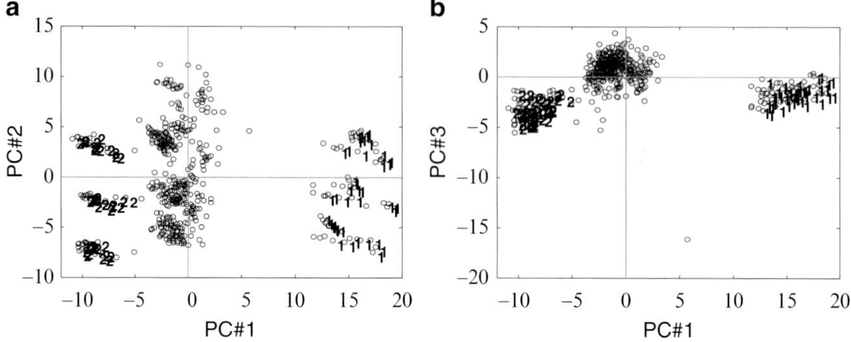

Fig. 5.11 Multivariate analysis (principal components scores plots of the first three principal components) of the response of the two-size CdSe nanocrystals sensor film: (**a**) PC#1 vs. PC#2 and (B) PC#1 vs. PC#3. Regions numbered 1 and 2 are data points of dynamic response from replicate ($n = 3$) film exposures to methanol and toluene, respectively. Unlabeled data points result from times when the film was exposed to a blank (dry air). Reprinted with permission from Leach and Potyrailo[26]; copyright 2006 Materials Research Society

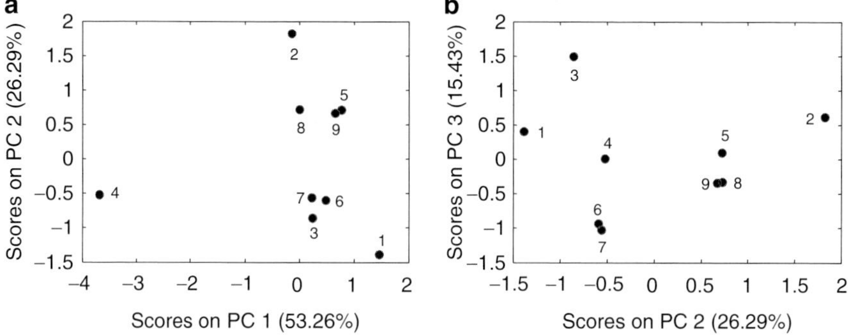

Fig. 5.12 PCA scores plots of the first three principal components of the gas response of nine polymer types with incorporated two-size CdSe nanocrystals: (**a**) PC#1 vs. PC#2 and (**b**) PC#2 vs. PC#3. Names of polymers 1–9 are provided in Table 5.1

that will incorporate molecular descriptors, new knowledge generated from high-throughput experiments may provide additional insights for the rational design of gas sensors based on incorporated semiconductor nanocrystals.

6 Response Stability

We further tested the stability of PL emission of new sensor films upon an extended irradiation with the laser light. Such response stability is critical in continuous monitoring applications. The parameters of interest in these evaluations were (1)

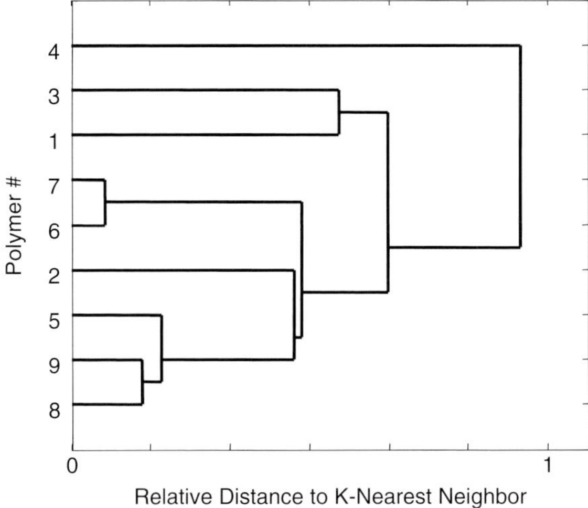

Fig. 5.13 Quantitative analysis of diversity in response to polar and nonpolar vapors of all screened polymers. Dendrogram of cluster analysis using Mahalanobis distance on three PCs of the PCA model. Names of polymers 1–9 are provided in Table 5.1

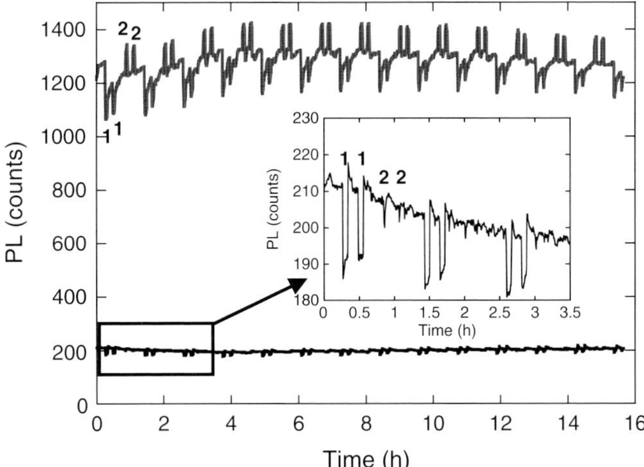

Fig. 5.14 Long-term stability of sensor response of 2.8 and 5.6-nm CdSe nanocrystals in poly(methyl methacrylate) polymer film upon repetitive exposures to methanol (1) and toluene (2)

the overall stability of the PL intensity and (2) the stability of the response pattern to methanol and toluene vapors.

Figure 5.14 shows response patterns from 2.8 and 5.6-nm nanocrystals in PMMA film over 16 h of continuous exposure of the film to laser radiation. During the laser exposure, the sensor film was periodically exposed to methanol and toluene vapors. These data indicated that the response of new selective sensor films was very stable upon extended exposure to laser excitation without significant degradation of both emission intensity and response pattern.

7   Conclusions

High-throughput experimentation has been applied for the evaluation of the responses of different-size CdSe nanocrystals incorporated in a variety of rationally selected polymer matrices. Automation of these experiments not only speeds up the screening process of candidate sensing materials, but also importantly reduces or eliminates variation sources that otherwise affect response of the sensing materials if measurements are done manually.

We have found that different polymeric matrices provide diverse response patterns of polymer-incorporated CdSe nanocrystals to vapors of different polarity. Such work promises to complement existing solvatochromic organic dye sensors with more photostable and reliable sensor materials. The stability of the sensor response pattern from 2.8 and 5.6-nm nanocrystals in PMMA film over more than 10 h of continuous exposure of the film to laser radiation is very attractive for diverse applications where continuous monitoring is needed.

Acknowledgment This work has been supported by GE Corporate long-term research funds. Authors are grateful to Fasila Seker for helpful discussions.

References

1. Potyrailo, R. A., Polymeric sensor materials: Toward an alliance of combinatorial and rational design tools?, *Angew. Chem. Int. Ed.* **2006**, *45*, 702–723
2. Wolfbeis, O. S., Fiber-optic chemical sensors and biosensors, *Anal. Chem.* **2004**, *76*, 3269–3284
3. Janata, J., Electrochemical sensors and their impedances: A tutorial, *Crit. Rev. Anal. Chem.* **2002**, *32*, 109–120
4. Bakker, E., Electrochemical sensors, *Anal. Chem.* **2004**, *76*, 3285–3298
5. Kooser, A.; Gunter, R. L.; Delinger, W. D.; Porter, T. L.; Eastman, M. P., Gas sensing using embedded piezoresistive microcantilever sensors, *Sens. Actuators B* **2004**, *99*, 474–479
6. Kröger, S.; Danielsson, B., Calorimetric biosensors, In *Handbook of Biosensors and Electronic Noses. Medicine, Food, and the Environment*; Kress-Rogers, E., Ed.; CRC: Boca Raton, FL, **1997**; 279–298
7. *Fiber Optic Chemical Sensors and Biosensors*; Wolfbeis, O. S., Ed.; CRC: Boca Raton, FL, **1991**
8. Potyrailo, R. A.; Hobbs, S. E.; Hieftje, G. M., Optical waveguide sensors in analytical chemistry: Today's instrumentation, applications and future development trends, *Fresenius' J. Anal. Chem.* 1998, *362*, 349–373
9. Hierlemann, A.; Weimar, U.; Kraus, G.; Schweizer-Berberich, M.; Göpel, W., Polymer-based sensor arrays and multicomponent analysis for the detection of hazardous organic vapours in the environment, *Sens. Actuators B* **1995**, *26*, 126–134

10. Potyrailo, R. A.; Sivavec, T. M., Boosting sensitivity of organic vapor detection with silicone block polyimide polymers, *Anal. Chem.* **2004**, *76*, 7023–7027

11. Wohltjen, H., A journey: From sensor ideas to sensor products, In *Plenary Talk at the 11th International Meeting on Chemical Sensors*, University of Brescia, Italy, July 16–19, 2006*** Elsevier Science: **2006**

12. Shortreed, M.; Monson, E.; Kopelman, R., Lifetime enhancement of ultrasmall fluorescent liquid polymeric film based optodes by diffusion-induced self-recovery after photobleaching, *Anal. Chem.* **1996**, *68*, 4015–4019

13. Bencic-Nagale, S.; Walt, D. R., Extending the longevity of fluorescence-based sensor arrays using adaptive exposure, *Anal. Chem.* **2005**, *77*, 6155–6162

14. Bruchez, M., Jr.; Moronne, M.; Gin, P.; Weiss, S.; Alivisatos, A. P., Semiconductor nanocrystals as fluorescent biological labels, *Science* **1998**, *281*, 2013–2016

15. Chan, W. C. W.; Nie, S., Quantum dot bioconjugates for ultrasensitive nonisotopic detection, *Science* **1998**, *281*, 2016–2018

16. Nazzal, A. Y.; Qu, L.; Peng, X.; Xiao, M., Photoactivated CdSe nanocrystals as nanosensors for gases, *Nano Lett.* **2003**, *3*, 819–822

17. Holtz, J. H.; Asher, S. A., Polymerized colloidal crystal hydrogel films as intelligent chemical sensing materials, *Nature* **1997**, *389*, 829–832

18. Convertino, A.; Capobianchi, A.; Valentini, A.; Cirillo, E. N. M., A new approach to organic solvent detection: High-reflectivity bragg reflectors based on a gold nanoparticle/teflon-like composite material, *Adv. Mater.* **2003**, *15*, 1103–1105

19. Jiang, P.; Smith, D. W. J.; Ballato, J. M.; Foujger, S. H., Multicolor pattern generation in photonic bandgap composites, *Adv. Mater.* **2005**, *17*, 179–184

20. Handley, J., Stretching the wire frontier, *Anal. Chem.* **2002**, *74*, 196A–199A

21. Zhou, Y.; Freitag, M.; Hone, J.; Staii, C.; Johnson, A. T.; Pinto, N. J.; MacDiarmid, A. G., Fabrication and electrical characterization of polyaniline-based nanofibers with diameter below 30 nm, *Appl. Phys. Lett.* **2003**, *83*, 3800–3802

22. Chakrabarti, R.; Klibanov, A. M., Nanocrystals modified with peptide nucleic acids (PNAS) for selective self-assembly and DNA detection, *J. Am. Chem. Soc.* **2003**, *125*, 12531–12540

23. Zhang, M.; Gorski, W., Electrochemical sensing based on redox mediation at carbon nanotubes, *Anal. Chem.* **2005**, *77*, 3960–3965

24. Potyrailo, R. A.; Leach, A. M., Gas sensor materials based on semiconductor nanocrystal/polymer composite films, In *Proceedings of Transducers '05, the 13th International Conference on Solid-State Sensors, Actuators and Microsystems*, Seoul, Korea, June 5–9, **2005**; 1292–1295

25. Potyrailo, R. A.; Leach, A. M., Selective gas nanosensors with multisize CdSe nanocrystal/polymer composite films and dynamic pattern recognition, *Appl. Phys. Lett.* **2006**, *88*, 134110

26. Leach, A. M.; Potyrailo, R. A., Gas sensor materials based on semiconductor nanocrystal/polymer composite films, In *Combinatorial Methods and Informatics in Materials Science. MRS Symposium Proceedings*, Wang, Q.; Potyrailo, R. A.; Fasolka, M.; Chikyow, T.; Schubert, U. S.; Korkin, A. Eds.; Materials Research Society: Warrendale, PA, **2006**; Vol. 894; 237–243

27. Murray, C. B.; Norris, D. J.; Bawendi, M. G., Synthesis and characterization of nearly monodisperse CdE (E = sulfur, selenium, tellurium) semiconductor nanocrystallites, *J. Am. Chem. Soc.* **1993**, *115*, 8706–8715

28. Kovalevskij, V.; Gulbinas, V.; Piskarskas, A.; Hines, M. A.; Scholes, G. D., Surface passivation in CdSe nanocrystal-polymer films revealed by ultrafast excitation relaxation dynamics, *Phys. Stat. Sol. B* **2004**, *241*, 1986–1993

29. Potyrailo, R. A.; Lemmon, J. P.; Leib, T. K., High-throughput screening of selectivity of melt polymerization catalysts using fluorescence spectroscopy and two-wavelength fluorescence imaging, *Anal. Chem.* **2003**, *75*, 4676–4681

30. Potyrailo, R. A.; Chisholm, B. J.; Olson, D. R.; Brennan, M. J.; Molaison, C. A., Development of combinatorial chemistry methods for coatings: High-throughput screening of abrasion resistance of coatings libraries, *Anal. Chem.* **2002**, *74*, 5105–5111

31. Potyrailo, R. A.; Ezbiansky, K.; Chisholm, B. J.; Morris, W. G.; Cawse, J. N.; Hassib, L.; Medford, G.; Reitz, H., Development of combinatorial chemistry methods for coatings: High-throughput weathering evaluation and scale-up of combinatorial leads, *J. Comb. Chem.* **2005**, *7*, 190–196

32. Potyrailo, R. A.; Pickett, J. E., High-throughput multilevel performance screening of advanced materials, *Angew. Chem. Int. Ed.* **2002**, *41*, 4230–4233

33. Potyrailo, R. A.; Wroczynski, R. J.; Pickett, J. E.; Rubinsztajn, M., High-throughput fabrication, performance testing, and characterization of one-dimensional libraries of polymeric compositions, *Macromol. Rapid Comm.* **2003**, *24*, 123–130

34. Potyrailo, R. A.; Hassib, L., Analytical instrumentation infrastructure for combinatorial and high-throughput development of formulated discrete and gradient polymeric sensor materials arrays, *Rev. Sci. Instrum.* **2005**, *76*, 062225

35. Sivavec, T. M.; Potyrailo, R. A. *Polymer coatings for chemical sensors*; US Patent 6,357,278 B1: **2002**

36. Nazzal, A. Y.; Wang, X.; Qu, L.; Yu, W.; Wang, Y.; Peng, X.; Xiao, M., Environmental effects on photoluminescence of highly luminescent CdSe and CdSe/ZnS core/shell nanocrystals in polymer thin films, *J. Phys. Chem. B* **2004**, *108*, 5507–5515

37. Wang, X.; Qu, L.; Zhang, J.; Peng, X.; Xiao, M., Surface-related emission in highly luminescent CdSe quantum dots, *Nano Lett.* **2003**, *3*, 1103–1106

38. Asami, H.; Abe, Y.; Ohtsu, T.; Kamiya, I.; Hara, M., Surface state analysis of photobrightening in CdSe nanocrystal thin films, *J. Phys. Chem. B.* **2003**, *107*, 12566–12568

39. Lisensky, G. C.; Meyer, G. J.; Ellis, A. B., Selective detector for gas chromatography based on adduct-modulated semiconductor photoluminescence, *Anal. Chem.* **1988**, *60*, 2531–2534

40. Seker, F.; Meeker, K.; Kuech, T. F.; Ellis, A. B., Surface chemistry of prototypical bulk ii–vi and iii–v semiconductors and implications for chemical sensing, *Chem. Rev.* **2000**, *100*, 2505–2536

41. Rabani, E.; Hetényi, B.; Berne, B. J.; Brus, L. E., Electronic properties of CdSe nanocrystals in the absence and presence of a dielectric medium, *J. Chem. Phys.* **1999**, *110*, 5355–5369

42. Leatherdale, C. A.; Bawendi, M. G., Observation of solvatochromism in CdSe colloidal quantum dots, *Phys. Rev. B* **2001**, *63*, 165315 165311–165316

43. Potyrailo, R. A.; Wroczynski, R. J.; Lemmon, J. P.; Flanagan, W. P.; Siclovan, O. P., Multivariate tools for real-time monitoring and optimization of combinatorial materials and process conditions, In *Analysis and Purification Methods in Combinatorial Chemistry*, B. Yan, Ed.; Wiley-Interscience: Hoboken, NJ, **2004**; 87–123

44. Wise, B. M.; Gallagher, N. B. *PLS_toolbox Version 2.1 for Use with Matlab*; Eigenvector Research, Inc.: Manson, WA, **2000**

45. Wise, B. M.; Gallagher, N. B.; Bro, R.; Shaver, J. M.; Windig, W.; Koch, R. S. *PLS_Toolbox Version 4.0 for Use with Matlab*; Eigenvector Research, Inc.: Manson, WA, **2006**

Section 3
Molecular Imprinting

Chapter 6
Computational Design of Molecularly Imprinted Polymers

Sreenath Subrahmanyam and Sergey A. Piletsky

Abstract Artificial receptors have been in use for several decades as sensor elements, in affinity separation, and as models for investigation of molecular recognition. Although there have been numerous publications on the use of molecular modeling in characterization of their affinity and selectivity, very few attempts have been made on the application of molecular modeling in computational design of synthetic receptors. This chapter discusses recent successes in the use of computational design for the development of one particular branch of synthetic receptors – molecularly imprinted polymers.

1 Introduction

Natural receptors are generally large protein molecules that form three-dimensional structures by highly specific intramolecular interactions. Although the recognition sites offer a precise configuration and exhibit very efficient recognition processes, this specific recognition is achieved at the expense of having complex and fragile structures with high molecular weight. An alternative to this is the choice of artificial receptors that incorporate a combination of medium-sized organic building blocks to which functional groups for molecular recognition can be attached. Rational design of artificial receptors, which possess very high affinity and selectivity, is currently one of the most researched topics in molecular recognition.

Several artificial receptors such as crown ethers, cyclodextrins, cyclophanes, and calixarenes find applications in molecular recognitions processes. Further, there have been several research publications that focus on computational design and analysis of recognition properties of artificial receptors such as cyclodextrins,[1–5] dendrimers,[6–10] crown ethers,[11,12] and calixarenes.[13–15] Typically computational

S. Subrahmanyam (✉) and S.A. Piletsky
Cranfield Biotechnology Center, Cranfield University at Silsoe, Bedfordshire, MK45 4DT, UK
sri@cranfield.ac.uk

R.A. Potyrailo and V.M. Mirsky (eds.), *Combinatorial Methods for Chemical and Biological Sensors*, DOI: 10.1007/978-0-387-73713-3_6,
© Springer Science + Business Media, LLC 2009

approaches were previously used for the design of small molecules.[16–36] The computational design of supramolecular synthetic receptors (analogues of natural protein receptors) still remains very challenging. Arguably the only class of supramolecular receptors that has been designed systematically using computational approach is molecularly imprinted polymers (MIPs). The present review will discuss in general these materials as well as different molecular modeling approaches that have been used to design them.

Molecular imprinting can be defined as the process of template-induced formation of specific recognition sites (binding or catalytic) in a material where the template directs the positioning and orientation of the material's structural components by a self-assembling mechanism.[37] The material itself could be oligomeric (a typical example is the DNA replication process), polymeric (organic MIPs and inorganic imprinted silica gels), or 2-dimensional surface assembly (grafted monolayer) (Fig. 6.1).

MIPs have several advantages when compared with other synthetic receptors:[39]

(a) High affinity and selectivity, which are similar to those of natural receptors
(b) Very high stability, which is superior to that of natural biomolecules
(c) Simplicity of their preparation and the ease of adaptation to different practical applications

A wide range of chemical compounds have been imprinted successfully, ranging from small molecules,[40–42] to large proteins and cells.[43] MIPs have been developed for a variety of applications including chromatography,[44,45] solid-phase extraction (SPE),[46,47] enzyme catalysis,[48] sensor technology,[44,49,50] biomimetic sensors,[51–53] and immunoassays.[54–56] MIPs are robust, inexpensive and, in many cases, possess affinity and specificity that are suitable for industrial applications. The high specificity and

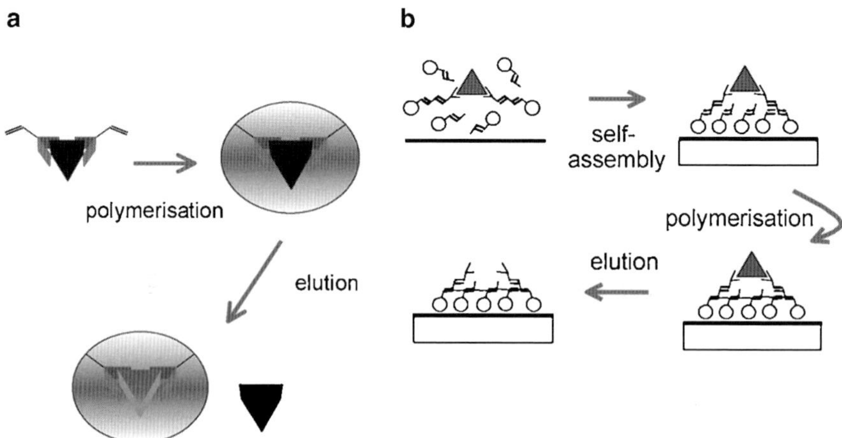

Fig. 6.1 (**a**) Three-dimensional and (**b**) two-dimensional imprinting polymerization, (courtesy of VTT)[38]

Table 6.1 Computational procedures for the rational design of MIPs

Molecular mechanics

Process	Application	Targets and references
LEAPFROG algorithm (Sybyl) was used to screen virtual library of functional monomers	Selection of best monomers leading to MIPs with high-binding capacity for the template	Abacavir[78]; biotin[79]; carbamate[80] cocaine, deoxyephedrine, methadone and morphine[81]; triazines[83,84]; tylosin[85]
Molecular mechanics (MM) calculations were done with AMBER7 Docking software to map the energetic interactions	Prediction of binding affinity and selectivity	Theophylline and its derivative[82]
MM and molecular dynamic (MD) calculations were performed for the template-monomer complex using HyperChem	Analysis of the complex formation between template and monomer and possible structure of imprinting sites	L- or D-tryptophan methyl ester[86]; N-α-t-boc-L-histidine[87]
Interactions between template and monomer in MIPs were analyzed using Amber MM method	Prediction of the ratios of template, functional monomer and solvent	Caffeine and theophylline[69]; ibuprofen[88]
3-D chemical structures of the labeled BLAs were modeled using molecular mechanics using (MOPAC, AM1 force field) using Chem3D Ultra 7.0 software	Analysis of recognition of the fluorescent analogues of template by the MIPs	Penicillin-G[73]
Energy minimization by molecular mechanics and quantum chemistry to estimate enthalpies of formation, bond orders, intermolecular distances and ionization potentials using PCModel for windows	Analysis of enthalpies of complex formation between functional monomer and template	Dibenzothiophene sulphone[89]
Calculations of interaction energies using PCMODEL 8.0, MMFF94 and force field	Selection of monomers for synthesis of MIPs based on the interaction energies	Paracetamol[90]
pKa calculations of template by Gaussian03W in vacuum	Study of correlation between molecular volumes and pKa of templates and the retention factors	Hydroxy polychlorinated biphenyls[91]

Molecular dynamics

Process	Application	Targets and references
Simulation of complex formation using simulated annealing (Sybyl)	Optimization of polymer composition	Creatinine[64]; microcystin LR[92]; ochratoxin A[93]
Molecular models of template and monomers were optimized by HyperChem 501; Simulated annealing process was applied to optimize the arrangement of the resulting structures	Selection of best monomers leading to MIPs with high-binding capacity for the template	N,O-dibenzylcarbamate[94]

(continued)

Table 6.1 (continued)

Molecular mechanics

Process	Application	Targets and references
Study of intermolecular interactions in molecular imprinting in complex monomeric systems using *Cerius* version 410 and Materials Studio	Prediction of monomers specific for the template	Theophylline and its derivatives[95,96]; chemical warfare agents[66]
Intermolecular Monte Carlo conformational analysis	Analysis of complex formation between template and monomer	Biotin[58]
Quantum mechanics		
HyperChemPro 60 software was used to calculate low energy confirmations and electronic distributions	Analysis of the effects of the electric charge distribution and of the size of the molecules on the retention mechanism in SPE	Terbutylazine and ametryn[71,97]
Energy minimization done by MOPAC. The structures were displayed using WebLab ViewerLite	Analysis of the possible interactions of the fluorescent monomer with a carboxamidrazone substrate	N1-benzylidene pyridine-2-carboxamidrazones[74]
A virtual library of the intermediates was constructed using CS Chem3D Pro software. The energy minimization was done using MOPAC	Optimization of monomer formulation for MIP	Transesterification[72]
Electronic energies were calculated through Density Functional Theory (DFT) using Gaussian 98	Choice of the best functional monomer and solvent	Homovanillic acid[98]
Calculating binding energy of a template molecule and a monomer in MIPs using Gaussian 98	Screening of functional monomers for preparation of MIPs	Theophylline and its derivatives[68]
Gaussian 03 and B3LYP used in calculation of interaction energies between the monomer and the template	Study of the influence of porogens on the affinity and selectivity of MIP	Nicotinamide[67]
Gaussian 03 and B3LYP used in calculation of interaction energies between the monomer and the template	The results of the interaction energies between the monomer and the template were correlated with the retention and imprinting factors	Nicotinamide and iso-nicotinamide[99]
MO calculations were carried out with CAChe and MOPAC; Molecular geometries were optimized by the AM1 method	Analysis of complex formation of template monomer mixture	(S)-Nilvadipine[100]
Chemometrics and neural network methods		
Chemometric Design Expert software was used to generate and manipulate the factorial data	Optimization of the template:monomer:cross-linker ratio	Sulfonamide[75]
Neural network was carried out using the back propagation algorithm (WEKA)	Prediction of imprinting factor of MIPs and study of template monomer complexes	Atropine and Boc-L-Trp d-Brompheniramine[76,77]

stability of MIPs render them as promising alternatives to enzymes, antibodies, and natural receptors for use in sensor technology.[37,48]

There have been several attempts on development of generic procedure for MIP preparation as mentioned below; however, the one that has been in prime focus in the recent years is computational design:

- Rational approaches that involve combinatorial methods, where an array of MIPs were prepared that could be analyzed in situ by binding assays[57–61]
- Use of a virtual library of functional monomers to assign and screen against the target template molecule[62–65]
- Rational approaches that involve computation of total energies (E), energy differences (DE), and distances (d) of closest approach between the monomers and template using molecular dynamics[66]
- The use of density functional theory (DFT) method to calculate the binding energy ΔE, between a template and monomers as a measure of their interaction that facilitates the selection of the monomers[67,68]
- Rational design that involve conformation of template–functional monomer complexes employing semi empirical methods[69–74]
- Chemometric approaches to optimize monomer, template, and cross linker ratios[75]
- Predicting template monomer complexes using neural network methods[76,77]

The following sections discuss each of the above-mentioned computational methods for the rational design of MIPs. Table 6.1 summarizes the different computational procedures adopted for the rational design of MIPs for a variety of templates or analysis of polymer properties.

2   Computational Methods for Rational Design of MIPs

2.1   Rational Approaches that Involve Molecular Mechanics

One of the most established rational approaches in design of imprinted polymers is combinatorial synthesis/screening.[57,60] However, combinatorial approach has its limitations. Considering even a simple two-component system utilizing 100 monomers, it would be a daunting task of preparing several thousand polymers. It gets further complicated when we look at the possibility of different ratios of the monomer mixtures further inflating the amount of time and resources required. One potential solution to the problem of rational design of polymer lies in molecular modeling and performing thermodynamic computations using a patented protocol developed within our group at Cranfield University.[62] Variations of this protocol are in use nowadays in many laboratories around the world.

When there is requirement of performing structural analysis in large molecular systems, comprising hundreds of molecules, there is an inherent demand on the computational time and resources. In these cases, molecular mechanics (MM) is used. MM refers to a system that can be used for qualitative descriptions that include only potential energy, which is essentially devoid of any quantum mechanical calculations. To facilitate calculations, MM considers atoms as balls of certain radius and the bonds between them as string. The exact values of atom sizes and bond geometry and strength originate from empirical data collected from X-ray crystallography and NMR experiments. Several MM softwares exist for a variety of general and specific applications. Some of the widely used MM softwares are AMBER, MOE, RasMol, QMol, Raster 3D, and AGM Build.

A major problem associated with the computational design of imprinted polymers is the difficulty of performing detailed thermodynamic calculations on multi-component systems. Although molecular modeling of complex systems and possible interactions of polymers with template, solvent, and other molecules are difficult because of the requirement of large computational workload, we could achieve this by simplifying the model. Since the structure of the monomers–template complexes formed in the monomer mixture is preserved in the synthesized polymer, instead of modeling the polymer, modeling the monomer mixture and the interactions taking place in solutions between monomers, cross-linker, template, and solvent would be possible, which substantially reduces computational load.[62,64]

The protocol developed at Cranfield University starts with the design of a virtual library of molecular models of functional monomers and template (Fig. 6.2). The next step is to screen the virtual library against a template to determine the monomers that strongly bind to the template. Calculations are performed to estimate how the monomers bind to template using simulated annealing to determine optimum ratios of template to monomers. In effect, the strength and type of interactions, existing between monomers and template in monomer mixture, which in theory, determining the recognition properties of the MIP will be analyzed and used for optimization of polymer composition. Each of the steps has been described below.

2.1.1 Modeling of the Template Molecule

A molecular model of the template molecule is made and charges for each atom are calculated and the structure of the template is refined using MM.

2.1.2 Construction of the Monomer Database

An example of monomer database is shown in Fig. 6.2. Although there are about 4,000 polymerizable compounds that have been reported that could be potentially used as functional monomers, in reality many of them have similar properties and functions and hence it is assumed that it is sufficient to test possible interactions between a minimal library of functional monomers and a target template.[101] Several different software packages can be used for creating virtual library of monomers

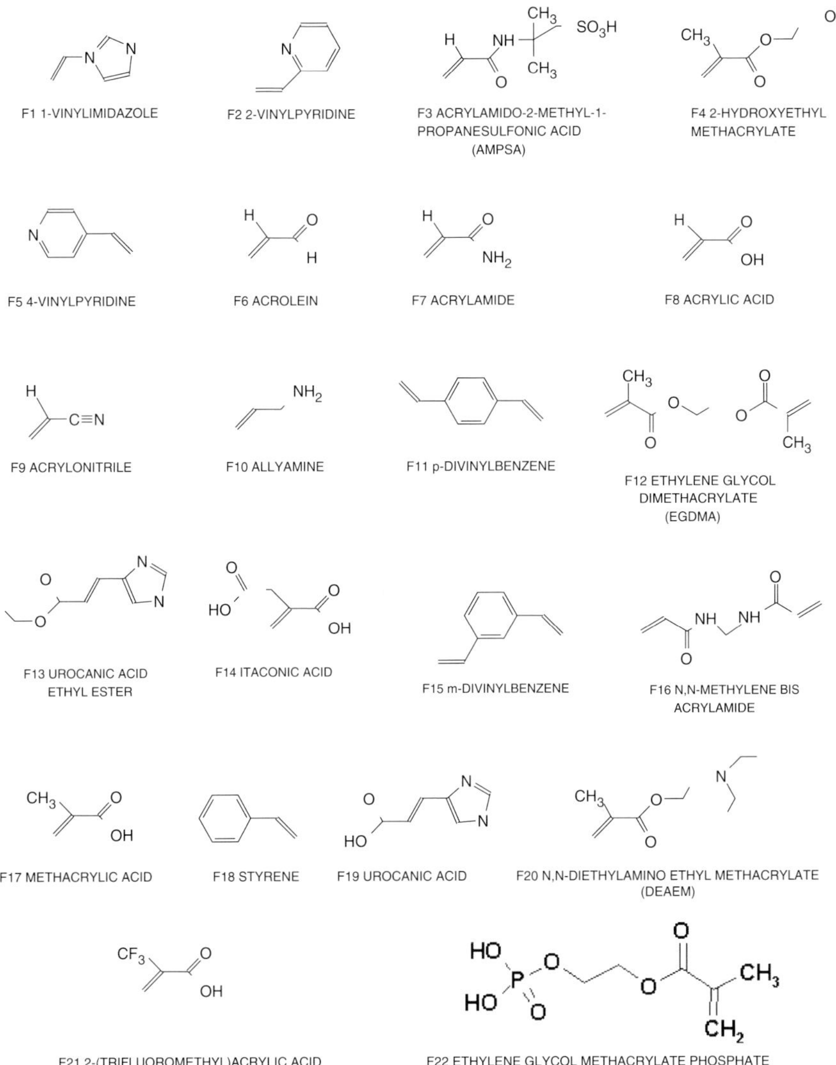

Fig. 6.2 Functional monomers in the virtual library

and molecular models of template. Examples include Agile Molecule, Sirius, Sybyl, Oscail X, and MOE.

2.1.3 Screening of the Virtual Library

The quantity and quality of MIP recognition sites that arise out of binding event is a direct function of the nature and extent of the monomer–template interactions present in the prepolymerization mixture. The previous research directed toward understanding physical basis of molecular recognition have shown that the extent of template complexation in equilibrium is governed by the change in Gibbs free energy of template–functional monomer interaction.[102–104] Calculation of the binding energies is one of the important requirements in this regard.[106–108] Andrews et al. (1984)[105] detailed an approach to calculate the average-binding energies of ten common functional groups based on analysis of structural factorization of the energetic contributions to binding. These approaches detailed the importance of each of the physical entities that govern a molecular recognition event. The general thermodynamic explanation that summarizes contribution of individual physical parameters in a binding event has been described by Williams (1). This equation can be used to describe template–monomers interactions as well as template–MIP binding events.[102,103,109]

$$\Delta G_{bind} = \Delta G_{t+r} + \Delta G_r + \Delta G_h + \Delta G_{vib} + \Sigma \Delta G_p + \Delta G_{conf} + \Delta G_{vdW}, \tag{1}$$

where the Gibbs free energy changes are: ΔG_{bind}, complex formation; ΔG_{t+r}, translational and rotational; ΔG_r, restriction of rotors upon complexation; ΔG_h, hydrophobic interactions; ΔG_{vib} residual soft vibrational modes; $\Sigma \Delta G_p$, the sum of interacting polar group contributions; ΔG_{conf}, adverse conformational changes; and ΔG_{vdW}, unfavorable van der Waals interactions.

This or similar equations lie in the cornerstone of practically all screening/modeling packages used in design of MIPs. In practical sense, the screening of virtual library is done by the means of Leapfrog algorithm (Tripos Inc). LeapFrog is used in drug development for screening of new, potentially active ligand molecules against known structure of receptor-binding sites. LeapFrog can also generate new compounds by repeatedly making small structural changes evaluating the binding energy of the new compound, and keeping or discarding the changes based on the results.[110,111.]

The first step in MIP design using LeapFrog is the identification of the binding sites on the surface of template molecule. LeapFrog samples the environment immediately surrounding the template and determines its average electrostatic, steric, and lipophilic characteristics. LeapFrog begins with placing each of the monomers in proximity of the template binding site. The second step is the calculation of binding energy. Once the binding site is well defined, the "fit" is assessed. Since many possible hits arise, each has to be scored to decide which one of those hits is most promising. There are a variety of scoring techniques employed by different programs that exist such as LEGEND,[112] LUDI,[113] SPROUT,[114] HOOK,[115] and PRO-LIGAND.[116] The scoring functions these programs employ, however, vary from (a) H-bond placement, (b) constraints that are due to steric effects, (c) explicit force fields, and (d) empirical or knowledge-based scoring methods. Programs such as GRID and LigBuilder set up a grid in the binding site and then assess interaction energies by placing probe atoms or fragments at each grid point.[117]

Scoring functions guide the growth and optimization of structures by assigning fitness values to the sampled space. Scoring functions attempt to approximate the binding free energy by substituting the exact physical model with simplified statistical methods. Force fields such as the one used by Leapfrog involve more computation than some other types of scoring functions. Leapfrog calculates major components of the binding energy such as steric, electrostatic, and hydrogen-bonding enthalpies. Other methods such as the one used in GRID4 program[118,119] also are in use that enhance the rate of calculation (Tripos, Inc).

2.1.4 Computation of Monomer Template Ratio

The next step in the protocol is the computation of monomer–template ratio performed by simulated annealing using a molecular dynamics approach (for detailed description see Sect. 2.2).

2.2 *Examples of Using MM Methods in MIP Design*

In this first example of rational design of MIP using the above-mentioned protocol,[43] we demonstrated the proof of concept where the screening of the virtual library of monomers led to an optimized MIP composition specific for creatinine. When this polymer was synthesized in the laboratory, it demonstrated superior selectivity in comparison to a MIP that was prepared using a traditional functional monomer methacrylic acid (MAA). In this work, we combined the above-described computational procedure for rational design of MIPs with a "Bite-and-Switch" approach for the detection of polymer–template interaction[120](Fig. 6.3).

In what could be considered as one of the best examples of the rational design using our protocol, a highly selective MIP for the cyanobacterial toxin microcystin-LR was designed and demonstrated.[65] Two MIPs for microcystin-LR were then synthesized, one using a functional monomer with the best binding score, 2-acryla-mido-2-methyl-1-propanesulfonic acid (AMPSA) (Fig. 6.4), and the other using a "traditional" functional monomer MAA. The optimal MIP formulation synthesized had affinity and sensitivity comparable with those of polyclonal antibodies and superior chemical and thermal stabilities compared with those of biological antibodies. The affinity of the computationally designed MIP, studied using ELISA (enzyme-linked competitive assay), was comparable to polyclonal antibodies (Table 6.2). The computationally designed MIP also showed higher affinity in comparison with the MAA-MIP. It was also found that MIPs had much lower cross-reactivity for microcystin-LR analogues than both polyclonal and monoclonal antibodies (Table 6.3).

Another polymer for Ochratoxin A (OTA) produced by several *Aspergillus* and *Penicillium* species, which is widespread in animal and human food, was designed[93] using this protocol. Two polymers were synthesized from MAA and acrylamide (AA) – monomers chosen computationally due to strong possibility of binding with OTA. Interestingly, MAA was used previously in design of polymer for OTA.[121] The polymer has shown complete binding in aqueous solutions. Binding mechanism

o-Phthaldialdehyde + R_1—SH + R_2—NH_2 ⟶ fluorescent isoindole   (I)

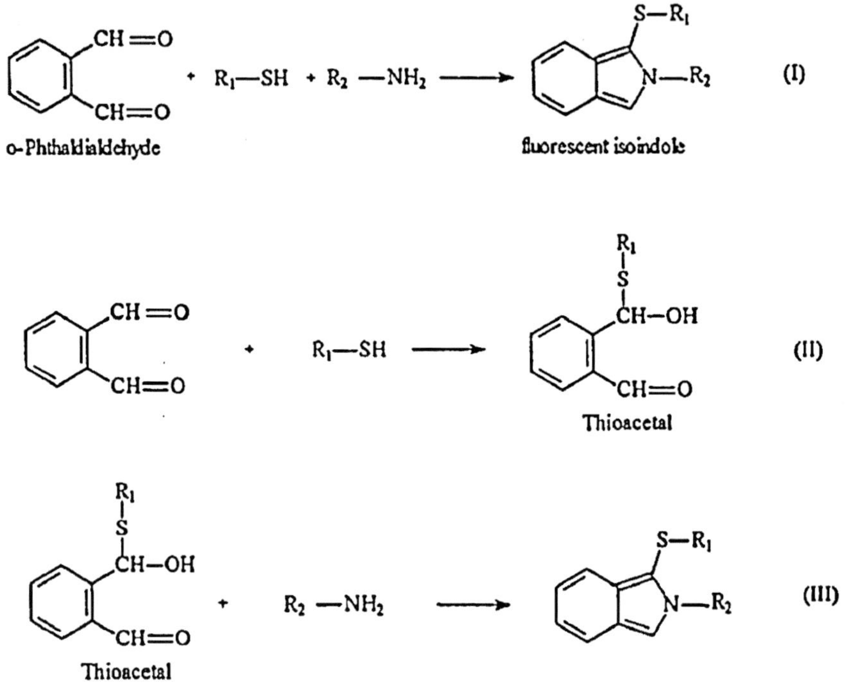

CH=O ... + R_1—SH ⟶ Thioacetal   (II)

Thioacetal + R_2—NH_2 ⟶   (III)

Fig. 6.3 Interaction between phthalic dialdehyde, mercaptan group (OPA reagents) and primary amine (I); thioacetal formation (II); formation of fluorescent complex between hemithioacetal and primary amine (III).[120] Reproduced with permission

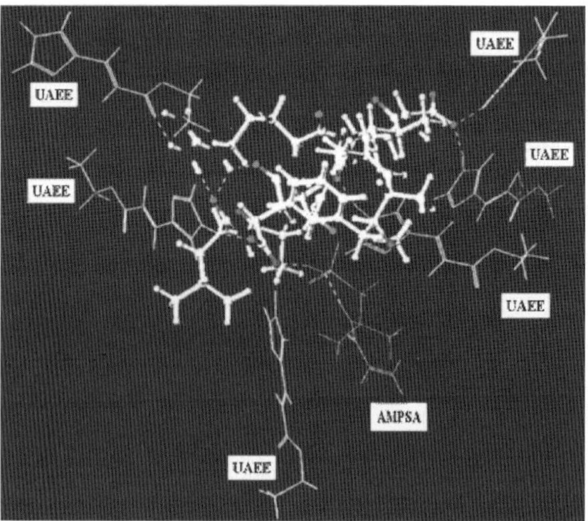

Fig. 6.4 Interactions between microcystin-LR and monomers. Microcystin-LR, in balls and sticks in the center of the picture, interacts with six molecules of urocanic acid ethyl ester (UAEE) and 1 molecule of AMPSA.[65] Reproduced with permission

Table 6.2 Affinity and sensitivity range of MIPs and antibodies for Microcystin-LR evaluated by competitive assay[65]

Receptor	K_d (nM)	Sensitivity range (μg/L)
Computational MIP	0.3 ± 0.08	0.1–100
MAA-MIP	0.9 ± 0.1	0.8–100
Monoclonal antibody	0.03 ± 0.004	0.025–5
Polyclonal antibody	0.5 ± 0.07	0.05–10

Reproduced with permission

Table 6.3 Cross-reactivity of MIPs and antibodies[65]

Receptor	MC-LR (%)	MC-RR (%)	MC-YR (%)	Nodularin (%)
Computational MIP	3333	21 ± 0.9	27 ± 2	22 ± 2
MAA-MIP	100	19 ± 0.8	30 ± 3	36 ± 0.5
Monoclonal antibody	100	106 ± 0.3	44 ± 2	18 ± 0.8
Polyclonal antibody	100	92 ± 2	142 ± 0.8	73 ± 1

Reproduced with permission

depended critically on the conformation of the polymeric binding pockets, which when combined with weak electrostatic interactions allows for specific recognition. Using this protocol, we developed efficient MIPs for drugs of abuse.[81] The polymers for four drugs of abuse: cocaine, deoxyephedrine, methadone, and morphine were developed. The best candidates for MIPs specific for cocaine were: IA, MAA, AA; for deoxyephedrine: IA, MAA, AA, and HEM; for methadone: IA, MAA, and HEM; and morphine: MAA, IA, and HEM. Quantity of monomer units able to form a complex with the template was further evaluated by saturating the space around the template with a combination of monomers selected. This step produced several complexes for each template (Fig. 6.5). The synthesized polymers possessed good recognition properties under the same conditions, which bring them a step closer to creation of a multisensor for drugs of abuse.

It is proven that MIPs perform well in organic solvents, while the practical application of MIPs is hindered due to their poor performance in polar media. Although it is desirable to achieve affinity separation and sensing in water, MIPs usually do not work equally well in aqueous media because of the disruption of hydrogen bonds and competition process between solvent and template molecules for their binding to the polymer functional groups. A significant contribution to the loss of polymer affinity originates from the potential difference in the structure of the polymer binding sites in organic solvent (traditionally used for polymer preparation) and in water due to differences in polymer swelling. In an effort to develop MIPs compatible with water, we imprinted biotin,[79] using the computational screening of a virtual library of functional monomers, and identified those that provide strong binding to the template in water. To mimic aqueous conditions, the energy minimization

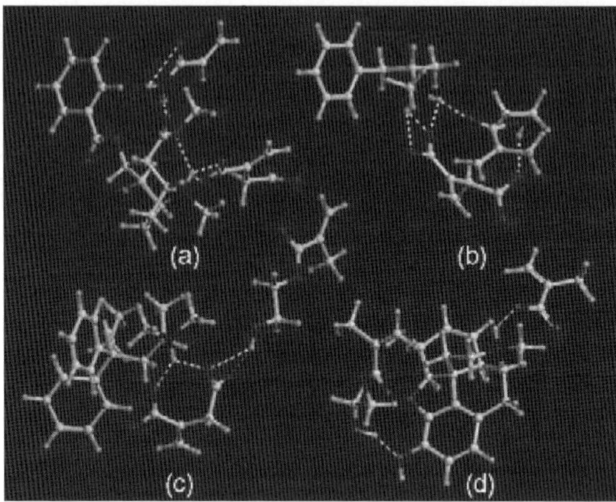

Fig. 6.5 Molecular complex formed between (**a**) cocaine, AA and IA; (**b**) deoxy-ephedrine, IA and HEM; (**c**) methadone, IA and HEM; (**d**) morphine, MAA and HEM, as predicted from molecular modeling.[81] Reproduced with permission

of monomers and template was performed using dielectric constant of water ($\varepsilon = 80$). The results of the modeling confirmed that monomers MAA, TFMAA, and AMPSA formed a strong complex with the template molecule in water through ionic and hydrogen bonds (Fig. 6.6). This was the first demonstration of the use of molecular modeling for rational selection of monomers capable of template recognition in water. The designed MIP was successfully grafted to the polystyrene surface in aqueous environment. The modified polymers demonstrated high affinity for biotin in water.

We also reported a design of MIP with high-binding capacity suitable for large scale extraction of abacavir, a HIV-1 reverse transcriptase inhibitor.[78] The MIP based on IA possessed very high-binding capacity for the template, up to 15.7% in 50 mM Na-acetate buffer (pH 4.0).

In another study, we synthesized a polymer specific for tylosin.[85] The synthesized polymer was examined for rebinding with the template and related metabolites such as tylactone, narbomycin, and picromycin. HPLC analysis showed that the computationally designed polymer is specific, and is capable of separating the template from its structural analogues. The MIP was capable of recovering tylosin from broth samples. This work demonstrated ability of MIP "dialing" – use of computational approach for prompt design of high-performance MIP for practical application. By using this approach, it is possible to design MIP composition in 2–3 weeks, which is substantially quicker than to perform combinatorial synthesis screening of multitude of polymers.

In other examples, computationally designed MIPs were prepared for herbicides: simazine,[84] carbamate,[80] and triazines.[83] The modeling approach was able to identify

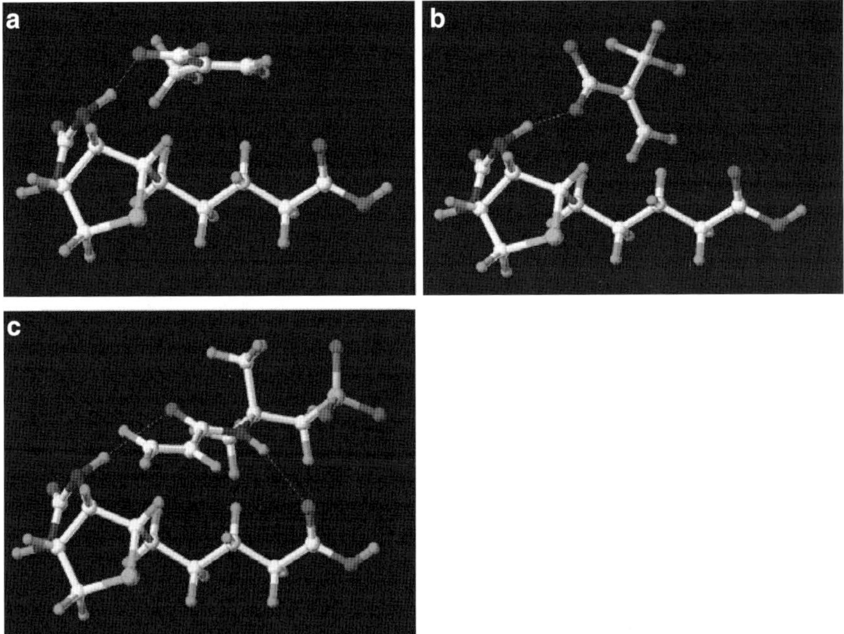

Fig. 6.6 Computationally derived structures of biotin–monomer complexes: (**a**) biotin–MAA; (**b**) biotin–TFMAA; (**c**) biotin–AMPSA.[80] Reproduced with permission

from 3 to 4 out of the 5 best monomers (Table 6.4). Interestingly, this work also showed that the computationally designed blank polymers have high affinity to the template, which potentially excludes the need for molecular imprinting. We further synthesized a MIP capable of controlled release of simazine in water.[84] Leapfrog identified a list of monomers that were used to produce polymers with different affinity and correspondingly different profile of the release of herbicides. The speed of release of herbicide correlated with the calculated binding characteristics. The high-affinity MAA-based polymer released ~2%, and the low-affinity HEM-based polymer released ~27% of the template over 25 days.

Chapuis et al.[97] studied the effects of the charge distribution and of molecular volume of the triazines on the selectivity of interactions between the analytes and the MIP using HyperchemPro 6.0. (Hypercube, Gainesville, FL, USA) Energy minimizations were performed using MM. The effects of the charge distribution and of molecular volume of the triazines on the selectivity of interactions between the analytes and the MIP were investigated. The synthesized materials were successfully applied for the class-selective extraction of triazines from industrial effluent and surface water samples.

In an interesting polymer design work that employs PM3 method, modeling studies were performed on caffeine (CAF) and theophylline (THO). Farrington et al.[69]

Table 6.4 Prediction of the performance of polymers from the modeling data[83]

Template	No. of polymers predicted out of 5 best	Monomers
Cyanazine	5	IA, MAA, TFMAA, AMPSA, ACM
Atrazine	4	IA, MAA, TFMAA, AMPSA
Simazine	4	MAA, IA, ACM, TFMAA
Bentazone	4	ALM, VI, ACM, MAA
Bromoxynil	3	IA, MAA, TFMAA
Propanil	4	TFMAA, MBAA, ACM, MAA
Tebuthiuron	4	MBAA, ACM, AMPSA, VI
Diuron	3	MBAA, ACM, TFMAA
Metribuzin	3	AMPSA, ACM, MAA
Hexazinone	2	ACM, ALM

Reproduced with permission

used Hyperchem 7.5 (Hypercube Inc., Gainsville, FL) to draw the structures of the templates and functional monomers (MAA and 2-VP), and minimize their conformation to the lowest energy using the semiempirical mechanic (PM3) method.[68,55] To analyze possible interactions between template and functional monomer and to calculate binding energies, the Amber MM method was used. The approach used here was able to predict the relative ratios of template to functional monomer and determine optimal solvent necessary for imprinting. The program also yielded information on thermodynamic stability of the prepolymerization complex.

Using the same modeling program, MIP specific for ibuprofen in aqueous media was developed.[88] Recoveries were typically >80% and good selectivity for ibuprofen over structurally related analogues was shown. In this study, calculation of the molecular volumes of the complexes (Fig. 6.7) was performed using Accerlys DS Viewer program (http://www.accerlys.com).

MM method is the fastest method available (least expensive), and hence is an ideal choice for studies on structural parameters and the most stable conformation molecules. Optimization steps are often carried out to confirm that the molecules are in their lowest energy state, so that calculated results can be compared with those made experimentally. However, since MM does not deal directly with electrons and orbitals, it cannot be used to study e.g. chemical reactivity of functional monomers.

2.3 Rational Approaches that Involve Molecular Dynamics (MD)

The MD simulations provides better description of interactions (generally electrostatic and van der Waals) because they reflect the effect that surrounding environment has on the properties of molecules. MD simulations in general are powerful tools to investigate complex systems made of thousands of atoms. A good understanding of intermolecular interactions, mechanism of imprinting, and properties in

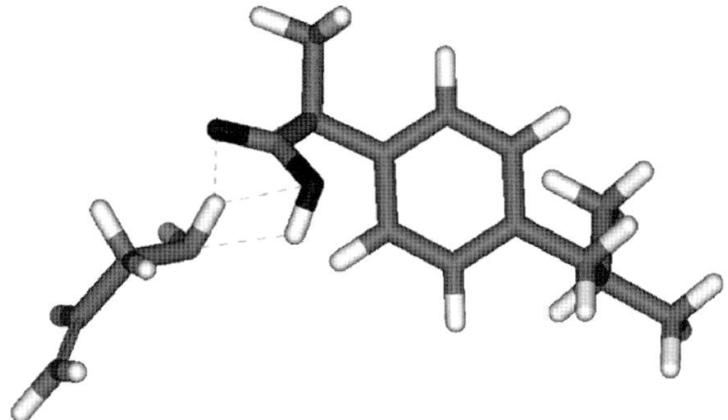

Fig. 6.7 The hyperchem derived energy minimized structure of ibuprofen and allylamine. The presence of hydrogen bonds is indicated by the *dashed lines*.[88] Reproduced with permission

molecular imprinting process requires advanced state of the art computational tools, which will help in investigations on the molecular clusters calling for NVT MD simulations and prediction of interaction energies.

One of the successful approaches that is used for the computation of monomer template ratio is simulated annealing. Simulated annealing is a Monte Carlo approach for minimizing multivariate functions. The term simulated annealing derives from a physical process of heating and then slowly cooling a substance to obtain a crystalline structure. In molecular modeling, a minima of the cost function corresponds to this ground state of the substance. The simulated annealing process lowers the temperature by slow stages until the system is "frozen or standstill" and no further changes occur. At each temperature, the simulation must proceed long enough for the system to reach a steady state or equilibrium. This is known as thermalization. The sequence of temperature changes and the number of iterations applied to thermalize the system at each temperature comprise an annealing schedule.[122] To apply simulated annealing, the system is initialized with a particular configuration. A new configuration is constructed by imposing a random displacement. If the energy of this new state is lower than that of the previous one, the change is accepted unconditionally and the system is updated. If the energy is greater, the new configuration is accepted probabilistically. This procedure permits an environment to proceed toward lower energy states, at the same time keeping open the option of escaping out of the local minima due to the probabilistic acceptance of some upward moves. Logarithmic decrease of temperature in simulated annealing generally assures an optimal solution.

2.4 Examples of Using MD Methods in MIP Design

Monti et al.[82] detailed a protocol that combined MD, MM, docking and site mapping to simulate the formation of possible imprints in acetonitrile solution for THO using MAA and MMA as monomers. MM calculations and MD simulations were carried out with AMBER7, as it previously showed satisfactory performance of this force field in the evaluation of the stability of hydrogen bonded as well as van der Waals adducts.[123–126] The structures of THO, MAA, and MMA molecules were first optimized using DFT,[127–129] and their atomic charges were determined with RESP.[130,131] All the simulations were performed in the NPT ensemble using Berendsen thermostat and barostat[132] with temperature set to 310 K and pressure to 1 atm. THO molecule was surrounded by functional monomers shells and solvated creating around it a rectangular parallelepiped acetonitrile box.[133,134] Docking procedures, used to find favorable orientations of the ligands inside the polymer cavity, were performed using the DOCK5.0 program[135–137] and the GRID program (GRID, 2004) was used to map the energetic interactions. The created model was able to predict binding affinity and selectivity when considering THO analogues, such as caffeine, theobromine, xanthine, and 3-methylxanthine.[82] The entire modeling study was performed in four different phases: first, a *non-covalent phase*, where the template and the functional monomers form noncovalent complexes in solution prior to polymerization; second, a *locking phase*, where the noncovalent monomers–template complexes are cross-linked and the binding site is generated with appropriately oriented functional monomers and model polymer structures are created and selected; third, *validation phase*, where the polymer specificity and recognition capabilities are tested, and finally the *mapping phase*, where the characteristics of the binding cavities are analyzed. This work showed that the MD simulations were able to predict the selectivity and the binding affinity and when complemented with experimental data gave a clearer picture of the system and the type of interactions in the complex (Fig. 6.8).

In another example of MD simulation Benito-Peña et al.[73] analyzed binding of seven novel fluorescent labeled β-lactams (BLAs) with a library of six polymers imprinted with penicillin G (PenG). The 3-D chemical structures of the labeled BLAs have been modeled followed by energy minimization by molecular dynamics (MOPAC, AM1 force field) using Chem3D Ultra 7.0 software (Cambridge-Soft, MA). The results of molecular modeling showed that recognition of the fluorescent analogues of PenG by the molecularly imprinted material is due to a combination of size and shape selectivity.

Yoshida et al.[86] employed HyperChem and performed MD calculations to verify the recognition mechanism of the MIP they synthesized for the separation of optically active tryptophan methyl ester. The computational modeling proved that the enantiomeric selectivity is conferred by the electrostatic and hydrogen bonding interactions between the functional molecule and the target tryptophan methyl ester along with the chiral space formed on the polymer surface.

Toorisaka et al.[87] studied the structure of the complex formed between a cobalt ion and alkyl imidazole that catalyses hydrolysis of an amino acid ester. By using HyperChem they calculated the lowest energy structure of the complex in vacuum. Then the complex was placed at the toluene–water interface by replacing 11 toluene molecules with molecules of water (Fig. 6.9). The MD simulation was performed in the (N, V, T) ensemble after MM calculation in the biphase system,

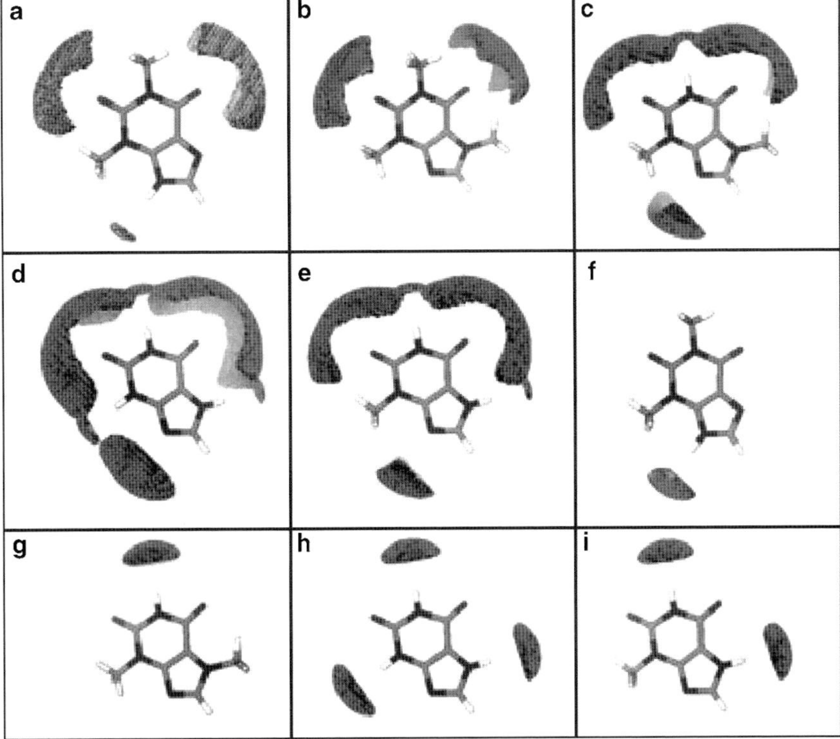

Fig. 6.8 Contour maps of the molecular interaction fields produced: by OH probe at −4.5 kcal/mol for THO (**a**), CAF (**b**), theobromine (**c**), xanthine (**d**), 3-methylxanthine (**e**), by O probe at −3.0 kcal/mol for THO (**f**), theobromine (**g**), xanthine (**h**), 3-methylxanthine (**i**).[82] Reproduced with permission

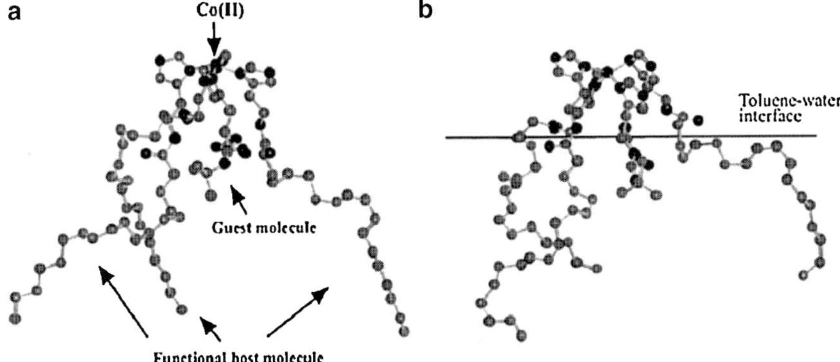

Fig. 6.9 The optimized structure for the active site of the MIP. The lowest energy structure of the complex (**a**) in vacuum before MD calculation; (**b**) at the oil–water interface after MD calculation (1 ps).[87] Reproduced with permission

to avoid a large strain energy. The computer model showed the structure of the imprinted sites formed on the polymer surface, which was in agreement with the structure predicted from the practical testing of synthesized polymer evaluated in the hydrolysis reaction.

Pavel and Lagowski[66,95,96] studied the intermolecular interactions in molecular imprinting of theophylline (THO). The minimized structures of five ligands, THO and its derivatives (theobromine, theophylline-8-butanoic acid, caffeine, and theophylline-7-acetic acid) were employed in MD simulation using Cerius2 version 4.10 software designed by Accelrys, Inc. (San Diego, CA, USA). The polymer consistent force field (PCFF) was employed, as it was found to be very suitable and reliable for the molecular simulation of organic molecular clusters of monomers and polymers.[138–140] The forces acting on each atom of a model polymer were calculated. The initial molecular clusters of the simulated monomers and polymers were optimized giving information about total energies (E), energy differences (DE), and distances (d) between the monomers and different ligands in a given cluster (Fig. 6.10).

Using the same MD simulations, Pavel et al.[66] designed monomers for MIPs specific for chemical warfare agents. They showed successful prediction of interaction energies, the closest approach, distances and the active site groups.

Several research groups have used PM3 method for analysis of template–functional monomer complexes.[141] From the computational perspective, PM3 method provides

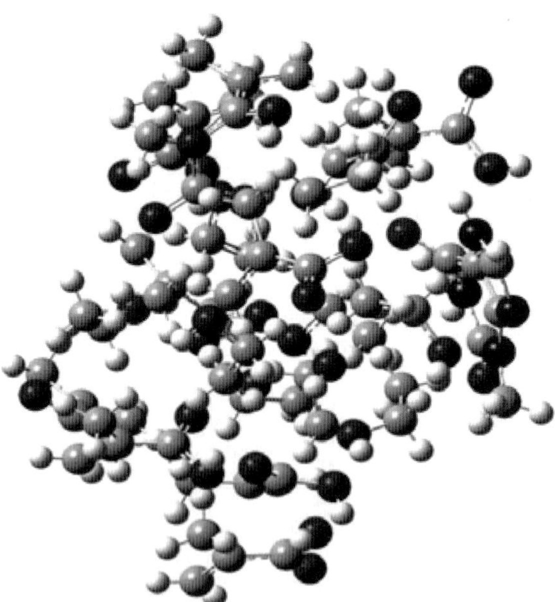

Fig. 6.10 A typical NVT-MD equilibrated conformation of the solvated molecular cluster (ten molecules of MAA, ten molecules of ethanol, and one molecule of THO).[96] Reproduced with permission

an improved modeling of noncovalent interactions such as hydrogen bonding and van der Waals interactions.

Baggiani et al.[94] used molecular graphic software HyperChem 5.01 (Hypercube Inc., Waterloo, Canada) to rationally design a MIP for recognition of the carbamate group. Functional monomers potentially able to form noncovalent interactions with the model molecule *N,O*-dibenzylcarbamate were computationally selected, describing possible interactions between the template and a small library of vinylic monomers. Molecular models of the template and a library of six possible functional monomers (ACM, AA, MAA, 2-HEM, 4- VP, DMAEM) were optimized by using a semiempirical quantum method (AM1). For each minimized structure, many combinations of template and monomers were assembled together after which a simulated annealing process was applied to optimize the arrangement of the resulting supramolecular structures. Annealing conditions were fixed as 300 K considering the dynamic equilibrium reached after 2,000 fs, with step of 0.1 fs. At the end of the annealing process, the position of functional monomers around the template was optimized. It was concluded that MAA is more efficient than ACM as a functional monomer, and that chloroform enhances polymer selectivity (Fig. 6.11). However, the drawback of this method is the fact that preliminary information on the relative stabilities of template-functional monomer(s) complexes were obtained by neglecting the presence of cross-linker (EDMA) and porogen molecules.

Since MD simulations are numerical and generally include large number of particles, the simulation time required for modeling is very substantial. However, contrary to MM, MD molecules and their complexes have ability of adapting to the environment, which provides more adequate representation of the reality.

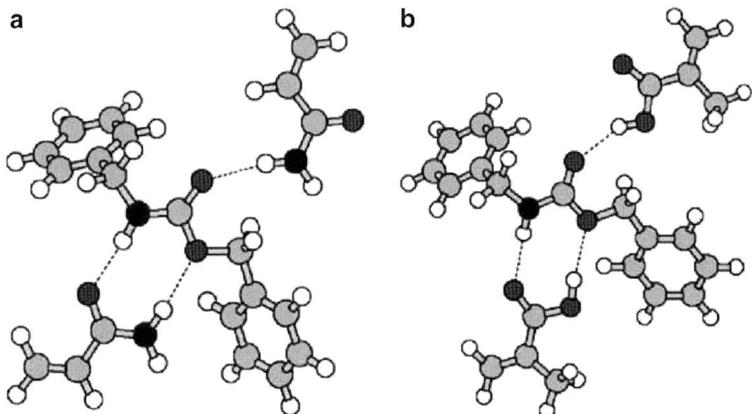

Fig. 6.11 Template–functional monomers complex between *N,O*-dibenzyl-carbamate and two molecules of acrylamide (**a**) and MAA (**b**).[94] Reproduced with permission

2.5 Rational Approaches that Involve Quantum Mechanics

Quantum mechanics (QM) is a field of quantum chemistry that uses mathematical basis to study chemical phenomena at a molecular level. It uses a complex mathematical expression called as a wave function with which energy and properties of atoms and molecules can be computed. For simple model systems wave functions can be analytically determined, while for complex systems such as those that involve molecular modeling, approximations have to be made. One of the commonly employed approximation methods is that of *Born and Oppenheimer*. This approximation exploits the idea that does not necessitate the development of a wave function description for both the electrons and the nuclei at the same time. The nuclei are heavier, and move much more slowly than the electrons, and therefore can be regarded as stationary, while electronic wave function is computed. By computing the QM of the electronic motion, the energy changes for different chemical processes, vibrations and chemical reactions can be understood.[142]

2.6 Examples of Using QM Methods in MIP Design

Dong et al.[68] employed this method to screen monomers using the binding energy, ΔE, of a template molecule and a monomer as a measure of their interaction. In this study, THO was chosen as the template molecule, and MAA, AA, and TFMAA were the functional monomers (Fig. 6.12). The calculation of ΔE was performed using DFT with the Gaussian 98 software.[143] First, the conformations of THO, MAA, AA, and TFMAA were optimized, and the energy of the molecules with the optimized conformation was calculated. Then the energy calculation was applied to the complex formed between THO and MAA, or AA or TFMAA, respectively. Finally, the binding energy of THO with the monomer was obtained from the following equation:

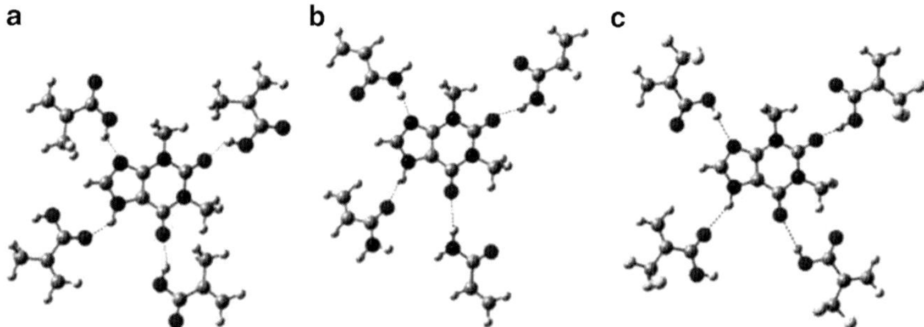

Fig. 6.12 The complex formed between THO and MAA (**a**), AA (**b**) and TFMAA (**c**), respectively.[68] Reproduced with permission

$$\Delta E = \left| E(\text{complex}) - E(\text{THO}) - E(\text{monomer}) \right|.$$

The MIP synthesized using TFMAA as monomer showed the highest selectivity to THO, while the MIP from AA gave the lowest, as predicted from the ΔE calculation.

Dineiro[98] deployed the same technique to select the best functional monomer and porogenic solvent for the construction of a recognition element for the dopamine metabolite homovanillic acid (HVA). The computational method[144] predicts that TFMAA and toluene are the monomer and solvent rendering the highest stabilization energy for the prepolymerization adducts (Fig. 6.13). MIP prepared using this formulation gave rise to a Freundlich binding isotherm. The stabilization energies in different solvents were calculated using the United Atom Hartree–Fock (UAHF) Polarizable Continuum Model (PCM)[145] to select the most stabilizing one.

Wu et al.[67] showed that the same technique could be employed to determine the influence of porogens on the affinity and selectivity of MIP. The interaction energy values between NAM and MAA were modeled with methanol, acetonitrile, chloroform, and toluene as the porogens. Gaussian 03[146] was adopted as the software to carry out the simulation, and B3LYP[147,148] was selected as the calculation method. B3LYP is a DFT method, which takes electronic correlation energy into consideration giving results of weak interaction system compared with Hartree–Fock method. The retention factors and selectivity factors of NAM and its analogues

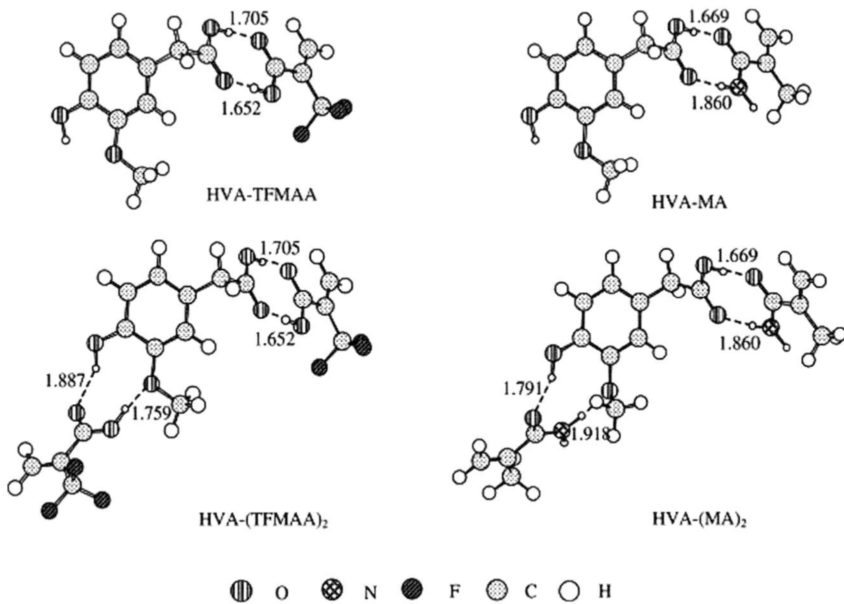

Fig. 6.13 Optimized geometries for the most stable adducts located between HVA and MA or TFMAA. Distances in Angstroms.[98] Reproduced with permission

were evaluated, and good correlations were found between the interaction energies and the retention factors. When the porogens had poor hydrogen bonding capacity, the interaction energy was mainly influenced by dielectric constant of the solvent, and when the porogen had strong capacity in forming hydrogen bond, both the dielectric constant of the solvent and the hydrogen bonding interference affected the formation of the template–monomer complex and the corresponding interaction energy.

Chiral recognition was examined for a MIP synthesized for (S)-nilvadipine using MAA, TFMAA, 2-VP, or 4-VP as a functional monomer and EGDMA as cross-linker.[100] Molecular computations were done with CAChe MOPAC version 94 implemented in CAChe programs23 run on a Windows 98-based desktop PC. Molecular geometries of (S)-nilvadipine and 4-VP were optimized by the AM1 method (Fig. 6.14). The simulation was performed on the hydrogen-bonding complex model with the dihydropyridine and pyridine rings of (S)-nilvadipine and 4-VP molecules, respectively. Molecular modeling revealed a one-to-one hydrogen-bonding-based complex formation of (S)-nilvadipine with 4-VP in chloroform and that (S)-nilvadipine imprinted EGDMA polymers should recognize the template molecule by its molecular shape, and that hydrophobic and hydrogen-bonding interactions seem to play important roles in the retention and chiral recognition of nilvadipine on the 4-VP-*co*-EGDMA polymers in hydro-organic mobile phases. The (S)-nilvadipine-imprinted 4-VPY-*co*-EGDMA polymers indeed gave the highest resolution for nilvadipine among the MIPs prepared.

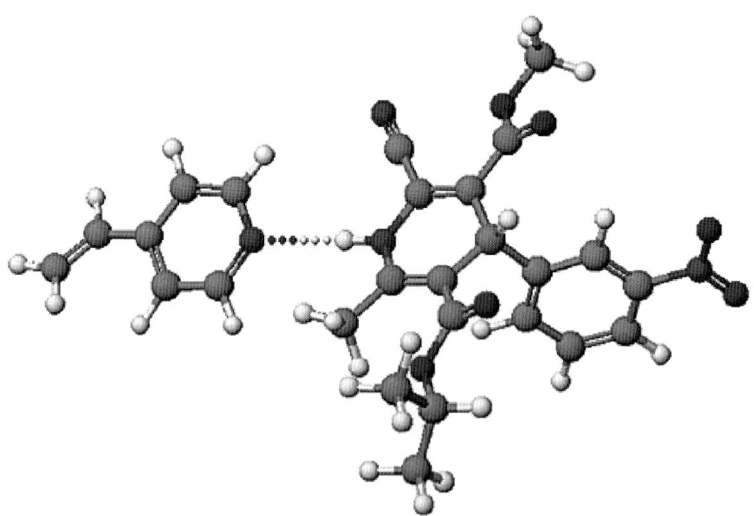

Fig. 6.14 AM1-optimized structure of the minimum energy complex.[100] Reproduced with permission

A computational optimization of the monomer formulation of molecularly imprinted catalysts (MIC) for lipase-catalyzed transesterification process was demonstrated.[72] Authors screened the intermediates of the lipase-catalyzed transesterification process commonly containing "catalytic triad" motif made up of a compounds such as serine, histidine, and aspartic acid.[149–151] To construct the virtual intermediates, p-nitrophenyl acetate was used as substrate, and monomers containing carboxylate moieties as molecular recognition elements. The energy of each intermediate was then minimized using the semiempirical MOPAC method with a minimum RMS gradient of 0.100, which specifies the convergence criteria for the gradient of the potential energy surface. AM1 theory was used with a closed shell function to calculate heat of formation (ΔHf) of the intermediates, which represents the gas-phase heat of formation at 298 K of 1 mol of the intermediate from its elements in their standard state. The result of this work has been utilized successfully for the design of artificial lipases.

In yet another study employing MOPAC AM1 calculations, Rathbone and Ge[74] computationally analyzed the possible interactions of the fluorescent monomer with a carboxamidrazone substrate (Fig. 6.15). MOPAC AM1 computations were carried out[152] within the program CAChe Work System Version 3.2 (Oxford Molecular Ltd). Energy minimizations were terminated after a gradient norm of 0.1 was achieved. The structures were displayed using the program WebLab ViewerLite Version 3.5 (Molecular Simulations Inc). GAMESS calculations were carried out[153] providing energy-minimized structures and Lowdin partial atomic charges. The results of simulations were used to explain the effects of the quenching of fluorescence of the template adsorbed by MIP.

Quantum methods are perhaps the most accurate approaches currently used in the field of molecular modeling, because the modeling method involves less "assumptions," and the results depend entirely on the accuracy of performed calculations.

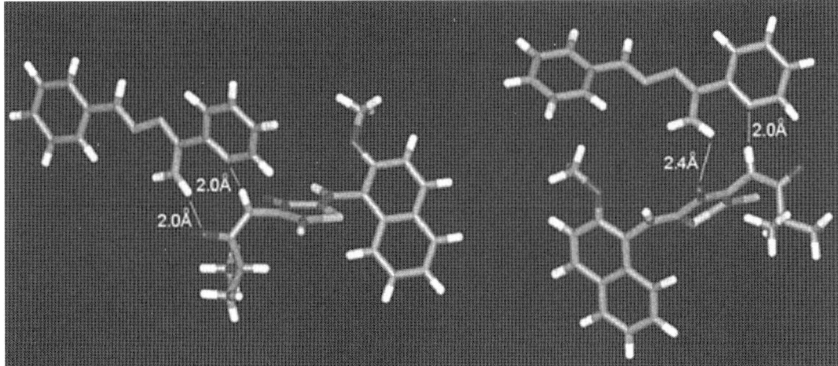

Fig. 6.15 Low energy conformation hydrogen bonding pairs of fluorescent monomer and substrate.[74] Reproduced with permission

However, the shear size of the required computations is huge and currently does not allow performance of realistic modeling of supramolecular systems.

2.7 Rational Approaches Involving Chemometrics and Neural Network Methods

Chemometrics uses mathematical and statistical methods for selecting optimal experimental procedures and extracting data for the analysis.[154] The design of experiments follows a mathematical framework for changing several selected factors simultaneously to predict the optimum conditions, thus reducing the number of experiments necessary.[155] In effect the goal is to plan and perform experiments to extract the maximum amount of information in the fewest number of trials. Chemometrics has various applications in scientific applications such as optimization of experimental parameters, design of experiments, data retrieval and statistical analysis, analysis of structure–property relationship estimations, signal processing, pattern recognition, and modeling.

2.8 Examples of Using Chemometrics Methods in MIP Design

Steinke et al.[156] have used the chemometrics approach to investigate the effect of variables such as type and quantity of monomers, cross-linker, porogens, initiator, type of initiation (UV or thermal), polymerization pressure, temperature, reaction time and reaction vial dimensions have on the properties of synthesized polymers.

Davies et al.[75] have used chemometrics design expert software to optimize the composition of MIP for sulfonamide, in particular the template/monomer/cross-linker ratio. MAA has been selected as functional monomer and EGDMA as cross-linker. They selected the amounts of template, monomer and cross-linker (T/M/X) in the MIP using a three-level full factorial design. The chemometrics design required synthesis of small number of MIPs with different ratio and using results of their testing for predicting polymer with optimum characteristics. The properties of polymer with predicted optimum ratio of 1:10:55 was compared with the properties of commonly used MIP with molar ratios such as 1:4:20[157] and 1:8:40.[158] The experiment proved that the predicted ratio generated the polymer with superior-binding properties.

In another study, Kempe and Kempe[159] employed multivariate data analysis (Modde 6.0 software, Umetrics, Umea, Sweden) for the optimization of monomer and cross-linker ratios in design of polymer specific for propranolol.

Mijangos et al.[160] used chemometrics (MODDE 6.0 software, Umetrics, Sweden) to optimize several parameters such as concentration of initiator (1,1′-azobis(cyclohexane-1-carbonitrile) and 2,2-dimethoxy-2-phenylacetophenone) and polymerization time required for design of high-performance MIP for ephedrine. A small set of (−) ephedrine-imprinted polymers was synthesized and tested by HPLC for their ability to interact with (+) and (−) ephedrine. This chemometric study provided evidence that to achieve high performance in chiral separation MIPs should be synthesized for a long period of time using low concentration of initiator and low temperature.

Prachayasittikul's group has used data from the literature such as monomer composition, retention factors of MIP and NIP, imprinting factor, mobile phase composition, etc. in calculations aimed at prediction of optimal template–monomer pairs.[76] The imprinting factor was predicted by artificial neural networks as a function of the calculated molecular descriptors and the mobile phase descriptors (the protocol is shown in Fig. 6.16). The quantum chemical descriptors were computed using Gaussian 03W. The model confirmed that the stronger template–functional monomer interactions lead to larger imprinting factors.

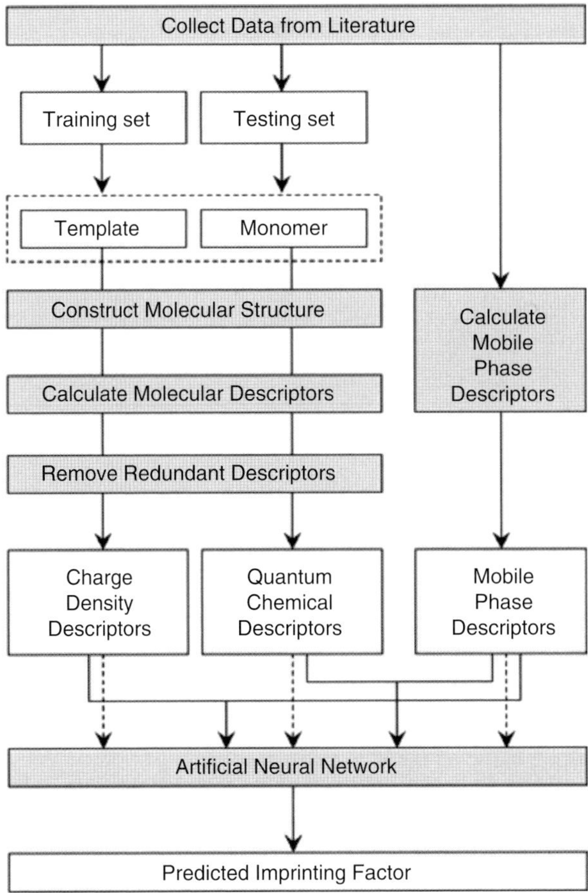

Fig. 6.16 The prediction procedure using artificial neural network.[77] Reproduced with permission

Most experiments are influenced by a variety of factors, and screening is often the first step in efficient assessment of which factors are important in influencing the desired outcome of the system under study.[161,162] Traditional factors that depend on monitoring the effects of changing one factor at a time on a response are extremely time consuming besides producing erroneous optimums in experiments.[161] However, chemometrics, often combined with artificial network simulations,[76] can help significantly in optimization and design of experiments for changing selected factors simultaneously to predict optimum conditions thus reducing the number of experiments necessary,[161] and providing error free analysis. There are a number of tools that are available to recognize patterns in the data and suggest predicted optimums while providing error analysis.[162]

Although chemometrics offers several advantages, complex numerical solutions that are generated by chemometrics approaches could often be misinterpreted unless a proper procedure is implimented. Since problem solving involves multivariate analysis, the interpreter needs to be fairly skilled in terms of analysis and interpretation. This approach also requires a number of experiments to be made, and as such cannot be performed entirely in silico.

3 Conclusion

Various methods such as MM, MD, or QM could be used to assist in MIP design. Each of these should, however, consider the various complex physical and chemical processes taking place during formation of monomer–template complexes as well as processes involved in polymer formation. It is important to note that although formation of monomer–template complexes in solution can be relatively easily modeled, the remaining challenge lies in modeling different stages of polymerization. Exact mechanism of events related to incorporation of the monomers into the polymer network and their effects on MIP recognition properties need to be fully understood. Future improvements in computational protocols would also throw light onto the reasons for discrepancy between the predicted and the experimental results of polymer performance that involve hydrophobic interactions. Future computational approaches would certainly help design artificial receptors and generate knowledge that could help in better understanding of very important molecular interactions in biological systems.

4 Acronyms and Further Descriptions

ΔE	Binding energy
2-VP	2-Vinyl pyridine
4-VP	4-Vinyl pyridine

"ab-initio"	Latin term meaning 'from the beginning'
AA	Acrylic acid
Acceryls DS Viewer	Modeling and simulation tools for drug discovery
Agile molecule	Is a 3 Dimensional molecular viewer which shows molecular models and provides geometry editing capabilities
ALM	Allylamine
AMBER	Assisted Model Building with Energy Refinement refers to a MM force field for the simulation of biomolecules and a package of molecular simulation programs.
AMPSA	2-acrylamido-2-methyl-1-propanesulfonic acid
B3LYP	Becke 3-Parameter, Lee, Yang and Parr, a density functional method
Bite and Switch	'Bite-and-Switch' is defined in terms of polymer's ability to bind the template (bite) and generate the signal (switch)
BLAs	β-lactams
B-Me	Biotin methyl ester
CAChe MOPAC	A general-purpose semiempirical molecular orbital package for the study of chemical structures and reactions
Cerius	A software to visualize structures, predict the properties and behavior of chemical systems, refine structural models, (Molecu lar Simulations Inc.)
Chem 3D	A software that provides visualization and display of molecular surfaces, orbitals, electrostatic potentials, charge densities and spin densities (http://www.cambridgesoft.com/)
DFT	Density functional theory
Dielectric constant	Is a measure of the ability of a material to store a charge from an applied electromagnetic field and then transmit that energy
DMAEM	Dimethyl aminoethyl methacrylate
DOCK	Program that addresses the problem of "docking" molecules to each other. It explores ways in which two molecules, such as a drug and an enzyme or protein receptor, might fit together
EGDMA	Ethylene glycol dimethacrylate
ELISA	Enzyme linked immuno sorbent assay
GAMESS	General Atomic and Molecular Electronic Structure System; a general ab initio quantum chemistry package that can compute wave functions ranging from RHF, ROHF, UHF, GVB, and MCSCF
Gibbs free energy	The chemical potential that is minimized when a system reaches equilibrium at constant pressure and temperature

GRID	Is a computational procedure for detecting energetically favorable binding sites on molecules of known structure. The energies are calculated as the electrostatic, hydrogen-bond and Lennard Jones interactions of a specific probe group with the target structure. *(Peter Goodford, Molecular Discovery Ltd)*
Guassian	"Ab initio" electronic structure program that originated in the research group of People at Carnegie-Melon. Calculate structures, reaction transition states, and molecular properties. (http://www.gaussian.com)
Guassview	Graphical user interface (GUI) designed for use with Gaussian for easier computational analysis
HEM	Hydroxyethyl methacrylate
His	Histidine
HOOK	Linker search for fragments placed by MCSS
HO-PCBs	Hydroxy polychlorinated biphenyls
HPLC	High performance liquid chromatography
HVA	Homovanillic acid
HyperChem	A molecular modeling package for windows
IA	Itaconic acid
k'	Retention factor
Leapfrog™	Is a component of the SYBYL™ software package (Tripos) and is a second-generation de novo drug discovery program that allows for the evaluation of potential ligand structures
LEGEND	Atom-based, stochastic search
Ligbuilder	A general purpose program for structure-based drug design
LUDI	Fragment-based, combinatorial search
MAA	Methacrylic acid
Materials Studio	A software for modeling and simulation of crystal structure, polymer properties, and structure–activity relationships (http://www.accelrys.com/products/mstudio)
MBAA	N,N′-methylenebisacrylamide
MD	Molecular dynamics
MIC	Molecularly imprinted catalysis
MIP	Molecularly imprinted polymer
MM	Molecular mechanics
MMA	Methylmethacrylate
MMFF94	A tool for conformational searching of highly flexible molecules

MOE	Molecular Operating Environment is a software system designed specifically for computational chemistry
Monte Carlo	An algorithm that computes based on repeated random sampling to arrive at results
MOPAC AM1	AM1 is used in the electronic part of the calculation to obtain molecular orbitals, the heat of formation and its derivative with respect to molecular geometry. MOPAC calculates the vibrational spectra, thermodynamic quantities, isotopic substitution effects and force constants for molecules, radicals, ions, and polymers
NAM	A scal able molecular dynamics code that can be run on the Beowulf parallel PC cluster used to run molecular dynamics simulations on selected molecular systems
NIP	Nonimprinted polymer
NVT-MD	Molecular dynamics performed under constant number of atom, volume, and temperature ensemble
OPA	*o*-phthalic dialdehyde
OscailX	A molecular modeling software from National University of Ireland. (http://www.ucg.ie/cryst/software.htm)
OTA	Ochratoxin A
PCFF	Polymer consistent force field
PCM	Polarizable continuum model
PCModel	Is a structure building, manipulation and display program which uses molecular mechanics and semiempirical quantum mechan ics to optimize geometry. Available on PC (DOS and Windows), Macintosh, SGI, Sun and IBM/RS computers. *(Kevin Gilbert, Serena Software)*
PenG	Penicillin G
pKa	Ionization constant
PRO-LIGAND	Fragment-based search
Qm	Mean absolute atomic charge
QM	Quantum mechanics
RECON	An algorithm for the rapid reconstruction of molecular charge densities and charge density-based electronic properties of molecules, using atomic charge density fragments precomputed from ab initio wave functions. The method is based on Bader's quantum theory of atoms in molecules.
RESP	Atomic partial charge assignment protocol
SDIM	Sulfadimethoxine
SHAKE	A molecular dynamics algorithm

Simulated annealing	A method that simulates the physical process of annealing, where a material is heated and then cooled leading to optimization.
SMZ	Sulfamethazine
SPROUT	Fragment-based, sequential growth, combinatorial search
SYBYL™	A molecular modeling and visualization package permitting construction, editing, and visualization tools for both large and small molecules (http://www.tripos.com)
T:M:X ratio	Template monomer crosslinker ratio
TAE	Transferable atom equivalent
TFMAA	2-(trifluoromethyl) acrylic acid
THO	Theophylline
UAHF	United Atom Hartree–Fock
Van-der Waals	Weak intermolecular forces that act between stable molecules
VI	1-vinylimidazole

References

1. Naidoo, K. J.; Chen, Y-J.; Jansson, J. L. M.; Widmalm, G.; Maliniak, A., Molecular Properties Related to the Anomalous Solubility of beta Cyclodextrin, A. *J. Phys. Chem. B* **2004**, *108*, 4236–4238

2. Chen, W.; Huang, J.; Gilson, M. K., Identification of symmetries in molecules and complexes, *J. Chem. Inf. Comput. Sci.* **2004**, *44*, 1301–1313

3. Zheng, X. M.; Lu, W. M.; Sun, D. Z., Enthalpy and entropy criterion for the molecular recognize of some organic compounds with beta cyclodextrin, *Acta. Phys-Chim. Sin.* **2001**, *17*, 343–347

4. Rekharsky, M. V.; Inoue, Y., Complexation thermodynamics of cyclodextrins, *Chem. Rev.* **1998**, *98*, 1875–1917

5. Liu, L.; Guo. Q. X., The driving forces in the inclusion complexation of cyclodextrins, *J. Incl. Phenom. Macro.* **2002**, *42*, 1–14

6. Cag&i;n, T.; Wang, G.; Martin, R.; Breen, N.; Goddard III, W.A., Molecular modelling of dendrimers for nanoscale applications, *Nanotechnology.* **2000**, *11*, 77–84

7. Cruz-Morales, J.A.; Guadarrama, P., Synthesis, characterization and computational modeling of cyclen substit uted with dendrimeric branches. Dendrimeric and macrocyclic moieties working together in a collective fashion, *J.Mol. Str.* **2005**, *779*, 1–10

8. Gorman, C.B., Dendritic encapsulation as probed in redox active core dendrimers *Comptes Rendus Chimie.* **2003**, *6*, 911–918

9. Amatore, C.; Oleinick A.; Svir, I., Diffusion within nanometric and micrometric spherical-type domains limited by nanometric ring or pore active interfaces. Part 1: conformal mapping approach, *J. Electroanal. Chem.* **2005**, *575*, 103–123

10. Cagin, T.; Wang, G; Martin, R.; Zamanakos, G.; Vaidehi, N.; Mainz, D. T.; Goddard, W. A., Multiscale modeling and simulation methods with applications to dendritic polymers, *Comp Theor. Polym. Sci.* **2001**, *11*, 345–356

11. Casnati, A.; Della Ca' N.; Sansone, F.; Ugozzoli, F.; Ungaro, R., Enlarging the size of calix[4] arene-crowns-6 to improve Cs^+/K^+ selectivity: a theoretical and experimental study, *Tetrahedron.* **2004**, *60,* 7869–7876

12. Majerski, K.M.; Kragol, G., Design, synthesis and cation-binding properties of novel adamantane and 2-oxaadamantane-containing crown ethers, *Tetrahedron.* **2001**, *57,* 449–457

13. Furer, V.L.; Borisoglebskaya, E.I.; Kovalenko, V. I., Modelling conformations and IR spectra of p-tert-butylthiacalix[4]arene tetraester using DFT method, *J. Mol. Str.* **2006**, *825,* 38–44

14. Inese, S. I.; Smithrud, D. B., Condensation reactions of calix[4]arenes with unprotected hydroxyamines, and their resulting water solubilities *Tetrahedron.* **2001**, *57,* 9555–9561

15. Korochkina, M.; Fontanella, M.; Casnati, A.; Arduini, A.; Sansone, F.; Ungaro, R.; Latypov, S.; Kataev, V.; Alfonsov, V., Synthesis and spectroscopic studies of isosteviol-calix[4]arene and -calix[6]arene conjugates, *Tetrahedron.* **2005**, *61,* 5457–5463

16. Chen,W.; Gilson, M. K., ConCept: de novo design of synthetic receptors for targeted ligands, *J. Chem. Inf. Mod.* **2007**, *47,* 425–434

17. Luo, R.; Gilson, M. K., Synthetic adenine receptors: Direct calculation of binding affinities. *J. Amer. Chem. Soc.* **2000**, *122,* 2934–2937

18. Gilson, M. K.; Given, J. A; Head, M. S.; A new class of models for computing receptor-ligand binding affinities, *Chem. Biol.* **1997**, *4,* 87–92

19. Gilson, M. K.; Given, J. A.; Bush, B.; McCammon, J. A., The statistical-thermodynamic basis for computation of binding affinities. A critical review, *Biophys. J.* **1997**, *72,* 1047–1069

20. Looger, L. L.; Dwyer, M. A.; Smith, J. J.; Hellinga, H. W., Computational design of receptor and sensor proteins with novel functions, *Nature.* **2003**, *423,* 185–190

21. Moore, G.; Levacher, V.; Bourguignon, J.; Dupas, G., Synthesis of a heterocyclic receptor for carboxylic acids, *Tetrahed. Lett.* **2001**, *42,* 261–263

22. Jaramillo, A.; Tortosa, P.; Rodrigo, G.; Suarez, M.; Carrera, J., Computational design of proteins with new functions, *BMC Syst. Biol.* **2007**, *1,* S15doi:10.1186/1752-05091-S1-S15

23. Evans, S.M.; Burrows, C.J.; Venanzi, C.A., Design of cholic acid macrocycles as hosts for molecular recognition of monosaccharides, *J. Mol. Str,* **1995**, *334,* 193–205

24. Cain, J. P.; Mayorov, A. V.; Cai, M.; Wang, H.; Tan, B.; Chandler, K.; Lee, Y.; Petrov, R. R.; Trivedi, D.; Hruby, V. J., Design, synthesis, and biological evaluation of a new class of small molecule peptide mimetics targeting the melanocortin receptors, *Bioorg. Med. Chem. Lett.* **2000**, *16,* 5462–5467

25. Biagi, G.; Bianucci, A. M.; Coi, A.; Costa, B.; Fabbrini, L.; Giorgi, I.; Livi, O.; Micco, I.; Pacchini, F.; Santini, E.; Leonardi, M.; Nofal, F.A.; Salerni, O.L.; Scartoni, V., 2,9-Disubstituted-N^6-(arylcarbamoyl)-8-azaadenines as new selective A_3 adenosine receptor antagonists: Synthesis, biochemical and molecular modelling studies, *Bioorg. Med. Chem.* **2005**, *13,* 4679–4693

26. Claramunt, R. M.; Herranz, F.; María, M. D. S.; Pinilla, E.; Torres, M. R.; Elguero, J., Molecular recognition of biotin, barbital and tolbutamide with new synthetic receptors, *Tetrahedron.* **2005**, *61,* 5089–5100

27. Cristalli, G.; Costanzi, S.; Lambertucci, C.; Taffi, S.; Vittori, S.; Volpini, R., Purine and deazapurine nucleosides: synthetic approaches, molecular modelling and biological activity, *Il Farmaco.* **2003**, *58,* 193–204

28. Albert, J. S.; Peczuh, M. W.; Hamilton, A. D., Design, Synthesis and evaluation of synthetic receptors for the recognition of aspartate pairs in an α-helical conformation, *Bioorg. Med. Chem.* **1997**, *5,* 1455–1467

29. Napolitano, E.; Fiaschi, R.; Herman, L. W.; Hanson, R. N., Synthesis and estrogen receptor binding of (17α,20E)- and (17α,20Z)-21-phenylthio- and 21-phenylseleno-19-norpregna-1,3,5(10),20-tetraene-3,17β-diols, *Steroids.* **1996**, *61,*384–389

30. Carrasco, M. R.; Still, W.C., Engineering of a synthetic receptor to alter peptide binding selectivity, *Chem. Biol.* **1995**, *2,* 205–212

31. Brizzi, A.; Cascio, M. G.; Brizzi, V.; Bisogno, T.; Dinatolo, M. T.; Martinelli, A.; Tuccinardi, T.; Marzo, V. D., Design, synthesis, binding, and molecular modeling studies of new potent ligands of cannabinoid receptors, *Bioorg. Med. Chem.* **2007**, *15*, 5406–5416
32. Menuel, S.; Joly, J.P.; Courcot, B.; Elysée, J.; Ghermani, N.E.; Marsura, A., Synthesis and inclusion ability of a bis-β-cyclodextrin pseudo-cryptand towards Busulfan anticancer agent, *Tetrahedron.* **2007**, *63*, 1706–1714
33. Cain, J.P.; Mayorov, A.V.; Cai, M.; Wang, H.; Tan, B.; Chandler, K.; Lee, Y.; Petrov, Y.R.; Trivedi, D.; Hruby, V.J., Design, synthesis, and biological evaluation of a new class of small molecule peptide mimetics targeting the melanocortin receptors, *Bioorg. Med. Chem. Lett.* **2006**, *16*, 5462–5467
34. Lassila, J.K.; Privett, H.K.; Allen, B.D.; Mayo, S.L., Combinatorial methods for small-molecule placement in computational enzyme design, *PNAS.* **2006**, *45*, 16710–16715
35. Vleeschauwer, M.D.; Vaillancourt, M.; Goudreau, N.; Guindon, Y.; Gravel, D., Design and synthesis of a new Sialyl Lewis X Mimetic: How selective are the selectin receptors? *Bioorg. Med. Chem. Lett.*, **2001**, *11*, 1109–1112
36. Moore, G.; Levacher, V.; Bourguignon, J.; Dupas, G., Synthesis of a heterocyclic receptor for carboxylic acids, *Tetrahed. Lett.* **2001**, *42*, 261–263
37. A New Generation of Chemical Sensors Based on MIPs, in Molecular Imprinting of Polymers; Piletsky, S.A.; Turner, A.P.F., Eds., Eurekah Publishers, 2006
38. VTT Information Technology, Tampere, Finland
39. Piletsky, S.A.; Piletska, E.V.; Chen, B.; Karim, K.; Weston, D.; Barrett, G.; Lowe, P.; Turner, A.P.F., Chemical grafting of molecularly imprinted homopolymers to the surface of microplates. Application of artificial adrenergic receptor in enzyme-linked assay for beta agonists determination, *Anal. Chem.* **2000**, *72*, 4381–4385
40. *Molecularly Imprinted Polymers, Man-Made Mimics of Antibodies and Their Practical Application in Analytical Chemistry*; Sellergren B.; Eds.; Elsevier, Amsterdam, The Netherlands, 2001, 59–70
41. Wulff, G., Molecular imprinting-a way to prepare effective mimics of natural antibodies and enzymes, *Stud. Surf. Sci. Catal.* **2002**, *141*, 35–44
42. Shea, K.J, Molecular imprinting of synthetic network polymers: the de-novo synthesis of macromolecular binding and catalytic sites, *Trends. Polym. Sci.* **1994**, *19*, 9–14
43. Janiak, D.S.; Kofinas, P., Molecular imprinting of peptides and proteins in aqueous media, *Anal. Bioanal. Chem. Anal. Bioanal. Chem.* **2007**, *389*, 399–404
44. Ansell, R.J.; Kriz, D.; Mosbach, K., Molecularly imprinted polymers for bioanalysis: chromatography, binding assays and biomimetic sensors, *Curr. Opin. Biotechnol.* **1996**, *7*, 89–94
45. Spivak, D.; Gilmore, M.A.; Shea, K.J., Evaluation of binding and origins of specificity of 9-ethyladenine imprinted polymers, *J. Am. Chem. Soc.* **1997**, *119*, 4388–4393
46. Lanza, F.; Sellergren B., The application of molecular imprinting technology to solid phase extraction, *Chromatographia.* **2001**, *53*, 599–611
47. Masque, N.; Marce R. M.; Borrull, F., Molecularly imprinted polymers: new tailor-made materials for selective solid-phase extraction. *Trends. Anal. Chem.*, **2001**, *20*,477–486
48. Wulff, G., Enzyme-like catalysis by molecularly imprinted polymer, *Chem. Rev.* **2002**, *102*, 1–27
49. Haupt, K.; Mosbach, K., Molecularly imprinted polymers and their use in biomimetic sensors, *Chem. Rev.* **2000**, *100*, 2495–2504
50. Kriz, D.; Ramstrom, O.; Mosbach, K., Molecular imprinting-new possibilities for sensor technology, *Anal. Chem.* **1997**, *69*, A345–A349
51. Kriz, D.; Ramström, O.; Svensson, A.; Mosbach, K., Introducing biomimetic sensors based on molecularly imprinted polymers as recognition elements, *Anal. Chem.* **1995**, *67*, 2142–2144
52. Jakusch, M.; Janotta, M.; Mizaikoff, B.; Mosbach, K.; Haupt, K., Molecularly imprinted polymers and infrared evanescent wave spectroscopy. A chemical sensors approach, *Anal. Chem.* **1999**, *71*, 4786–4791
53. Greene, N. T.; Shimizu, K. D., Colorimetric molecularly imprinted polymer sensor array using dye displacement, *J. Am. Chem. Soc.* **2005**, *127*, 5695–5700

54. Hennion, M.C.; Pichon, V., Immuno-based sample preparation for trace analysis, *J. Chromatogr. A.* **2003**, *1000*, 29–52
55. Wei, S.; Molinelli, A.; Mizaikoff, B., Molecularly imprinted micro and nanospheres for the selective recognition of 17β-estradiol, *Biosens. Bioelectron.* **2006**, *21*, 1943–1951
56. Vidyasankar S.; Ru M.; Arnold F.H., Molecularly imprinted ligand-exchange adsorbents for the chiral separation of underivatized amino acids, *J. Chromatogr A.* **1997**, *775*, 51–63
57. Takeuchi T.; Fukuma D.; Matsui J., Combinatorial molecular imprinting: an approach to synthetic polymer receptors, *Anal. Chem.* **1999**, *71*, 285–290
58. Takeuchi, T.; Dobashi, A.; Kimura, K., Molecular imprinting of biotin derivatives and its application to competitive binding assay using nonisotopic labeled ligands, *Anal. Chem.* **2000**, *72*, 2418–2422
59. Takeuchi, T.; Fukuma, D.; Matsui, J.; Mukawa, T., Combinatorial molecular imprinting for formation of atrazine decomposing polymers, *Chem. Lett.* **2001**, *1*, 530–531
60. Lanza, F.; Sellergren B., Method for synthesis and screening of large groups of molecularly imprinted polymers, *Anal. Chem.* **1999**, *71*, 2092–2096
61. Lanza, F.; Hall, A.J.; Sellergren, B.; Bereczki, A.; Horvai, G.; Bayoudh, S.; Cormack, P. A. G.; Sherrington D.C., Development of a semiautomated procedure for the synthesis and evaluation of molecularly imprinted polymers applied to the search for functional monomers for phenytoin and nifedipine, *Anal. Chim. Acta.* **2001**, *435*, 91–106
62. Chen, B.; Day, R. M.; Piletsky, S. A.; Piletska, O.; Subrahmanyam, S.; Turner, A. P.F., *Rational design of MIPs using computational approach.* UK Patent, GB0001513–1, 2000
63. Piletsky, S.; Karim, K.; Piletska, E. V.; Day, C. J.; Freebairn, K. W.; Legge, C.; Turner, A. P. F., Recognition of ephedrine enantiomers by molecularly imprinted polymers designed using a computational approach, *Analyst.* **2001**, *126*, 1826–1830
64. Subrahmanyam, S.; Piletsky, S.; Piletska, E.; Chen, B.; Karim, K.; Turner, A. P. F., Bite-and-Switch approach using computationally designed molecularly imprinted polymers for sensing of creatinine, *Biosens. Bioelectron.* **2001**, *16*, 631–637
65. Chianella I.; Lotierzo M.; Piletsky S. A.; Tothill I. E.; Chen B.; Karim K.; Tuner A. P. F., Rational design of a polymer specific for microcystin-LR using a computational approach, *Anal. Chem.* **2002**, *74*, 1288–1293
66. Pavel, D.; Lagowski, J.; Lepage, C. J., Computationally designed monomers for molecular imprinting of chemical warfare agents e Part V, *Polymer.* **2006**, *47*, 8389–8399
67. Wu, L.; Zhu, K.; Zhao, M.; Li, Y., Theoretical and experimental study of nicotinamide molecularly imprinted polymers with different porogens, *Anal. Chim. Acta.* **2005**, *549*, 39–44
68. Dong, W.; Yan, M.; Zhang, M.; Liu, Z.; Li Y., A computational and experimental investigation of the interaction between the template molecule and the functional monomer used in the molecularly imprinted polymer, *Anal. Chim. Acta.* **2005**, *542*, 186–192
69. Farrington, K; Magner, E.; Regan, F., Predicting the performance of molecularly imprinted polymers, Selective extraction of caffeine by molecularly imprinted solid phase extraction, *Anal. Chim. Acta.* **2006**, *566*, 60–68
70. Baggiani, C.; Baravalle, P.; Anfossi, L.; Tozzi, C., Comparison of pyrimethanil-imprinted beads and bulk polymer as stationary phase by non-linear chromatography, *Anal. Chim. Acta.* **2005**, *542*, 125–134
71. Chapuis, F.; Pichon, V.; Lanza, F.; Sellergren, S.; Hennion, M. C., Optimization of the class-selective extraction of triazines from aqueous samples using a molecularly imprinted polymer by a comprehensive approach of the retention mechanism, *J. Chromatogr A.* **2003**, *999*, 23–33
72. Meng, Z.; Yamazaki, T.; Sode, K., A molecularly imprinted catalyst designed by a computational approach in catalysing a transesterification process, *Biosens. Bioelectron.* **2004**, *20*, 1068–1075

73. Benito-Peña, E.; Mar;' C. Bondi,M.; Aparicio, S.; Orellana, G.; Cederfur, J.; Kempe, M., Molecular Engineering of Fluorescent Penicillins for Molecularly Imprinted Polymer Assays, *Anal. Chem.* **2006**, *78*, 2019–2027

74. Rathbone, D.L.; Ge, Y., Selectivity of response in fluorescent polymers imprinted with *N*1-benzylidene pyridine-2-carboxamidrazones, *Anal. Chim. Acta.* **2001**, *435*, 129–136

75. Davies, M. P.; De Biasi, V.; Perrett, D., Approaches to the rational design of molecularly imprinted polymers, *Anal. Chim. Acta.* **2004**, *504*, 7–14

76. Nantasenamat, C.; Naenna, T.; Ayudhya, C. I. N.; Prachayasittikul, V., Quantitative prediction of imprinting factor of molecularly imprinted polymers by artificial neural network, *J. Comp-Aided Mol.Des.* **2005**, *19*, 509–524

77. Nantasenamat, C.; Ayudhya, C. I.N.; Naenna, T.; Prachayasittikul, V., Quantitative structure-imprinting factor relationship of molecularly imprinted polymers, *Biosens Bioelectron.* **2007**, *22*, 3309–3317

78. Chianella, I.; Karim, K.; Piletska, E. V.; Preston, C.; Piletsky, S. A., Computational design and synthesis of molecularly imprinted polymers with high binding capacity for pharmaceutical applications-model case: Adsorbent for abacavir, *Anal. Chim. Acta.* **2006**, *559*, 73–78

79. Piletska, E. V.; Piletsky, S. A.; Karim, K.; Terpetschnig, E.; Turner, A. P. F.; "Biotin-specific synthetic receptors prepared using molecular imprinting", *Anal. Chim. Acta.* **2004**, *504*, 179–183

80. Barragán, I. S.; Karim, K.; Fernández, J. M. C.; Piletsky, S. A.; Medel, A. S., A molecularly imprinted polymer for carbaryl determination in water, *Sens. Actuators B.* **2007**, *123*, 798–804

81. Piletska, E.V.; Romero-Guerra, M.; Chianella, I.; Karim, K.; Turner, A. P. F.; Piletsky, S. A., Towards the development of multisensor for drugs of abuse based on molecular imprinted polymers, *Anal.Chim. Acta.* **2005**, *542*, 111–117

82. Monti, S.; Cappelli, C.; Bronco, S.; Giusti, P.; Ciardelli, G., Towards the design of highly selective recognition sites into molecular imprinting polymers: A computational approach, *Biosens. Bioelectron.* **2006**, *22* 153–163

83. Breton, F.; Rouillon, R.; Piletska, E.V.; Karim, K.; Guerreiro, A.; Chianella, I.; Piletsky, S. A.; Virtual imprinting as a tool to design efficient MIPs for photosynthesis-inhibiting herbicides, *Biosens. Bioelectron.* **2007**, *22*, 1948–1954

84. Piletska, E. V.; Turner, N. W.; Turner, A. P. F.; Piletsky, S. A, Controlled release of the herbicide simazine from computationally designed molecularly imprinted polymers, *J. Controlled Release.* **2005**, *108*, 132–139

85. Piletsky, S.; Piletska, E.V.; Karim, K.; Foster, G.; Legge, C.; Turner, A. P. F., Custom synthesis of molecular imprinted polymers for biotechnological application Preparation of a polymer selective for tylosin, *Anal. Chim. Acta.* **2004**, *504*, 123–130

86. Yoshida, M.; Hatate, Y.; Uezu, K.; Goto, M.; Furusaki, S., Chiral-recognition polymer prepared by surface molecular imprinting technique, *Colloids Surf A.* **2000**, *169*, 259–269

87. Toorisaka, E.; Uezua, K.; Goto, M.; Furusaki, S., A molecularly imprinted polymer that shows enzymatic activity, *Biochem. Eng. J.* **2003**, *14*, 85–91

88. Farrington, K.; Regan, F., Investigation of the nature of MIP recognition: The development and characterisation of a MIP for Ibuprofen, *Biosens Bioelectron.* **2007**, *22*, 1138–1146

89. Castro, B.; Whitcombe, M. J.; Vulfson, E. N.; Duhalt, R. V.; Bárzana,E., Molecular imprinting for the selective adsorption of organosulphur compounds present in fuels, *Anal. Chim. Acta.* **2001**, *435*, 83–90

90. Liu, Y.; Wang, F.; Tan, T.; Lei, M., Study of the properties of molecularly imprinted polymers by computational and conformational analysis, *Anal. Chim. Acta.* **2007**, 581, 137–146

91. Kubo, T.; Matsumoto, H.; Shiraishi, F.; Nomachia, M.; Nemotoa, K.; Hosoya, K.; Kaya, K., Selective separation of hydroxy polychlorinated biphenyls (HO-PCBs) by the structural recognition on the molecularly imprinted polymers: Direct separation of the thyroid hormone active analogues from mixtures, *Anal. Chim. Acta.* **2007**, *589*, 180–185

92. Chianella, I.; Lotierzo, M.; Piletsky, S. A.; Tothill, I. E.; Chen, B.; Karim, K.; Turner, A. P. F., Rational design of a polymer specific for microcystin-LR using a computational approach, *Anal. Chem.* **2002**, *74*, 1288–1293

93. Turner, N. W.; Piletska, E. V.; Karim, K.; Whitcombe, M.; Malecha, M.; Magan, N.; Baggiani, C.; Piletsky, S. A., Effect of the solvent on recognition properties of molecularly imprinted polymer specific for ochratoxin A, *Biosens. Bioelectron.* **2004**, *20*, 1060–1067

94. Baggiani, C.; Anfossi, L.; Baravalle, P.; Giovannoli. C.; Tozzi, C., Selectivity features of molecularly imprinted polymers recognising the carbamate group, *Anal. Chim. Acta.* **2005**, *531*, 199–207

95. Pavel, D.; Lagowski, J., Computationally designed monomers and polymers for molecular imprinting of theophylline and its derivatives, Part I, *Polymer.* **2005**, *46*, 7528–7542

96. Pavel, D.; Lagowski, J., Computationally designed monomers and polymers for molecular imprinting of theophylline, Part II, *Polymer,* **2005**, *46*, 7543–7556

97. Chapuis, F.; Pichon, V.; Lanza, F.; Sellergren, B.; Hennion, M.C., Retention mechanism of analytes in the solid-phase extraction process using molecularly imprinted polymers:-Application to the extraction of triazines from complex matrices, *J. Chromatogr. B* **2004**, *804*, 93–101

98. Dineiro, Y.; Menendez, M. I.; Lopez, M. C. B.; Jesus, M.; Lobo.C.; Ordieres, A. J. M.; Blanco, P. T., Computational predictions and experimental affinity distributions for a homovanillic acid molecularly imprinted polymer, *Biosens. Bioelectron.* **2006**, *22*, 364–371

99. Wu, L.; Li, Y., Study on the recognition of templates and their analogues on molecularly imprinted polymer using computational and conformational analysis approaches, *J. Mol. Recognit.* **2004**, *17*, 567–574

100. Fu, Q.; Sanbe, H.; Kagawa, C.; Kunimoto K.; Jun H., Uniformly Sized Molecularly Imprinted Polymer for (S)-Nilvadipine. Comparison of Chiral Recognition Ability with HPLC Chiral Stationary Phases Based on a Protein, *Anal. Chem.* **2003**, *75*, 191–198

101. Karim K.; Breton, F.; Rouillon, R.; Piletska E. V.; Guerreiro, A.; Chianella, I.; Piletsky, S. A., How to find effective functional monomers for effective molecularly imprinted polymers? *Adv. Drug Del. Rev.* **2005**, *57*, 1795–1808

102. Nicholls, I.A., Thermodynamic considerations for the design of and ligand recognition by molecularly imprinted polymers, *Chem. Lett.* **1995**, *24*, 1035–1036

103. Nicholls, I. A.; Andersson, H. S., Thermodynamic principles underlying molecularly imprinted polymer formulation and ligand recognition. In Molecularly Imprinted Polymers, Man-Made Mimics of Antibodies and Their Practical Application in Analytical Chemistry, Sellergren, Ed.; Elsevier, Amsterdam, The Netherlands 2001, 59–70

104. Nicholls, I. A.; Adbo, K.; Andersson, H. S.; Andersson, P. O.; Ankarloo, J.;.Hedin-Dahlstro(m, J.; Jokela, P.; Karlsson, J. G.; Olofsson, L.; Rosengren, J. P.; Shoravi, S.; Svenson, J.; Wikman, S., Can we rationally design molecularly imprinted polymers? *Anal. Chim. Acta.* **2001**, *435*, 9–18

105. Andrews, P. R.; Craik D. J.; Martin, J. L., Functional group contributions to drug-receptor interactions, *J. Med. Chem.* **1984**, *27*, 1648–1657

106. Williams, D. H.; Cox, J. P. L.; Doig, A. J.; Gardner, M.,; Gerhard, U.; Kaye, P. T.; Lal, A. R.; Nicholls, I. A.; Salter, C. J.; Mitchell, R. C.; Toward the semiquantitative estimation of binding constants. Guides for peptide–peptide binding in aqueous solution, *J. Am. Chem. Soc.* **1991**, *113*, 7020–7030

107. Searle, M.; Williams, D. H.; Gerhard, U., Partitioning of free energy contributions in the estimation of binding constants: Residual motions and consequences for amide–amide hydrogen bond strengths, *J. Am. Chem. Soc.* **1992**, *114*, 10704–10710

108. Holroyd, S. E.; Groves, P.; Searle, M. S.; Gerhard, U.; Williams, D. H, Rational design and binding of modified cell-wall peptides to vancomycin-group antibiotics: Factorising free energy contributions to binding, *Tetrahedron.* **1993**, *49*, 9171–9182

109. Nicholls, I. A., Towards the rational design of molecularly imprinted polymers, *J. Mol. Recognit.* **1998**, *11*, 79–82

110. Dixon, S.; Blaney, J.; Weininger, D., Characterizing and Satisfying the Steric and Chemical Restraints of Binding Sites. *Presented at the Third York Meeting*, **1993**

111. Payne, A. W. R.; Glen, R. C., Molecular recognition using a binary genetic search algorithm, *J. Mol. Graph.* **1993**, *11*, 74–91

112. Nishibata, Y.; Itai, A., Automatic creation of dug candidate structures based on receptor structure. Starting point for artificial lead generation, *Tetrahedron.* **1991**, *47*, 8985–8990

113. Böhm, H. J., The computer program LUDI: a new simple method for the de-novo design of enzyme inhibitors, *J. Comput. Aided Mol. Des.* **1992**, *6*, 61–78

114. Gillett, V. J.; Myatt, G.; Zsoldos Z.; Johnson, A. P. SPROUT, HIPPO and CAESA: tools for de novo structure generation and estimation of synthetic accessibility, *Perspect. Drug Discov. Des.* **1995**, *3*, 34–50

115. Eisen, M. B.; Wiley, D. C.; Karplus, M.; Hubbard, R. E., HOOK: a program for finding novel molecular architectures that satisfy the chemical and steric requirements of a macromolecule binding site, *Proteins* **1994**, *19*, 199–221

116. Clark, D.E.; Frenkel, D.; Levy, S. A.; Li, J.; Murray, C.W.; Robson, B.; Waszkowycz, B.; Westhead, D.R., PRO LIGAND: an approach to de novo molecular design. Application to the design of organic molecules, *J. Comput. Aided Mol. Des.* **1995**, *9*, 13–32

117. Luthra, R., Computer Based *De Novo* Design of Drug like Molecules, Lecture Seminar Report; **2006**, Department of Chemistry, University of Alabama, United States

118. Goodsell, D. S.; Olson, A. J., Automated docking of substrates to proteins by simulated annealing, *Proteins.* **1990**, *8*, 195–202

119. Pattabiraman, N.; Levitt, M.; Ferrin, T. E.; Langridge, R. Computer graphics in real-time docking with energy calculation and minimization, *J. Comput. Chem.* **1985**, *6*, 432–436

120. Piletska, E. V.; Piletsky, S. A.; Subrahmanyam, S.; Karim, K.; Turner, A. P. F., A new reactive polymer suitable for covalent immobilization and monitoring of the primary amines, *Polymer.* **2000**, *42*, 3603–3608

121. Baggiani, C.; Girauldi, G.; Vanni, A., A molecular imprinted polymer with recognition properties towards the carcinogenic mycotoxin ochratoxin A, *Bioseperation.* **2001**, *10*, 389–394

122. Gould, H.; Tobochnik, J., An Introduction to Computer Simulation Methods: Applications to Physical Systems Parts 1 and 2. Addison-Wesley, Reading MA, 1988, Translated into Russian

123. Case, D. A.; Pearlman, D. A.; Caldwell, J. W.; Cheatham III, T.E. C.; Wang, J.; Ross, W.S.; Simmerling, C. L.; Darden, T. A.; Merz, K. M.; Stanton, R. V.; Cheng, A. L.; Vincent, J. J.; Crouley, M.; Tsui, V.; Gohlke, H.; Radmer, R. J.; Duan, Y.; Pitera, J.; Massova, I.; Seibel, G. L.; Singh, U. C.; Weiner, P. K.; Kollman, P. A., **2002**, AMBER 7, University of California, San Francisco. US

124. Bronco, S.; Cappelli, C.; Monti, S., Understanding the interaction between collagen and modifying agents: a theoretical perspective, *J. Phys. Chem. B.* **2004**, *108*, 10101–10112

125. Monti, S.; Bronco, S.; Cappelli, C., Toward the supramolecular structure of collagen: a molecular dynamics approach, *J. Phys. Chem. B.* **2005**, *109*, 11389–11398

126. Hobza, P.; Kabelac, M.; Sponer, J.; Mejzlik, P.; Vondrasek, J., Performance of empirical potentials (AMBER, CFF95, CVFF, CHARMM, OPLS, POLTEV), semiempirical quantum chemical methods (AMI, MNDO/M, PM3, and ab initio Hartree–Fock method for interaction of DNA bases: comparison with nonempirical beyond Hartree–Fock results, *J. Comp. Chem.* **1997**, *18*, 1136–1150

127. Becke, A.D., Density functional thermochemistry. III. The role of exact exchange, *J. Chem. Phys,* **1993**, *98*, 5648–5652

128. Koch, W.; Holthausen, M. A. C., A Chemists Guide to Density functional Theory. Wiley-VCH, Berlin, 2001

129. Hehre, W. J.; Radom, L.; Schleyer, P. V.; Pople, J. A., Ab initio Molecular Orbital Theory. Wiley, New York, 1986

130. Cieplak, P.; Cornell, W. D.; Bayly, C. I.; Kollman, P. A., Application of the multi-molecule and multi-conformational RESP methodology to biopolymers: charge derivation for DNA, RNA, and proteins, *J. Comp. Chem.* **1995**, *16*, 1357–1377

131. Bayly, C. I.; Cieplak, P.; Cornell, W. D.; Kollman, P. A., A well-behaved electrostatic potential based method using charge restraints for deriving atomic charges: the RESP model, *J. Phys. Chem.* **1993**, *97*, 10269–10280

132. Berendsen, H. J.C.; Postma, J. P. M.; VanGunsteren, W. F.; DiNola, A.; Haak, J. R., Molecular dynamics with coupling to an external bath, *J. Chem. Phys.* **1984**, *81*, 3684–3690

133. Grabuleda, X.; Jaime, C.; Kollman, P. A., Molecular dynamics simulation studies of liquid acetonitrile: new six-site model, *J. Comput. Chem.* **2000**, *21*, 901–908

134. Ryckaert, J. P.; Ciccotti, G.; Berendsen, H. J. C., Numerical integration of the cartesian equations of motion of a system with constraints: molecular dynamics of *n*-alkanes, *J. Comp. Phys.* **1977**, *23*, 327–341

135. Ewing, T. J. A.; Kuntz, I. D., Critical evaluation of search algorithms used in automated molecular docking, *J. Comput. Chem.* **1997**, *18*, 1175–1189

136. Kuntz, I. D.; Blaney, J. M.; Oatley, S. J.; Langridge, R.; Ferrin, T. E., A geometric approach to macromolecule–ligand interactions, *J. Mol. Biol.* **1982**, *161*, 269–288

137. Kuntz, I. D.; Brooijmans, N., Molecular recognition and docking algorithms, *Annu. Rev. Biophys. Biomol. Struct.* **2003**, *32*, 335–373

138. Sun, H. J., Force field for computation of conformational energies, structures, and vibrational frequencies of aromatic polyesters, *Comput. Chem.* **1994**, *15*,752–768

139. Sun, H.; Mumby, S. J.; Maple, J. R.; Hagler, A. T., An ab initio CFF93 all-atom force field for polycarbonates, *J. Am. Chem. Soc.* **1994**, *116*, 2978–2987

140. Brostow W., Science of Materials. Wiley, New York, 1979

141. Stewart, J. J. P., Optimization of parameters for semiempirical methods II. Applications, *J. Comput. Chem.* **1989**, *10*, 221–264

142. Sherwood, P., 2006, Unpublished data on the website, Daresbury Laboratory, U.K

143. Frisch, M. J.; Trucks, G. W.; Schlegel, H. B.; Scuseria, G. E.; Robb, M. A.; Cheeseman, J. R.; Zakrzewski, V. G.; Montgomery, J. A.; Stratmann, R. E.; Burant, J. C.; Dapprich, S.; Millam, J. M.; Daniels, A. D.; Kudin, K. N.; Strain, M. C.; Farkas, O.; Tomasi, J.; Barone, V.; Cossi, M.; Cammi, R.; Mennucci, B.; Pomelli, C.; Adamo, C.; Clifford, S.; Ochterski, J.; Petersson, G. A.; Ayala, P. Y.; Cui, Q.; Morokuma, K.; Malick, D. K.; Rabuck, A. D.; Raghavachari, K., Foresman, J. B.; Cioslowski, J.; Ortiz, J. V.; Stefanov, B. B.; Liu, G.; Liashenko, A.; Piskorz, P.; Komaromi, I.; Gomperts, R.; Martin, R. L.; Fox, D. J.; Keith, T.; Al-Laham, M. A.; Peng, C. Y.; Nanayakkara, A.; Gonzalez, C.; Challacombe, M.; Gill, P. M. W.; Johnson, B.; Chen, W.; Wong, M. W.; Andres, J. L.; Gonzalez, C.; Head-Gordon, M.; Replogle, E. S.; Pople, J.A., Gaussian Inc., Pittsburgh, PA, 1998

144. Dineiro, Y.; Menendez, M. I.; Blanco-Lopez M. C.; Lobo-Castanon M. J.; Miranda-Ordieres A. J.; Tunon-Blanco P., Computational approach to the rational design of molecularly imprinted polymers for voltammetric sensing of homovanillic acid, *Anal. Chem.* **2005**, *77*, 6741–6746

145. Cossi, M.; Barone, V.; Cammi, R.; Tomasi, J., Ab initio study of solvated molecules: a new implementation of the polarizable continuum model, *Chem. Phys. Lett.* **1996**, *255*, 327–335

146. Frisch, M. J.; Trucks, G. W.; Schlegel, H. B.; Scuseria, G. E.; Robb, M. A.; Cheeseman, J. R.; Montgomery Jr., J. A.; Vreven, T.; Kudin, K. N.; Burant, J. C.; Millam, J. M.; Iyengar, S. S.; Tomasi, J.; Barone, V.; Mennucci, B.; Cossi, M.; Scalmani, G.; Rega, N.; Petersson, G. A.; Nakatsuji, H.; Hada, M.; Ehara, M.; Toyota, K.; Fukuda, R.; Hasegawa, J.; Ishida, M.; Nakajima, T.; Honda, Y.; Kitao, O.; Nakai, H.; Klene, M.; Li, X.; Knox, J. E.; Hratchian, H. P.; Cross, J. B.; Adamo, C.; Jaramillo, J.; Gomperts, R.; Stratmann, R. E.; Yazyev, O.; Austin, A. J.; Cammi, R.; Pomelli, C; Ochterski, J. W.; Ayala, P. Y.; Morokuma, K.; Voth, G.A.; Salvador, P.; Dannenberg, J. J.; Zakrzewski, V. G.; Dapprich, S.; Daniels, A. D.; Strain, M. C.; Farkas, O.; Malick, D. K.; Rabuck, A. D.; Raghavachari, K.; Foresman, J. B.; Ortiz, J. V.; Cui, Q.; Baboul, A. G.; Clifford, S.; Cioslowski, J.; Stefanov, B. B.; Liu, G; Liashenko, A. Nanayakkara, P.; Challacombe, M.; Gill, P. M. W.; Johnson, B.; Chen, W.; Wong, M. W.; Gonzalez, C.; Pople, J. A., Gaussian 03, Revision B. 05, Gaussian, Inc., Pittsburgh, PA, **2003**

147. Becke, A. D., Density-functional thermochemistry. III. The role of exact exchange, *J. Chem. Phys.* **1993**, *98*, 5648–5652

148. Lee, C. T.; Yang, W. T.; Parr, R. G., Development of the Colle-Salvetti correlation-energy formula into a functional of the electron density, *Phys. Rev.* **1988**, *B 37*,785–789

149. Alexander, C.; Davidson, L.; Hayes, W., Imprinted polymers: artificial molecular recognition materials with applications in synthesis and catalysis, *Tetrahedron.* **2003**, *59*, 2025–2057

150. Brady, L.; Brzozowski, A. M.; Derewenda, Z. S.; Dodson, E.; Dodson, G.; Tolley, S.; Turkenburg, J. P.; Christiansen, L.; Huge-Jensen, B.; Norskov, L.; Thim, L.; Menge, U., A serine protease triad forms the catalytic centre of a triacylglycerol lipase, *Nature* **1990**, *343*, 767–770

151. Schrag, J. D.; Li, Y. G.; Wu, S.; Cygler, M., Ser-His-Glu triad forms the catalytic site of the lipase from *Geotrichum candidum*, *Nature* **1991**, *351*, 761–764

152. Dewar, M. J. S.; Zoebisch, E. G.; Healy, E. F.; Stewart, J. I. P., AM1: A new general purpose quantum mechanical molecular model, *J. Am. Chem. Soc.* **1985**, *107*, 3902–3909

153. Schmidt, M. W.; Baldridge, K. K.; Boatz, J. A.; Elbert, S. T.; Gordon, M. S.; Jensen, J. H.; Koseki, S.; Matsunaga, N.; Nguyen, K. A.; Su, S. J.; Windus, T. L.; Dupuis, M.; Montgomery, J. A., General atomic and molecular electronic structure system, *J. Comput. Chem.* **1993**, *14*, 1347–1363

154. Araujo, P.W.; Brereton, R.G., Experimental design III. Quantification, *Trends Anal. Chem.* **1996**, *15*, 156–163

155. Box, G. E. P.; Hunter, W. G.; Hunter, J.S., Statistics for Experimenters, Wiley, New York, **1978**

156. Steinke, J.; Sherrington, D. C.; Dunkin, I. R., Imprinting of Synthetic Polymers Using Molecular Templates, Advances in Polymer Science, Springer, Berlin, **1995**

157. Koeber, R.; Fleischer, C.; Lanza, F.; Boos, K. S.; Sellergren, B.; Barcelo, D., *Anal. Chem.* **2001**, *73*, 2437–2444

158. Molinelli, A.; Weiss, R.; Mizaikoff, B., Advanced solid phase extraction using molecularly imprinted polymers for the determination of quercetin in red wine, *J. Agric. Food Chem.* **2002**, *50*, 1804–1808

159. Kempe, H.; Kempe, M., Novel methods for the synthesis of molecularly imprinted polymer bead libraries, *Macromol. Rapid Commun.* **2004**, *25*, 315–320

160. Mijangos, I.; Villoslada, F. N.; Guerreiro, A.; Piletska, E.V.; Chianella, I.; Karim, K.; Turner, A. P. F.; Piletsky, S. A., Influence of initiator and different polymerisation conditions on performance of molecularly imprinted polymers, *Biosenors Bioelectron.* **2006**, *22*, 381–387

161. Araujo, P. W.; Brereton. R. G, Experimental design I. Screening, *Trends Anal. Chem.* **1996**, *15*, 26–31

162. Araujo P. W.; Brereton R. G., Experimental design II. Optimization, *Trends Anal. Chem.* **1996**, *15*, 63–77

Chapter 7
Experimental Combinatorial Methods in Molecular Imprinting

Börje Sellergren, Eric Schillinger, and Francesca Lanza

Abstract Combinatorial techniques to synthesize and evaluate molecularly imprinted polymer sorbents for various molecular recognition based applications are reviewed. Various techniques to screen for binding and imprinting effects are discussed with reference to the application for which the sorbents are intended. State of the art high throughput synthesis techniques are outlined, and examples of how these have been used to identify suitable polymer building blocks or to optimize prepolymerization compositions to improve the molecular recognition properties in a given matrix are given.

1 Introduction

Molecular imprinting is an increasingly adopted technique for solving selectivity problems in a number of chemistry disciplines. In particular, the ability of the imprinted polymers to recognize their targets with receptors like affinities and selectivities are attractive in this regard.[1,2] During the last decade, imprinting has developed into a mature discipline allowing polymeric receptors to be prepared for small and more recently large molecules with a range of molecular recognition-based applications.

Molecular imprinting comprises the following key steps (Fig. 7.1): (1) A template molecule, exhibiting some resemblance with the target molecule, is mixed with functional monomer(s) in solution; (2) Polymerization is then initiated in presence of an excess of a cross-linking monomer resulting in a three-dimensional cross-linked porous network polymer; (3) The template is extracted or cleaved from the polymer matrix leading to a molecularly imprinted polymer (MIP) capable of sophisticated molecular recognition. Prior to step (3), the MIP is typically crushed and sieved to obtain a desired size fraction of particulate material. This particulate material can be packed into a chromatographic column and used for chromatographic separation of the template from other components of a mixture with similar structure or functionality.

B. Sellergren (✉), E. Schillinger, and F. Lanza
INFU, Universität Dortmund, Otto Hahn Strasse 6, 44221 Dortmund, Germany

R.A. Potyrailo and V.M. Mirsky (eds.), *Combinatorial Methods for Chemical and Biological Sensors*, DOI: 10.1007/978-0-387-73713-3_7,
© Springer Science+Business Media, LLC 2009

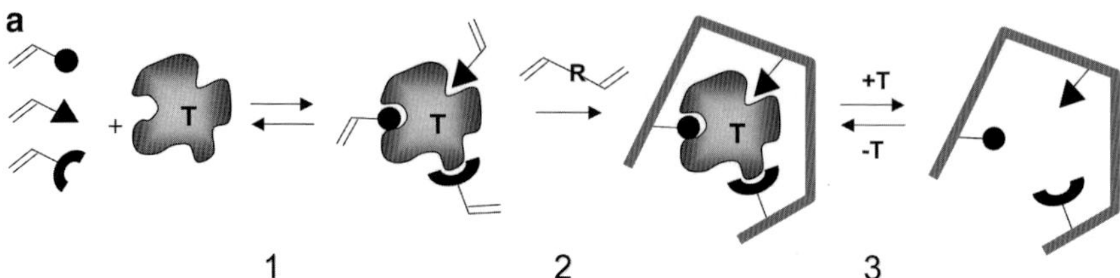

Fig. 7.1 Principle of molecular imprinting. *T* = template

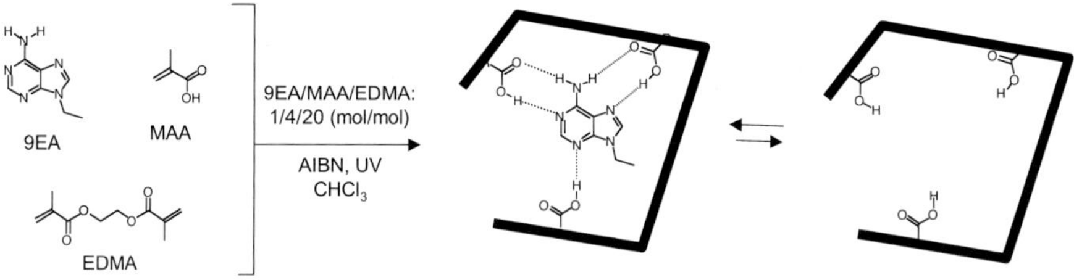

Fig. 7.2 Imprinting protocol based on methacrylic acid (MAA) as the functional monomer and ethylenglycoldimethacrylate (EDMA) as the cross-linking monomer. Template = 9-ethyladenine (9EA)

Currently, the most widely applied technique to generate molecularly imprinted binding sites is represented by the "noncovalent" route. This makes use of noncovalent self-assembly of the template with the functional monomer(s) prior to polymerization. After free radical polymerisation with a cross-linking monomer, followed by template extraction, rebinding is achieved via noncovalent interactions.

The polymerization is performed in the presence of a pore-forming solvent called a porogen. To stabilize the electrostatic interactions between the functional monomers and the template, the porogen is often chosen to be aprotic and of low to moderate polarity. This restriction limits the ability to fully control the morphology of the materials and excludes a number of two phase polymerization techniques to make beads or nanoparticles. One difficulty in the process is thus to find a compromise between these two aspects: materials morphology and molecular recognition. Methacrylic acid (MAA) is probably the most widely used functional monomer in imprinting, and it has been used to generate MIPs showing high selectivity for a large number of target molecules (Fig. 7.2).[3] Unfortunately, it does not provide a universal solution to imprinting problems and for several targets it fails to produce sufficient binding affinity and selectivity. In this case, other monomer or combinations of monomers need to be found. Another fundamental problem in imprinting concerns the selection or design of a template that is likely to produce a sufficient cross-reactivity for the desired target. Apart from functional and shape complementarity with the target, the template needs to be stable and soluble in the prepolymerization mixture and during polymerization. Commonly, the polymer matrix of the MIP needs to be optimized to minimize nonspecific binding of the target or matrix components in the sample matrix for which they are intended. This requires fine tuning of the type and amount of cross-linking monomer as well as the

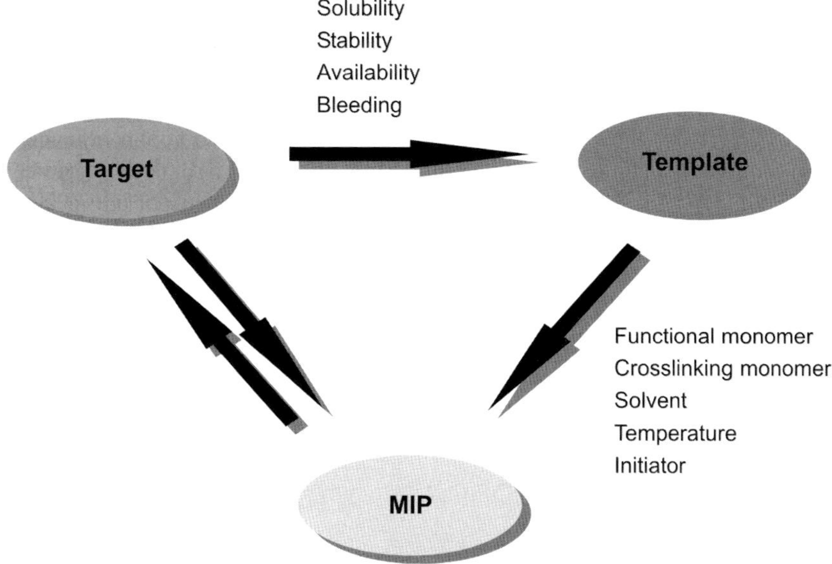

Fig. 7.3 Factors influencing the quality of the molecular recognition properties of MIPs

porogenic solvent. The factors influencing the binding site properties are summarized in Fig. 7.3.

Other urgent improvements concern the mass transfer properties of the materials and bleeding of residual traces of template remaining entrapped in the material.[4] The latter properties are influenced by the polymer morphology, which in turn is affected by the type and amount of cross-linking monomer, and the type of porogenic solvent.

Thus many parameters are influencing the materials' recognition and structural properties suggesting that finding an optimum MIP is not an easy task. Because of the cumbersome nature of the conventional method to prepare MIPs, only a very limited set of MIPs are synthesised and tested. This implies that for most reported MIPs considerably better performing materials could probably have been found given a practical and reliable way of producing and testing polymer libraries.

Thus, combinatorial synthesis approaches[5–11] or computational techniques,[12] allowing the main factors to be rapidly screened, have offered valuable tools in the development of new MIPs. Here, a large collection of MIPs (based on different functional monomers, comonomers, cross-linkers, solvents, initiators) are generated in a format facilitating the selection of individual members possessing promising properties. In both cases, the result will be a starting point for optimisation. Because of the many variables that control the imprinting process, the optimisation then needs to be achieved by a multivariate strategy such as Experimental Design or Modeling.

2   Parameters Influencing the Performance of MIPs

In the self-assembly approach to molecular imprinting (Fig. 7.1) studies have indicated that the solution structure of the monomer–template assemblies defines the subsequently formed binding sites.[3] In other words, the amount and quality of recognition sites in the MIP depends on the number and strength of specific interactions occurring between the template and the monomers in the prepolymerisation mixture (Fig. 7.4). These are in turn influenced by the quality of the solvent, cross-linking monomer, temperature, and pressure used in the polymerisation.

An important initial criterion is thus to enhance the stability of these interactions.[13] At the same time, the number of nonspecific binding sites will be minimized since there will be a reduction in the amount of free nonassociated functional monomer. However, varying these synthesis-related factors will also affect the morphology of the materials at a meso and macroscopic level,[14] which in turn determine their kinetic properties, e.g., diffusional mass transfer limitations, size exclusion effects, bleeding.

2.1 Choice of Template

The majority of templates used so far exhibit moderate to high solubility in the resulting polymerisation media and they (or their structural analogues) can therefore be used directly in the conventional procedure. However, for several reasons, a structural analogue to the target molecule is commonly preferred as template:

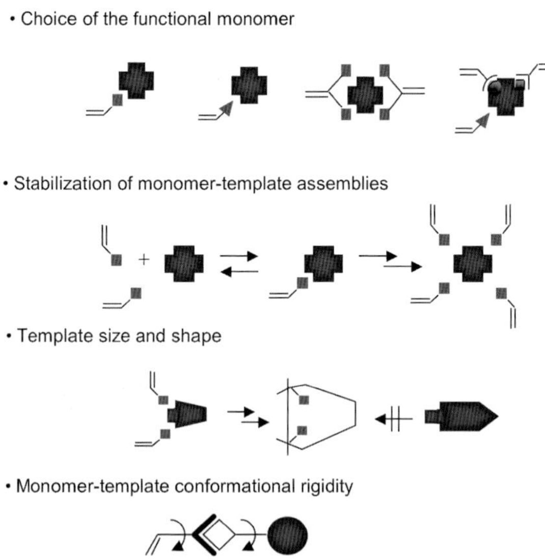

Fig. 7.4 Factors influencing the quality of the imprinted sites related to the monomer–template assemblies

- The target is unstable under the polymerisation conditions or inhibits the polymerisation.[15]
- The target is not available in sufficient quantities to make imprinting worthwhile.[16]
- To avoid template bleeding, i.e., the bleeding of low levels of template from the MIP, thus precluding SPE applications in trace analysis.[16–20]
- The target is insoluble or poorly soluble in the prepolymerization mixture.[21]
- The target is toxic or hazardous in other ways, e.g., explosive.

Therefore, a close target analogue is often chosen as template, leading to binding sites that also can capture the target analyte.[16–21] In addition to being a closely related analogue to the target, the template should possess functionalities that interact with the functional monomer, it should be stable and inert during the polymerization, and it should not interfere during the subsequent quantification of the analyte. The cross-reactivity of MIPs can be very hard to predict as shown by several surprising results. Screening for cross-reactive MIPs has therefore become an important part in the development of new receptors (vide infra).

2.2   Choice of Functional Monomers

The guiding principle in the choice of functional monomer is functional group complementarity (Fig. 7.4).[13] Thus, for templates containing acidic groups, Brönsted basic functional monomers (e.g., 2- or 4-vinylpyridine (VPy);diethyla minoethylmethacrylate (DEAEMA)) are preferably chosen whereas acidic functional monomers (e.g., methacrylic acid (MAA), trifluoromethylacrylic acid (TFM), itaconic acid (ITA)) are used to target Brönsted bases. For carboxylic acids and amides, high selectivities have also been observed with primary amide containing monomers (e.g., methacrylamide (MAAM)). Other neutral solvating monomers that commonly enhance the imprinting effect are N-vinylpyrrolidone (NVP) and hydroxyethylmethacrylate (HEMA).

In the case of poorly polar to apolar templates, with few polar interaction sites, it may instead be beneficial to use amphiphilic monomers, stabilizing the monomer–template assemblies by hydrophobic and Van der Waals forces or, in the case of extended π-systems, through charge transfer stabilization. Similarly to biological systems, a large number of complementary interactions is expected to increase the strength and fidelity of recognition. Thus, the use of comonomers that may orthogonally target different subunits of a complex template is often a successful strategy.[22,23] This leads to terpolymers or higher polymers and the screening and optimization task is considerably more complex here. As discussed later, comonomers can also be chosen to impart compatibility of the material with a certain solvent, i.e., to reduce nonspecific binding of target analogues or matrix components.[21,24]

The selection of suitable monomers may also be aided by molecular modeling and computational methods.[12] A virtual library of functional monomers is created and screened for all possible interactions the individual monomers may engage in with the template. Monomers with the highest binding scores are subsequently selected to produce full scale MIPs with hopefully superior recognition properties.

4-VP   2-VP   DEAEMA

MAA   TFM   ITA

MAAM   HEMA   NVP

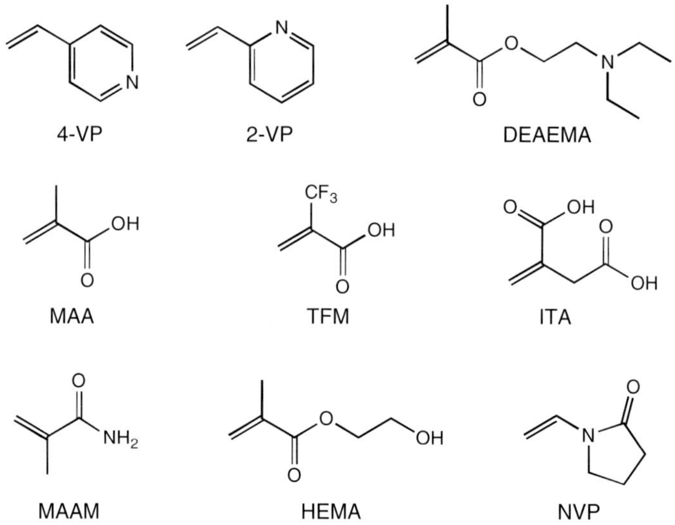

Finally, the search for the optimal structural motif, which can complement the template functionality, may also be guided by results from the areas of host–guest chemistry. Polymerizable host monomers designed to strongly complex a given guest in solution have been used to successfully imprint similar guests.[25,26] In some cases, this has resulted in a quantitative yield of high energy binding sites (stoichiometric imprinting), which can recognize the guest also under aqueous conditions.

2.3   Choice of Cross-Linking Monomer and Solvent

Ethyleneglycoldimethacrylate (EDMA) (Fig. 7.2) is the most commonly used cross-linker for the methacrylate-based systems, primarily because it provides materials with mechanical and thermal stability, good wettability in most rebinding media, and rapid mass transfer with good recognition properties. Except for trimethacrylate monomers, such as trimethylolpropane trimethacrylate (TRIM),[27] no other cross-linking monomer provides similar recognition properties for such a large variety of target templates. Divinylbenzene (DVB) has proven to be a superior matrix monomer for some templates but commonly in combination with other polymerization formats (e.g., emulsion, precipitation, or suspension). In the imprinting of peptides, TRIM has been used to prepare resins possessing a higher sample load capacity and a better performance than similar resins prepared using EDMA as cross-linker. Polar protic cross-linking monomers such as methylenediacrylamide (MDA)[28] and pentaerythritoltrimethacrylate (PETRA)[21] have found use for imprinting and applications in more polar solvents. In this context, the amide-based cross-linkers such as methylenediacrylamide (MDA) are interesting since they provide a polar, more protein like microenvironment. This has proven beneficial in the imprinting of polar templates with limited solubility in organic solvents.

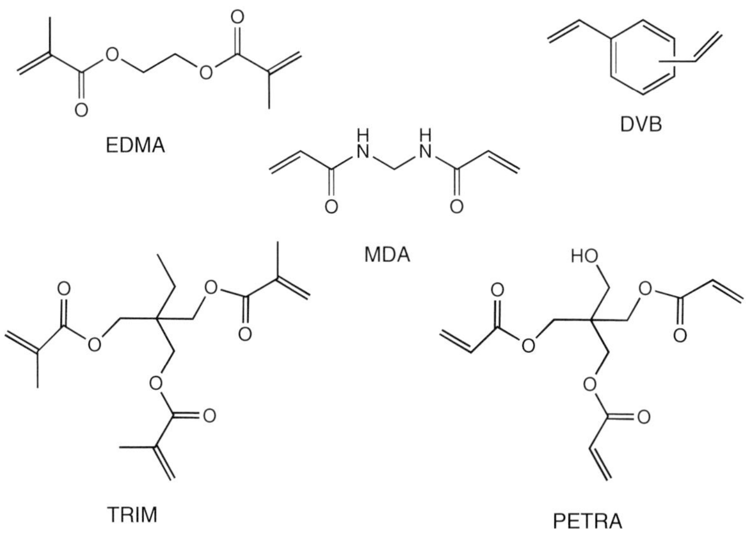

EDMA

DVB

MDA

TRIM

PETRA

The solvent should be capable of fully solubilizing the monomers and template in one pot. For monomer–template interactions stabilized by polar forces, non-protic solvents of low polarity should be chosen, since they are less likely to compete with the monomers for the template. The functional monomer–template complexes are often based on hydrogen bond interactions. If the solvent is a good hydrogen bond donor or acceptor, it will compete with the monomers and destabilize the complexes. For monomer–template systems stabilized by solvatophoic forces, more polar solvents and higher temperatures may be favorable.

2.4 Choice of Temperature and Initiator

In the case of electrostatic interactions, lower temperatures of polymerization are known to increase the stability of monomer–template assemblies. Therefore, low temperature initiators are commonly used (for example V-70, V-65 by Wako) or the polymerization is performed using photolabile initiators at lower temperature.[29] However, thermally initiated polymerization has a more general scope and may be preferable in certain monomer–template systems. After these initial considerations, initial screening experiments can be designed.

3 MiniMIPs and How to Evaluate Them

By reducing the batch size of conventional imprinting, the rapid synthesis and screening of large groups of MIPs becomes possible. Thus, significant time saving can be made by preparing only ~50 mg of imprinted polymer (MiniMIPs) on the bottom of small vials or of 96-well plates followed by in situ testing of the monoliths by equilibrium batch rebinding experiments.[29] The general procedure shown in Fig. 7.5 is essentially a scaled down version of the traditional monolith procedure (Fig. 7.2). Once a library of miniMIPs has been prepared, different options are available for assessing the imprinting effect and molecular recognition properties. Generally, we are interested in knowing how strongly and selectively our target compound interacts with the MIP in a given medium. Thus measurements reflecting these properties are needed.

3.1 Measuring the Imprinting Effect

Three different approaches to estimate the imprinting effect have been proposed (Fig. 7.6):

Measurement of template release: The rate or equilibrium value of the release of template from the freshly prepared polymer is measured, typically using a solvent identical to the one used as porogen during preparation of the polymer. This anticipates that the rate and amount of released template directly reflects the affinity of the polymer for the template in a subsequent rebinding step and thus the success of imprinting. The approach has accurately predicted imprinting effects[6] but lacks generality due to the influence of secondary morphology effects and disturbances from released unreacted monomers.

Comparison of binding to a NIP: This technique is commonly used to estimate the imprinting effect and has the benefit of allowing the effect to be estimated based on single component analysis.[6,24,30] This in turn allows nonselective quantification techniques to be used such as simple readings of UV-absorbances, emission intensities, or scintillation. The validity of this approach assumes, however, that the NIP is truly

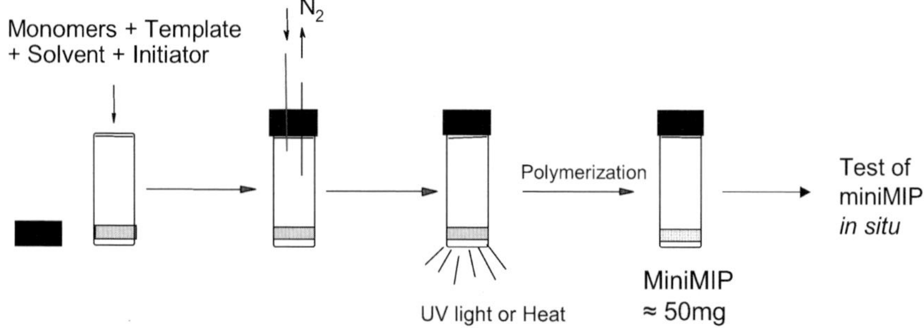

Fig. 7.5 Synthesis and evaluation of miniMIPs

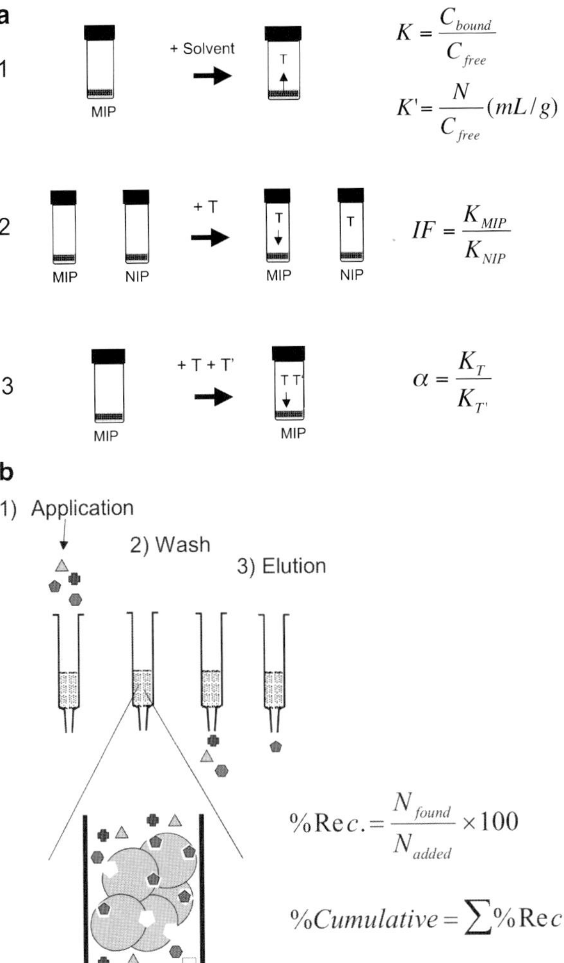

Fig. 7.6 Options to assess imprinting effects and corresponding response factors. (**a**) Static mode assessment. (1) Template release. K = partitioning coefficient. (2) Comparison of template binding to a MIP with that to a NIP. IF = imprinting factor. (3) Comparison of binding of template with that of a close analog to the MIP. α = selectivity factor. (**b**) Flow through SPE mode assessment. Response factors can here be the % recovered template and for gradient elution, the cumulative recovery

comparable with the MIP in terms of morphology, i.e., pore size distribution, surface area, and other physical properties such as polarity, pKa, and functional properties.

Competitive binding experiments: This may allow imprinting effects to be accurately assessed using the MIP only and relies on the measurement of the binding of the target in presence of a structural analog for which the MIP should discriminate against.[5] Obviously to minimize the influence of nonspecific binding, this is valid only for competitive binding of closely related analogs. The use of two enantiomers of a racemate, with their identical physical properties in an achiral environment, is the perfect model system for estimating imprinting effects by this approach.

3.2 Measuring Binding

The options available for measuring analyte binding can be divided into those giving quantitative data on the amount of bound analyte at equilibrium (Fig. 7.6a) and those reflecting binding under nonequilibrium conditions (Fig. 7.6b) e.g. in the chromatographic mode. The former technique is the most widely used and consists in the measurement of the amount of nonbound analyte, either directly or indirectly whereafter the amount of bound analyte is obtained from the mass balance. The latter technique has been less exploited but will become important when optimizing MIPs for chromatographic applications.

The amount of nonbound analyte can be quantified directly, especially if it contains a strong chromophore, fluorophore, or isotope label. Alternatively, it can be postlabeled or the evaluation can take place in a competitive or displacement assay format using a labeled probe.[31,32] A final option is to incorporate a signaling group in the imprinted site leading to a chromogenic response reflecting the amount of bound analyte.[33,34]

The direct measurement of the nonbound fraction is attractive because of its simplicity and since it, in principle, allows the derivation of binding energies. The measurements can here take place in a parallel or serial mode (Fig. 7.7). The

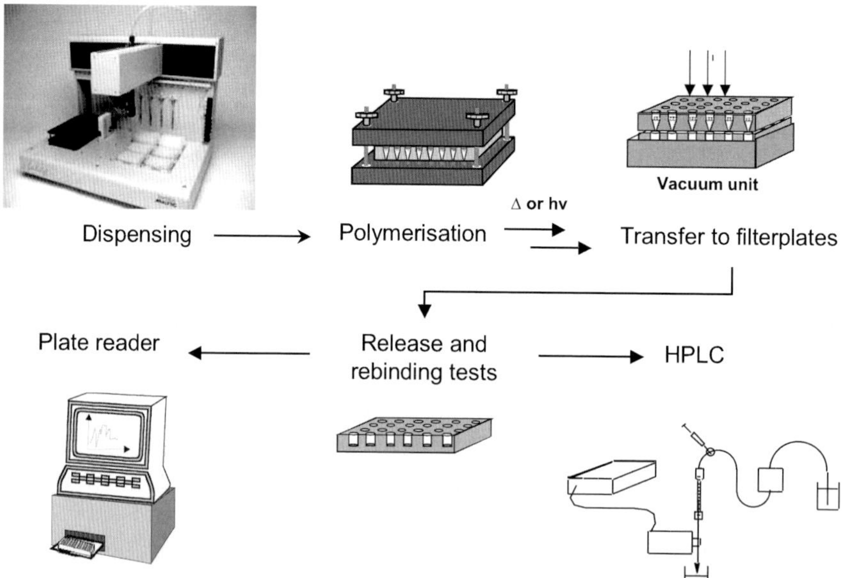

Fig. 7.7 Semiautomated high throughput procedure for the synthesis of miniMIP libraries in 96-well microtiter plate format and evaluation of the library via serial (HPLC) or parallel (multifunctional plate reader) techniques

parallel mode typically makes use of plate readers and is attractive due to the higher throughput it allows, whereas the serial mode makes use of slower chromatographic analysis but is more precise allowing the assessment of competitive binding events.

3.3   Response Factors in the Assessment of miniMIPs under Equilibrium Conditions

After polymerization, the polymers are first incubated with the porogen in a release step. It is here assumed that the rate of release and the amount of released template correlates with the quality of the templated sites. Hundred percent release in this step can be taken as a criterion for discarding a particular MiniMIP. Following the release step, the template is exhaustively extracted from the polymer with a series of washing steps. After template removal, all the polymers are exposed to a solution of the template in the porogen or in another solvent (rebinding step) (Fig. 7.6). The amount of the nonbound template left in solution (C_{free}) is quantified at definite time intervals by any analytical method (HPLC-UV, MS, fluorescence, radioactivity measurements). In the rebinding step, the partition coefficient (K) and the imprinting factor (IF = K_{MIP}/K_{NIP}) (where $K = n/C_{free}$ and n is the amount of template rebound per gram of dry polymer) are calculated based on the weight of the polymer and the original concentration of the template in the rebinding solution. Thus the calculated partition coefficients and imprinting factors are the most important responses used to evaluate the polymers.

4   Techniques for Generating miniMIP Libraries

Owing to the low consumption of reagents and the absence of any manipulation of the polymers, the small scale protocol can be partly automated. In particular, the use of modern plate technology (plate readers, pipetting robots) can be exploited for the preparation and screening of 96-well MIPs (Fig. 7.7).[24] The most recent technique incorporates a liquid sample handling robot for the rapid dispensing of monomers, templates, solvents, and initiator into the reaction vessels of a 96-well plate. A library of polymers, each ca. 50mg, can thus be prepared in 24h. The polymers are thereafter transferred to solvent resistant filter plates for accelerated template removal and improved handling. The template release and MIP rebinding capacity and selectivity can be assessed in batch rebinding experiments by quantifying nonbound fractions using serial analysis by an automated HPLC-system or parallel analysis using a UV monochromator plate reader. Using the parallel analysis mode, the complete polymer library can be evaluated in one to two weeks time. The use of solvent resistant filter-plates allows the solvent to be rapidly removed and exchanged. Thereby it is possible to test real solid phase extraction conditions on the entire library albeit under equilibrium conditions. The problem in this context is to generate a chromatographic bed

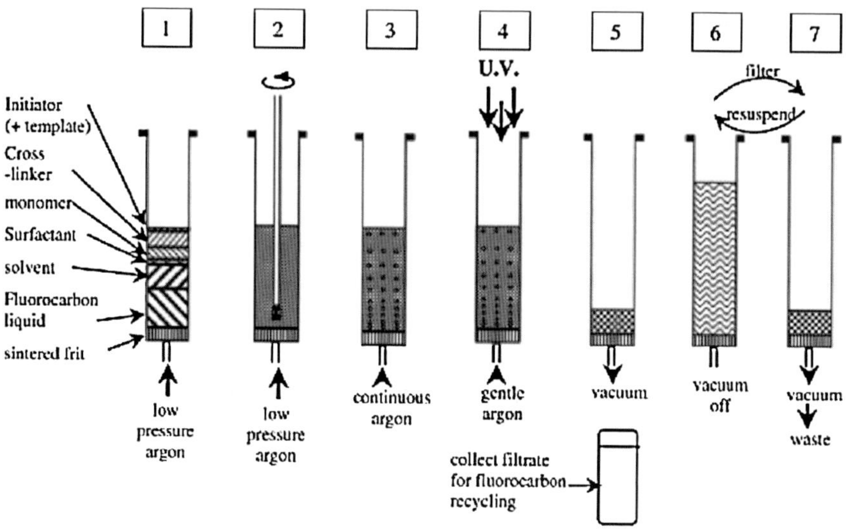

Fig. 7.8 Procedure for the parallel synthesis of beaded MIPs in principle permitting flow through mode assessment of the recognition properties

allowing flow through testing of each member of the library. A recent contribution from the group of Mayes seems promising in this regard (Fig. 7.8).[35] They introduced a parallel method for generating MIPs in a beaded form based on suspension polymerization in liquid fluorocarbons. The polymers were directly polymerized under UV light in solid phase extraction (SPE) cartridges, then washed and extracted in the same cartridges where they had been synthesized, resulting in a rapid and automatable process that required no transfer or manipulation of the polymer particles.

5 Screening of Functional Monomers for Small Target Molecules

The HPLC-vial based approach has been successfully applied to the optimization of imprinted polymers for a range of low molecular weight targets. Most of these studies have focused on the search for optimal functional monomers, although some describe the optimization of monomer ratios, type of cross-linker, template, and solvent.

In the absence of literature precedents, the first screen is often aimed at identifying the most promising functional monomers for targeting a given template. A low polar, non-protic solvent capable of dissolving all the components can be used as polymerization diluent, and EDMA is usually the first choice of cross-linker for this initial group of MiniMIPs. The most successful monomer combinations will be taken as starting points for other screening experiments (i.e., optimization of cross-linker and/or porogen) and finally for the fine-tuning that will involve the variation of the ratios of all the components of the imprinting protocol.

5.1 Triazines

Triazines are a group of herbicides, which due to their persistancy and health hazards require environmental monitoring. MIPs could contribute to selectively enrich these from environmental waters.[19,36] To optimize a MIP for this purpose, a small library of miniMIPs was synthesized by manually pipetting solutions containing different functional monomers into 2 mL vials followed by photoinitiated polymerization of the mixtures at 15°C.[6] The MiniMIPs were then incubated with the solvent (CH_2Cl_2) overnight, and the supernatants were analyzed by HPLC-UV. The MiniMIPs prepared using MAA and TFM showed the lowest concentration of template in the supernatant solutions. This agrees with predictions based on observed solution complexation between substituted pyridines and carboxylic acids and is in agreement with other reports. The MiniMIPs prepared using TFM and MAA were selected for the rebinding test. After complete extraction of the template, the polymers were submitted to equilibrium batch rebinding with a 1 mM solution of terbutylazine in CH_2Cl_2. The IF:s of the MIPs prepared using MAA and TFM as functional monomers were the highest with values of 11 and 6, respectively.

A similar experiment was performed in a 96-well plate format. The components of the polymerization mixture were the same except for the solvent (CH_3CN was used instead of CH_2Cl_2). The polymerization was thermally initiated. The solutions were dispensed using a pipetting robot (Fig. 7.7) and the supernatants analyzed in series by HPLC-UV or in parallel with a plate reader. Eleven WellMIPs were prepared and screened for each monomer. As seen in Fig. 7.9, the reproducibility of the automated procedure is good and the fast parallel assessment delivers similar results as the serial analysis by HPLC.

To optimize MIPs for the two triazines ametryn (an S-triazine) and atrazine (a chloro triazine) Takeuchi et al. prepared a library of imprinted terpolymers, using

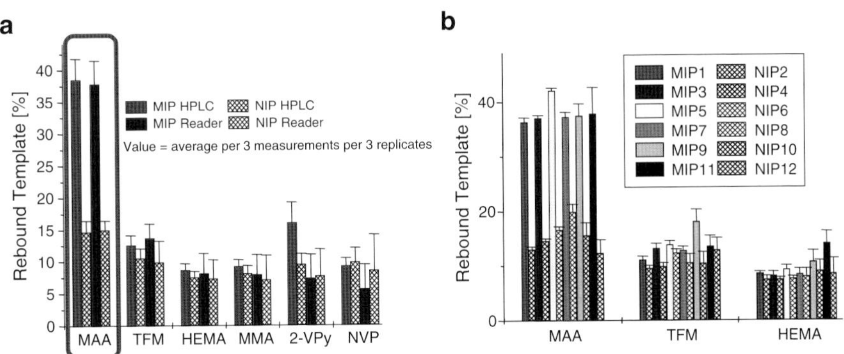

Fig. 7.9 (a) Percentage of template rebound to MIPs and NIPs prepared using different functional monomers after 24 h incubation with 1 mM terbutylazine in acetonitrile. MiniMIPs and MiniNIPs were prepared with a liquid sample handling robot in a 96-well plate and evaluated by a serial (HPLC-UV) or a parallel (plate reader) technique as depicted in Fig. 7.7. (b) Results for 6 replicate MIPs and NIPs

MAA and the stronger acid TFM as functional monomers.[5] The polymers were different in the used molar ratio of total functional monomer to template and the molar ratio of MAA to TFM. The polymer library was subjected to a competitive binding experiment by incubating each member with an equimolar mixture of ametryn and atrazine. The templates exhibited a different preference for the two monomers. Although the more weakly basic chlorotriazine prefered the weaker acid MAA, ametryn was best imprinted using TFM as functional monomer.

5.2 *Nifedipine*

Nifedipine is the archetype of the dihydropyridine calcium entry blockers, increasingly used as a probe drug for the assessment of cytochromeP-450 IIIA4 enzyme activity in humans. To assess the usefulness of MIPs for sample pretreatment of this drug, a min- iMIP monomer screening was performed (Fig. 7.10).[29] Monomers were chosen to target the few functionalities of the molecule: acidic monomers such as MAA, TFM to target the ring nitrogen or the ester groups, basic monomers such as AS, 4-VP, 2-VP, DEAEMA to target the slightly acidic exocyclic proton, hydrogen bonding monomers such as HEMA, AS, MAA, TFM to target the ester groups, and finally π-donors/acceptors such as AnthraMMA, NaphtMA or NVC to target the π-electron poor aromatic ring.

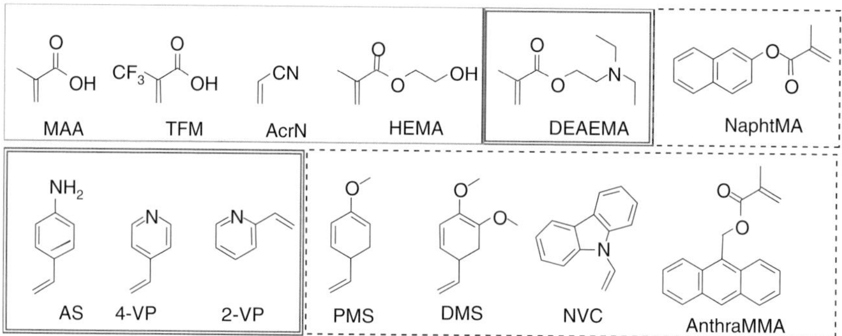

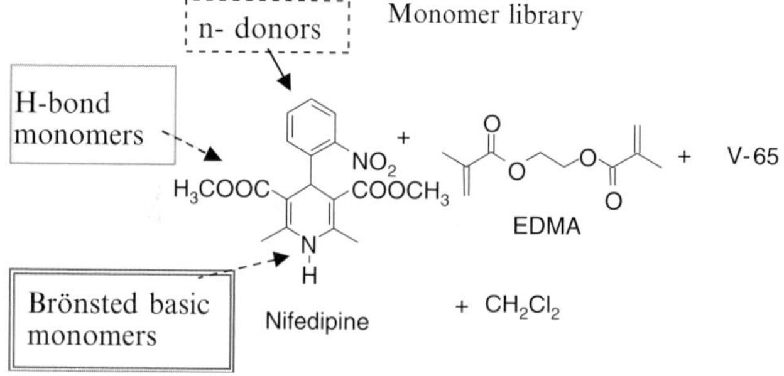

Fig. 7.10 Screening of functional monomers for imprinting of nifedipine

From the results of the incubation with the porogen (CH_2Cl_2), the basic monomers 4-VP, 2-VP, the acidic monomer TFM, and the π-donor AnthraMMA could be identified as the most promising monomers (Fig. 7.11a). The results of the equilibrium batch rebinding test partially confirmed the first findings. AS, MAA, HEMA, and AcrN could also be identified as good monomers (Fig. 7.11b). In this context note that the π-donor monomers are compatible with the acidic ones, and thus they could be combined in a combinatorial approach. An important issue concerns the choice between a static or dynamic mode assessment of the rebinding selectivity. In these studies, the promising selectivity observed in the batch mode was not reflected in chromatographic retention results. This highlights the need to evaluate the polymers in a dynamic situation (vide supra), particularly if the final material is to be implemented in solid phase extraction or chromatography.

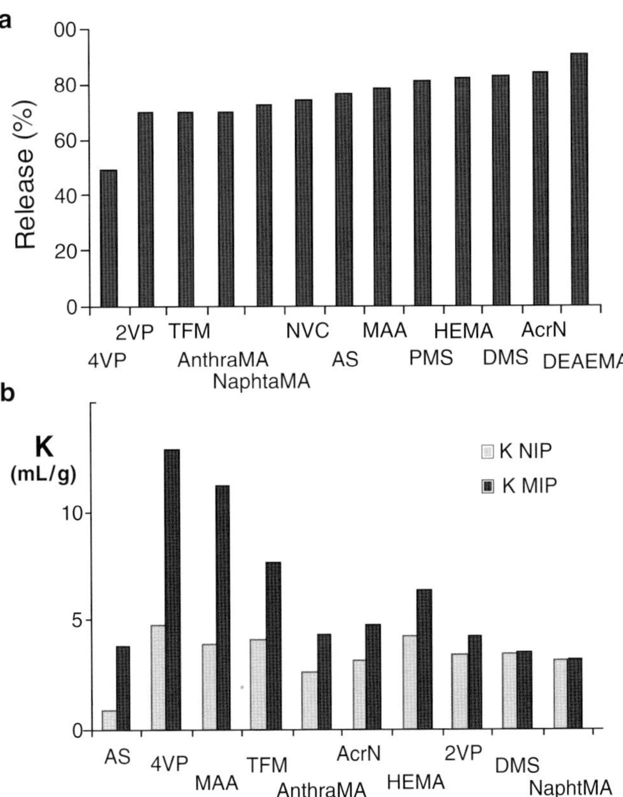

Fig. 7.11 Percentage of Nifedipine detected in the miniMIP supernatants after 30 min sonication in the presence of 1 ml CH_2Cl_2. (**a**).and partition coefficients K_{MIP} and K_{NIP} of MiniMIPs after overnight incubation with 1 mM nifedipine in dichloromethane (**b**)

5.3 *Estradiol*

17β-Estradiol

One target that has proven difficult to recognize with MIPs is the natural hormone 17β-Estradiol.[37-40] This belongs to the class of endocrine disrupters, which mimic or anatagonize the effects of endogenous hormones. Because of its widespread use and to its persistance, it has become an important environmental pollutant. The need for corresponding sensitive and selective analytical methods has led to a search for efficient extraction techniques allowing enrichment and cleanup of the target analyte.[41] Molecularly imprinted solid phase extraction (MISPE) belongs to one of the tools assessed for this purpose.[42] It is, therefore, important to search for monomer combinations that would lead to MIPs exhibiting higher affinity and selectivity for this analyte. For this purpose, we made use of 96-well plate technique described above including 96-well filter plates and a multifunctional plate reader.[30]

On the basis of previous literature, a limited set of functional monomers was tested for β-estradiol imprinting using two different porogens (ACN and THF) and two different cross-linking monomers (EDMA and DVB). Thus eight functional monomers were used encompassing acidic monomers (MAA, TFMAA), basic monomers (2-VP, 4-VP, DEAEMA, AJH242), and neutral monomers (MAAM, NVP, AS) all of which can engage in weak to moderate hydrogen bonds through the OH groups of the template molecule. After polymerization, the PTFE plate containing four identically prepared MIPs and NIPs per functional monomer was dried in a drying oven. Thereafter, the polymers were weighed and transferred to the solvent-resistant filter plates. The gravimetric yield of the polymers in this step was as expected quantitative. However, after exhaustive washing of the polymers, it dropped somewhat but exceeded 84% for all polymers whereby no significant difference was measured between MIP and NIP. After exhaustive extraction of the template, the library was incubated with a solution of β-estradiol in ACN (0.5 mM) for 16 h. Thereafter, the supernatants were vacuum sucked through the filter plate and transfered to quartz microtiter plates for fluorescence readings, or to an HPLC autosampler for HPLC quantification.

Template solubility is a particularly critical factor in the selection of conditions for MIP synthesis. Although it is most convenient if the template exhibit solubility in the porogenic solvent alone, monomer assisted solvation is a good indicator for a successful outcome of the imprinting step. Best results are commonly obtained using porogens capable of solvating the template only by assistance of the functional monomer.[43] As a good illustration of the above effect, the library prepared using THF as porogen, in which β-estradiol is perfectly soluble, generally displayed small imprinting factors (Table 7.1). We noted, however, that the basic functional monomer DEAEMA performed best in this system with an imprinting factor of 1.36 and that both of the cross-linking monomers, DVB and EDMA, performed similarly. This contrasted with the results using acetonitrile (ACN) as porogen.

Acetonitrile belongs to the poor solvents for β-estradiol but perhaps of its intermediate solvating property, it has been the solvent of choice for imprinting β-estradiol in the past. After initial poor results using THF as porogen, acetonitrile was therefore used as porogen in the further optimization work.

The amount of β-estradiol bound to each MIP and NIP (as an average of four replicas) is seen in Fig. 7.12 together with the corresponding imprinting factors.

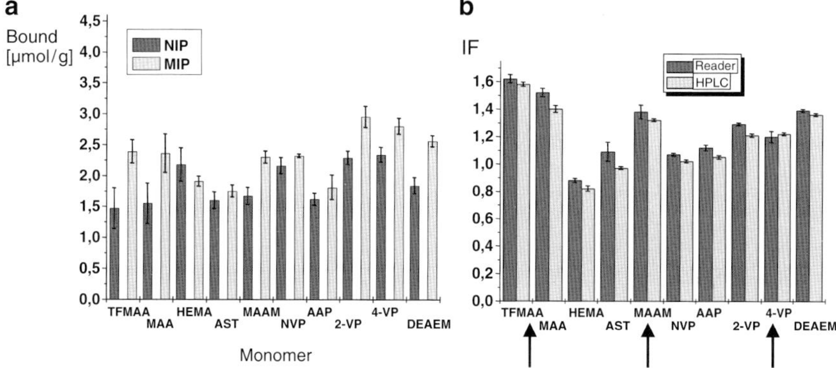

Fig. 7.12 (**a**) Amount of β-estradiol bound to MIPs and NIPs of an 80-polymer library. The results are averages of four replicas with the standard deviations given as error bars. (**b**) Average imprinting factors (IF = n_{MIP}/n_{NIP}) obtained for the monomer library using the parallel evaluation technique with reader or the serial technique with HPLC

Table 7.1 Imprinting factors for polymers prepared using EDMA or DVB as cross-linking monomers and THF as porogen

Monomer	IF (EDMA)	IF (DVB)
MAA	0.75	0.81
MAAM	1.03	0.87
2-VP	0.95	0.97
4-VP	1.03	1.02
DEAEMA	1.36	1.37
AJH242	0.88	0.95
EDMA	0.79	–
DVB	–	0.84

In view of the relatively small errors, the batchwise reproducibility is acceptable and the effects reliable. Furthermore, in agreement with results reported in the literature,[38,40] the acidic and the basic functional monomers give polymers showing significant imprinting effects. Among these, MAA and TFMAA show the best results followed by DEAEMA. Among the neutral functional monomers, MAAM also showed a significant imprinting effect. It is interesting to compare these results with the recognition mechanism of the biological receptor for estradiol. In agreement with the finding that both basic and acidic functional monomers give enhanced imprinting factors, β-estradiol is bound to the receptor assisted by hydrogen bond interactions involving a basic amino acid (His) at the 17 position and an acidic amino acid (Glu) at the 3 position.

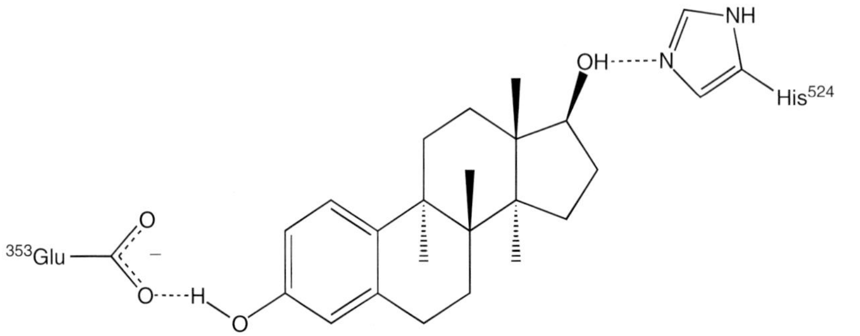

Estradiol recepton

6   Optimization of Prepolymerization Composition

Because of the complexity of the imprinting process, effective optimization is unlikely to be obtained by changing one separate factor at a time (COST strategy). Indeed if the factors (components and parameters) are not independent, a COST strategy will not lead to the real optimum and will not allow the real effects to be separated from "noise." Instead, all relevant factors should be changed together over a set of experimental runs and the results charted and connected by means of a semiempirical mathematical model. This chemometry strategy thus contains experimental design and modeling,[44] and has been successfully used to optimize MIPs for a variety of applications.

The choice of the response factors depends on the field of application, which is foreseen for the materials. If, for instance, the MIPs are to be used as chromatographic stationary phases, retention and imprinting factors should be used as responses. If the MIPs will be used as sensing elements, the response will be dependent on the nature of the transducer (intensity of absorption, fluorescence, voltage, conductivity, resistance) and can be expressed as the equilibrium value of the signal or as the rate of signal generation. If the MIPs will be used in solid phase extraction, the recovery of the target analyte, as well as that of closely-related analogues, will be the natural variable to be chosen as response.

Thus, as an alternative to expressing the imprinting effect as an imprinting factor, the NIPs can be neglected and binding selectivity directly assessed in a competitive mode by adding a mixture of the target compound and a structurally-related analogue.

6.1 *Watercompatible MIP for Bupivacaine*

Under aqueous conditions, the hydrophobic surface of EDMA based imprinted polymers leads to substantial nonspecific binding of nonrelated lipophilic compounds.[20] In addition, biofluid constituents such as plasma-proteins and lipids are strongly adsorbed to the polymer surface. Elaborate wash protocols are required to remove these matrix components commonly leading to lower recoveries of the target analyte as well.[45] Another issue concerns column reuse. In some cases, these can be restored by suitable washing schemes but often, however, the only resort is a frequent change to fresh columns or, alternatively, the use of additional sample pretreatment procedures to remove harmful matrix components. We, therefore, embarked on the development of imprinted sorbents capable of selective extraction of the analyte from pure aqueous buffer with minimum nonspecific binding of other matrix components.[24] This would obviate the need for organic solvent-based washing steps. The starting point for the optimization was the well characterized MIP consisting of poly(MAA-*co*-EDMA) imprinted with bupivacaine.[46] This polymer was chosen as the reference for subsequent comparisons.

Bupivacaine

The library was constructed by slightly modifying the procedure used to make the reference polymer. The modifications comprised (1) the use of the hydrophilic comonomer 2-hydroxyethylmethacrylate (HEMA), known to impact water compatibility in a number of unrelated systems, (2) the use of four porogens (DCM, TCE, toluene, and MTBE), and (3) the relative ratios of (a) HEMA/MAA and (b) (HEMA + MAA)/EDMA. For each porogen, the latter factors were optimized by a 2^2 factorial design experiment including one center point (Fig. 7.13). The chosen response factors were the partition coefficients (K) and the imprinting factor (IF). The imprinting factors for the four sets of polymers are shown in Fig. 7.14. As indicated by the arrows, polymers synthesized using low cross-linking levels (50%) and high HEMA concentrations, in combination with the use of two of the porogens (toluene or 1,1,1-trichloroethane), showed high imprinting factors. These were high mainly as a result of low nonspecific binding.

$F_2 =$ Crosslinking % $F_1 =$ HEMA/MAA	HEMA / MAA 0 / 1	HEMA / MAA 1 / 1	HEMA / MAA 2 / 1
83.3 %	IP 1	–	IP 2
66.6%	–	IP 5	–
50 %	IP 3	–	IP 4

Fig. 7.13 2^2 full factorial design used for optimizing bupivacaine (BV) imprinted polymers (poly(MAA-*co*-HEMA-*co*-EDMA) for the selective extraction of bupivacaine from blood plasma samples

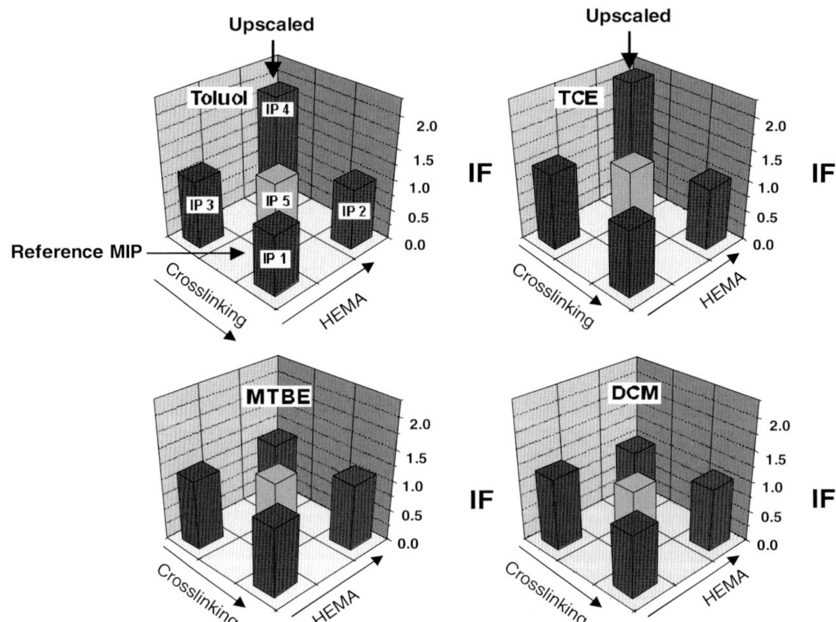

Fig. 7.14 3D-representation of imprinting factors (IF) obtained from the rebinding of BV in phosphate buffer pH 7.4 to the polymer library prepared from the design shown in Fig. 7.13. *Arrows* indicate compositions selected for upscaling as well as the conventional composition used to produce the previously reported MIP

The best performing polymers were upscaled for assessment as sorbents for extractions of bupivacaine from blood plasma samples.[47] As can be seen in Table 7.2 an SPE procedure involving aqueous washes containing maximum 10% organic solvent led to effective removal of bound bupivacaine from the watercompatible nonimprinted polymer contrasting with the quantitative recovery obtained using the corresponding MIP. Meanwhile the recoveries measured using the reference MIP and NIP were quantitative indicating a pronounced nonspecific binding in this case.

Table 7.2 Recoveries (%) of bupivacaine from imprinted (MIP) and nonimprinted (NIP) water compatible polymer (AquaMIP) and reference polymers (reference) in SPE in absence of a solvent switch step using less than 10% organic solvent in water

Polymer	MIP	NIP
Reference	105	105
AquaMIP	104	4

Protein binding to the optimized polymers was also lower compared with that for the reference polymer.

6.2 *Watercompatible MIP for Sildenafil*

A similar approach was used for achieving a polymer that works in water but targeted toward the lipophilic drug sildenafil (Viagra).[18] An extended 2 × 2 experimental design led to a library of 90 polymers. This library was assessed in the SPE mode by completely adsorbing the target to the polymers in water, followed by stepwise desorption by applying wash fractions containing increasing amounts of MeCN. This produced cumulative recovery curves for each library member, from which a polymer hydrophilicity index (HI) and a cumulative imprinting factor (= 50/Y) could be calculated as shown in Fig. 7.15. As examplified for the library subset prepared using toluene as porogen (Fig. 7.16), some compositions resulted in high hydrophilicities and cumulative imprinting factors simultaneously. By using the hydrophilicity and imprinting factor criteria together, polymers were identified which could be successfully implemented in SPE protocols for the target analyte.

Sildenafil

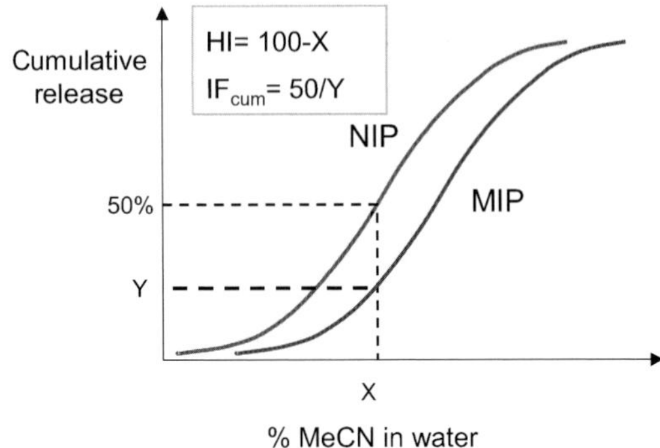

Fig. 7.15 Approach to assess a MIP/NIP library in the SPE mode. After quantitative nonspecific adsorption of the analyte from water the analyte is gradually eluted from the cartridge by increasing the acetonitrile content in the wash steps. The cumulative recovery is plotted against % acetonitrile in the wash step. The hydrophilicity index is defined as the % water present in the wash solvent leading to 50% release of the analyte from the NIP. The cumulative imprinting factor is defined as the ratio of the recovery from the NIP at that point (=50%) over that from the MIP (= *Y*)

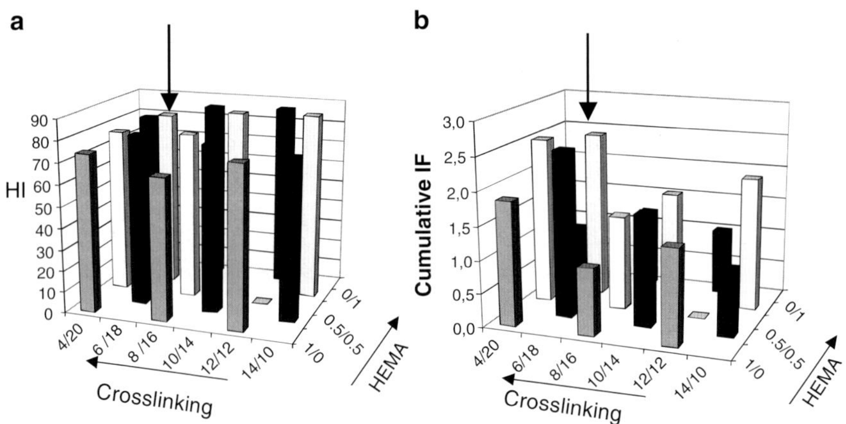

Fig. 7.16 Hydrophilicity index (**a**) and cumulative imprinting factors (**b**) measured for a Sildenafil imprinted poly(MAA-*co*-HEMA-*co*-EDMA) library produced using toluene as porogen. The *arrows* indicate a composition leading simultaneously to a high HI and IF

7   Exploring MIP Cross-Reactivity

The cross-reactivity of MIPs can be very hard to predict as shown by several sometimes surprising results. Table 7.3 shows some examples of template substructures providing good sites for indicated target structures. Screening

Table 7.3 Examples of templates used to produce MIPs with a cross-reactivity for a given target analyte

Target analyte	Template	Ref.
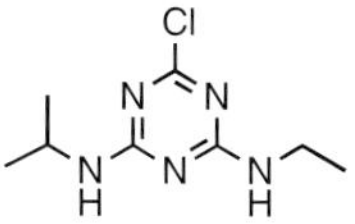	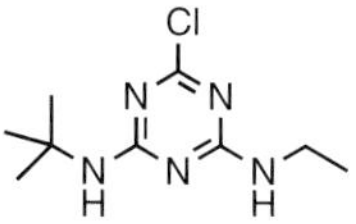	36
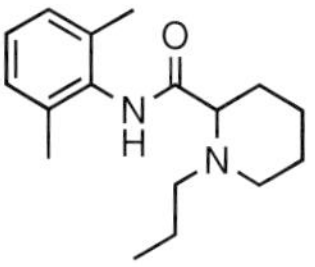	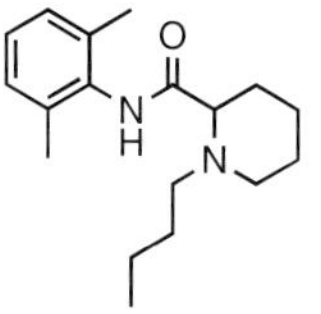	46
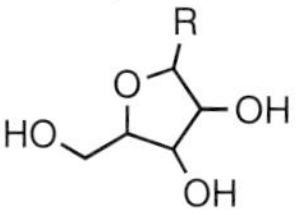	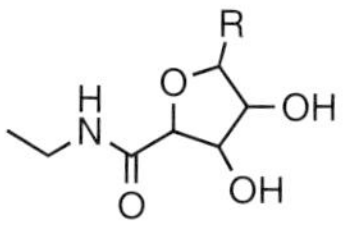	16

for cross-reactive MIPs has therefore become an important part in the development of new receptors (vide supra). On the basis of the so-called Explorasep plates marketed by MIP Technologies (http://www.miptechnologies.com), screening of analytes against a continuously evolving library of molecularly imprinted polymer (MIP) sorbents may accelerate the discovery of new MIPs for a given target. Depending on the solvent used during the interaction MIPs can be highly selective toward the imprinted molecule or show a broader cross-reactivity with other compounds, which possess the same or similar chemical substructures or functionalities.

8 Conclusions

High throughput and combinatorial synthesis techniques have proven valuable for the rapid development and testing of new MIPs. Most use of the technique has aimed at finding suitable monomers for a given template or target analyte (e.g., dihydropyridines, sulfonamides, pteridines, triazines, drug lead compounds) but equally important is a second stage optimization with the objective of finding optimum monomer ratios and porogens in terms of affinity and selectivity for the target analyte in a given matrix. Because of the high throughput and the reliability of the present method, it is now an important tool in the development and optimization of new MIPs. Applications of the resulting materials encompass selective solid phase extraction for quantitative analysis, chemical sensors, sorbents for preparative separations, screening tools in drug discovery and enzyme like catalysis.

References

1. Molecularly imprinted polymers. Man made mimics of antibodies and their applications in analytical chemistry; Sellergren, B., Ed.; Elsevier Science B.V.: Amsterdam, 2001; Vol. 23
2. Alexander, C.; Andersson, H. S.; Andersson, L. I.; Ansell, R. J.; Kirsch, N.; Nicholls, I. A.; O'Mahony, J.; Whitcombe, M. J., Molecular imprinting science and technology: A survey of the literature for the years up to and including 2003, *J. Molec. Recogn.* **2006**, 19, 106–180
3. Sellergren, B.; Lepistoe, M.; Mosbach, K., Highly enantioselective and substrate-selective polymers obtained by non-covalent interactions. NMR and chromatographic studies on the nature of recognition, *J. Am. Chem. Soc.* **1988**, 110, 5853–5860
4. Ellwanger, A.; Karlsson, L.; Owens, P. K.; Berggren, C.; Crecenzi, C.; Ensing, K.; Bayoudh, S.; Cormack, P.; Sherrington, D.; Sellergren, B., Evaluation of methods aimed at complete removal of template from molecularly imprinted polymers, *Analyst* **2001**, 126, 784–792
5. Takeuchi, T.; Fukuma, D.; Matsui, J., Combinatorial molecular imprinting: An approach to synthetic polymer receptors, *Anal. Chem.* **1999**, 71, 285–290
6. Lanza, F.; Sellergren, B., Method for synthesis and screening of large groups of molecularly imprinted polymers, *Anal. Chem.* **1999**, 71, 2092–2096
7. Cederfur, J.; Pei, Y.; Meng, Z.; Kempe, M., Synthesis and screening of a molecularly imprinted polymer library targeted for penicillin G, *J. Combin. Chem.* **2003**, 5, 67–72
8. Takeuchi, T.; Fukuma, D.; Matsui, J.; Mukawa, T., Combinatorial molecular imprinting for formation of atrazine decomposing polymers, *Chem. Lett.* **2001**, 6, 530–531
9. Davies, M. P.; De Biasi, V.; Perrett, D., Approaches to the rational design of molecularly imprinted polymers, *Anal. Chim. Acta* **2004**, 504, 7–14
10. Navarro-Villoslada, F.; San Vicente, B.; Moreno-Bondi, M. C., Application of multivariate analysis to the screening of molecularly imprinted polymers for bisphenole, *Anal. Chim. Acta* **2004**, 504, 149–162
11. Davidson, L.; Blencowe, A.; Drew, M. G. B.; Freebairn, K. W.; Hayes, W., Synthesis and evaluation of a solid supported molecular tweezer type receptor for cholesterol, *J. Mater. Chem.* **2003**, 13, 758–766
12. Piletsky, S. A.; Karim, K.; Piletska, E. V.; Day, C. J.; Freebairn, D. W.; Legge, C.; Turner, A. P. F., Recognition of ephedrine enantiomers by molecularly imprinted polymers designed using a computational approach, *Analyst* **2001**, 126, 1826–1830
13. Sellergren, B., Polymer- and template-related factors influencing the efficiency in molecularly imprinted solid-phase extractions, *Trends Anal. Chem.* **1999**, 18, 164–174
14. Sellergren, B.; Shea, K. J., On the influence of polymer morphology on the ability of imprinted polymers to separate enantiomers, *J. Chromatogr.* **1993**, 635, 31
15. Quaglia, M.; Chenon, K.; Hall, A. J.; De Lorenzi, E.; Sellergren, B., Target analogue imprinted polymers with affinity for folic acid and related compounds, *J. Am. Chem. Soc.* **2001**, 123, 2146–2154
16. Dirion, B.; Lanza, F.; Sellergren, B.; Chassaing, C.; Venn, R.; Berggren, C., Selective solid phase extraction of a drug lead compound using molecularly imprinted polymers prepared by the target analogue approach, *Chromatographia* **2002**, 56, 237–241
17. Andersson, L. I.; Paprica, A.; Arvidsson, T., A highly selective solid-phase extraction sorbent for preconcentration of sameridine made by molecular imprinting, *Chromatographia* **1997**, 46, 57–62
18. Dzygiel, P.; O'Donnell, E.; Fraier, D.; Chassaing, C.; Cormack, P. A. G., Evaluation of water-compatible molecularly imprinted polymers as solid-phase extraction sorbents for the selective extraction of sildenafil and its desmethyl metabolite from biological samples, *J. Chromatogr.* B **2007**, 853(1–2), 346–353
19. Köber, R.; Fleischer, C. T.; Lanza, F.; Boos, K.-S.; Sellergren, B.; Barcelo, D., Evaluation of a multidimensional solid phase extraction platform (six-spe) for highly selective on-line clean-up and high throughput lc-ms analysis of triazines in river water samples using molecularly imprinted polymers (mips), *Anal. Chem.* **2001**, 73, 2437–2444

20. Karlsson, J. G.; Andersson, L. I.; Nicholls, I. A., Probing the molecular basis for ligand-selective recognition in molecularly imprinted polymers selective for the local anesthetic bupivacaine, *Anal. Chim. Acta* **2001**, 435, 57–64

21. Manesiotis, P.; Hall, A. J.; Courtois, J.; Irgum, K.; Sellergren, B., An artificial riboflavin receptor prepared by a template analogue imprinting strategy, *Angewandte Chemie, Intern. Ed.* **2005**, 44, 3902–3906

22. Ramström, O.; Andersson, L. I.; Mosbach, K., Recognition sites incorporating both pyridinyl and carboxy functionalities prepared by molecular imprinting, *J. Org. Chem.* **1993**, 58, 7562–7564

23. Lübke, C.; Lübke, M.; Whitcombe, M. J.; Vulfson, E. N., Imprinted polymers prepared with stoichiometric template–monomer complexes: Efficient binding of ampicillin from aqueous solutions, *Macromolecules* **2000**, 33, 5098–5105

24. Dirion, B.; Cobb, Z.; Schillinger, E.; Andersson, L. I.; Sellergren, B., Water-compatible molecularly imprinted polymers obtained via high-throughput synthesis and experimental design, *J. Amer. Chem. Soc.* **2003**, 125, 15101–15109

25. Wulff, G.; Knorr, K., Stoichiometric imprinting, *Bioseparation* **2002**, 10, 257

26. Hall, A. J.; Manesiotis, P.; Emgenbroich, M.; Quaglia, M.; De Lorenzi, E.; Sellergren, B., Urea host monomers for stoichiometric molecular imprinting of oxyanions, *J. Org. Chem.* **2005**, 70, 1732–1736

27. Kempe, M.; Mosbach, K., Receptor binding mimetics: A novel molecularly imprinted polymer receptor, *Tetrahedron Lett.* **1995**, 36, 3563–3566

28. Hart, B. R.; Shea, K. J., Synthetic peptide receptors: Molecularly imprinted polymers for the recognition of peptides using peptide-metal interactions, *J. Am. Chem. Soc.* **2001**, 123, 2072–2073

29. Lanza, F.; Hall, A. J.; Sellergren, B.; Bereczki, A.; Horvai, G.; S; Cormack, P. A. G.; Sherrington, D. C., Development of a semiautomated procedure for the synthesis and of molecularly imprinted polymers applied to the search for functional monomers for phenytoin and nifedipine, *Anal. Chim. Acta* **2001**, 435, 91–106

30. Dirion, B.; Schillinger, E.; Sellergren, B., Development of a high throughput synthesis technique for the optimization of mips for 17b-estradiol, Mat. Res. Soc. 787 (Mat. Res. Soc. Fall Meeting) Boston, 53–59 (2003)

31. Hunt, C. E.; Ansell, R. J., Use of fluorescence shift and fluorescence anisotropy to evaluate the re-binding of template to (s)-propranolol imprinted polymers, *Analyst* **2006**, 131, 678–683

32. Greene, N.; Lee, J.-D.; Hong, J.-I.; Shimizu, K., Employing molecular imprinted polymers for sensor discrimination, *Polymer Preprints* (Amer. Chem. Soc., Division of Polym. Chemistry) **2005**, 46, 139–140

33. Ye, L.; Mosbach, K., Polymers recognizing biomolecules based on a combination of molecular imprinting and proximity scintillation: A new sensor concept, *J. Am. Chem. Soc.* **2001**, 123, 2901–2902

34. Manesiotis, P.; Hall, A. J.; Emgenbroich, M.; Quaglia, M.; de Lorenzi, E.; Sellergren, B., An enantioselective imprinted receptor for z-glutamate exhibiting a binding induced color change, *Chem. Comm.* **2004**, 2278–2279

35. Perez-Moral, N.; Mayes, A. G., Direct rapid synthesis of mip beads in spe cartridges, *Biosens. Bioelectron.* **2006**, 21, 1798–1803

36. Ferrer, I.; Lanza, F.; Tolokan, A.; Horvath, V.; Sellergren, B.; Horvai, G.; Barcelo, D., Selective trace enrichment of chlorotriazine pesticides from waters and sediment samples using terbuthylazine, *Anal. Chem.* **2000**, 72, 3934–3941

37. Rachkov, A.; McNiven, S.; Cheong, S.-H.; El'Skaya, A.; Yano, K.; Karube, I., Molecularly imprinted polymers selective for Beta-estradiol, *Supramol. Chem.* **1998**, 9, 317–323

38. Haginaka, J.; Sanbe, H., Uniform-sized molecularly imprinted polymers for.Beta.-estradiol, *Chem. Lett.* **1998**, 1089–1090

39. Idziak, I.; Benrebouh, A., A molecularly imprinted polymer for 17.Alpha.-ethynylestradiol evaluated by immunoassay, *Analyst* **2000**, 125, 1415–1417

40. Ye, L.; Yu, Y.; Mosbach, K., Towards the development of molecularly imprinted artificial receptors for the screening of estrogenic chemicals, *Analyst* **2001**, 126, 760–765

41. Special issue on molecular imprints and related approaches for solid-phase extraction and sensors in chemical analysis; Barceló, D., Ed.; 1999; Vol. 18
42. Lanza, F.; Sellergren, B., The application of molecular imprinting technology to solid phase extraction, *Chromatographia* **2001**, 53, 599–611
43. Sellergren, B., Imprinted dispersion polymers: A new class of easily accessible affinity stationary phases, *J. Chromatogr.* A **1994**, 673, 133–141
44. Eriksson, L.; Johansson, E.; Kettaneh-Wold, N.; Wikström, C.; Wold, S. Design of experiments. Principles and applications; Umetrics AB: Umeå, 2000
45. Andersson, L. I.; Schweitz, L., Solid-phase extraction on molecularly imprinted polymers, Handbook of Analytical Separations **2003**, 4, 45–71
46. Andersson, L. I., Efficient sample pre-concentration of bupivacaine from human plasma solid-phase extraction on molecularly imprinted polymers, *Analyst* **2000**, 125, 1515–1517
47. Cobb, Z.; Sellergren, B.; Andersson, Water-compatible molecularly imprinted polymers for efficient direct injection on-lin solid phase extraction of ropivacaine and bupivacaine from human plasma. *Analyst* **2007**, 132, 1262–71

Section 4
Biological Receptors

Chapter 8
Combinatorially Developed Peptide Receptors for Biosensors

Chikashi Nakamura and Jun Miyake

Abstract Various combinatorial libraries were screened for short peptides of 4–10 mer, which were used as sensor molecules for capturing target chemicals or biomolecules. Immuno-antibodies can be synthesized in the living bodies of higher animals even for low-molecular-weight nonnatural chemical compounds, such as dioxins or PCBs. Recently, some peptide ligands that can even bind to inorganic crystals have been reported. This indicates that the 20 natural amino acids have the potential to recognize almost all types of molecules and substances. The question arises whether one should design a "rational" mini library of peptides consisting of a limited number of amino acids according to the motifs in epitopes or paratopes or the binding pocket sequences in receptors, or a completely "random" combinatorial library containing all sequences. If one wants to obtain a peptide binder to target a small chemical compound, the answer is a "random" library, since the molecular interaction between the target compound and an amino acid cannot be precisely predicted beforehand. In this section, we discuss the possibility of using short combinatorial peptides as binders for biosensors to detect chemical compounds.

1 Peptides as Materials for Molecular Recognition

Almost all molecular recognition in living organisms is performed by proteins such as immuno-antibodies and receptors. The molecular recognition elements of the proteins are 20 natural amino-acids. Amazingly, the variety of side chains of natural amino-acids enables recognition of not only biomolecules and organic compounds but also inorganic materials such as fullerenes,[1] solid titanium,[2] and carbon nanotubes.[3,4] The wide target range and robust binding are exceptional properties that have led

C. Nakamura (✉) and J. Miyake
Research Institute for Cell Engineering, National Institute of Advanced Industrial Science and Technology (AIST), Central 6, 1-1-1 Higashi, Tsukuba, Ibaraki 305-8566, Japan
chikashi-nakamura@aist.go.jp

R.A. Potyrailo and V.M. Mirsky (eds.), *Combinatorial Methods for Chemical and Biological Sensors*, DOI: 10.1007/978-0-387-73713-3_8,

to the use of polymers of amino-acids as binders in sensors. In the development of sensors using polypeptides, immuno-antibodies have commonly been used because of their high selectivity and sensitivity. However, there are some limitations to the use of antibodies. Antibodies cannot be obtained for all targets. There are difficulties in the preparation of antibodies, for example, with highly toxic compounds, low molecular weight compounds, highly conserved protein motifs in animals, sugar chains and highly hydrophobic materials, etc. Moreover, immuno-antibodies must be used under conditions that do not cause denaturation of proteins. We can retain the molecular recognition properties of immuno-antibodies and overcome their disadvantages by using a combinatorial peptide. Polypeptides of less than 20–30 mer can be handled easily because they do not have a rigid conformation, and they do not denature even under physiologically stringent conditions. Furthermore, short peptides can be designed and chemically synthesized depending on the target and the purpose.

First of all, a peptide with a binding sequence that recognizes the target is needed for the development of a sensor. In future, it will be possible to utilize quantitative structure–activity relationship (QSAR) methodology, by which the sensor peptide sequence can be predicted and de novo designed from the structure of the target substance. However, this is not possible at the present time, since both the information about molecular interactions and the computing methodology for molecular docking have not yet been established. So, in 2007, combinatorial screening of the peptide sequence from a library is the most common method still used. There are two approaches to the screening of a sensor peptide. One is the rational design and screening of a mini library based on the motif sequences of natural antibodies or receptors. The other is random full library screening. A pentapeptide library consisting of natural amino acids has a total of 20^5 (3.2 million) different sequences. This means that full library screening requires a huge effort. Better sensor performance can be expected from longer peptides since they can form highly-dimensional and rigid conformations, which form stable bindings with targets. Because of these considerations, a mixed method of rational design and random screening is generally chosen. The CDR sequence of an antibody or the binding pocket of a receptor is usually used as the motif and part of the sequence is randomized, thereby constructing a mini library. For example, a dioxin-binding peptide screened from a library based on the binding domain sequence of the Ah receptor has been reported.[5] Moreover, randomized domains can be minified by employing rigid base bone structures, e.g., a alpha-helix bundle [6–8] or a circular peptide.[9–11] Stable binding can also be expected when using such rigid backbones.

2   Porphyrin Binding Peptide

We proposed the use of a short peptide consisting of 4–5 amino acids as the binder molecule in a biosensor for targeting low-molecular weight molecules of less than 1,000 Da. In preparing a random full library by solid phase synthesis, it is thought that a pentapeptide library is the largest possible in a usual laboratory. In a pentapeptide library, at least 195 (2.5 million) sequences of resins, from which cysteine has been excluded, must be treated. It is not expected that peptides of 4–5 amino acids have a secondary structure since pentapeptides cannot form alpha-helices. However, recently, native unfolded proteins have been reported. Such intrinsically-disordered proteins can interact with a target ligand and undergo a structural transition to a folded form when bound.[12] It is probable that we can expect such flexible conformational changes occurring when short peptides bind to a target.

2.1 Screening of Porphyrin-Binding Peptide

A porphyrin molecule was used to confirm that pentapeptides bind sufficiently to small chemical compounds. This was demonstrated by Sugimoto and his colleagues.[13] The porphyrin-binding peptide, HASYS, 5,10,15,20-tetrakis(*N*-methyl-pyridinium-4-yl)-21H,23H-porphine (TMpyP) was obtained from a M13 phage library (Fig. 8.1, and the figure was made based on the personal communication with Prof. Kawakami and Prof. Sugimoto who are the authors of reference 13). Although the molecular weight of TMpyP is only 680 Da, a single phage body has a weight of over 16 MDa. They proposed that a phage could not be retained on the target immobilized on a solid phase since the large binding energy needed to hold such a huge phage body could not be expected with low molecular weight compounds. Therefore, they developed a plaque screening method. In this case, phage plaques were transferred to nitrocellulose membranes, a target molecule solution was added to the membranes and TMpyP-bound plaques selected by fluorescence imaging. The binding constant of the peptide obtained was 10^5 M^{-1}, which is not very high compared with those of immuno-antibodies. The peptide was chosen as a model peptide binder for detecting small chemicals.

2.2 Apparatus to Detect Nondescript Target Bound by Peptide

The nondescript target compounds, for example, environmental persistent organic pollutants (POPs) such as dioxins, PCBs, DDT, are not chemically or physically reactive, and do not have typical photo-spectroscopic features. So, apparatuses that can directly detect the event of binding target molecules have been required. A surface plasmon resonance (SPR) sensor, a quartz crystal microbalance (QCM), and an atomic force microscope (AFM) are candidates to detect such null substances. The SPR, QCM, and AFM apparatus can monitor the refractive index change, the mass change, and the molecular interaction itself, respectively. The porphyrin-binding peptide was used as an example short peptide binder targeting small compound. Cystein was added to the original peptide sequence HASYS to immobilize the peptide.

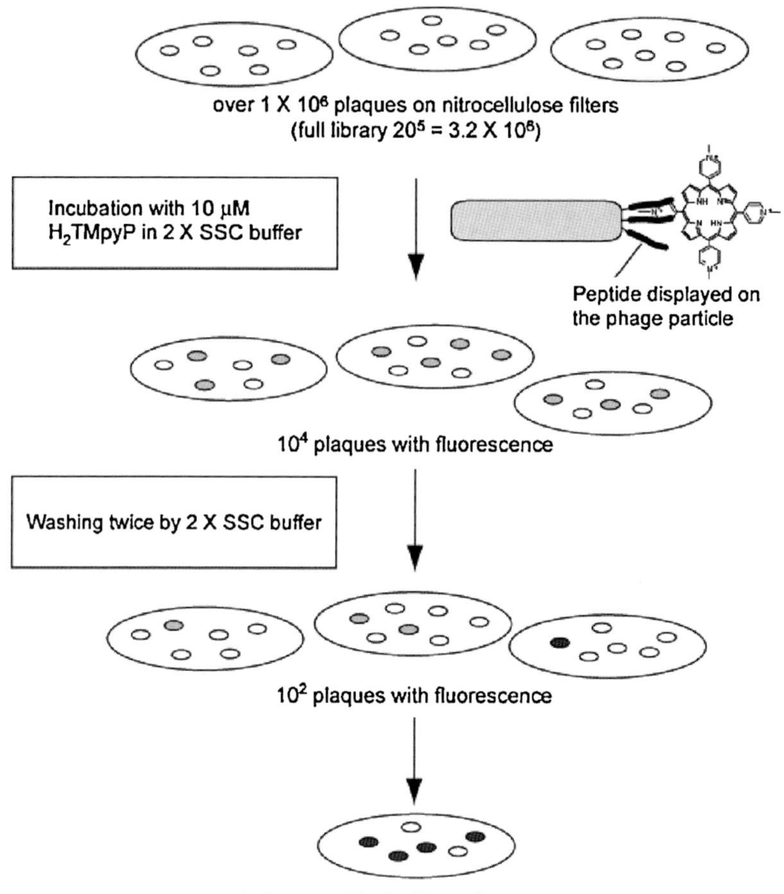

Fig. 8.1 Strategy for screening of porphyrin binding peptide

2.3 SPR Sensor

HASYSC peptides (MW: 667) were immobilized on sensor chips, CM5 and J1, of the BIACORE system (Biacore AB, Uppsala, Sweden). The CM5 chip, which has a 100-nm-thick carboxylated dextran layer on a 50-nm-thick gold layer, was activated by amine coupling of the N-hydroxysuccinimide groups or by thiol coupling of 2-(2-pyridinyldithio)ethaneamine. For the CM5 chip, the peptide was immobilized via the thiol group of cystein by thiol coupling method using 2-(2-pyridinyldithio)etha-neamine hydrochloride (Fig. 8.2). For BIACORE X, a 0.1 milli-degree change of the SPR angle corresponds to 1 resonance unit (RU) and an SPR signal of 1,000 RU corresponds to about $1.0\,ng/mm^2$ of bound protein.[14] The amounts of immobilized HASYSC were 1.6 ng for the thiol coupling[15,16]. About 1.4×10^{12} peptide molecules (2.4 pmol) were immobilized on the SPR chip. The porphyrin solution was introduced onto each SPR chip, and calibration curves were obtained. As shown in Fig. 8.3,

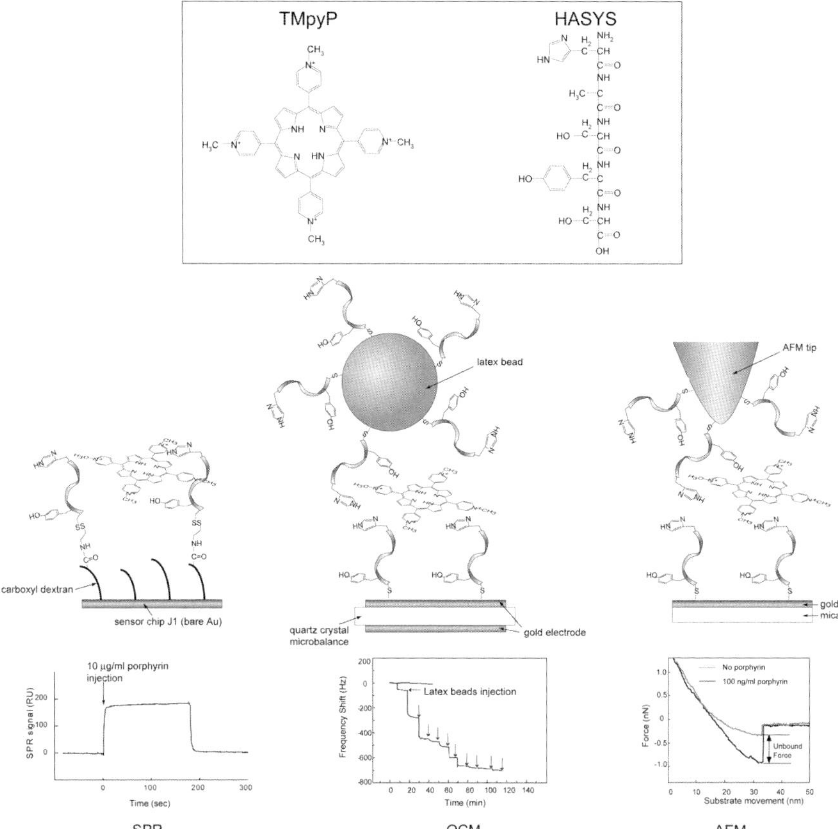

Fig. 8.2 Porphyrin-binding peptide and sensor devices

a linear correlation was obtained in the range from 100 ng/ml to 10 µg/ml with a correlation coefficient of over 0.999.[16] In the saturated condition, a signal of 1,570 RU was obtained indicating that approximately 2.3 pmol of H_2TMpyP was bound on the CM5 chip and almost all the peptide was associated with porphyrin. The detection limit of 0.3 µg/ml is not sensitive enough for detecting hazardous chemical compounds at the ppb or ppt level. However, the SPR sensor is a versatile flow-injection apparatus, and the short peptide can act as a ligand on an SPR chip in spite of the low-binding constant of 10^5 M^{-1}.[13]

2.4 QCM Sensor

The QCM system is a sensor that monitors the mass change caused by adsorption of a substance on the electrode surface. In this case, the crystal was made from AT-cut quartz and had a resonance frequency of 9 MHz. The area of the electrode was 20 mm². The theoretical frequency shift caused by adsorption of 1 ng of a substance

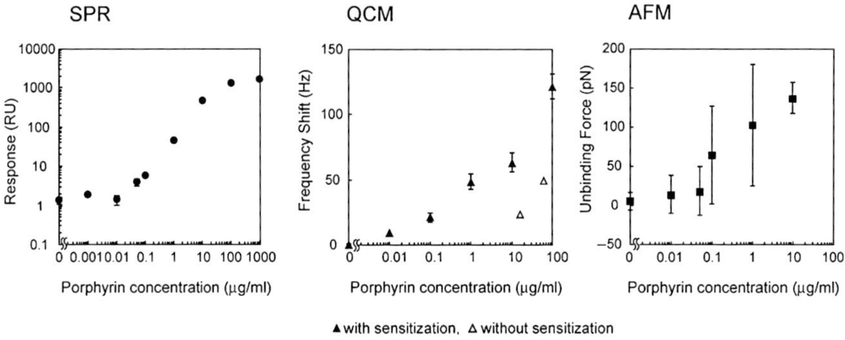

Fig. 8.3 Calibration curves for porphyrin detection using peptide-coupled devices

is about 1 Hz obtained using Kanazawa and Gordon's equation.[17] To oscillate in an aqueous condition, one electrode must be covered and the other functions as a sensor surface. A peptide SAM was formed by dipping the quartz crystal in 1 ml of a peptide solution for 2 h. As a result, the frequency shift indicating peptide immobilization was about 50 Hz, and thus, about 50 ng (75 pmol) of the peptide was assembled on the gold electrode.[18] Because of the roughness of the gold electrode, more peptide than predicted by theory was probably immobilized on the surface. Unfortunately, a specific frequency shift could not be observed even when 1 µg/ml of porphyrin was added to the solution (opened triangle plots). However, a clear shift occurred when polystyrene latex beads, to which HASYSC covalently immobilizes, were added to the measurement vessel. The YAGY[19] or HASYS[13] peptides were selected from a biased combinatorial library or phage plaque library in which two aromatic amino acids were inserted into positions i and $i+3$. Sugimoto et al. has suggested that the stacking of aromatic amino acids and porphyrin is the most likely manner by which a short peptide will interact with a porphyrin molecule. The pi electron orbital of porphyrin is probably large enough to interact with more than two aromatic rings of histidine or tyrosine by stacking. The beads were bound to the electrode surface by sandwich binding of porphyrin to the peptides on the electrode and to those on the beads as shown in the schematic diagram of Fig. 8.2. This implies that at least two porphyrin-binding peptides can associate with one porphyrin molecule. Such sandwich binding is not generally possible for small targets when an immuno-antibody is used. This is an advantage in using short peptides. Thus, peptide immobilized beads can be used as a mass-sensitizer in QCM systems (Fig. 8.3). The detection limit of the sensitized QCM system was 0.1 µg/ml,[18] which was ten times better than that of the SPR. Of course, similar sensitization can be applied to the SPR system but a strong affinity that can maintain sandwich binding in a flow shear stress is required.

2.5 *AFM Sensor with Combinatorial Peptide*

Mechanical force measurements have been carried out to evaluate various interactions using atomic force microscopy. The so-called rupture force or unbinding force, which is the force required to break a covalent bond [20] or break an interaction, [21,22] can be measured by molecules immobilized between a substrate and the apex of an AFM tip as shown in Fig. 8.2. The unbinding forces observed in a force distance-curve inform us of how many bonds or complexes there are between the surfaces. Therefore, an unbinding force measurement using a binder-modified AFM tip can be applied for molecular detection. Since it was demonstrated in the QCM measurement that porphyrin and HASYS can interact with one or two molecules, HASYSC peptide SAMs were formed on both the gold surface of a mica substrate and a gold coated AFM tip. In the porphyrin solution, the tip approached and made contact to the substrate and was then retracted. If sandwich binding occurred when the apex of the tip made contact to the substrate, an unbinding force could be observed in the force curve of the retraction process. The unbinding force increased with the force loading rate, therefore, the unbinding force does not represent a specific characteristic of the molecular interaction. However, the value of the unbinding force correlated with the number of molecules in the sample solution. As shown in Fig. 8.3, the standard deviation in the measurement was large. Nevertheless, the averaged unbinding force increases with the concentration of H_2TMpyP. It is difficult to evaluate the detection limit because the deviations are too large. However, if the deviation could be smaller by improvement of AFM apparatus, the expected detection limit would be about 0.1 µg/ml, the same as that of the SPR sensor.[23]

In this measurement, we calculated the number of complexes involved in an unbinding event to be seven in a solution of 1 µg/ml of H_2TMpyP.[23] The unbinding force at this concentration was 100 pN, which divided by seven gives a value of 14 pN for the unbinding force of a single complex of porphyrin with the peptide. It is proposed that the interaction of the HASYS peptide and TMpyP is a sandwich pi-stacking interaction by histidine and tyrosine, and it is suggested that the unbinding force of such an interaction is around 14 pN.

The three apparatus for detecting molecular interactions were investigated using the same peptide binder, HASYSC, and the peptides were immobilized in the same manner. All the apparatus show the ability for detection. Of course, porphyrins can be detected by a spectroscopic or florescence detection method; however, these apparatus can be applied to substances that do not have spectroscopic or electro-chemical properties.

3 Herbicide-Binding Peptide

We tried to obtain a herbicide binder from a tetrapeptide library. The herbicides diuron (DCMU) [3-(3,4-dichlorophenyl)-1, 1-dimethylurea] and atrazine (2-chloro-4-(ethylamino)-6-(isopropylamino)-s-triazine) are widely used for crop protection (Fig. 8.4). However, due to their widespread use, they contribute greatly to water pollution on a worldwide scale.[24,25] Many risk characterizations or epidemiological studies on triazine herbicides (i.e., atrazine, simazine, cyanazine) have been reported including the clastogenic potentials of atrazine and simazine to Chinese hamster ovary cells[26] or their carcinogenicity.[27,28] The standard level of atrazine in

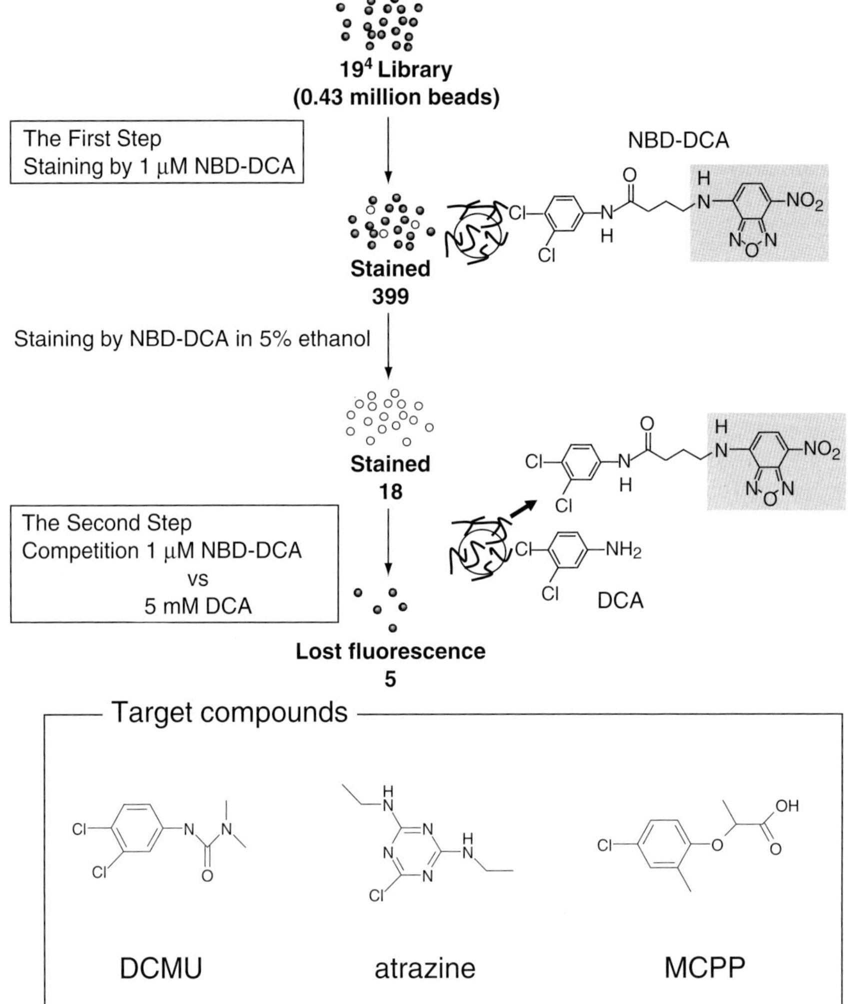

Fig. 8.4 Strategy for herbicide binder screening

drinking water is 3 µg/l in the US. The Environment Agency of Japan sets the basic limits for water pollutants and the standard for simazine is set at 3 ng/ml. Although such herbicidal pollution must be monitored to conserve the safety of the environment, it is impossible to check huge areas of farmland by standard chemical analysis techniques such as gas chromatography mass spectrometry (GC-MS). Some simple methods such as ELISA have been developed; however, detection methods using immuno-antibodies are expensive. The use of short peptides as binders for targeting herbicides is a more cost-effective solution than antibodies. The targets, DCMU and atrazine, are very small compounds, and screening for a tetrapeptide binder for such small chemicals is a very challenging task.

3.1   Screening Strategy to Obtain an Herbicide Binder

A library consisting of 19 natural amino acids, excluding cysteine, was synthesized on resin using a solid phase split synthesis approach. The reactions were based on Fmoc solid-phase peptide synthesis.[29] TentaGel® resins (TGS-NH2) were obtained from Rapp Polymere GmbH (Tübingen, Germany). All building blocks and coupling reagents were purchased from HiPep Laboratories (Kyoto, Japan). One hundred milligrams of resin per vessel were subjected to synthesis such that the number of peptide beads surpassed the number of sequences in the tetrapeptide library (1.3×10^5). The mixing and pooling method was performed to produce the library, using a peptide synthesizer PSSM-8 (Shimadzu, Kyoto, Japan).

We screened peptides that recognize and bind to a partial structure of DCMU from a combinatorial library of tetrapeptides on solid-phase. We used the 3,4-dichloroaniline (DCA) group as bait after taking into consideration the chemical structure of DCMU and its simple synthesis. To screen peptides that bind to the DCA group, a fluorescent-labeled dichloroaniline molecule (NBD-DCA) was synthesized by conjugating a dye with the dichloroaniline through a linker (Fig. 8.4). We selected a NBD moiety for the dye because of its small chemical structure and large Stokes shift.

Two step screening was done for the 430,000 library beads. In the first step, the beads stained by $1\,\mu$M NBD-DCA were selected. The fluorescent beads were picked up by a capillary using an inverted fluorescent microscope for observation. The 399 peptide beads obtained were then washed with ethanol. After washing, the beads still fluorescing were excluded since they had bound irreversibly to the NBD-DCA. Thus, only 18 beads were tested in the second step. In the second step, the beads were subjected to a competitive reaction to select only the peptides binding to the DCA moiety of the NBD-DCA. They were suspended in a $5\,$mM DCA solution containing 5% ethanol for $3\,$h and only those five beads for which the fluorescence decreased were selected. Finally, the sequences of the five beads obtained were determined by a standard Edman degradation method.

3.2   Sequences and Characteristics
of Herbicide-Binding Peptides

The sequences of the five screened peptides were DTYY, DFYA, DNIY, DVIV, and DQFL. Interestingly, all sequences contain an Asp residue at the N terminus but had no further charged residues throughout the rest of the sequence. It seems to be purely coincidental that Asp terminus beads only were screened. N terminus substituted derivatives, DTYYC, ETYYC, and NTYYC, were synthesized and their binding abilities investigated. These three sequences were immobilized on a SPR chip by thiol coupling, and the affinities to DCMU were investigated as described in the previous section. ETYYC showed the same level of SPR response to the

DCMU solution as DTYYC, while NTYYC showed little increase in response (Table 8.1). These results strongly suggest that an N-terminal amino acid bearing a carboxyl group is crucial for binding of DCMU. The carboxyl group may play a role in controlling peptide conformation by electrostatic interactions with the N-terminal amine or by forming hydrogen bonds with the ligand or peptide chain. Thus, the question remains as to why peptide sequences such as EXXX could not survive. The SPR signal of ETYYC to NBD-DCA (4.8 RU) was much smaller than that of DTYYC (12.6 RU). The small affinity to the bait is probably the reason why EXXX could not be screened. This is an unexpected result for which we do not have a certain explanation. However, we can say that if we had chosen only glutamic acid, an acidic amino acid, in the construction of the library, we would not have obtained a positive result. All the sequences were found to have a negative charge on the side-chain of the Asp residue and a positive charge at the N terminus, giving a net charge of zero on the resins. Their hydrophobicities varied greatly and ranged from −0.43 (DTYY) to 2.64 (DVIV) on a conventional hydrophobic scale.[30] These results suggest that the hydrophobicity of the overall polypeptide chain is not the sole factor in determining the binding ability of the peptides. Structural analysis by molecular dynamics was done for the DTYY-DCA complex but no useful information was obtained. The short peptide is not structurally rigid, and it is very difficult to predict its structure under aqueous condition. Docking studies for such low affinity and flexible structure binders are meaningless.

3.3   Sensor Using Herbicide-Binding Peptides

The parameters of the binding kinetics of the five peptides to chlorinated aromatic herbicides, DCMU, atrazine, and MCPP were studied using SPR sensors, and detailed data have been presented in a previous paper.[31] The cysteine-added peptides were synthesized and immobilized on CM5 chips. The DVIVC peptide, which is the most hydrophobic of the peptides, did not show reproducible responses to any of the herbicide solutions. The largest SPR signal ratio of DCMU/MCPP was obtained from the chip with DTYYC. The incremental response to DCMU was greater than to either atrazine or 2-(2-methyl-4-chlorophenoxy) propionic acid (MCPP). In particular, the response to MCPP was very small, suggesting that the peptides bind to molecules that contain a chlorinated six-member aromatic ring, such as DCMU or atrazine. The screening process carried out in this study was

Table 8.1 SPR responses of the N terminus modified derivatives

Peptide	DCMU	fishing bait
	SPR signal (RU)/pmol (peptide)	
DTYYC	26.4	12.6
ETYYC	24.6	4.8
NTYYC	N.D.	N.D.

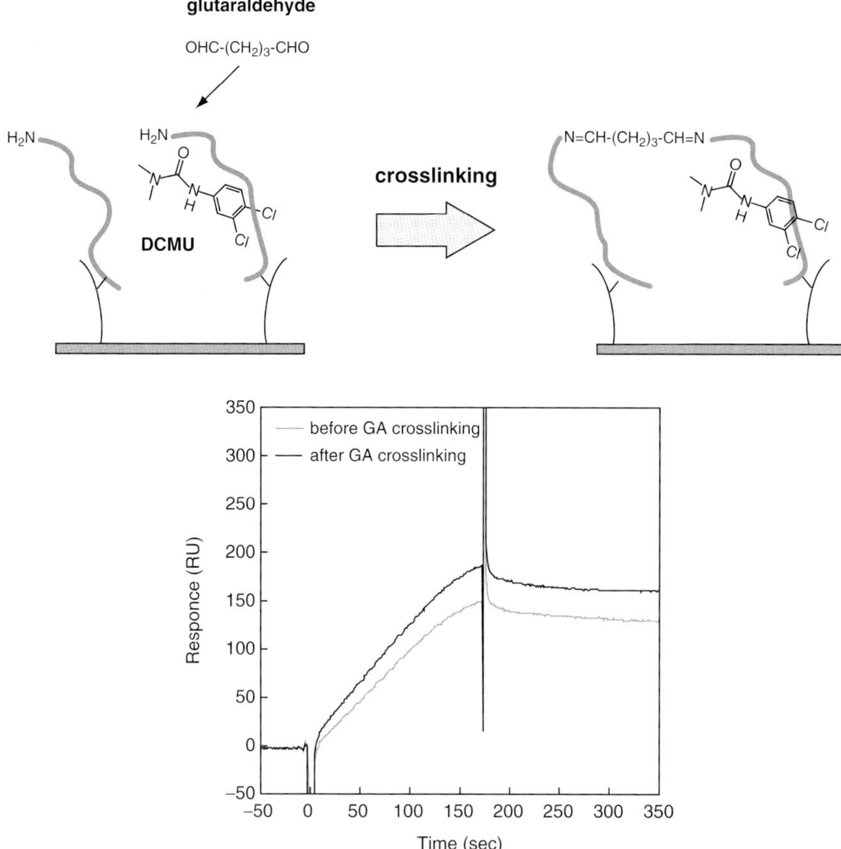

Fig. 8.5 Molecular imprinting on the SPR chip using the obtained binder peptide

successful, since the peptide DTYY was able to recognize the chemical structure of the target molecule. A linear correlation between the herbicide concentration and the SPR signal was observed from 0.25 to 1.0 mM, and the detection limit for both DCMU and atrazine was about 0.25 mM (50 mg/l), which is, of course, insufficient to detect the standard level in drinking water (3 μg/l). The dissociation constant to DCMU was about 0.4 mM. In the screening, 5 mM DCA was needed for the competitive exclusion of the bound NBA-DCA. The screening condition dominated the binding ability of the screened peptides. If one has an objective concentration, the target concentration should be considered in the screening. If the detection method or type of sensor has been chosen before screening, the screening should be performed using the sensor.

The low affinities of the peptides to the targets are due to the flexible structure of peptides. We tried to make the peptide structures rigid on the SPR chip by molecular imprinting. DTYYC was immobilized on the SPR chip by the same thiol coupling as above. Then DCMU solution was introduced to the chip. After confirming an incremental change in the SPR signal, glutalaldehyde (GA) solution was injected to produce intramolecular links between the amines of the peptides to make the peptide structure rigid. The bound DCMU and the excess GA was washed out, and a new DCMU solution was injected into the flow cell. The apparent SPR signal increased when compared with that before cross-linking (Fig. 8.5). The terminal amine group of DTYY is presumably not involved in the molecular recognition. As a result, the binding constant was doubled. Even on carboxydextran functionalized chips, such imprinting methods are effective in increasing the binding ability.

4   Dioxin-Binding Peptide

Dioxin is a well-known toxic chemical in the environment and must be strictly monitored, since it is not a commercial product but is generated from incinerators when burning waste. The method used to accurately determine the concentration of dioxin in the environment is chemical analysis using high-resolution gas chromatography mass spectrometry (HR-GC/MS). The Japanese environmental quality standard for dioxin in the soil is 1,000 pg-TEQ/g. An ELISA or methods derived from ELISAs using immuno-antibodies are the prime candidates as alternatives for GC/MS analysis. In previous reports, proof has been given that antibodies or receptors that consist of natural amino acids have the potential to recognize dioxin molecules.[32] However, denaturation of the antibodies caused by the presence of the organic solvent in the aqueous solutions required in order to dissolve hydrophobic dioxin is a serious problem in the use of antibodies in dioxin assays. If the concentration of organic solvent in the assay solution is lowered, high concentrations of dioxin cannot be dissolved and are not detected precisely. Some synthetic peptides that bind dioxins have been reported: their design is based on the binding pocket of the aryl hydrocarbon receptor (AhR)[5] or CDR domain of a monoclonal anti-dioxin antibody.[33] We tried to screen dioxin-binding peptides using almost the same strategy used to select a herbicide binder. Detailed information about the screening and application of dioxin-binding peptides has been described in previous papers.[34,35]

4.1   Screening Strategy to Obtain a Dioxin Binder

For the dioxin binder screening, some changes were made as shown in Fig. 8.6. A new pentapeptide library was prepared because a strong affinity is required, to detect dioxins at the ppb level. The design of the screening solution is very important since it determines which peptides can be screened. Also, the composition of the screening solution may restrict the detection conditions of the sensor. For the herbicide binder screening, ethanol was used to dissolve the herbicide in an aqueous buffer. The refractive index of ethanol (1.3623) is not very different from that of water (1.3335); therefore, the ethanol-based buffer solution could be used in the

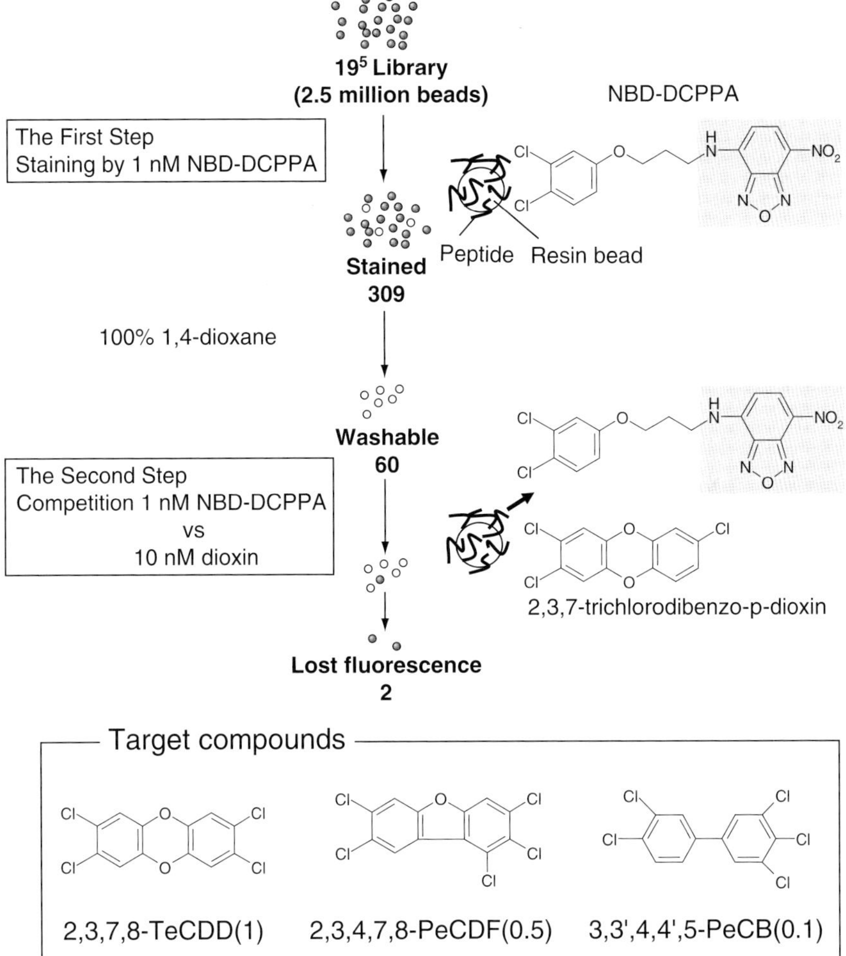

Fig. 8.6 Strategy for dioxin binder screening

SPR measurement, which is carried out with a refractive index between 1.33 and 1.40. Such requirements for the apparatus must be considered when preparing the screening solution. 1,4-Dioxane was chosen to dissolve dioxins since the most toxic target dioxin, 2,3,7,8-tetrachlorodibenzodioxin (2,3,7,8-TeCDD), can be dissolved at high concentrations, and the boiling point of 1,4-dioxane, 101.1 °C, is sufficiently high taking autonomous vaporization during screening into consideration. 1,4-Dioxane's refractive index is 1.4224 and the maximum concentration of 1,4-dioxane in the buffer to measure the SPR signal is 30% (v/v). 20% 1,4-dioxane in a 10 mM phosphate buffer (pH 8.2) was used for this screening.

The dye NBD was again used to synthesize the fluorescent conjugate mimicking the dioxin structure. Its high fluorescent quantum yield in a hydrophobic environment contributed to the low background measurement. In this screening, 3,4-dichlorophenol was employed as a moiety of 2,3,7,8-TeCDD after consideration of the chemical structure of dioxin and its synthetic simplicity. N-NBD-3-(3′,4′-dichlorophenoxy)-1-propylamine (NBD-DCPPA) was synthesized as a labeled compound competing against dioxin. All possible 19^5 (2.5×10^6) sequences were synthesized on TentaGel® resins. Two step competitive screening was carried out in this dioxin binder screening. In the second step, 2,3,7-trichlorodibenzodioxin (2,3,7-TriCDD), which is a nontoxic dioxin derivative and has the most similar structure to 2,3,7,8-TeCDD, was added for competitive binding against NBD-DCPPA. From the 2.5 million beads, only two were selected that bound 2,3,7-TriCDD (10nM) competitively with 1nM NBD-DCPPA in a buffer containing 20% 1,4-dioxane.

4.2 Sequences and Characteristics of Dioxin-Binding Peptides

The only two sequences obtained from the possible 2.5 million beads in the library were FLDQV and FLDQI. Surprisingly, apart from the C terminus, four of the amino acids were identical in the two peptides obtained. Unfortunately, in our investigations using SPR, the peptide sequences separated from the resin beads did not function as binders to dioxins or NBD-DCPPA. Figure 8.7 shows the association and dissociation of NBD-DCPPA with time for FLDQV beads of various diameter from 10 to 130 μm. TentaGel® beads of 130 μm were used for the screening. Association (A) and dissociation (B) curves were fitted to the data using simple one-to-one binding equations.[36] Both the association and dissociation rate constants increase with decreasing resin bead size and the binding constant (K_A) obtained for 30 μm beads is five times lower than that of 130 μm beads, which is 8.3×10^9 M^{-1}.

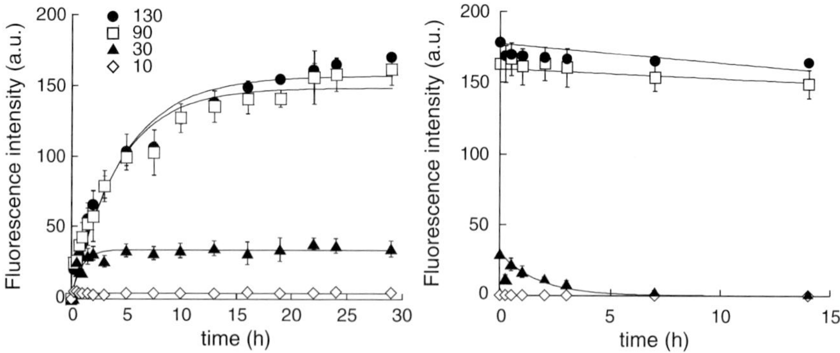

Fig. 8.7 Association and dissociation of NBD-DCPPA with time for various peptide resin bead diameters

This result implies that the resin itself also makes a contribution to the target binding. In this case, the peptide cannot be used as a binder element for the sensor since the entire very complex structure consisting of polystyrene, PEG linker and peptide sequence must be assembled on the sensor platform. In combinatorial screening, such phenomena are not unusual. A peptide and its support, which is a resin or phage body, sometimes work cooperatively to bind the target.

Unfortunately, after cleavage from the resins, neither FLDQV nor FLDQI could bind target dioxins absolutely. Hereby, detection of dioxins using the sensor apparatuses such as SPR and the peptide immobilized on the SPR chip was impossible. However, 130 μm diameter peptide resin beads showed very high affinity to NBD-DCPPA, so we investigated the potential of fluorescence microscopy for detecting dioxins using a bead-synthesized intact peptide. Competitive binding with 2,3,7-TriCDD and 2,3,7,8-TeCDD was investigated. As shown in Fig. 8.8, a typical sigmoidal decrease representing the competitive binding of 2,3,7-TriCDD to DB2 beads against NBD-DCPPA (open circles) was obtained. The IC_{50} of 2,3,7-TriCDD against 1 nM NBD-DCPPA was 1.3 nM. This result indicates that both 2,3,7-TriCDD and NBD-DCPPA have almost equal affinity to the peptide with a 20% 1,4-dioxane solution (same condition as that used in the screening process). With this solution, the correlation for 2,3,7,8-TeCDD was poor; however, increasing the concentration of the organic solvent to 30% was effective for clear competitive binding (closed circles). This phenomenon might be ascribed to poor solubility owing to the higher hydrophobicity of 2,3,7,8-TeCDD than 2,3,7-TriCDD. We

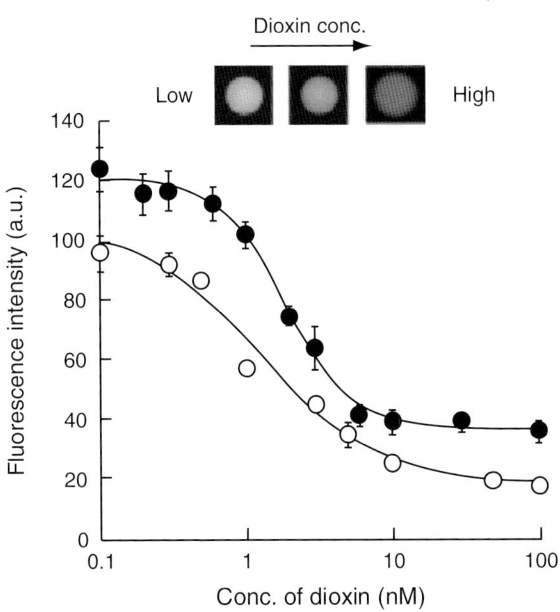

Fig. 8.8 Calibration curves for detecting 2,3,7-TriCDD (*open circle*) and 2,3,7,8-TeCDD (*filled circle*)

chose nontoxic 2,3,7-TriCDD, taking screening safety into account, and a concentration of 20% was adopted, taking the solubility of 2,3,7-TriCDD into account. Changing the organic solvent concentration or other conditions may sometimes be effective in expanding the target variety. These results suggest that there are different optimum concentrations of 1,4-dioxane for each dioxin congener. However, the highly hydrophobic congeners 1,2,3,7,8,9-HxCDD and 1,2,3,4,6,7,8-HpCDD could be bound to FLDQV beads with a 30% 1,4-dioxane solution. Thus, 30% 1,4-dioxane was used for further investigations.

The IC_{50} of 2,3,7,8-TeCDD against 4 nM NBD-DCPPA was about 3 nM. The coefficient of variation for points on the curve of 2,3,7,8-TeCDD was between 1 and 8%. The deviation in fluorescence intensity might be due to irregularities in the resin beads. The detection limit for 2,3,7,8-TeCDD was about 1 nM (0.3 ng/ml). Surprisingly, this sensitivity is comparable to that obtained with ELISAs using anti-dioxin immuno-antibodies.[32,37,38] Mascini et al. developed piezoelectric sensors (QCM) for dioxins using the pentapeptides NFQGI, NFQGQ, and NFQGF. The design was based on a binding pocket of the aryl hydrocarbon (Ah) receptor (Phe-Gln-Gly).[5] They evaluated the potential applicability of the peptides in solid–gas analysis, using a QCM sensor array and succeeded in detecting dioxins at the ppb level. The similarity between the amino acids is discussed in Sect. 4.3.

When using a simple method, the specificity for binding is the most important factor in detection. It was shown that FLDQV binds to chlorinated dibenzo-*p*-dioxin, chlorinated dibenzofuran, and some polycyclic aromatic hydrocarbons (PAHs), and does not bind to nonchlorinated dibenzo-*p*-dioxin and polychlorinated biphenyls (PCBs). These results suggest a process of molecular recognition by the FLDQV peptide. It is clear that FLDQV binds specifically to coplanar aromatic structures similar to 2,3,7-TriCDD. Interestingly, 3,4-dichlorophenol or 3-(3′,4′-dichlorophenoxy)-1-propylamine were not recognized by FLDQV. Furthermore, the fluorescent dye of the conjugate was not exchanged from NBD to the other dyes. The NBD domain of the NBD-DCPPA is probably involved in the molecular recognition, and it demonstrates that the dye or tag-compound used to prepare the bait molecule in the screening strongly affects the result.

4.3   Second Screening to Improve the Sensing Capability of the Peptides

The result of the alanine scan indicates that the components of FLDQV are all important to dioxin binding. In particular, substituting similar amino acids, L2I, D3N, D3E, Q4E, or Q4N, in place of the three internal ones all gave negative results suggesting the structure of the three amino acids sequence LDQ is an essential element in binding dioxin to this peptide. Mascini et al. refers to a sequence on which Kobayashi et al. had proposed a model for a dioxin-binding site composed of NFQGR as the backbone.[39] Kobayashi et al. reported that the central glutamine residue plays an important role in the formation of an energetically stable complex with dioxin. The amine of the glutamine is thought to form a hydrogen bond with the chlorine atom of dioxin [40] or one of its oxygen atoms.[39] The glutamine involved in the screened peptides is not a coincidence. Acidic and amide residues were found in all these dioxin-binding sequences. At the N terminus, aspartic acids were involved in all five herbicide-binding sequences. The sequence, NERF, has been found in a CDR2 of the anti-dioxin monoclonal antibody.[41] At present, we have no clear explanation as to the requirement for an acidic residue for recognizing chlorinated aromatic hydrocarbons.

To optimize the peptide sequence, a library with a single amino acid substitute was prepared using amino acids such as the nonnatural amino acids, naphthyla-lanine (Nal), cyclohexylalanine (Cha), and phenylglycine (Phg), etc. The N and C terminus amino acids were replaced by certain nonnatural amino acids, and some peptides showed stronger affinities to NBD-DCPPA than the original FLDQV (FLDQV's affinity is stronger than FLDQI's). Especially, FLDQ-Phg was used for detection of 2,3,7,8-TeCDD. The sequence FLDQL, which revealed an affinity comparable to FLDQI, was presumably lost in the screening. Only three sequences, FLDQV, FLDQI, and FLDQL, consisting of natural amino acids might have the ability to bind dioxins in this screening condition. The widest fluorescent signal range was observed for the original peptide, FLDQV; however, the steep slope of the calibration curve demonstrates that the detection range is narrow. The detection limits were defined as being 3SD below the average maximum value or 3SD above the average minimum value. The lowest detection limit obtained, 50 pg/ml of 2,3,7,8-TeCDD, was with FLDQ-Phg, and this is about ten times better than the original sequence, FLDQV. Such a screening strategy using a natural amino acid library and optimization using nonnatural amino acids provides us with a superior binder candidate. The target sample of this research project is dioxin in soil. We consider the sensitivity of FLDQ-Phg in soil samples to be high enough to determine whether or not the concentration is above the Japanese environmental quality standard (1,000 pg-TEQ/g).

Cha-LDQV showed less cross reactivity to 1,2,3,4,6,7,8-HpCDD, OCDD, *o*-dichlorobenzene, naphthalene, anthracene, benzo[a]anthracene than did FLDQV. As we predicted, molecular recognition by pentapeptides is not so high; however, the difference in molecular recognition by derivative peptides can be used to distinguish the target.

4.4 Detecting Dioxin in Practical Environmental Soil Samples

Thirty soil samples extracted and purified using the official method employed in Japan were subjected to an on-bead assay, and the values obtained were compared with the toxic equivalent quantity (TEQ) obtained by GC/MS analysis.[35] Dioxin is a generic term for a group of chemicals, which consists of many congeners and the toxicity of each congener is different. Environmental samples contain mixtures of such congeners. To estimate the total toxicity of dioxin congeners in practical samples by the official GC/MS method, the TEQ is calculated. The toxic equivalency factor (TEF) is the relative toxicity of congeners where 2,3,7,8-TeCDD is defined as unity. The TEQ of each congener is calculated as the product of its TEF and the observed concentration from GC/MS. The sum of the TEQ of each congener is calculated as the total TEQ. No clear correlation between the values obtained by on-bead assays and those by GC/MS (TEQ) was observed. The results showed that it was difficult to use the on-bead assay method as a technique for determining the dioxin TEQ concentration. This result is a consequence of the fact that the affinities of dioxin congeners to peptides do not correlate with the TEF. For example, when the peptide affinity to 2,3,7,8-TeCDD (1) is 100%, that to 2,3,4,7,8-PeCDF (0.5) must be 50%. For all dioxin congeners, the correlations must be guaranteed, which is probably not possible with a single peptide sequence. If we can use multiple sequences, which have a variety of affinities to the dioxin congeners, the TEQ concentration can be evaluated using a peptide chip. Recently, peptide arrays have been developed to screen-binding molecules.[42,43] Such a peptide array can be assembled using peptide derivatives for quantitative and qualitative analysis.

5 Importance of Full Library Screening

Our answer to the question of whether a mini "rational" library or a full "random" library of peptides is needed to target small chemical compounds is that full "random" library screening is needed. By selecting amino acids and preparing a mini library, we were unable to find a peptide for neither herbicides nor dioxins, particularly without the knowledge that aspartic acid was the key amino acid for screening both herbicide and dioxin binders. Minimizing the size of the library requires choosing an amino acid from ones similar to those found in antibodies. In selecting glutamic acid as an acidic amino acid based on the sequence NERF found in a CDR2 of the anti-dioxin antibody,[41] we were unable to hit the dioxin binder. This fact could not have been predicted; therefore, we strongly recommend random library screening when constructing a short peptide library, such as a pentapeptide library.

When developing a biosensor using a combinatorially screened peptide, the peptide should, if possible, be screened on the sensor device to be used. It is best to measure the binding signal using the device so that the peptide can be screened directly. For example, the SPR-based biosensor screening methodology developed by Tseng and Chu provides an efficient approach with which to search for ligands in libraries of small organic molecules.[44] The main advantage of this direct screening technology is that the sensor and peptide combination is used directly as a sensor for detecting the target. Moreover, SPR array systems have recently become commercially available, e.g., the Flexchip of Biacore. The peptide binder screened by using such a system can be assembled on the SPR chip and used directly to detect targets in an injected sample solution.

References

1. Morita, Y.; Ohsugi, T.; Iwasa, Y.; Tamiya, E., A screening of phage displayed peptides for the recognition of fullerene (c60). *J. Mol. Catal. B: Enzym.* **2004**, *28*, 185–190

2. Sano, K.; Shiba, K., A hexapeptide motif that electrostatically binds to the surface of titanium. *J. Am. Chem. Soc.* **2003**, *125*, 14234–14235

3. Su, Z.; Leung, T.; Honek, J. F., Conformational selectivity of peptides for single-walled carbon nanotubes. *J. Phys. Chem. B Condens. Matter. Mater. Surf. Interfaces Biophys.* **2006**, *110*, 23623–23627

4. Wang, G.; De, J.; Schoeniger, J.; Roe, D.; Carbonell, R., A hexamer peptide ligand that binds selectively to staphylococcal enterotoxin b: Isolation from a solid phase combinatorial library. *J. Pept. Res.* **2004**, *64*, 51–64.

5. Mascini, M.; Macagnano, A.; Monti, D.; Del Carlo, M.; Paolesse, R.; Chen, B.; Warner, P.; D'Amico, A.; Di Natale, C.; Compagnone, D., Piezoelectric sensors for dioxins: A biomimetic approach. *Biosens. Bioelectron.* **2004**, *20*, 1203–1210

6. Boon, C. L.; Frost, D.; Chakrabartty, A., Identification of stable helical bundles from a combinatorial library of amphipathic peptides. *Biopolymers* **2004**, *76*, 244–257

7. Obataya, I.; Kotaki, T.; Sakamoto, S.; Ueno, A.; Mihara, H., Design, synthesis and peroxidase-like activity of 3alpha-helix proteins covalently bound to heme. *Bioorg. Med. Chem. Lett.* **2000**, *10*, 2719–2722

8. Obataya, I.; Sakamoto, S.; Ueno, A.; Mihara, H., Design and synthesis of 3alpha-helix peptides forming a cavity for a fluorescent ligand. *Biopolymers* **2001**, *59*, 65–71

9. Fukumori, T.; Morita, Y.; Tamiya, E.; Yokoyama, K., Design of peptide that recognizes double-stranded DNA. *Anal. Sci.* **2003**, *19*, 181–183

10. Meyer, S. C.; Gaj, T.; Ghosh, I., Highly selective cyclic peptide ligands for neutravidin and avidin identified by phage display. *Chem. Biol. Drug. Des.* **2006**, *68*, 3–10

11. Qin, C.; Bu, X.; Zhong, X.; Ng, N. L.; Guo, Z., Optimization of antibacterial cyclic decapeptides. *J. Comb. Chem.* **2004**, *6*, 398–406

12. Fink, A. L., Natively unfolded proteins. *Curr. Opin. Struct. Biol.* **2005**, *15*, 35–41

13. Kawakami, J.; Kitano, T.; Sugimoto, N., A selection of short peptides that interact with a porphyrin as a small target by immobilized phage display. *Chem. Commun.* **1999**, 1765–1766

14. Karlsson, R.; Stahlberg, R., Surface plasmon resonance detection and multispot sensing for direct monitoring of interactions involving low-molecular-weight analytes and for determination of low affinities. *Anal. Biochem.* **1995**, *228*, 274–280

15. Nakamura, C.; Inuyama, Y.; Shirai, K.; Nakano, S.; Sugimoto, N.; Miyake, J., Analysis for peptide binding to porphyrin using surface plasmon resonance. *Synthetic Met.* **2001**, *117*, 127–129

16. Nakamura, C.; Inuyama, Y.; Shirai, K.; Sugimoto, N.; Miyake, J., Detection of porphyrin using a short peptide immobilized on a surface plasmon resonance sensor chip. *Biosens. Bioelectron* **2001**, *16*, 1095–1100

17. Kanazawa, K.; Gordon, J. G., The oscillation frequency of a quartz resonator in contact with liquid. *Anal. Chem. Acta* **1985**, *175*, 99–105

18. Nakamura, C.; Song, S.-H.; Chang, S.-M.; Sugimoto, N.; Miyake, J., Quartz crystal microbalance sensor targeting low molecular weight compounds using oligopeptide binder and peptide-immobilized latex beads. *J. Anal. Chim. Acta.* **2002**, *469*, 183–188

19. Sugimoto, N.; Nakano, S., Sandwiching interaction of peptides with a porphyrin. *Chem. Lett.* **1997**, 939–940

20. Grandbois, M.; Beyer, M.; Rief, M.; Clausen-Schaumann, H.; Gaub, H. E., How strong is a covalent bond? *Science* **1999**, *283*, 1727

21. Florin, E. L.; Moy, V. T.; Gaub, H. E., Adhesion forces between individual ligand-recepter pairs. *Science* **1994**, *264*, 415–417

22. Yuan, C.; Chen, A.; Kolb, P.; Moy, V., Energy landscape of streptavidin-biotin complexes measured by atomic force microscopy. *Biochemistry* **2000**, *39*, 10219–10223

23. Nakamura, C.; Takeda, S.; Kageshima, M.; Ito, M.; Sugimoto, N.; Sekizawa, K.; Miyake, J., Mechanical force analysis of peptide interactions using atomic force microscopy. *Biopolymers.* **2004**, *76*, 48–54

24. Solomon, K.; Bake, D. B.; Richards, R. P.; Dixon, K. R.; Klaine, S. J.; Point, T. W. L.; Kendall, R. J.; Weisskopf, C. P.; Giddings, J. M., About atrazine. *Environ. Toxicol. Chem.* **1997**, *15*, 31–76

25. Wauchope, R. D.; Buttler, T. M.; Hornsby, A. G.; Augustijn-Beckers, P. M. W.; Burt, J. P., Scs/ars/ces pesticide properties database for environmental decision making. *Rev. Environ. Contam. Toxicol.* **1992**, *123*, 1–157

26. Taets, C.; Aref, S.; Rayburn, A. L., The clastogenic potential of triazine herbicide combinations found in potable water supplies. *Environ. Health Perspect* **1998**, *106*, 197–201

27. Sathiakumar, N.; Delzell, E., A review of epidemic studies of triazine herbicides and cancer. *Crit. Rev. Toxicol.* **1997**, *27*, 599–612

28. Van Leeuwen, J. A.; Waltner-Toews, D.; Abernathy, T.; Smit, B.; Shoukri, M., Associations between stomach cancer incidence and drinking water contamination with atrazine and nitrate in ontario (canada) agroecosystems, 1987–1991. *Int. J. Epidemiol.* **1999**, *28*, 836–840

29. Chan, W. C.; White, P. D.; Chan, W. C. White, P. D. ed.; Oxford University Press: New York, NY, 2000, pp 41–7630.

30. Eisenberg, D.; Schwarz, E.; Komarony, M.; Wall, R., Amino acid scale: Normalized consensus hydrophobicity scale. *J. Mol. Biol.* **1984**, *179*, 125–142.

31. Obataya, I.; Nakamura, C.; Enomoto, H.; Hoshino, T.; Nakamura, N.; Miyake, J., Development of a herbicide biosensor using a peptide receptor screened from a combinatorial library. *J. Mol. Catal. B: Enzym* **2004**, *28*, 265–271

32. Shan, G.; Leeman, W. R.; Gee, S. J.; Sanborn, J. R.; Jones, A. D.; Chang, D. P. Y.; Hammock, B. D., Highly sensitive dioxin immunoassay and its application to soil and biota samples. *Anal. Chim. Acta.* **2001**, *444*, 169–178

33. Morita, Y.; Murakami, Y.; Yokoyama, K.; Tamiya, E., Synthesis and analysis of peptide ligand for biosensor application using combinatorial chemistry. *Biol. Sys. Eng.* **2002**, *(ACS Symposium Series 830)*, 210–219

34. Nakamura, C.; Inuyama, Y.; Goto, H.; Obataya, I.; Kaneko, N.; Nakamura, N.; Santo, N.; Miyake, J., Dioxin-binding pentapeptide for use in a high-sensitivity on-bead detection assay. *Anal. Chem.* **2005**, *77*, 7750–7757

35. Inuyama, Y.; Nakamura, C.; Oka, T.; Yoneda, Y.; Obataya, I.; Santo, N.; Miyake, J., Simple and high-sensitivity detection of dioxin using dioxin-binding pentapeptide., *Biosens. Bioelectron.* **2007**, *22*, 2093–2099

36. O'Shannessy, D. J.; Brigham-Burke, M.; K., S. K.; Hensley, P.; Brooks, I., Determination of rate and equilibrium binding constants for macromolecular interactions using surface plasmon resonance: Use of nonlinear least squares analysis methods. *Anal. Biochem.* **1993**, *212*, 457–468

37. Roy, S.; Mysior, P.; Brzezinski, R., Comparison of dioxin and furan teq determination in contaminated soil using chemical, micro-erod, and immunoassay analysis., *Chemosphere* **2002**, *48*, 833–842

38. Shimomura, M.; Nomura, Y.; Lee, K. H.; Ikebukuro, K.; Karube, I., Dioxin detection based on immunoassay using a polyclonal antibody against octa-chlorinated dibenzo-p-dioxin (ocdd). *Analyst* **2001**, *126*, 1207–1209

39. Kobayashi, S.; Kitadai, M.; Sameshima, K.; Ishii, Y.; Tanaka, A., A theoretical investigation of the conformation changing of dioxins in binding site of dioxin receptor model; role of absolute hardness-electronegativity diagrams for biological activity. *J. Mol. Struct.* **1999**, *475*, 203–217

40. Procopio, M.; Lahm, A.; Tramontano, A.; Bonati, L.; Pitea, D., A model for recognition of polychlorinated dibenzo-p-dioxins by the aryl hydrocarbon receptor. *Eur. J. Biochem.* **2002**, *269*, 13–18

41. Recinos, A., III; Silvey, K. J.; Ow, D. J.; Jensen, R. H.; Stanker, L. H., Sequences of cDNAs encoding immunoglobulin heavy- and light-chain variable regions from two anti-dioxin monoclonal antibodies. *Gene* **1994**, *149*, 385–386

42. Takahashi, M.; Nokihara, K.; Mihara, H., Construction of a protein-detection system using a loop peptide library with a fluorescence label. *Chem. Biol.* **2003**, *10*, 53–60

43. Wenschuh, H.; Volkmer-Engert, R.; Schmidt, M.; Schulz, M.; Schneider-Mergener, J.; Reineke, U., Coherent membrane supports for parallel microsynthesis and screening of bioactive peptides. *Biopolym. (Peptide Sci.)* **2000**, *55*, 188–206

44. Tseng, M. C.; Chu, Y. H., Using surface plasmon resonance to directly identify molecules in a tripeptide library that bind tightly to a vancomycin chip. *Anal. Biochem.* **2005**, *336*, 172–177

Chapter 9
Combinatorial Libraries of Arrayable Single-Chain Antibodies

Itai Benhar

Abstract Antibodies that bind their respective targets with high affinity and specificity have proven to be essential reagents for biological research. Antibody phage display has become the leading tool for the rapid isolation of single-chain variable fragment (scFv) antibodies in vitro for research applications, but there is usually a gap between scFv isolation and its application in an array format suitable for high-throughput proteomics. In this chapter, we present our antibody phage display system where antibody isolation and scFv immobilization are facilitated by the design of the phagemid vector used as platform. In our system, the scFvs are fused at their C-termini to a cellulose-binding domain (CBD) and can be immobilized onto cellulose-based filters. This made it possible to develop a unique filter lift screen that allowed the efficient screen for multiple binding specificities, and to directly apply library-derived scFvs in an antibody spotted microarray.

1 Introduction

With a long track record in diagnostics and therapy, antibodies are currently the leading biopharmaceuticals.[1,2] For many years, and with increasing pace more recently, antibodies had also served as the "scalpel of the molecular surgeon" in providing opportunities for elucidating the molecular details of protein–protein interactions.[3–5] The rapid rate of gene discovery by large-scale genome projects has generated an unprecedented wealth of data. Unfortunately, subsequent analyses of the resulting sequences have not been able to keep pace. One particular area of concern is the functional study of putatively encoded gene products. Although DNA arrays continue to improve and provide information regarding changes at the

I. Benhar
Department of Molecular Microbiology and Biotechnology,
The George S. Wise Faculty of Life Sciences, Tel-Aviv University, Israel
ben-har@post.tau.ac.il

R.A. Potyrailo and V.M. Mirsky (eds.), *Combinatorial Methods for Chemical and Biological Sensors*, DOI: 10.1007/978-0-387-73713-3_9,

mRNA level,[6] experimental evidence clearly shows a disparity between the relative expression levels of mRNA and their corresponding protein products.[7,8] Therefore, biological function, posttranslational modifications, changes in cellular location, or interactions within complexes resulting from environmental stimuli or disease-state can only be assessed at the protein level. Methods to systematically examine proteins on a genomic scale that are known as "Functional genomics" and "Proteomics" have only recently begun to be developed, and the scope of analysis has thus far been limited.[9–11] As a result, only a fraction of the available informational content from each genomic sequencing effort is currently being extracted for subsequent experimentation. Antibodies that bind their respective targets with high affinity and specificity have proven to be essential reagents for biological research. Although systems for the isolation and production of "whole" IgG antibodies led the quest for therapeutic antibodies, phage display has become the leading tool of making single-chain variable fragment (scFv) antibodies in vitro for more "investigational" applications.[4,12–15] This technology links the antigen-specific immunoglobulin variable domains from both heavy (V_H) and light (V_L) chains into a single DNA coding sequence through a flexible peptide linker.[16,17] Assembled scFv antibodies, fused to a minor coat protein of a bacteriophage particle containing the gene encoding the antibody, can be isolated against any desired target from sufficiently large libraries. The availability of large antibody-phage libraries[18–22] has provided an unlimited source of binders to almost any antigen of choice.

Phage libraries are enriched for specific binding clones by subjecting the phage to repetitive rounds of affinity-selection (bio-panning).[23] Many variations exist to phage selection protocols, which in most cases involve capturing of antibody-displaying phages on immobilized antigen, washing out nonbinders, eluting the bound phages by chemical or enzymatic means and reinfecting host bacteria with the eluted phages. This "panning cycle" is repeated several times (usually 3–6 cycles) until sufficient enrichment of binders allow the identification of individual antigen-specific phage clones by means of an immunoassay such as enzyme-linked-immunosorbent-assay (ELISA).[4,15,24] With most panning approaches, applying repetitive selection cycles result in the diminution of diversity with a few dominant phage clones (usually those having the highest affinity or display efficiency or those directed against immunodominant epitopes, or others that are less "harassing" to the host bacteria and are produced in higher amounts) being recovered.

To isolate the maximum diversity of binders, it is therefore advantageous to analyze clones as early as possible in the selection procedure. Depending on the library size, the number of input phages and the selection scheme, the first round of phage selection generally yields 10^3–10^7 clones. Thus, conventional ELISA screening using 96-well plates allows only a small fraction of the selected clones to be screened. It has been suggested that to increase the screening throughput, one should use larger filter-based assays, or antibody chip approaches that allow up to 10^6 clones to be simultaneously screened on a single filter or array.[19,25–29] Basically, two types of filter-based or array-based screens for antibody libraries have been described. One involved capture of specific binders on a filter coated with specific target antigens that was subsequently probed with a generic ligand.[25,30–32] The second type involved capture of the entire population on a filter coated with a generic ligand such as anti immunoglobulins[27,33] and subsequently probing with the specific antigen. Recently, we described the construction of a human synthetic single-chain Fv antibody library of 2×10^9 independent clones where in vivo formed human complementarity-determining region (CDR) gene fragments were combinatorially incorporated onto the appropriate positions of variable-region master frameworks.[22] The repertoire of scFvs represented in this "Ronit 1" library were cloned into our pCC_{16} phagemid vector that was derived from our prototype pCC vector.[34] As such, the scFvs may be displayed on phage by infection of the host *E. coli* with a helper phage or may be expressed as soluble scFvs as can most phage displayed antibodies of the 3 + 3 type.[15,35] In our system, both (phage displayed and soluble antibody) the scFvs are fused at their C-termini to a cellulose-binding domain (CBD) and can be immobilized onto cellulose-based filters either when displayed on phage or when expressed as soluble scFv-CBD fusions.[34,36] This made it possible to develop an unique filter lift screen that allowed the efficient simultaneous screen for multiple binding specificities, and to apply library-derived scFvs in an antibody spotted microarray.[37] The "Ronit 1" library is a large high quality library that had already been screened against peptide and several proteins or against protein–peptide complexes and the selected scFvs exhibited dissociation constants in the low nanomolar range. In this chapter, I discuss the construction and applications of that library in comparison to alternative solutions that are offered by antibody engineering professionals.[38]

2   Design of the Human Combinatorial "Ronit 1" Library

Combinatorial assembly of antibody-coding DNA is the fulcrum of antibody libraries, since the earliest ones that were reported.[12–14] These libraries were derived from pools of immune cells of immunized donors. Since the antibody heavy and light chains are encoded by separate mRNAs, the pooling of isolated nucleic acids from the source cells resulted in the loss of the heavy-light chain pairs that were present in the individual B-cells. V_H–V_L pair (for scFv libraries) or Fd-light chain pairs (for Fab' libraries) were assembled at random during the cloning steps of library construction. This loss of original pairs made it challenging to recreate such a pair within the library, but this challenge was met with ease owing to the size and efficiency of the affinity-selection process that is inherent of antibody phage display. Hence, little attempt was made to isolate original heavy chain-light chain pairs by applying a "single-cell PCR" approach.[39–41]

The "second generation" of antibody libraries were the naïve, the semi and the fully-synthetic antibody libraries. These were significantly larger than the immune libraries, and the source for antibody coding sequences were large pools of B-cells derived from nonimmunized donors. This made it easier to construct libraries based on human antibody genes, hence most modern libraries are human libraries. In the semisynthetic format, parts of the variable domains (usually CDR3 sequences) were fully or partially randomized by PCR-based processes. These "single-pot" libraries compensated by size and antibody diversity for the lack of in-vivo affinity selection were incorporated in the immune libraries.[18–20,42]

To more deeply explore the sequence space beyond the limits of natural diversity, Borrebaeck and coworkers shuffled in-vivo formed CDRs to generate scFvs in which every CDR originated from a different antibody.[21,43,44] According to this approach, named n-CoDeR,[45] the single-chain antibody fragments are constructed based on a single master framework (FR) for each single-domain using shuffled CDR 1–3 sequences that originate from many different in vivo formed V-genes. The purpose of this approach was to introduce genetic diversity in antibody libraries, where diverse natural CDRs that were subject to "proof-reading" in vivo (since they originated from functional antibodies and may have superior functional qualities compared with CDRs whose sequence diversity is produced in vitro) are combined into a single antibody framework. That single framework is chosen based on parameters of structural robustness and proved performance in recombinant antibody technology, principles that were used by other professional library builders.[20,46] A high degree of functional variation was achieved by means of simultaneous and random combination of six biologically derived CDRs, which can extend the genetic diversity beyond what is naturally created in the immune system.

We constructed the "Ronit 1" antibody phage-display library using this n-CoDeR principle.[44,45] The human germline genes DP-47, DPL-3, and DPK-22 were used as master frameworks for V_H, V_K, and V_λ, respectively. As illustrated schematically in Fig. 9.1, we designed our primer set to amplify all possible human V_H's/V_L's CDRs that belong to the master framework's antibody gene family. The combination of synthetic framework regions that partially overlap with the CDR-amplifying primers allowed their assembly into intact antibody domains in the context of each master framework. The following criteria were considered in the primer set design: The oligonucleotides were designed essentially as described by Jirholt et al.,[43] using sequence information from the V-Base database of human antibody germline sequences (http://vbase.mrc-cpe.cam.ac.uk/) as was present in 2003. The primers were designed to allow as much different CDRs as possible to be recovered, by using degenerate nucleotide positions, and by avoiding restricted 3′ residues.

DP-47[47] that was chosen as a master framework for V_H domains belongs to the V_HIII subgroup. Accordingly, PCR primers were designed to amplify all possible in vivo formed CDRs that belong to that subgroup. The primers were basically identical to the set used by[43] with a few modifications. We included mixed bases at several positions to guarantee full coverage of the entire V_HIII family. Primers

---

Fig. 9.1 (continued) antibody domain in the second overlap-extension PCR. This second PCR yields cloning-ready anti body domains with flanking restriction sites and combinatorially shuffled CDRs. This scheme is for V_H domains and identical scheme with the appropriate primer sets was used to prepare the light chain domains. RE 5′ represents a restriction site for *Nco*I for V_H domains or *Eco*RV for V_L domains. RE 3′ represents a restriction site for *Blp*I for V_H domains or *Not*I for V_L domains

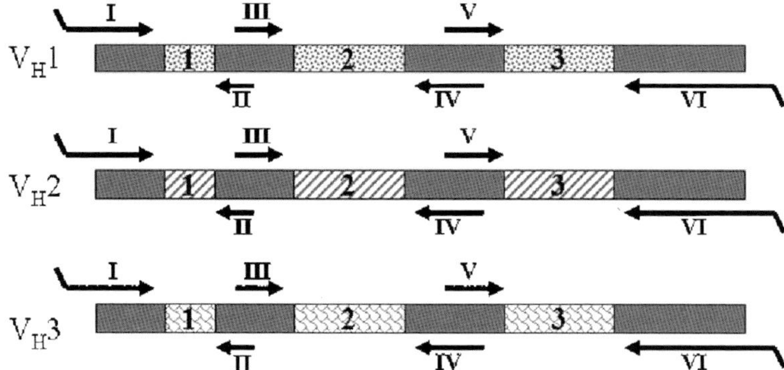

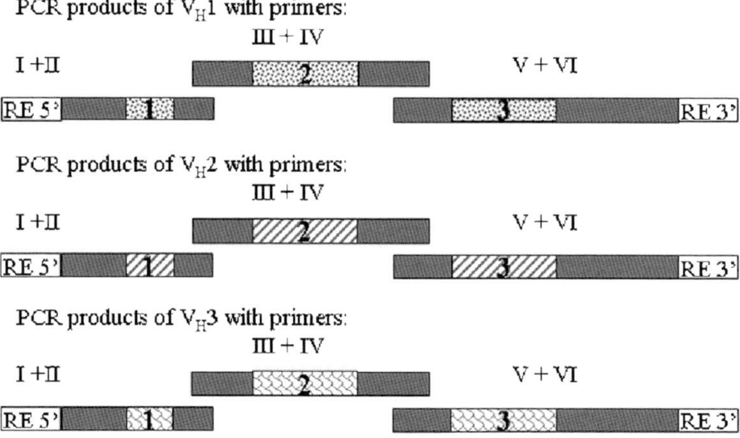

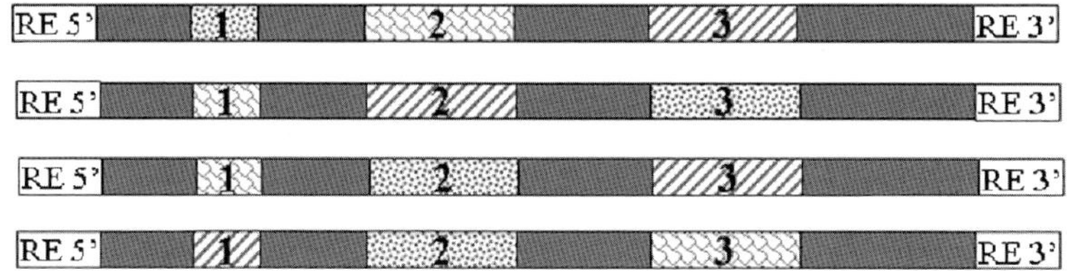

Fig. 9.1 CDR shuffling into a master framework according the n-CoDeR concept. A primer set that was designed according to the master framework was used to amplify individual CDRs (that are numbered 1–3) with flanking framework sequences from the human cDNA libraries that were used as template (V_H1, V_H2, and V_H3 represent a sample of antibody genes that are present in the natural human repertoire that contains millions of such genes). The resulting PCR products contain CDRs with flanking framework sequences that overlap with each other, allowing assembly of intact

located at the termini introduced *Nco*I site at the 5′ end and *Blp*I at the 3′ end of the assembled V_Hs to make them compatible for cloning into our pCC_{16} vector (marked RE 5′ and RE 3′, respectively in Fig. 9.1).

The advantages of using single master frameworks, chosen based on stability and expression/display levels for antibody libraries, have been discussed elsewhere.[20,42,44,48] The DP-47 germline gene was selected as a master framework for V_H domains since it folds easily and is well expressed in *E. coli*.[43,49] DP47 belongs to the V_HIII family that represents about 40% of the human repertoire and is well represented in functional repertoires of antigen-specific antibodies.[20,47] Similarly, the germline genes we selected as master frameworks for V_L domains, DPL-3 and DPK-22, share similar properties and were thus incorporated into several antibody libraries.[20,21,48,50–52]

3   Construction of the Library

As described earlier, we used the germline sequences of DP-47 (3–23)[47] and DPL3 (1 g)[53] as master frameworks for V_H and V-lambda domains, respectively, as was done previously by Jirholt et al.[43] In addition, our library employed DPK22 (A27)[50] as a master framework for V-kappa domains. Our CDR pools were amplified by PCR using human spleen (pool of 14 adults), lymph nodes (pool of 12 adults) and peripheral blood lymphocytes (large unspecified number of adult donors), cDNA pools (Clontech, USA) as template, and the primer set we designed. As shown schematically in Fig. 9.1, the first set of PCR reactions yielded pools of cloning ready antibody domains. The library construction is shown schematically in Fig. 9.2. Our pCC vector system was applied for display and secretion of scFvs.[34] In this system, the scFvs are expressed as cellulose-binding domain (CBD) fusions allowing the immobilization of displaying phages or of secreted soluble scFv-CBD fusions onto cellulose or cellulose-based matrices. We took advantage of this property during library construction and later during scFv isolation and downstream application, as will be described in detail later.

First, we took advantage of being able to immobilize phages that display an intact scFv-CBD fusion on cellulose during the construction of the library. The library was constructed in a two-step process; we first constructed three individual single-domain libraries in our phagemid vector pCC_{16}–gal6(Fv) (Fig. 9.2). This

Fig. 9.2 (continued) cells and rescued with helper phage to recover scFv-CBD-displaying phages. The rescued phage particles were immobilized onto cellulose followed by washing out the unbound phages that represent truncated or otherwise nondisplaying phage clones. Step 4: Cellulose-immobilized phages were used directly as template in a PCR reaction, where synthesized intact variable heavy and light domains were joined together to form complete cassettes encoding full-length scFvs. These were cloned into the pCC_{16} phagemid vector, yielding scFvs-CBD libraries. Pooling together the V_H–V_K and the V_H–V_λ libraries yielded the final product: the "Ronit 1" library

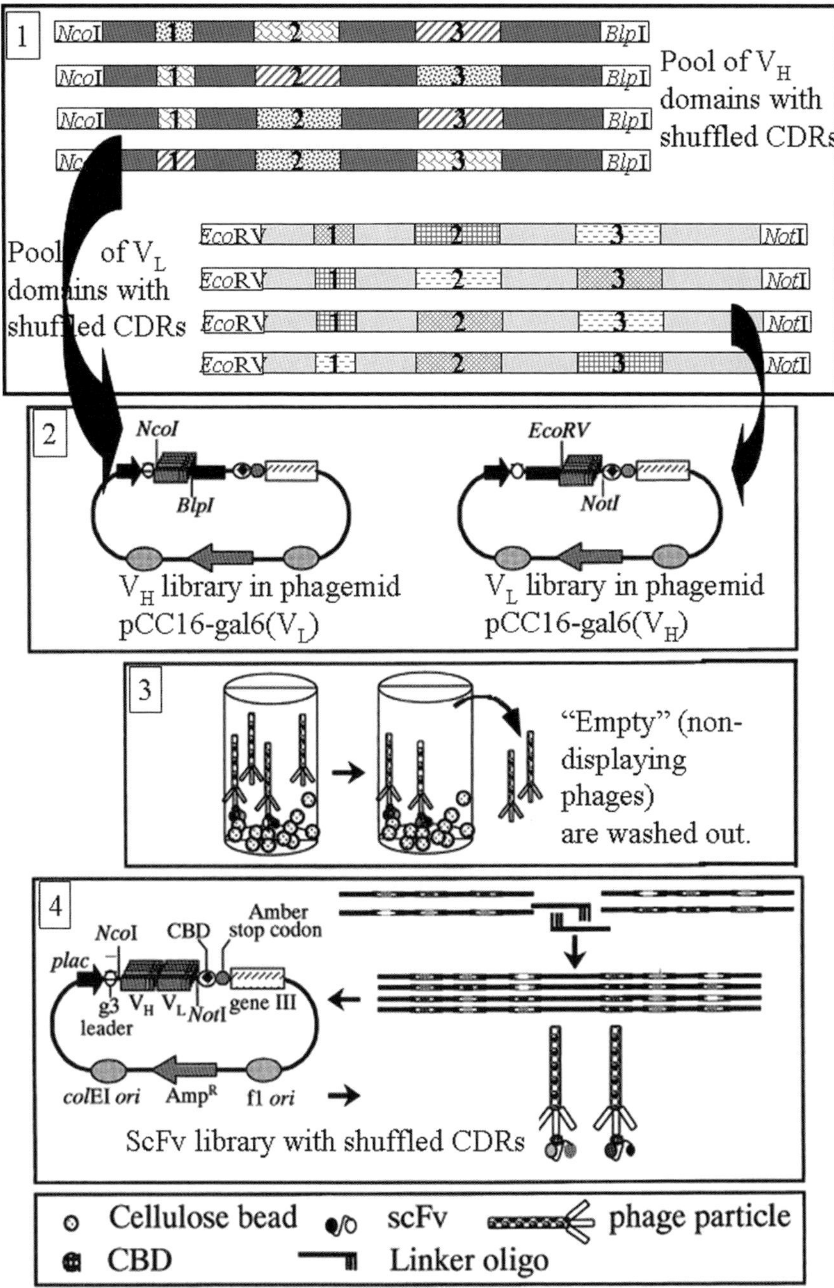

Fig. 9.2 The "Ronit 1" library construction. Step 1: The amplified pools of single antibody domains V_H and V_L domains were prepared as shown in Fig. 9.1. Step 2: following digestion with the appropriate restriction enzymes (*Nco*I and *Blp*I for V_H, *Eco*RV and *Not*I for V_L), the V_H and V_L domains were cloned into the pCC$_{16}$ gal6(Fv)[34] phagemid vector, replacing the resident corresponding gal6 domain. Step 3: The resulting single-domain libraries were introduced into *E. coli*

resulted in three single-domain phage libraries of 10^7–10^8 clones each. The pCC$_{16}$ phagemid vector described herein was modified from that described by us previously,[34] as it carries a 18 residue linker enclosed between *Blp*I and *Eco*RV sits, allowing easy replacement of antibody domains for e.g., in vitro affinity maturation. In addition, the CBD domain carries the R76G mutation obtained by in vitro evolution for efficient display (on phage) or secretion (as free scFv-CBDs).[54] The assembled and purified antibody domains were digested with the appropriate restriction enzymes and were cloned into the phagemid vector backbone replacing the resident corresponding domain (gal6 V_H or V_L). Following DNA ligation, the phagemid DNA was introduced into TG-1 *E. coli* cells by electroporation. The resulting single-domain libraries were "rescued" by infection with M13KO7 helper phage essentially as described[55] and 10^{11} phage particles were immobilized in 1 ml PBS containing 2% crystalline cellulose (SIGMACell 20, Sigma, Israel) for 15 min followed by three washes with PBS as described.[34] The immobilized phages (corresponding to phages that display intact, in-frame scFv-CBD clones) were used directly as template in a PCR reaction to recover the intact single domains followed by their assembly into complete scFv cassettes using the scFv-assembly primers. The scFv PCR products were purified from 1 agarose gels, digested with *Nco*I and *Not*I and ligated into pCC$_{16}$ *Nco*I and *Not*I digested vector DNA. The ligated DNA was introduced into TG-1 *E. coli* cells by electroporation. In that process, the intact single-domain clones that were attached to cellulose via the CBD moiety were then amplified by PCR and assembled together (V_H and V_L) to form two complete scFv libraries, V_H–V_K and V_H–V_λ of about 10^9 clones each. The libraries were mixed at 1:1 ratio with a single library, the "Ronit 1" library being the end product with a final complexity of 2×10^9 clones. It is sometimes debated if the number of transformants truly represents the actual size of the library. Personally, I believe that by following the two-step process in library construction we were never close to exhaust to available sequence space, thus estimating library size from the number of transformed bacteria seems a reasonable estimate.

4   Quality Control of the Library After Construction

To evaluate the quality of the library, we carried our assays of functionality (expression of full length scFv-CBD proteins) and DNA sequence analysis of randomly selected library clones that were not subject to any affinity selection step. Known variable gene sequences available in the March 1997 release of the V-Base sequence directory were used to compare and determine the origin of the CDRs. Our primer set was designed maximized diversity by amplifying all possible human V_H's and V_L's CDRs that belong to the master framework gene families. Indeed, the sequence analysis of random library clones showed the appearance of more than a single residue at these particular positions; therefore, diversity was present at framework positions that flank the CDRs and not restricted to the CDRs only, as was done in preparing the original n-CoDeR libraries.[21,43]

Our success in isolating diverse CDRs can be appreciated from the sequence analysis that was carried out. The following properties of the "Ronit 1" library are evident from these sequences: (a) a wide distribution of CDR lengths was incorporated into the library (Fig. 9.3); (b) The CDRs originated from a wide collection of germline sequences and in many cases correspond to affinity-selected CDRs (no perfect match to a germline CDR sequence).[22] The CDR shuffling that was done according to the n-CoDeR concept was successful, as in all analyzed clones, each CDR originated from a different donor.[22]

Library size and diversity are critical variables for the isolation of valuable antigen binders. Since the probability of finding antibodies with certain specificity is higher in an antibody library that contains a large number of individual clones, the generation of diversity is a key element. It has been reported that libraries that exceed 10^9 clones are sufficient for the isolation of nanomolar affinity binders.[56]

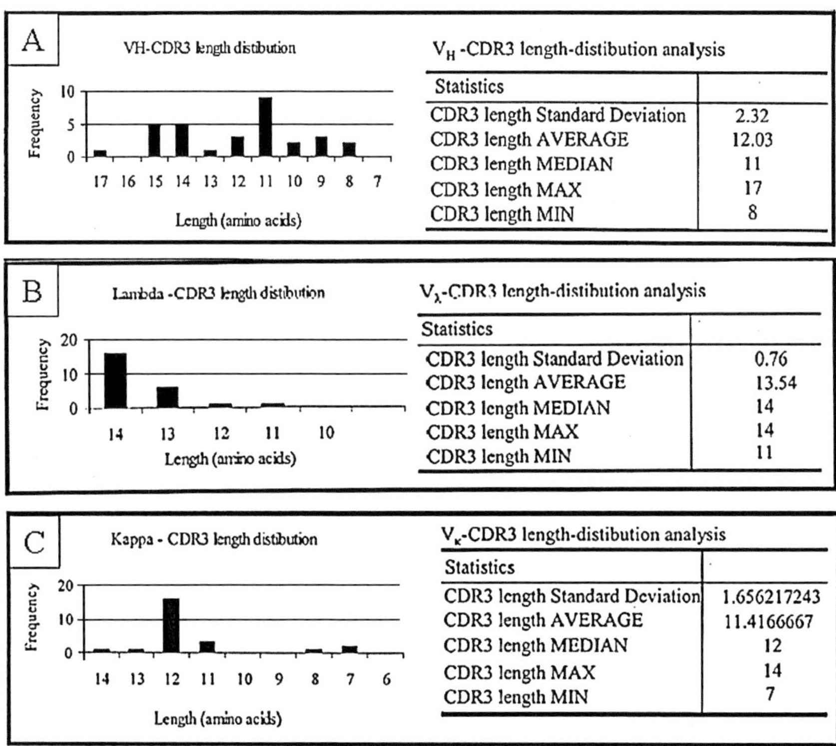

Fig. 9.3 Diversity of randomly selected library clones – V_H genes. A unique computer algorithm was written in "visual basic" and incorporated as a module into Microsoft Excel® for the purpose of sequence alignment and statistical analysis of "Ronit 1" library clones (available from the authors upon request). In addition to identifying the source of each CDR1 and CDR2 and how much it had diversified (shown in[22]) the algorithm also calculated, plotted, and presented a statistical analysis the length distribution of V_H (**a**), V_λ (**b**) and V_K (**c**) CDR3 of all the sequenced clones

To obtain such a library, we constructed it in two steps as described above since stepwise library construction is more likely to yield a large and diverse library.[57] We obtained the desired library of 2×10^9 clones by first constructing single-domain libraries, enriching them for integrity via cellulose binding and eventually mixing them in a predetermined ratio. Indeed, sequence analysis of randomly selected library clones showed that the library is composed of >90% intact clones. The remaining clones contained mostly short in-frame deletions that resulted in a truncated but operational (in terms of cellulose binding) CBD fusion being generated. However, no such clone was recovered following antigen-affinity selection. This further demonstrates the utility of scFv-CBD display[34] that makes it possible to eliminate nondisplaying clones at any desired step. In addition, as predicted from the library size, we successfully isolated nanomolar and even subnanomolar affinity binders from the library. This further shows that a library of about 10^9 intact scFv-CBD clones, which is not too cumbersome to handle, is diverse enough to allow the isolation of high affinity binders without having to resort to in vitro affinity maturation. It should be noted that larger libraries exist, particularly such that were made by companies.[18,20,46,58,59] Libraries in excess of 10^{10} clones require fermentors for growth and require the expertise of highly trained professionals for routine maintenance and use. On the one hand, one should bear in mind that the larger the library – the more likely it is to yield binders with affinities below even 10^{-10} M for any imaginable antigen. On the other hand, such very large libraries are usually not easily available to researchers at academia who wishes to isolate an antibody against his/her favorite antigen without the hassles of negotiating cumbersome Material transfer agreements (MTAs) with industry.

5 Isolation of Specific Binders from the Library: "Standard" Bio-Panning

To demonstrate the potential of the new library to yield antibodies against a diverse set of antigens, we used several different protein and peptide antigens for affinity selection (panning) (Table 9.1). Affinity selection was done by mixing phages and antigen in solution followed by capturing antigen–phage complexes by magnetic beads.[19,55] It seems that it is an effective panning strategy as we were successful in isolating binders for most attempted antigens following two panning cycles. Specific binders were initially identified using phage and soluble scFv ELISA as described[55] and subsequently by the filter lift screen described later. As shown, we were successful in isolating specific binders against all target antigens. The binding properties of library-derived scFvs were evaluated using ELISA, Flow cytometry, and BIAcore analyses. Virtually any library-derived scFv we tested was found to be exclusively specific to its cognate antigen. The affinities of library-derived scFvs ranged between 0.9 and 200nM, indicating that in some cases library derived scFvs will be appropriate for downstream application without further affinity improvement. Concerning the anti HLA-A2-PAP complex scFv we isolated, the rank order of their apparent binding affinities (Fig. 9.9b) perfectly corresponds to their potency of binding as analyzed by

Table 9.1 Affinity selection of scFvs from the Ronit-1 library

Antigen name	Antigen description	No. of panning cycles	Positive binders/ screened clones	No. of uniqne clones[a]	Affinities of selected scFvs[b]
GDEP peptide	38 a.a. putative prostate cancer-associated antigen	2	34/94	14/34	0.9–200 nM
		3	69/87	7/20	
		1[c]	15/500[c]		
		2[c]	80/300[c]		
CD25	1L2 receptor alpha subunit	3	2/48	1/2	
Streptavidin		2	50/94	3/20	
		2[c]	76/674[c]		
		3[c]	231/540[c]		
EGFP	Enhanced green fluorescent protein	3	24/47	1/9	
MBP	The *E. coli* Maltose-binding protein	3 / 2[c]	None/96 3/500[c]	2/2	
Herceptin	Humanized therapeutic antibody	4	36/96	8/24	
NS3	Serine protease of the hepatitis C virus	3[c]	25/300[c]		
HLA-A2-PAP	Prostatic acid phosphatase (PAP) peptide in complex with HLA-A2	3	30/96	6/20	5–50 nM
HLA-A2-HBV	HBV peptide in complex with HLA-A2	3	12/48	7/12	
HLA-A2-G9-209	Peptide derived from melanoma differentiation antigen gp100 in complex with HLA-A2	3	4/48	¼	<5 nM

[a]Determined by *Bst*NI restriction fingerprinting that was carried out on part of the candidates for binders that were identified in the initial screen
[b]Affinity was not determined for all the selected antibodies
[c]Clones isolated by the colony lift screen

flow cytometry (Fig. 9.9a). This reaffirmed an assertion made by others that flow cytometry is a legitimate tool for affinity comparison between different antibodies.[60]

6 Isolation of Specific Binders from the Library: Cellulose Filter Colony Lift Screen

The second time we took advantage of being able to immobilize phages that display an intact scFv-CBD fusion on cellulose was in developing a colony-lift screen for identifying binders. Most of the scFvs described in Table 9.1 were isolated by the standard practice of picking individual colonies after affinity selection and testing them individually in a phage ELISA.[55] As a substitute to this laborious process, our phagemid system employing scFv-CBD fusions made it possible to develop a novel filter-based colony lift screen. Two types of filter-based screens for antibody libraries have been described in the literature. One involved the capture of scFv on an antigen-coated filter that was subsequently probed with a generic ligand[25] screened an array of 27648 immobilized human fetal brain proteins with 12 well-expressed antibody fragments that had not previously been exposed to any antigen. They identified four highly specific antibody–antigen pairs. They suggested that the specificity and sensitivity of binding demonstrates that this "naive" screening approach could be applied to the high throughput isolation of specific antibodies against many different targets in the human proteome. Giovannoni et al.[32] described a method, based on iterative colony filter screening, for the rapid isolation of binding specificities from a large synthetic repertoire of human scFvs. In their system, *E. coli* cells, expressing the library of scFvs, were grown on a porous master filter in contact with a second filter coated with the antigen, onto which scFvs that were secreted by the bacteria were able to diffuse. Detection of antigen binding on the second filter (with a secondary antibody that recognized the scFvs) allowed the recovery of a number of *E. coli* cells, including those expressing the binding specificity of interest from the master filter. Those could be submitted to a second round of screening for the isolation of specific scFvs.

The second approach for filter screening antibodies involved capture of the entire antibody library on a filter coated with a generic ligand and probing with the specific antigen. Skerra et al.[33] grew bacteria harboring plasmid vectors that direct the secretion of Fab fragments into the bacterial periplasm on one membrane. The secreted fragments were allowed to diffuse to a second "capture" membrane that was coated with anti-globulin, and that second membrane was probed with labeled antigen. Watkins et al.[27] devised an approach in which nitrocellulose filters were coated with an anti-immunoglobulin reagent and overlaid with phage-infected bacterial lawn. Subsequently, capture lifts were incubated with biotinylated antigen, and reactive Fabs were identified with streptavidin conjugates. de Wildt et al.[31] used both approaches: they compared the signals derived by direct capture of scFvs on antigen-coated filters followed by probing with a labeled generic ligand with signals derived by capture on a generic ligand-coated filter (in their case the generic ligands were protein A and protein L) followed by probing with labeled antigen. They found that direct capture on antigen gave a consistently higher signal-to-noise ratio. They further claimed that coating the antigen on the filter removes the need to label every antigen, which is time consuming and can be difficult for certain antigens.

Our screen, shown schematically in Fig. 9.4, may be regarded as a combination between the approaches that were undertook by Skerra et al.[33] and by Giovannoni et al.,[32] but is unique in comparison to those colony lift screens in that the scFvs are immobilized virtually irreversibly to the cellulose filter and are then probed with labeled or tagged antigen. Thus our screen could, in principle, allow the reprobing of the same filter with different antigens making it possible to isolate binders of different specificities from one filter. More than that, we could in fact probe the same filter-immobilized scFv-CBD fusions with two probes (labeled differently) mixed together, each recognizing its cognate scFvs. An example is provided in Fig. 9.5, where both streptavidin binders and MBP-NS3 binders were identified on a single filter.

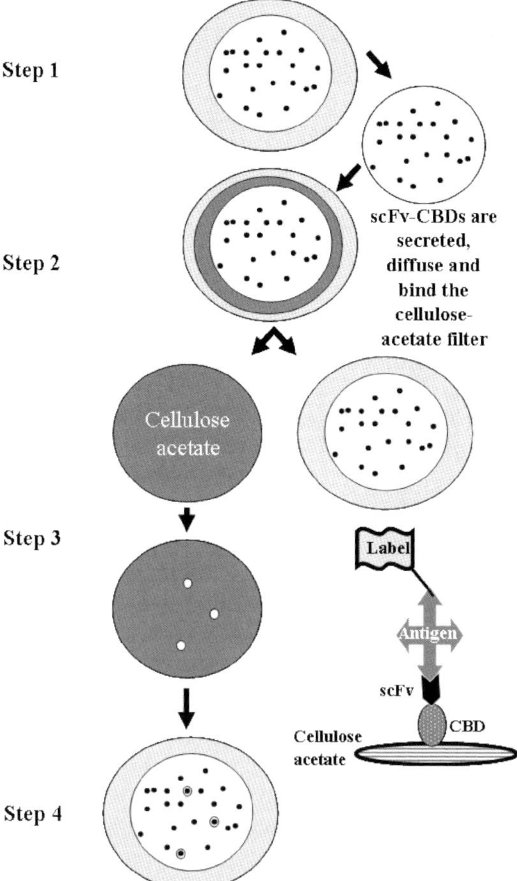

Fig. 9.4 The colony lift screen. Step 1: Bacteria are spread on a Supor (low protein binding) filter on YTG –agar plate and grown for 10 h at 30°C to form microcolonies. Step 2: The Supor filter is placed on top of cellulose-acetate filter on IPTG containing plate at 30°C for 16 h (induction of scFv-CBD expression). During the induction period the scFv-CBD fusion are secreted, diffuse and bind tightly to the cellulose acetate filter. Step 3: The colonies on Supor filter are saved for recovery later. The cellulose acetate filter is processed with labeled antigen as illustrated in the cartoon. Step 4: Probable binders are identified on the cellulose-acetate filter and colonies picked from the master Supor filter. These candidates are later verified by ELISA for specificity

Since the colony lift screen allows the screening of a large number of scFv-CBD fusions on a single filter, it raises the chances of isolating rare binders, as shown here with MBP as target. We failed to identify MBP binders by "traditional" picking and testing 96 random clones of the third cycle output (Table 9.1). However, screening a filter of >500 colonies allowed the identification of three unique MBP binders. When "traditional" screening approaches are used, successive panning that is necessary to achieve the enrichment level required for identification of binders results in the diminution of diversity with a few dominant scFv clones being recovered.

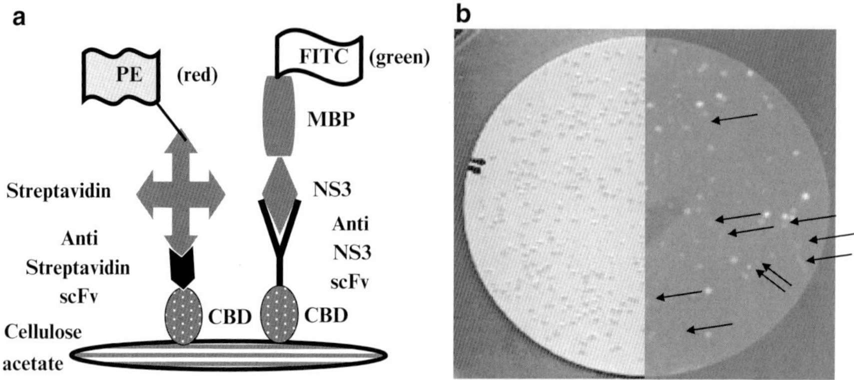

Fig. 9.5 Simultaneous identification of binders of two antigens on a single filter. (**a**) scheme of the screen with two antigens each labeled with a different fluorophore. (**b**). Similar numbers of colonies from two independent selections (done on each antigen separately) were plated on the Supor master filter (the left half in (**b**)). The antibodies that diffused down into the cellulose acetate filter were developed with the fluorescently labeled antigens mixed together at 10 μg/ml each. The filters were washed and analyzed using a Typhoon 8600 reader (Molecular Dynamics). The right-hand half in (**b**) is a mirror image of the left filter with bright spots were green and dimmer spots marked by *arrows* were *red*

It is therefore desirable to isolate binders following fewer, preferably a single panning cycle that would maximize the diversity of unique binders isolated against any particular antigen. Our filter lift screen that allows the screening of a large number of scFvs has that additional advantage in making it possible to screen after fewer panning cycles (even a single cycle, as shown in Fig. 9.6).

One may go further as to contemplate screening an entire library on an array or a chip without resorting to any affinity selection whatsoever. With current protein array manufacturing abilities, less than 10^5 different antibodies may be printed on the most advanced arrays.[62] It is thus clear that large libraries cannot be accessed in their entirety using current array technology. A possible solution to isolating binders without any affinity selection was put forward by Dario Neri and coworkers using their own filter lift screen,[32] which like ours uses a two membrane sandwich. This group poured an entire library (ETH-2 library of 7×10^8 clones) on a porous filter that was placed on top of an antigen coated PVDF filter. Following induction of scFv expression and their secretion, diffusion and binding at antigenic sites on the PVDF filter was processed with secondary antibodies that recognized the FLAG epitope present on their scFvs. This resulted in signals that allowed them to isolate a densely-populated spot from the master porous filter, expand the bacteria in fresh medium, dilute them to a lower density, and repeat the entire process. After three such cycles, they could identify unique binders as single colonies, all by screening, without ever resorting to affinity maturation.

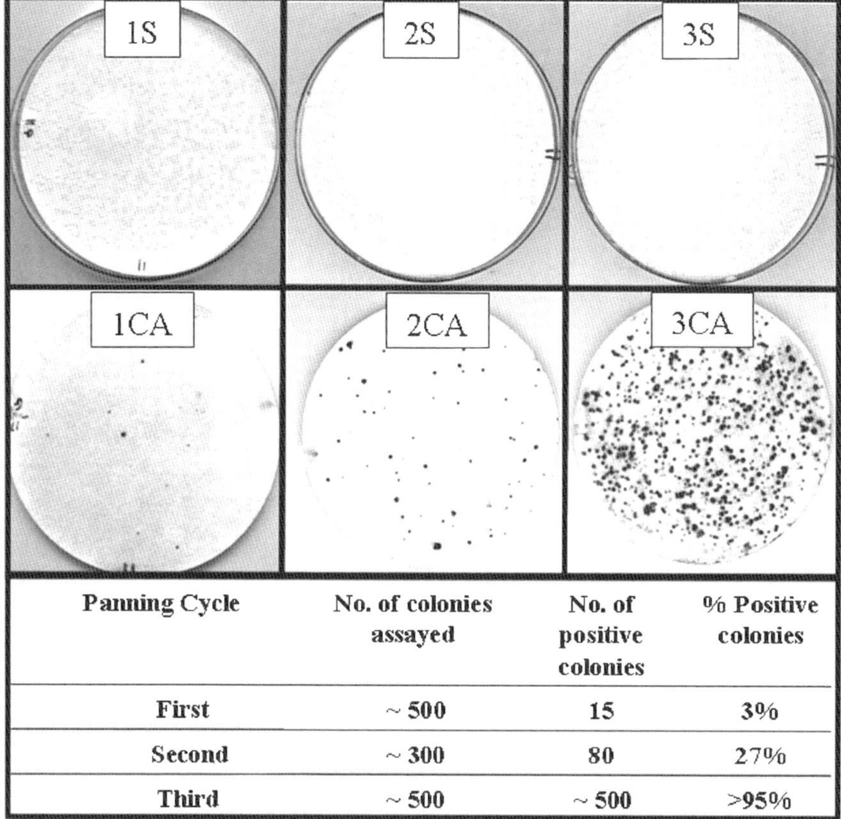

Panning Cycle	No. of colonies assayed	No. of positive colonies	% Positive colonies
First	~ 500	15	3%
Second	~ 300	80	27%
Third	~ 500	~ 500	>95%

Fig. 9.6 Identification of antigen binders by the colony lift screen. *E. coli* TG-1 colonies that were infected with phages from panning cycles I, II, and III outputs, where the GDEP peptide[61] was used in affinity selection, are shown in 1S, 2S, and 3S, respectively. Cellulose acetate membranes were used for the detection of scFvs that specifically recognise the GDEP peptide. The membranes were reacted with 5 µg/ml of biotinylated-GDEP peptide together with HRP-streptavidin. 1CA, 2CA, and 3CA shows the membranes after the development with the "TMB-membrane" (Sigma, Israel) HRP-substrate. Note that candidates for GDEP binders can be identified already after the first panning cycle (filter 1CA)

We adapted that approach to our own filter-lift screen, and could isolate binders of a MBP-EGFP fusion protein, as shown in Fig. 9.7. We named it "pa(i)nless" filter screening.

Since current robotics do not allow full coverage of large libraries in a single screen, the iterative screen described by Giovannoni et al.[32] and by us may represent an attractive alternative. We used the filter lift screen also to further evaluate the quality of the "Ronit 1" library, by probing filter lifts with rabbit polyclonal anti CBD antibodies[36] (data not shown). A positive signal necessarily indicated that the corresponding colony expresses and secrets an intact scFv-CBD, and is not the

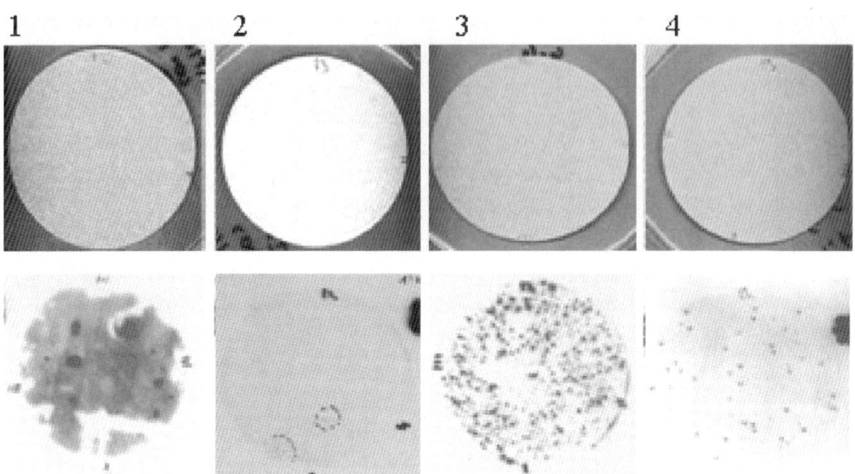

Fig. 9.7 Isolation of EGFP binders by "pa(i)nless" filter screening. TG-1 *E. coli* cells carrying some 10⁸ unselected "Ronit 1" library phagemids were streaked at high density on filter 1 that was probed with antigen + secondary antibodies. A "positive" region was picked from the master Supor filter, diluted and streaked on filter 2 (about 10⁶ cells). The process was repeated, diluting ×100 each step, until well-isolated unique binders were isolated after the fourth cycle. Filters were processed with 5 μg/ml MBP-EGFP fusion protein mixed with HRP-conjugated anti MBP mAb and ECL reagents. Upon later verification by ELISA, they were all found as EGFP binders

result of fortuitous capture of a truncated protein. Indeed, the results from such filters, of homogeneous signals in >95% of the plated colonies were in agreement with our estimation of the library quality in terms of full-length scFv-CBD clones. Furthermore, all antigen positive clones thus probed were indeed intact clones, as was subsequently shown by DNA sequencing.

7 Applications of Library-Derived scFvs

Antibody phage display library-derived scFvs can be used for a multitude of applications as can any antigen-specific scFv derived through various avenues of antibody engineering, as have been extensively reviewed elsewhere, a very nonexhaustive list of reviews.[63–71] The scFvs that we and our collaborators isolated from the "Ronit 1" library were not exception, and we could use the phage-displayed scFvs or soluble forms of the scFvs following subcloning into overexpression systems for ELISA, immunoblotting, flow cytometry, immunostaining of cells, as T cell receptor-like antibodies, as the targeting moieties of recombinant immunotoxins, and for recognitions of crystalline facets of a gallium arsenide semiconductor.[22,72–76]

A case in point are five scFvs we isolated against the MHC-peptide complex HLA-A*0201-PAP.[77] Eisenbach and coworkers previously reported the identification of this highly immunogenic HLA-A*0201-restricted prostatic acid phosphatase-derived peptide (PAP-3) by a two-step in vivo screening in an HLA-transgenic (HHD) mouse system and discussed its potential as a tumor-associated antigen for prostate cancer.[78] In the more recent study, we aimed to elucidate the efficiency of PAP-3-based vaccine upon active antitumor immunization, and applied scFv that are specific to the HLA-A*0201-restricted PAP complex in the process. Following their isolation, the scFvs were evaluated for apparent binding affinity to the recombinant MHC-peptide complex in ELISA, as shown in Fig. 9.8a. The apparent affinities ranged between 5 and 50 nM.

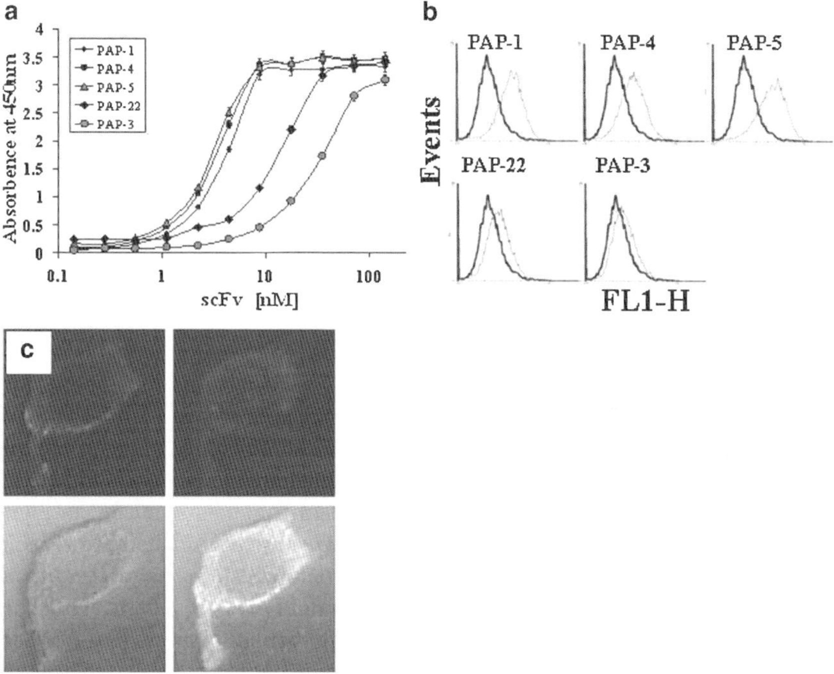

Fig. 9.8 Analysis of HLA-A2-PAP complex-specific scFvs. Anti HLA-A*0201-PAP complex scFvs were assayed by (**a**) ELISA of five HLA-A2-PAP complex-specific scFvs. Twofold dilutions of purified scFvs were used in ELISA-plate coated with 5 µg/ml of recombinant HLA-A*0201-PAP complex protein. (**b**) Flow cytometry: antigen-presenting JY cells were loaded with PAP peptide and stained with the same five purified scFvs selected against the HLA-A2-PAP complex, followed by FITC-labeled anti-myc antibody. *Thick line* represents cells incubated with the secondary FITC-labeled anti-myc antibody alone. (**c**) Immunostaining of D122-HHD-PAP antigen-presenting cells with the PAP-3 scFv. Cells were fixed and stained with 10 mg/ml of myc-tagged soluble PAP-5 scFv followed by FITC-conjugated rabbit myc antibodies (*upper left panel*). The expression of the HLA-A*0201 molecule on the cells was tested with monoclonal antibody BB7.2 followed by rhodmaine-conjugated goat anti mouse antibodies (*upper right panel*) the *lower left panel* is the Nomarski image of the cell while the *lower right panel* is the merge of the *upper panels*

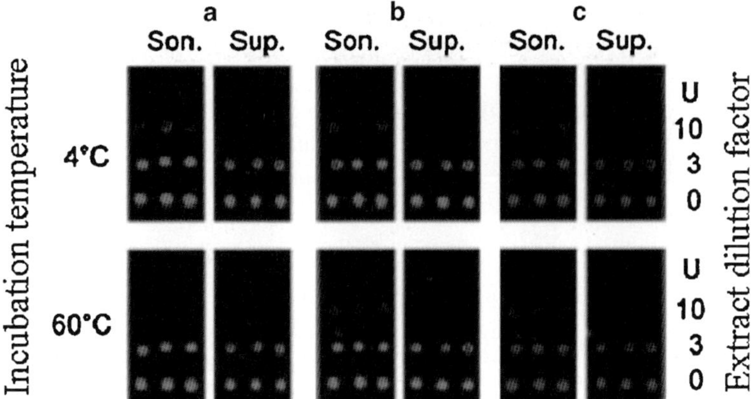

Fig. 9.9 Fluorescence-labeling of anti-SA scFv-CBD on a cellule-based spotted microarray. *E. coli* BL21 cell-extract (Son) and growth medium (Sup) after expression of SA-specific scFv-CBD were incubated for 30 min at 4°C or 60°C and centrifuged for 10 min at 20,800 × *g*. The supernatant fluids were fabricated onto cellulose-coated slides directly (0) or after dilution 3× or 10× with PBS. PBS only was used as negative control (U). The slides were treated for 60 min with 0.5 µg/ml Alexa Fluor 546-labeled streptavidin (**a**) (direct labeling) and scanned. Indirect labeling was demonstrated by incubating the slides with Cy5-labeled chicken anti-CBD (**c**). Double-labeled slides were first incubated for 40 min with Alexa Fluor 546-labeled streptavidin followed by addition of Cy5-labeled chicken anti-CBD (indirect labeling) (**b**)

To explore whether such library-derived scFvs would bind MHC–peptide complexes presented on antigen-presenting cells (APCs) (and therefore may eventually be used to visualize these complexes on the surface of tumor cells), and to estimate their binding affinities in that assay, we performed flow cytometry analysis on peptide loaded antigen presenting cells (Fig. 9.9b). As shown, all five selected scFvs were able to bind JY cells presenting the HLA-A*0201-PAP complex but not JY cells presenting a control irrelevant HLA-A2-HBV complex. Interestingly, the rank order of their apparent binding affinities (Fig. 9.8a) perfectly corresponded to their potency of binding as analyzed by flow cytometry (Fig. 9.8b). This reaffirmed an assertion made by others that flow cytometry is a legitimate tool for affinity comparison between different antibodies.[60] The most potent binder (PAP-5 scFv) was further applied for immunostaining of APCs, generating the expected membrane staining pattern as shown in Fig. 9.8c. The weaker binders failed to generate clear staining patterns on these APCs. The probable reason being that while the flow cytometry analysis was done with JY APCs, artificially loaded with the PAP peptide, and hence displaying the HLA-A*0201-PAP complex at high level, the D122-HHD-PAP cells are APCs derived from a mouse transgenic for HLA-A*0201, which were transfected with the PAP protein. As a result, these cells process the antigen internally and present it at more "natural" levels that are much lower than those presented by the artificial JY system.[79]

8 Applications of Library-Derived scFvs in a Cellulose-Based Spotted Microarray

An additional application of "Ronit 1" library-derived scFvs was in a proof-of-principle incorporation of such scFvs in a cellulose-based, spotted protein microarray. The inclusion of scFvs in arrays and microarrays is a slowly evolve field, mainly due to that fact that in general, scFvs are less stable and robust than whole IgG antibodies that are preferred for inclusion in protein (antibody) arrays. Still, scFvs that are derived from libraries that were designed for robustness are being applied for such applications.

Some recent examples for scFv-based antibody (micro)arrays include the report by Gilbert et al.,[80] who addressed the problem of cross-reactivity that limits the degree of multiplexing in parallel sandwich immunoassays (using monoclonal antibodies (mAbs)). They claimed that antibodies must be prescreened to reduce false positives. In contrast, they applied a second chip surface for the local application of detection antibodies, thereby efficiently eliminating antibody cross-reactions, while illustrating the potential advantages of using single-chain Fv fragments rather than mAbs as capture and detection molecules with this double chip technology. Borrebaeck et al.[81] demonstrated the possibilities of designing protein chips with a capacity of about 8,000 spots/chip, which were based on specially constructed scFvs using nanostructure surfaces with biocompatible characteristics, resulting in sensitive detection in the 600-amol range.[31] They have developed a novel technique for high-throughput screening of recombinant antibodies, based on the creation of antibody arrays based on robotic picking and high-density gridding of bacteria containing antibody genes. This was followed by filter-based ELISA screening to identify clones that express binding antibody fragments. By eliminating the need for liquid handling, they could screen up to 18,342 different antibody clones at a time and, because the clones are arrayed from master stocks, the same antibodies could be double spotted and screened simultaneously against 15 different antigens. Angenendt et al.[82] presented an approach to reduce time, material, and to extend automation beyond the selection process by application of conventional microarray machinery. They were able to express recombinant scFvs in a single inoculation and expression step and subjected them without purification directly to an automated high throughput screening procedure based on "multiple spotting technique." While obtaining comparable sensitivities to enzyme-linked immunosorbent assays, they could minimize manual interaction steps and stream-line the technique to be accessible within the automated selection procedure. Wingren et al.[28] discussed several years of experience in constructing scFv-based microarrays. They have evaluated a protein microarray platform based on nonpuri-fied affinity-tagged scFvs to generate proof-of-principle and to demonstrate the specificity and sensitivity of the array design. To this end, they used their human recombinant scFv antibody library genetically constructed around one robust framework, the n-CoDeR library containing 2×10^{10} clones, as a source for probes. The probes were immobilized via engineered C-terminal affinity tags, his-

or myc-tags, to either Ni^{2+}-coated slides or anti-tag antibody coated substrates. Their results showed that highly functional microarrays could be generated and that nonpurified scFvs could be readily applied as probes, providing a limit of detection in the pM to fM range.

Our example of arraying scFvs[37] had common features with some of the above-mentioned examples, and some features that were unique. For one, the immobilization of scFvs was in some reported cases, a random process of covalent conjugation to a chemically-activated surface, which could harm some of the antibodies used. In other examples[28,31], immobilization of scFvs was through a generic ligand or a peptide tag, which is noncovalent and subject to leaching of the arrayed scFvs or their dissociation under extreme pH or elevated temperatures. In our study, scFv-CBD fusion proteins were expressed directly from library-derived phagemid-carrying bacteria, either from a clarified whole cell extract, or from the crude medium that contains secreted scFv-CBD protein. These crude preparations were applied by a commercial arrayer onto a spotted microarray platform made by coating a conventional microscope slide with cellulose. Hence, at the fabrication level, we could use nonpurified scFv-CBD fusions that bound virtually irreversibly to the microarray surface (very harsh conditioned are required to elute scFv-CBD fusion protein from cellulose.[36]).

ScFv robustness was also tested, by heating the crude scFv-CBD preparations to 60°C before printing them onto the slide and comparing their performance to those kept at 4°C. As shown in Fig. 9.9, no substantial loss of reactivity could be observed. We could take the credit for that stability in our choice of robust frameworks in the design of the library, but it could also be a feature unique to the two anti-streptavidin scFvs we tested, and not common to any imaginable library derived clone.

9 Conclusion

In conclusion, the "Ronit 1" combinatorial antibody phage library described herein is a valuable source of antibodies to essentially any target. To date, we and our collaborators have been able to select specific antibodies to over 20 antigens (Table 9.2). The antibodies may be used as research reagents or as a starting point for the development of therapeutic antibodies. As the list of sequenced genomes and disease-related gene products is expanding rapidly, there will be a growing need for an in vitro and eventually automated method for antibody isolation. The arrayable nature of our scFv-CBD fusions, which may be immobilized onto cellulose-coated surfaces of any size and shape (including cellulose-coated glass slides that are readable by commercial chip readers), should facilitate such automation. As antibodies have been and will be ideal probes for investigating the nature, localization, and purification of novel gene products, our library is envisaged to play an important role in target validation and target discovery in the area of functional genomics and proteomics.

Table 9.2 Selected scFvs from the "Ronit 1" library

Antigen name	Antigen description	Selection methods
GDEP peptide	Synthetic biotinylated 25 amino-acids peptide	Panning in solution followed by capture on streptavidin-magnetic beads
Mutant p53 peptide	Synthetic biotinylated 7 amino-acids peptide	Panning in solution followed by capture on streptavidin-magnetic beads
HLA-A2-PAP complex	Recombinant HLA-A2 in complex with ILLWQPIPV; Peptide derived from prostatic acid phosphatase (PAP)	Panning in solution followed by capture on streptavidin-magnetic beads
HLA-A2-HBV complex	Recombinant HLA-A2 in complex with FLPSDFFPSV; Peptide derived from Hepatitis B Virus core protein	Panning in solution followed by capture on streptavidin-magnetic beads
HLA-A2- G9–209 complex	Recombinant HLA-A2 in complex with YLEPGPVTA; Peptide derived from melanoma differentiation antigen gp100	Panning in solution followed by capture on streptavidin-magnetic beads
CD25	Biotinylated IL-2 receptor α subunit (Tac)	Panning in solution followed by capture on streptavidin-magnetic beads
CD4	Biotinylated CD4	Panning in solution followed by capture on streptavidin-magnetic beads
EGFP (biotinylated)	Biotinylated Enhanced green fluorescent protein	Panning in solution followed by capture on streptavidin-magnetic beads
EGFP (biotinylated)	Biotinylated Enhanced green fluorescent protein	Filter-lift screening
EGFP	Enhanced green fluorescent protein	Panning on plastic-immobilized protein
MBP (biotinylated)	Biotinylated *E. coli* maltose-binding protein	Panning in solution followed by capture on streptavidin-magnetic beads
MBP (biotinylated)	Biotinylated *E. coli* maltose-binding protein	Filter-lift screening
MBP	*E. coli* maltose-binding protein	Panning on plastic-immobilized protein
Streptavidin		Panning in solution followed by capture on streptavidin-magnetic beads
Streptavidin		Filter-lift screening
BSA	Bovine serum albumin	Panning on plastic-immobilized protein
Cetuximab	Herceptin® therapeutic monoclonal antibody	Panning on plastic-immobilized protein

(continued)

Table 9.2 (continued)

Antigen name	Antigen description	Selection methods
Human Fc	Commercial Fc fragment of human IgG	Panning on plastic-immobilized protein
NS3	NS3 serine protease of the Hepatitis C virus	Panning on plastic-immobilized protein
NS3	NS3 serine protease of the Hepatitis C virus	Filter-lift screening
ASGPR	Human asialoglycoprotein receptor	Panning on plastic-immobilized protein
Tumor-specific scFv-1	Undefined antigen on A549 NSCLC cells	Panning on cells following depletion
Tumor-specific scFv-2	Undefined antigen on A549 NSCLC cells	Panning on cells following depletion
Tumor-specific scFv-3	Undefined antigen on A549 NSCLC cells	Panning on cells following depletion

Acknowledgments and Notes The work reported in this chapter was supported in part by a research grant from the Israel Science Foundation, administered by the Israel National Academy for Sciences and Humanities (Jerusalem, Israel). The n-CoDeR® technology is covered by IP rights held by BioInvent, Sweden (http://www.bioinvent.com).

References

1. Hale, G., Therapeutic antibodies–delivering the promise? *Adv. Drug Deliv.* Rev. **2006**, 58, 633–639

2. Baker, M., Upping the ante on antibodies, *Nat. Biotechnol.* **2005**, 23, 1065–1072

3. Wingren, C.; Borrebaeck, C. A., High-throughput proteomics using antibody microarrays, *Expert Rev. Proteomics* **2004**, 1, 355–364

4. Hoogenboom, H. R., Selecting and screening recombinant antibody libraries, *Nat. Biotechnol.* **2005**, 23, 1105–1116

5. Bradbury, A. R.; Velappan, N.; Verzillo, V.; Ovecka, M.; Marzari, R.; Sblattero, D.; Chasteen, L.; Siegel, R.; Pavlik, P., Antibodies in proteomics, *Methods Mol. Biol.* **2004**, 248, 519–546

6. DeRisi, J. L.; Iyer, V. R.; Brown, P. O., Exploring the metabolic and genetic control of gene expression on a genomic scale, *Science* **1997**, 278, 680–686

7. Anderson, L.; Seilhamer, J., A comparison of selected mrna and protein abundances in human liver, *Electrophoresis* **1997**, 18, 533–537

8. Gygi, S. P.; Rochon, Y.; Franza, B. R.; Aebersold, R., Correlation between protein and mRNA abundance in yeast, *Mol. Cell Biol.* **1999**, 19, 1720–1730

9. Martzen, M. R.; McCraith, S. M.; Spinelli, S. L.; Torres, F. M.; Fields, S.; Grayhack, E. J.; Phizicky, E. M., A biochemical genomics approach for identifying genes by the activity of their products, *Science* **1999**, 286, 1153–1155

10. Ross-Macdonald, P.; Coelho, P. S.; Roemer, T.; Agarwal, S.; Kumar, A.; Jansen, R.; Cheung, K. H.; Sheehan, A.; Symoniatis, D.; Umansky, L.; Heidtman, M.; Nelson, F. K.; Iwasaki, H.; Hager, K.; Gerstein, M.; Miller, P.; Roeder, G. S.; Snyder, M., Large-scale analysis of the yeast genome by transposon tagging and gene disruption, *Nature* **1999**, 402, 413–418

11. Carr, K. M.; Rosenblatt, K.; Petricoin, E. F.; Liotta, L. A., Genomic and proteomic approaches for studying human cancer: Prospects for true patient-tailored therapy, *Hum. Genomics* **2004**, 1, 134–140

12. McCafferty, J.; Griffiths, A. D.; Winter, G.; Chiswell, D. J., Phage antibodies: Filamentous phage displaying antibody variable domains, *Nature* **1990**, 348, 552–554

13. Marks, J. D.; Hoogenboom, H. R.; Bonnert, T. P.; McCafferty, J.; Griffiths, A. D.; Winter, G., By-passing immunization. Human antibodies from v-gene libraries displayed on phage, *J. Mol. Biol.* **1991**, 222, 581–597

14. Barbas, C. F. d., Recent advances in phage display, *Curr. Opin. Biotechnol.* **1993**, 4, 526–530

15. Benhar, I., Biotechnological applications of phage and cell display, *Biotechnol. Adv.* **2001**, 19, 1–33

16. Bird, R. E.; Hardman, K. D.; Jacobson, J. W.; Johnson, S.; Kaufman, B. M.; Lee, S. M.; Lee, T.; Pope, S. H.; Riordan, G. S.; Whitlow, M., Single-chain antigen-binding proteins, *Science* **1988**, 242, 423–426

17. Huston, J. S.; Levinson, D.; Mudgett-Hunter, M.; Tai, M. S.; Novotny, J.; Margolies, M. N.; Ridge, R. J.; Bruccoleri, R. E.; Haber, E.; Crea, R.; et al., Protein engineering of antibody binding sites: Recovery of specific activity in an anti-digoxin single-chain fv analogue produced in *Escherichia coli*, *Proc. Natl. Acad. Sci. (USA)* **1988**, 85, 5879–5883

18. Vaughan, T. J.; Williams, A. J.; Pritchard, K.; Osbourn, J. K.; Pope, A. R.; Earnshaw, J. C.; McCafferty, J.; Hodits, R. A.; Wilton, J.; Johnson, K. S., Human antibodies with sub-nanomolar affinities isolated from a large non-immunized phage display library, *Nat. Biotechnol.* **1996**, 14, 309–314

19. de Haard, H. J.; van Neer, N.; Reurs, A.; Hufton, S. E.; Roovers, R. C.; Henderikx, P.; de Bruine, A. P.; Arends, J. W.; Hoogenboom, H. R., A large non-immunized human fab fragment phage library that permits rapid isolation and kinetic analysis of high affinity antibodies, *J. Biol. Chem.* **1999**, 274, 18218–18230

20. Knappik, A.; Ge, L.; Honegger, A.; Pack, P.; Fischer, M.; Wellnhofer, G.; Hoess, A.; Wolle, J.; Plückthun, A.; Virnekas, B., Fully synthetic human combinatorial antibody libraries (hucal) based on modular consensus frameworks and cdrs randomized with trinucleotides, *J. Mol. Biol.* **2000**, 296, 57–86

21. Soderlind, E.; Strandberg, L.; Jirholt, P.; Kobayashi, N.; Alexeiva, V.; Aberg, A. M.; Nilsson, A.; Jansson, B.; Ohlin, M.; Wingren, C.; Danielsson, L.; Carlsson, R.; Borrebaeck, C. A., Recombining germline-derived cdr sequences for creating diverse single-framework antibody libraries, *Nat. Biotechnol.* **2000**, 18, 852–856

22. Azriel-Rosenfeld, R.; Valensi, M.; Benhar, I., A human synthetic combinatorial library of arrayable single-chain antibodies based on shuffling in vivo formed cdrs into general framework regions, *J. Mol. Biol.* **2004**, 335, 177–192

23. Coomber, D. W., Panning of antibody phage-display libraries. Standard protocols, *Methods Mol. Biol.* **2002**, 178, 133–145

24. Barbas, C. F.; Burton, D. R.; Scott, J. K.; Silverman, G. J., Phage display: A laboratory manual; Cold Spring Harbor Laboratory Press: Cold Spring Harbor, NY, **2001**, pp 736

25. Holt, L. J.; Bussow, K.; Walter, G.; Tomlinson, I. M., By-passing selection: Direct screening for antibody-antigen interactions using protein arrays, *Nucleic Acids Res.* **2000**, 28, E72

26. Radosevic, K.; Voerman, J. S.; Hemmes, A.; Muskens, F.; Speleman, L.; de Weers, M.; Rosmalen, J. G.; Knegt, P.; van Ewijk, W., Colony lift assay using cell-coated filters: A fast and efficient method to screen phage libraries for cell-binding clones, *J. Immunol. Methods* **2003**, 272, 219–233

27. Watkins, J. D.; Beuerlein, G.; Wu, H.; McFadden, P. R.; Pancook, J. D.; Huse, W. D., Discovery of human antibodies to cell surface antigens by capture lift screening of phage-expressed antibody libraries, *Anal. Biochem.* **1998**, 256, 169–177

28. Wingren, C.; Steinhauer, C.; Ingvarsson, J.; Persson, E.; Larsson, K.; Borrebaeck, C. A., Microarrays based on affinity-tagged single-chain fv antibodies: Sensitive detection of analyte in complex proteomes, *Proteomics* **2005**, 5, 1281–1291

29. Kwon, Y.; Han, Z.; Karatan, E.; Mrksich, M.; Kay, B. K., Antibody arrays prepared by cutinase-mediated immobilization on self-assembled monolayers, *Anal. Chem.* **2004**, 76, 5713–5720

30. Rodenburg, C. M.; Mernaugh, R.; Bilbao, G.; Khazaeli, M. B., Production of a single chain anti-cea antibody from the hybridoma cell line t84.66 using a modified colony-lift selection procedure to detect antigen-positive scfv bacterial clones, *Hybridoma* **1998**, 17, 1–8

31. de Wildt, R. M.; Mundy, C. R.; Gorick, B. D.; Tomlinson, I. M., Antibody arrays for high-throughput screening of antibody-antigen interactions, *Nat. Biotechnol.* **2000**, 18, 989–994

32. Giovannoni, L.; Viti, F.; Zardi, L.; Neri, D., Isolation of anti-angiogenesis antibodies from a large combinatorial repertoire by colony filter screening, *Nucleic Acids Res.* **2001**, 29, E27

33. Skerra, A.; Dreher, M. L.; Winter, G., Filter screening of antibody fab fragments secreted from individual bacterial colonies: Specific detection of antigen binding with a two-membrane system, *Anal. Biochem.* **1991**, 196, 151–155

34. Berdichevsky, Y.; Ben-Zeev, E.; Lamed, R.; Benhar, I., Phage display of a cellulose binding domain from clostridium thermocellum and its application as a tool for antibody engineering, *J. Immunol. Methods* **1999**, 228, 151–162

35. Hayashi, N.; Kipriyanov, S.; Fuchs, P.; Welschof, M.; Dorsam, H.; Little, M., A single expression system for the display, purification and conjugation of single-chain antibodies, *Gene* **1995**, 160, 129–130

36. Berdichevsky, Y.; Lamed, R.; Frenkel, D.; Gophna, U.; Bayer, E. A.; Yaron, S.; Shoham, Y.; Benhar, I., Matrix-assisted refolding of single-chain fv- cellulose binding domain fusion proteins, *Protein Expr. Purif.* **1999**, 17, 249–259

37. Ofir, K.; Berdichevsky, Y.; Benhar, I.; Azriel-Rosenfeld, R.; Lamed, R.; Barak, Y.; Bayer, E. A.; Morag, E., Versatile protein microarray based on carbohydrate-binding modules, *Proteomics* **2005**, 5, 1806–1814

38. Maynard, J.; Georgiou, G., Antibody engineering, *Annu. Rev. Biomed. Eng.* **2000**, 2, 339–376

39. Jiang, X.; Suzuki, H.; Hanai, Y.; Wada, F.; Hitomi, K.; Yamane, T.; Nakano, H., A novel strategy for generation of monoclonal antibodies from single b cells using rt-pcr technique and in vitro expression, *Biotechnol. Prog.* **2006**, 22, 979–988

40. Coronella, J. A.; Telleman, P.; Truong, T. D.; Ylera, F.; Junghans, R. P., Amplification of igg vh and vl (fab) from single human plasma cells and b cells, *Nucleic Acids Res.* **2000**, 28, E85

41. Wang, X.; Stollar, B. D., Human immunoglobulin variable region gene analysis by single cell RT-PCR, *J. Immunol. Methods* **2000**, 244, 217–225

42. Nissim, A.; Hoogenboom, H. R.; Tomlinson, I. M.; Flynn, G.; Midgley, C.; Lane, D.; Winter, G., Antibody fragments from a 'single pot' phage display library as immunochemical reagents, *Embo J.* **1994**, 13, 692–698

43. Jirholt, P.; Ohlin, M.; Borrebaeck, C. A. K.; Soderlind, E., Exploiting sequence space: Shuffling in vivo formed complementarity determining regions into a master framework, *Gene* **1998**, 215, 471–476

44. Borrebaeck, C. A.; Ohlin, M., Antibody evolution beyond nature, *Nat. Biotechnol.* **2002**, 20, 1189–1190

45. Carlsson, R.; Soderlind, E., N-coder concept: Unique types of antibodies for diagnostic use and therapy, *Expert Rev. Mol. Diagn.* **2001**, 1, 102–108

46. Rothlisberger, D.; Pos, K. M.; Plückthun, A., An antibody library for stabilizing and crystallizing membrane proteins – selecting binders to the citrate carrier cits, *FEBS Lett.* **2004**, 564, 340–348

47. Tomlinson, I. M.; Walter, G.; Marks, J. D.; Llewelyn, M. B.; Winter, G., The repertoire of human germline vh sequences reveals about fifty groups of vh segments with different hypervariable loops, *J. Mol. Biol.* **1992**, 227, 776–798

48. Pini, A.; Viti, F.; Santucci, A.; Carnemolla, B.; Zardi, L.; Neri, P.; Neri, D., Design and use of a phage display library. Human antibodies with subnanomolar affinity against a marker of angiogenesis eluted from a two-dimensional gel, *J. Biol. Chem.* **1998**, 273, 21769–21776.

49. Kobayashi, N.; Soderlind, E.; Borrebaeck, C. A., Analysis of assembly of synthetic antibody fragments: Expression of functional scfv with predefined specificity, *Biotechniques* **1997**, 23, 500–503

50. Cox, J. P.; Tomlinson, I. M.; Winter, G., A directory of human germ-line v kappa segments reveals a strong bias in their usage, *Eur. J. Immunol.* **1994**, 24, 827–836

51. Griffiths, A. D.; Williams, S. C.; Hartley, O.; Tomlinson, I. M.; Waterhouse, P.; Crosby, W. L.; Kontermann, R. E.; Jones, P. T.; Low, N. M.; Allison, T. J.; et al., Isolation of high affinity human antibodies directly from large synthetic repertoires, *Embo J.* **1994**, 13, 3245–3260

52. Ohlin, M.; Borrebaeck, C. A., Characteristics of human antibody repertoires following active immune responses in vivo, *Mol. Immunol.* **1996**, 33, 583–592

53. Huang, S. C.; Jiang, R.; Glas, A. M.; Milner, E. C., Non-stochastic utilization of ig v region genes in unselected human peripheral b cells, *Mol. Immunol.* **1996**, 33, 553–560

54. Benhar, I.; Tamarkin, A.; Marash, L.; Berdichevsky, Y.; Yaron, S.; Shoham, Y.; Lamed, R.; Bayer, E. A., Phage display of cellulose binding domains for biotechnological application, In Glycosyl hydrolases for biomass conversion; M. E. Himmel; J. O. Baker and J. N. Saddler, Ed.; American Chemical Society: Washington, DC, 2001; Vol. 769; 168–189

55. Benhar, I.; Reiter, Y., Phage display of single-chain antibodies (scfvs), In Current protocols in immunology; J. E. Coligan. Ed.; Wiley, New York, NY, 2002; 10.19B.11–10.19B.39

56. Feldhaus, M. J.; Siegel, R. W.; Opresko, L. K.; Coleman, J. R.; Feldhaus, J. M.; Yeung, Y. A.; Cochran, J. R.; Heinzelman, P.; Colby, D.; Swers, J.; Graff, C.; Wiley, H. S.; Wittrup, K. D., Flow-cytometric isolation of human antibodies from a nonimmune saccharomyces cerevisiae surface display library, *Nat. Biotechnol.* **2003**, 21, 163–170

57. Sheets, M. D.; Amersdorfer, P.; Finnern, R.; Sargent, P.; Lindqvist, E.; Schier, R.; Hemingsen, G.; Wong, C.; Gerhart, J. C.; Marks, J. D., Efficient construction of a large nonimmune phage antibody library: The production of high-affinity human single-chain antibodies to protein antigens, *Proc Natl Acad Sci U S A* **1998**, 95, 6157–6162

58. Steukers, M.; Schaus, J. M.; van Gool, R.; Hoyoux, A.; Richalet, P.; Sexton, D. J.; Nixon, A. E.; Vanhove, M., Rapid kinetic-based screening of human fab fragments, *J. Immunol. Methods* **2006**, 310, 126–135

59. Soderlind, E.; Carlsson, R.; Borrebaeck, C. A.; Ohlin, M., The immune diversity in a test tube–non-immunised antibody libraries and functional variability in defined protein scaffolds, *Comb. Chem. High Throughput Screen.* **2001**, 4, 409–416

60. Shimizu, T.; Oda, M.; Azuma, T., Estimation of the relative affinity of b cell receptor by flow cytometry, *J. Immunol. Methods* **2003**, 276, 33–44

61. Olsson, P.; Bera, T. K.; Essand, M.; Kumar, V.; Duray, P.; Vincent, J.; Lee, B.; Pastan, I., Gdep, a new gene differentially expressed in normal prostate and prostate cancer, *Prostate* **2001**, 48, 231–241

62. Holt, L. J.; Enever, C.; de Wildt, R. M.; Tomlinson, I. M., The use of recombinant antibodies in proteomics, *Curr. Opin. Biotechnol.* **2000**, 11, 445–449

63. Ward, E. S., Antibody engineering using *Escherichia coli* as host, *Adv. Pharmacol.* **1993**, 24, 1–20

64. Adams, G. P.; Schier, R., Generating improved single-chain fv molecules for tumor targeting, *J. Immunol. Methods* **1999**, 231, 249–260

65. Worn, A.; Plückthun, A., Stability engineering of antibody single-chain fv fragments, *J. Mol. Biol.* **2001**, 305, 989–1010

66. Cohen, P. A., Intrabodies. Targeting scfv expression to eukaryotic intracellular compartments, *Methods Mol. Biol.* **2002**, 178, 367–378

67. Bilbao, G.; Contreras, J. L.; Curiel, D. T., Genetically engineered intracellular single-chain antibodies in gene therapy, *Mol. Biotechnol.* **2002**, 22, 191–211

68. Leath, C. A., III; Douglas, J. T.; Curiel, D. T.; Alvarez, R. D., Single-chain antibodies: A therapeutic modality for cancer gene therapy (review), *Int. J. Oncol.* **2004**, 24, 765–771

69. Huhalov, A.; Chester, K. A., Engineered single chain antibody fragments for radioimmunotherapy, *Q. J. Nucl. Med. Mol. Imaging* **2004**, 48, 279–288

70. Holliger, P.; Hudson, P. J., Engineered antibody fragments and the rise of single domains, *Nat. Biotechnol.* **2005**, 23, 1126–1136

71. Kreutzberger, J., Protein microarrays: A chance to study microorganisms? *Appl. Microbiol. Biotechnol.* **2006**, 70, 383–390

72. Denkberg, G.; Lev, A.; Eisenbach, L.; Benhar, I.; Reiter, Y., Selective targeting of melanoma and apcs using a recombinant antibody with tcr-like specificity directed toward a melanoma differentiation antigen, *J. Immunol.* **2003**, 171, 2197–2207
73. Artzy Schnirman, A.; Zahavi, E.; Yeger, H.; Rosenfeld, R.; Benhar, I.; Reiter, Y.; Sivan, U., Antibody molecules discriminate between crystalline facets of a gallium arsenide semiconductor, *Nano Lett.* **2006**, 6, 1870–1874
74. Machlenkin, A.; Azriel-Rosenfeld, R.; Volovitz, I.; Vadai, E.; Lev, A.; Paz, A.; Goldberger, O.; Reiter, Y.; Tzehoval, E.; Benhar, I.; Eisenbach, L., Preventive and therapeutic vaccination with pap-3, a novel human prostate cancer peptide, inhibits carcinoma development in HLA transgenic mice, *Cancer Immunol. Immunother.* **2007**, 56, 217–226
75. Azriel-Rosenfeld, R. Ph.D. Thesis, Tel-Aviv University, 2005
76. Valensi, M. M.Sc. Thesis, Tel-Aviv University, 2005
77. Machlenkin, A.; Azriel-Rosenfeld, R.; Volovitz, I.; Vadai, E.; Lev, A.; Paz, A.; Goldberger, O.; Reiter, Y.; Tzehoval, E.; Benhar, I.; Eisenbach, L., Preventive and therapeutic vaccination with pap-3, a novel human prostate cancer peptide, inhibits carcinoma development in hla transgenic mice, *Cancer Immunol. Immunother.* **2007**, 56, 217–226
78. Machlenkin, A.; Paz, A.; Bar Haim, E.; Goldberger, O.; Finkel, E.; Tirosh, B.; Volovitz, I.; Vadai, E.; Lugassy, G.; Cytron, S.; Lemonnier, F.; Tzehoval, E.; Eisenbach, L., Human ctl epitopes prostatic acid phosphatase-3 and six-transmembrane epithelial antigen of prostate-3 as candidates for prostate cancer immunotherapy, *Cancer Res.* **2005**, 65, 6435–6442
79. Pascolo, S.; Bervas, N.; Ure, J. M.; Smith, A. G.; Lemonnier, F. A.; Perarnau, B., Hla-a2.1-restricted education and cytolytic activity of cd8(+) t lymphocytes from beta2 microglobulin (beta2m) hla-a2.1 monochain transgenic h-2db beta2m double knockout mice, *J. Exp. Med.* **1997**, 185, 2043–2051
80. Gilbert, I.; Schiffmann, S.; Rubenwolf, S.; Jensen, K.; Mai, T.; Albrecht, C.; Lankenau, A.; Beste, G.; Blank, K.; Gaub, H. E.; Clausen-Schaumann, H., Double chip protein arrays using recombinant single-chain fv antibody fragments, *Proteomics* **2004**, 4, 1417–1420
81. Borrebaeck, C. A.; Ekstrom, S.; Hager, A. C.; Nilsson, J.; Laurell, T.; Marko-Varga, G., Protein chips based on recombinant antibody fragments: A highly sensitive approach as detected by mass spectrometry, *Biotechniques* **2001**, 30, 1126–1132
82. Angenendt, P.; Wilde, J.; Kijanka, G.; Baars, S.; Cahill, D. J.; Kreutzberger, J.; Lehrach, H.; Konthur, Z.; Glokler, J., Seeing better through a mist: Evaluation of monoclonal recombinant antibody fragments on microarrays, *Anal. Chem.* **2004**, 76, 2916–2921

Chapter 10
A Modular Strategy for Development of RNA-Based Fluorescent Sensors

Masatora Fukuda, Tetsuya Hasegawa, Hironori Hayashi, and Takashi Morii

Abstract Fluorescent biosensors that directly transduce binding events of small molecules into optical signals are valuable tools in the areas of therapeutics and diagnostics. However, construction of fluorescent biosensors from macromolecular receptors with desired characteristics, such as detection wavelengths and concentration ranges for ligand detection, is not a straightforward task. A ribonucleopeptide (RNP) receptor was easy to convert to a fluorescent RNP sensor without chemically modifying the nucleotide in the ligand-binding RNA. The strategy of converting the ligand-binding RNP receptor to a fluorescent RNP sensor was applied to generate fluorescent ligand-binding RNP libraries by utilizing a pool of RNA subunits obtained from the in vitro selection of ATP-binding RNPs and various fluorophore-modified peptide subunits. Simple screening of the fluorescent RNP library based on the fluorescence emission intensity changes in the absence and presence of the ligand afforded a wide variety of fluorescent RNP sensors with emission wavelengths ranged from 390 to 670 nm. Screening of the fluorescence emission intensity changes in the presence of increasing concentrations of ligand provided RNP sensors responding at wide concentration ranges of ligand. The combinatorial strategy using the modular RNP receptor enables tailoring of a fluorescent sensor for a specific ligand without knowledge of detailed structural information for the macromolecular receptor.

1 Introduction

Biosensors that directly transduce binding events into measurable optical signals are valuable tools in the areas of therapeutics and diagnostics.[1,2] Especially, the fluorescent biosensors have been the most successful, and still have the most promising future for practical application. Macromolecular receptors, such as

M. Fukuda, T. Hasegawa, H. Hayashi, and T. Morii (✉)
Institute of Advanced Energy, Kyoto University, Uji, Kyoto 611-0011, Japan
t-morii@iae.kyoto-u.ac.jp

R.A. Potyrailo and V.M. Mirsky (eds.), *Combinatorial Methods for Chemical and Biological Sensors*, DOI: 10.1007/978-0-387-73713-3_10,
© Springer Science + Business Media, LLC 2009

proteins[3-10] and nucleic acids,[11-28] are modified with reporter groups by chemical method to transduce ligand binding into measurable optical signals.

A variety of optical sensors for small ligands has been successfully constructed by the site-specific incorporation of fluorophores to proteins.[3-10] For example, a biosensor designed from a natural protein domain optically responds to the concentration change of D-*myo*-inositol-1,4,5-trisphosphate (IP$_3$),[10] a second messenger that controls the cellular Ca^{2+} concentration. As a natural protein domain with high affinity and specificity to IP$_3$, the pleckstrin homology (PH) domain, a −120 amino acid protein domain identified in several key regulatory proteins,[29,30] was utilized as a scaffold for the construction of IP$_3$ biosensors. The PH domain of phospholipase C (PLC) δ_1 binds to IP$_3$ with a submicromolar affinity and associates with lipid vesicles containing phosphatidylinositol-4,5-bisphosphate, but weakly binds to other inositol phosphates and phosphatidylinositol-4-phosphate.[31,32] To convert the PH domain to a fluorescent IP$_3$ sensor, invariant cysteine residues of the PH domain from rat PLC δ_1 were mutated to serine residues, followed by an introduction of unique cysteine near the IP$_3$ binding site. The three-dimensional structure of the PH domain from PLC δ_1 and IP$_3$ complex[33] was utilized to identify locations where a single cysteine mutation could be introduced for the covalent attachment of environmentally-sensitive fluorophores without greatly affecting the affinity to the ligand. Three kinds of mutant PH domains with an unique cysteine residue were constructed, and four thiol reactive fluorophores were attached at each of these positions. These 12 different fluorophore-labeled PH domains were purified, then evaluated for their optical IP$_3$ responses to afford usable fluorescent IP$_3$ sensors.

Construction of fluorescent biosensors responding at desirable wavelengths and expedient ligand concentration ranges is not a straightforward task. The first step to construct a fluorescent biosensor requires an effort to find a macromolecular receptor with appropriate affinity and specificity to the target, which is not always available. The second step converts the macromolecular receptor to a fluorescent sensor usually by chemical modification of the receptor with a fluorophore with desired optical characteristics. Because the direct coupling between the ligand binding and the optical signaling events of fluorescent biosensors relies on the ligand binding-induced conformational change unique to each macromolecular receptor,[3-27] the fluorophore-labeled receptor is not always guaranteed to execute the expected optical signals. The chemical modification of the receptors to fluorescent sensors involves following difficulties to overcome (1) prediction of an appropriate position for the fluorophore modification and choice of a suitable fluorophore to respond to desired emission properties of biosensors are challenging even with detailed structural information; and (2) mutations or chemical modifications of biomolecular receptors often cause a loss of binding property or stability of the parent receptor. Many mutants of the receptor are constructed and modified with a series of fluorophores, and then resulting fluorophore-labeled receptors have to be evaluated for the fluorescence responses to obtain macromolecular-based sensors with desired optical sensing properties.

The availability of macromolecular receptors with appropriate affinity and specificity to the target has been expanded by in vitro selection[34] of RNA or DNA aptamers for various targets ranging from small molecules to proteins or even cell membranes.[35] Aptamers or the RNA or DNA receptors obtained by SELEX (systematic evolution of ligands by exponential enrichment)[34] are suitable for the structural base for biosensors because various three-dimensional shapes characterized by stems, loops, bulges, hairpins, pseudoknots, triplexes, or quadruplexes allow high affinity and selectivity. RNA or DNA aptamers have been utilized as macromolecular receptors to construct fluorescent aptamer sensors (sensing aptamers). Fluorescent aptamer sensors for adenosine-triphosphate (ATP) have been reported by utilizing ATP-binding DNA or RNA aptamers. These fluorescent aptamer sensors were constructed based on three-dimensional structures of the ATP-binding RNA aptamer or DNA aptamer.[36–38] Various constructs were made to place fluorescent dyes adjacent to the ligand-binding site, and the signaling abilities of the resultant fluorescent aptamers were evaluated by determining whether changes in fluorescence intensity occurred in the presence of the ATP. Among the fluorophore-labeled aptamers, only a few aptamers revealed marked increases in fluorescence intensity in the presence of ATP. In the case of constructing fluorescent aptamer sensors by chemically incorporating fluorophore-labeled nucleotides into the aptamers,[11,12,14–20,22–26,28] the difficulties described above also remained to overcome.

A new strategy has been reported that enables isolation of fluorescent ribonucleopeptide (RNP) sensors with a variety of binding and signal-transducing characteristics, i.e., high signal-to-noise ratios, detection wavelengths, and concentration ranges for the ligand detection.[39] A modular structure of RNP is ideal for construction of a fluorescent sensor without a chemical modification of the ligand-binding region wihin the RNA aptamer. This section describes a strategy to construct fluorescent ATP sensors by utilizing an RNP complex (Fig. 10.1). ATP-binding RNP receptors were obtained by in vitro selection of a library of stable RNP complexes[40,41] composed with HIV Rev-peptide and its target RNA sequence (RRE).[42] The RNA subunit of RNP was utilized to construct a ligand-binding cavity by in vitro selection, while further functionalization of the peptide subunit was possible. The ATP-binding

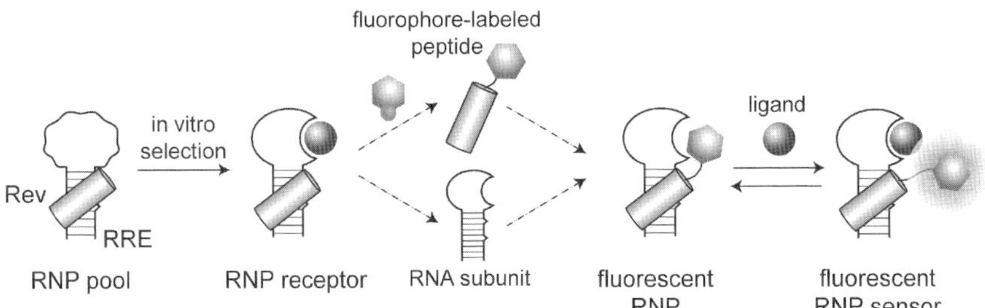

Fig. 10.1 Direct conversion of a ribonucleopeptide (RNP) receptor to a fluorescent sensor. A complex of the HIV-Rev peptide and RRE RNA was used as a framework for the RNP receptor and RNP sensor

RNP receptor can be converted to a fluorescent ATP sensor without chemically modifying the ATP-binding RNA region. The RNA subunit of the ATP-binding RNP and a Rev peptide modified with a fluorophore formed a stable fluorescent RNP complex that showed an increase in the fluorescence intensity with binding to ATP. The step to convert the ATP-binding RNP receptor to a fluorescent ATP sensor was applied to generate fluorescent ATP-binding RNP libraries. Combination of a pool of RNA subunits of ATP-binding RNPs and a variety of fluorophore-modified peptide subunits afforded fluorescent RNP libraries, from which RNP sensors with expedient optical and binding properties were selected in a convenient manner. The selection strategy of fluorescent RNP sensors described here would potentiate generation of an array of fluorescent sensors for analytes with a variety of detection wavelengths and wide concentration ranges.

2 Modular Strategy for Tailoring Fluorescent Biosensors from RNP

2.1 Conversion of an ATP-Binding RNP Receptor to a Fluorescent ATP Sensor

An ATP-binding RNP receptor can be converted to a fluorescent ATP sensor to replace the Rev peptide of an ATP-binding RNP receptor with a fluorophore-labeled Rev peptide (Fig. 10.1). The ATP-binding RNP was obtained by in vitro selection method.[39,40] The size of random nucleotides within the RNA subunit was 30 nucleotides (RRE30N) (Fig. 10.2a). The RNP receptors obtained by the in vitro selection against the ATP-agarose resin were assessed for the ATP-binding characteristics. An RNP complex A28/Rev had ATP-binding activity with a dissociation constant (K_D) of 15.2 μM (Fig. 10.2b) and discriminated ATP over other ribonucleotides GTP, CTP, and UTP efficiently (Fig. 10.2c).

The RNP receptor of A28/Rev was converted to a fluorescent ATP sensor by complexation of pyrene-labeled Rev (Pyr-Rev) and A28 RNA. Fluorescence responses of A28/Pyr-Rev were evaluated in the absence or presence of ATP. Addition of ATP resulted in a two increase of the fluorescence intensity of A28/Pyr-Rev (Fig. 10.3a). A significant change in the fluorescence intensity of A28/Pyr-Rev was induced only with ATP, not with other NTPs, indicating that the specificity of A28/Pyr-Rev to ATP is parallel to that of the parent A28/Rev receptor (Fig. 10.3b). Titration analysis of the changes in fluorescence intensities of A28/Pyr-Rev by ATP gave a K_D value of 6.6 μM for the A28/Pyr-Rev-ATP complex, which was in good agreement with the K_D value for the complex of parent A28/Rev and ATP (Fig. 10.3b). The results indicate that reconstitution of the ATP-binding RNP receptor with a fluorophore-labeled Rev peptide simply and effectively convert ATP receptors into fluorescent ATP sensors without diminishing the affinity and selectivity of the parent ATP receptors.

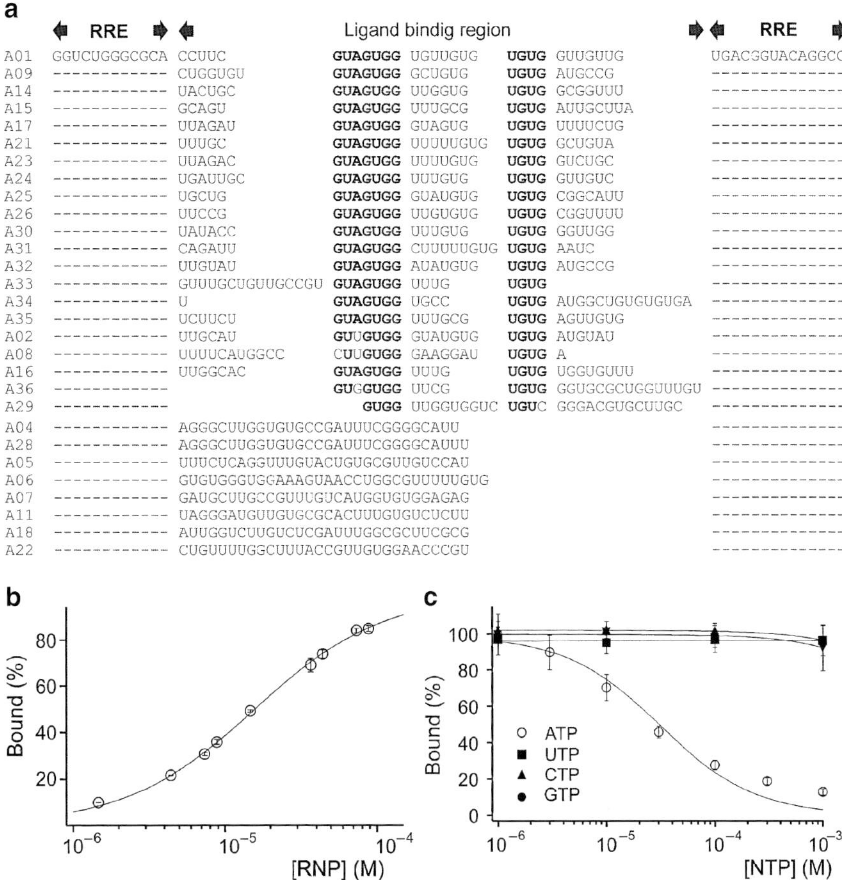

Fig. 10.2 (**a**) Nucleotide sequences of the ligand-binding region (N30) in the RNA subunit selected against ATP revealed possible consensus nucleotide sequences (*shown in bold face*). RRE denotes the Rev-binding RNA sequence. (**b**) A saturation curve for the binding of A28/Rev (*open circles*) to immobilized ATP. (**c**) Competition binding assays of the ATP complex of the RNP receptor A28/Rev with ATP (*open circles*), UTP (*filled squares*), CTP (*filled triangles*), or GTP (*filled circles*) indicate a selective ATP binding of A28/Rev

2.2 Construction of Fluorescent RNP Libraries and Screening of Fluorescent RNP Sensors

In vitro selection of the ATP-binding RNP receptors generated a variety of RNA sequences varying in the location of the consensus sequence within the randomized nucleotide region, affording a library of RNA subunits (Fig. 10.2a). Each RNA subunit forms a ligand-binding pocket that has a unique geometry to the *N*-terminal of the Rev peptide and an individual affinity to the ligand in the RNP complex. Combination of the RNA subunit library and a fluorophore-labeled Rev peptide

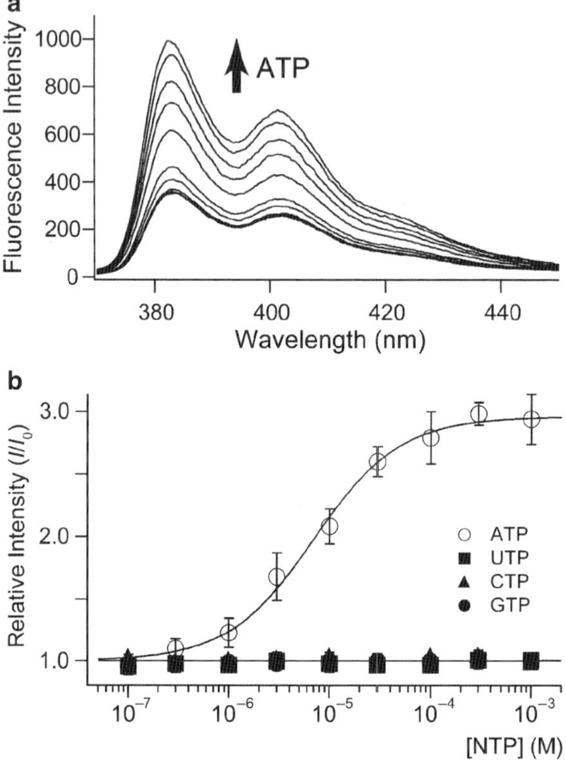

Fig. 10.3 (**a**) Direct titration of a fluorescent RNP complex (0.1 μM) of the A28 RNA subunit and a Rev peptide modified with pyrene at the amino terminal (Pyr-Rev) with ATP (1, 3, 10, 30, 100, 300, 1,000, and 3,000 μM) shows an increase in fluorescence intensity. A spectrum in the absence of ATP is shown in *bold line*. (**b**) A saturation curve for the fluorescence emission intensity of A28/Pyr-Rev to ATP (*open circles*), UTP (*filled squares*), CTP (*filled triangles*), or GTP (*filled circles*) indicates A28/Pyr-Rev responds selectively to the addition of ATP

generates a library of fluorescent RNP receptors with a series of affinities that emit at the wavelength expected from the labeled fluorophore (Fig. 10.4). Combination of another fluorophore-labeled Rev peptide and the RNA subunit library affords a new fluorescent RNP library that exhibits an emission wavelength different from the former library. Such fluorescent RNP libraries are ideal for obtaining a fluorescent sensor tailor-made for a given target.

The feasibility of the screening scheme of fluorescent RNP sensors (Fig. 10.4) was demonstrated by utilizing libraries of fluorescent RNP receptors constructed from the 29 different ATP-binding RNP, which obtained from RRE30N RNAs (Fig. 10.2a) and Rev peptides modified with various fluorophores, 7-methoxycoumarin-3-carboxylic acid (7mC-Rev), 4-fluoro-7-nitrobenz-2-oxa-1,3-diazole (NBD-Rev), and Cy5 mono NHS ester (Cy5-Rev) at the *N*-terminal. Complex formation of the 29 different RNA subunits and Pyr-Rev, 7mC-Rev, NBD-Rev, or Cy5-Rev afforded four independent fluorescent RNP libraries.

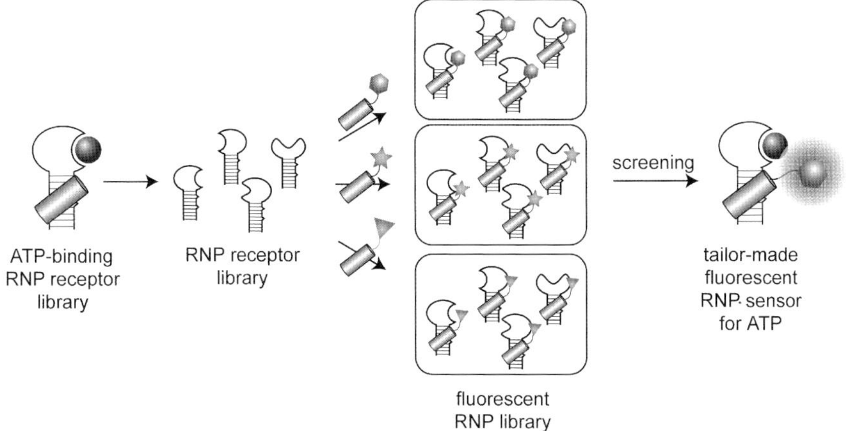

Fig. 10.4 A screening strategy of a tailor-made RNP fluorescent sensor. Combination of the RNA subunit pool of the RNP receptor and Rev peptide subunits labeled with different fluorophores generates combinatorial fluorescent RNP receptor libraries, from which RNP sensors with desired optical and/or binding properties are screened

Each complex of 7mC-Rev and RNA was placed individually on a multiwell plate, and was measured by the change of fluorescence intensities in the absence or presence of ATP (1 mM) by using a microplate reader. A scanned image of multiwell plate assay of a fluorescent ATP-binding RNP library constructed from the RNA subunits of the ATP-binding RNP (Fig. 10.2a) and 7mC-Rev peptide is shown in Fig. 10.5. In most of the cases, the fluorescence emission of 7mC-Rev was quenched upon the formation of RNP complexes when compared with that of 7mC-Rev alone (the lane marked "no RNA"). As typically shown in lanes marked 14, 23, 32, 34, 35, 02, 28, 07, and 06, the fluorescence emission increased in the intensity upon addition of ATP, indicating that these fluorescent RNPs could be utilized as ATP sensors. The fluorescence intensities in the absence or presence of ATP were evaluated in the similar manner for fluorescent RNP libraries constructed by combination of the RNA subunits of the ATP-binding RNP and Pyr-Rev, NBD-Rev and Cy5-Rev. Relative ratios of fluorescence intensity (I/I_0) in the absence (I_0) and the presence (I) of ATP for fluorescent RNPs with 7mC-, Pyr-, NBD-, and Cy5-Rev monitored at 390, 390, 535, and 670 nm, respectively, are shown in Fig. 10.6. Each sensor showed a variety of I/I_0 ratio from 0.8 to 6. More than a half of fluorescent ATP sensors responded with the I/I_0 ratio over 2, as typically shown for the libraries with 7mC-, Pyr-, and NBD-Rev. The fluorescent RNP library with Cy5-Rev provided ATP sensors with an emission wavelength of 670 nm and with I/I_0 ratios of over 2 (Fig. 10.6d).

The above simple screening method efficiently provided fluorescent ATP sensors emitting from 390 to 670 nm with excitation wavelengths ranging from 340 to 650 nm. The degree to which the fluorescence intensity changed upon binding to ATP varied with each fluorescent RNP. A26 and A30 RNAs afforded ATP sensors with high I/I_0 ratios with 7mC-Rev and NBD-Rev (Fig. 10.6a, c). In contrast, even

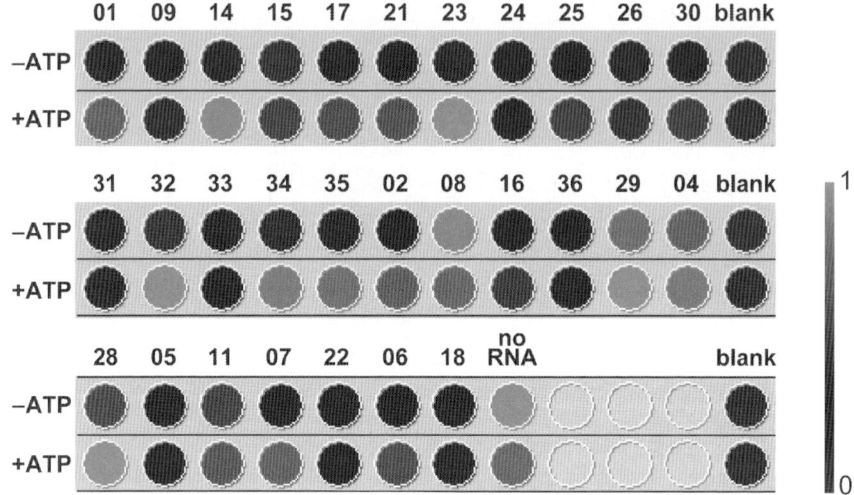

Fig. 10.5 Microplate assay for the combinatorial screening of the fluorescent RNP library. Fluorescence intensities of 7mC-Rev derived RNPs (1 µM) were evaluated in the absence or presence of 1 mM of ATP with excitation at 355 nm and emission 390 nm with intensities being weak in *black color* and strong in *white*. The fluorescent RNP complexes are in the same order as the RNA sequences presented in Fig. 10.2a

the sensors were constructed from the same RNP receptor, the I/I_0 ratios varied with the fluorophore attached to the Rev peptide. A25/Pyr-Rev showed the highest I/I_0 value of 4.5 within the Pyr-Rev sensors (Fig. 10.6b), while A25/7mC-Rev showed a moderate I/I_0 ratio of 2.5 within the 7mC-Rev sensors (Fig. 10.6a). These results again emphasized the difficulty in predicting the efficiency of optical response of the fluorescence sensor, and demonstrated the advantage of the above strategy to obtain usable fluorescence sensors. In the case of fluorescent RNP sensors, the fluorophore at the *N*-terminal of Rev does not necessarily locate in the ligand-binding pocket composed by the RNA subunit to exhibit large fluorescence intensity changes upon ligand binding. The binding of ATP to the RNP sensor will cause conformational changes of the RNA subunit that can sufficiently alter the fluorescence intensity of the fluorophore.

2.3 Screening of ATP Sensors Responding Within Desired Concentration Range

On the one hand, the screening scheme based on the relative changes in emission intensities of fluorescent RNPs with or without the ligand ATP (I/I_0) is useful for obtaining RNP sensors with desired emission wavelength and with high I/I_0 ratio. On the other hand, screening of the fluorescence emission intensities in the presence of increasing

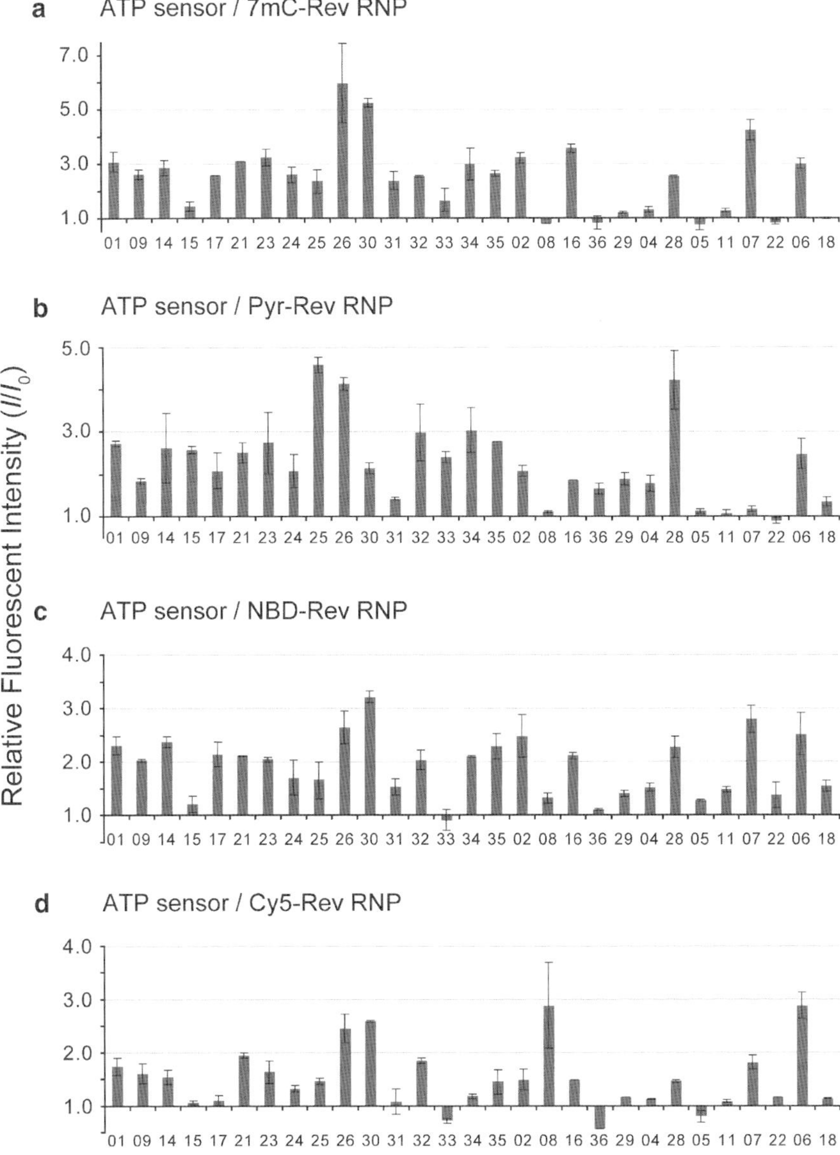

Fig. 10.6 Relative fluorescence intensity changes (I/I_0) are shown in the bar graphs for (**a**) 7mC-Rev RNP, (**b**) Pyr-Rev RNP, (**c**) NBD-Rev RNP, and (**d**) Cy5-Rev RNP

concentrations of ATP (from 10 nM to 10 mM) allowed titration analysis of the fluorescent RNP library, which enabled screening of ATP sensors responding within certain concentration ranges of ATP (Fig. 10.7a). Evaluation of the fluorescence emission pattern vs. the added ATP concentration instantly revealed a concentration range at which each fluorescent RNP responses. For example, saturation midpoints of three fluorescent

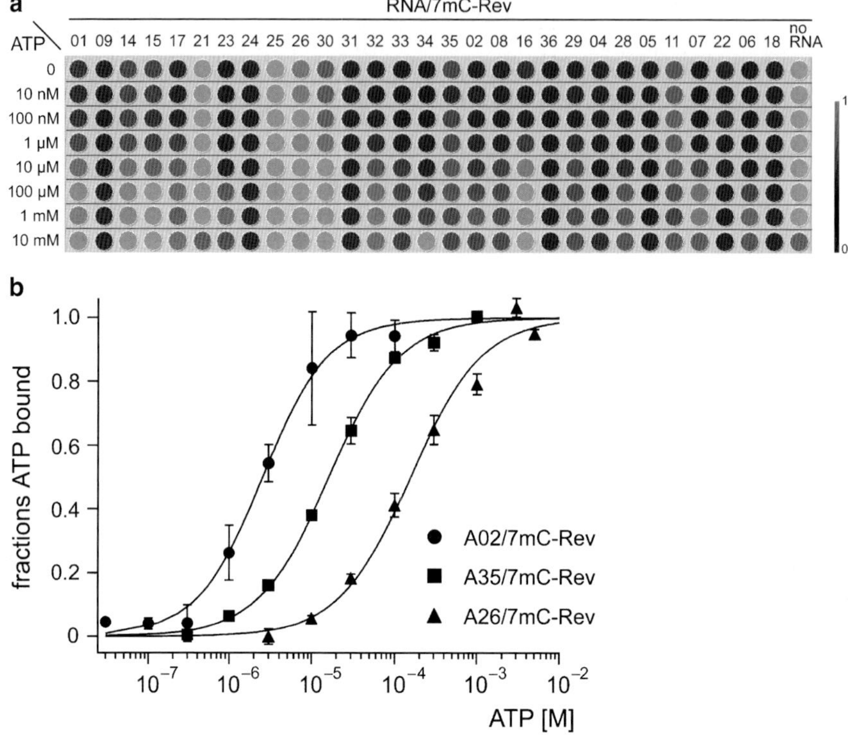

Fig. 10.7 (**a**) Combinatorial screening of fluorescent RNP library with 7mC-Rev in the presence of 10 nM to 10 mM ATP visualize a concentration range of ATP for each fluorescent RNP to respond effectively. The fluorescence intensities are weak in *black color* and strong in *white*. (**b**) Histograms show titrations of fluorescence intensity changes of fluorescent RNPs consisting of A02 (*filled circles*), A35 (*filled squares*), and A26 (*filled triangles*) RNAs in the presence of 10 nM to 10 mM ATP

RNPs, A02/7mC-Rev, A35/7mC-Rev, and A26/7mC-Rev, were spaced at approximately one order of magnitude intervals (Fig. 10.7b), with K_D values being 2.2, 15.7, and 156 µM, respectively. A31/7mC-Rev appeared to show fluorescence response at the ATP concentration of 1–10 mM (Fig. 10.7a). Simultaneous application of these four sensors covers ATP concentration ranges from ~10^{-7} to ~10^{-2} M.

2.4 Sensing Multiple Ligands at Different Wavelengths

The adaptability of in vitro selection to a panel of ligands[34,35] is applicable to the RNP-based selection scheme. RNP receptors for GTP were isolated from the pool of RNP library by the in vitro selection method described for the selection of ATP-binding RNP.[39] The sequences of the RNA subunit of GTP-binding RNPs revealed that there

```
        ◄ RRE ►  ◄              ligand binding region              ►◄  RRE  ►
G03  GGUCUGGGCGCA  UAAU     GCGG CGUUGUUUAUCG UGUCUAC AUG                UGACGGUACAGGCC
G08  ------------  UCGUU    GCGG UGUAUUUACA   UGUCUAC AUGG               --------------
G09  ------------  UUGGC    GCGC UGUGACUAUA   GGUCUAC GUGA               --------------
G20  ------------  GUGUC    GCGA AGUGUUGGCU   UGUCUAC GUGC               --------------
G23  ------------  UGGA     GCGG CUUGUUGCGGAG UGUCUAC UUG                --------------
G26  ------------  UUAAU    GCGG CCCUGAAGGG   CGUCUAC AUGU               --------------
G28  ------------  UUGUU    GCGG CGUUAUGGCG   UGUCUAC AUAG               --------------
G31  ------------  GUCGUU   GCGG AUCUUGGUGGAU UGUCUAC A                  --------------
G33  ------------  UUGUU    GCGG CGUUAUGGCG   UGUCUAC AUAG               --------------
G35  ------------  CGUG     GCGG GGACUCUUGUCU UGUCUAC CUG                --------------
G02  ------------       UGUCUAC CUGUGCAUGCUUGCGCGUG GCGA                --------------
G04  ------------       UGUCUAC CUGUGCAUGCUUGCGCGUG GCGG                --------------
G27  ------------  AGGA UGUCUAC UCGUUUGGCGG            GCGG UUUU         --------------
G10  ------------  UUGUAGGUAC AGCUGUUUCUGCU GCGG UUU                    --------------
G16  ------------  UUGUAGGUAC GGAGUGGGUUUCC GCGG UUU                    --------------
G07  ------------  GUGGUUGGCGGGCU              GCGGUC CGUGCUUGCUG        --------------
G13  ------------  CUGUCUUACGGCGUGGUUG        GCGGUC GCGGG              --------------
G17  ------------  CAAGGGUUUGGU               GCGGUC GGUGUGGCGUUG        --------------
G25  ------------  UGUUUUGGUUUCGGUAGUCCU GCGGUC CUUGU                   --------------
G19  ------------  GUGGCCUGGGGUAU             GUGGUC GAGUUGUUGC          --------------
G30  ------------  GUCGUAGCGCGCUGUUUGUA       GUGGUC AACU               --------------
G12  ------------  UGCCGUCGUA                      AUACCUG AUUUUUUCGUGGC --------------
G29  ------------  CGUGGCUGGCCUUGUUGGGUUGU AUACCUG                      --------------
G05  ------------  UAGUCGGCGGUUCGAUUGAACGUUAUCCCG                       --------------
G14  ------------  CUUUACACGGCUUCGUGCGCGUCUUGGGUG                       --------------
G34  ------------  GAUGGCGCGGCCUUGAUGGUGUUUACGUAUU                      --------------
```

Fig. 10.8 Nucleotide sequences of the ligand-binding region for the RNA subunit of RNP selected against GTP revealed consensus nucleotide sequences (*shown in bold*). RRE denotes the Rev-binding RNA sequence

are consensus 5′-GCGG-3′ and 5′-UGUCUAC-3 sequences (Fig. 10.8). As with the case for the ATP-binding RNP, combination of the RNA subunit pool of the GTP-binding RNP and several fluorophore-labeled Rev peptide subunits can form combinatorial fluorescent RNP receptor libraries. The fluorescence intensities in the absence and presence of GTP were assessed in the similar manner for fluorescent RNP libraries obtained by combination of the RNA subunits of the ATP-binding RNP and the fluorophore-labeled Rev peptides. A series of fluorescent GTP sensors with I/I_0 ratios of over 2 were obtained from the fluorescent RNP libraries derived from the RNA subunit of GTP-binding RNP and 7mC-Rev, Pyr-Rev or NBD-Rev (Fig. 10.9). In the case of 7mC-Rev derived GTP sensor library, G23/7mC-Rev, G26/7mC-Rev, G05/7mC-Rev, and G14/7mC-Rev had high I/I_0 ratios of over 4. As observed for the screening of the ATP sensor libraries, the observed I/I_0 ratios of GTP sensors depended on the nature of fluorophore even using the same RNA subunit. For example, The I/I_0 ratio of G05/7mC-Rev showed almost 6 (Fig. 10.9a), whereas that of G05/Pyr-Rev was measured to be 1 (Fig. 10.9b).

2.5 Selective Fluorescence Responses of the ATP and GTP Sensors

These fluorescent GTP-binding RNPs showed distinct selectivity to GTP over other NTPs (Fig. 10.10). With ATP and GTP sensors emitting at different wavelengths, concentrations of ATP and GTP can be monitored independently in the same

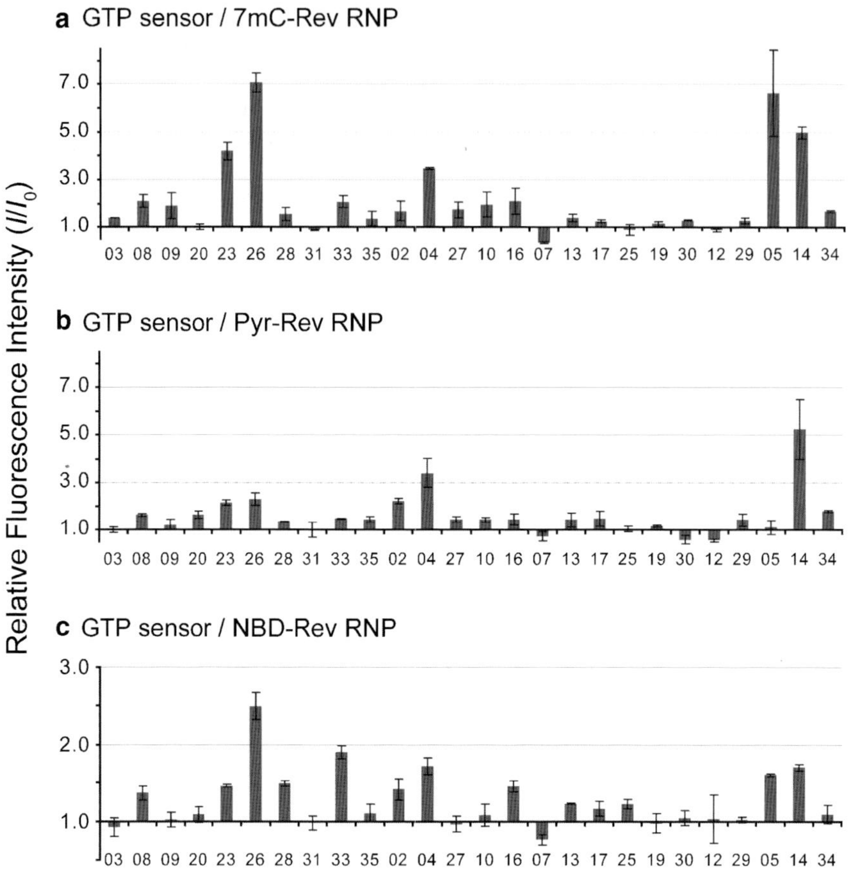

Fig. 10.9 Relative fluorescence intensities (I/I_0) of fluorescent GTP-binding RNPs upon GTP binding are shown in the bar graphs: (**a**) 7mC-Rev RNP, (**b**) Pyr-Rev RNP, (**c**) NBD-Rev RNP

solution. The ATP sensor A32/Cy5-Rev could monitor ATP from a submicromolar to millimolar concentration range at 670 nm (Fig. 10.11, lane 1). Even when the ATP titration was carried out in the presence of GTP (500 µM), the fluorescence intensity of A32/Cy5-Rev increased (Fig. 10.11, lane 3) with the affinity similar to that in the absence of GTP. Increasing the concentration of ATP did not affect the fluorescence intensity of the GTP sensor G33/NBD-Rev (Fig. 10.11, lane 2). G33/NBD-Rev showed constant fluorescence intensity in solutions containing 500 µM GTP and increasing concentrations of ATP (Fig. 10.11, lane 4). The GTP sensor G33/NBD-Rev also responded to GTP from a submicromolar to millimolar concentration range at 535 nm in the absence (Fig. 10.11, lane 6) and presence (lane 8) of ATP (500 µM). The affinity of the GTP sensor G33/NBD-Rev to GTP in the presence of 500 µM ATP was almost equal to that in the absence of ATP.

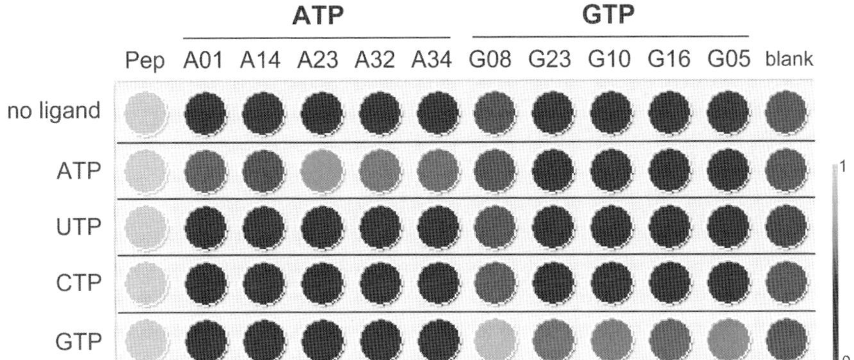

Fig. 10.10 Selective fluorescence responses of the ATP and GTP sensors. Experiment was conducted on a microplate for fluorescent ATP and GTP-binding RNPs generated by mixing 7mC-Rev (1 μM) and the RNA subunits (1 μM) of ATP-binding RNP (A01, A14, A23, A32, and A34) and GTP-binding RNP (G08, G23, G10, G16, and G05) in the presence of 0.1 mM ATP, UTP, CTP, or GTP. Lanes marked pep and blank contained 7mC-Rev and only a reaction buffer, respectively. Fluorescence intensities of samples were measured with excitation at 355 nm and emission 390 nm. Images of the fluorescence intensity of wells were shown with intensities being weak in *black color* and strong in *white*

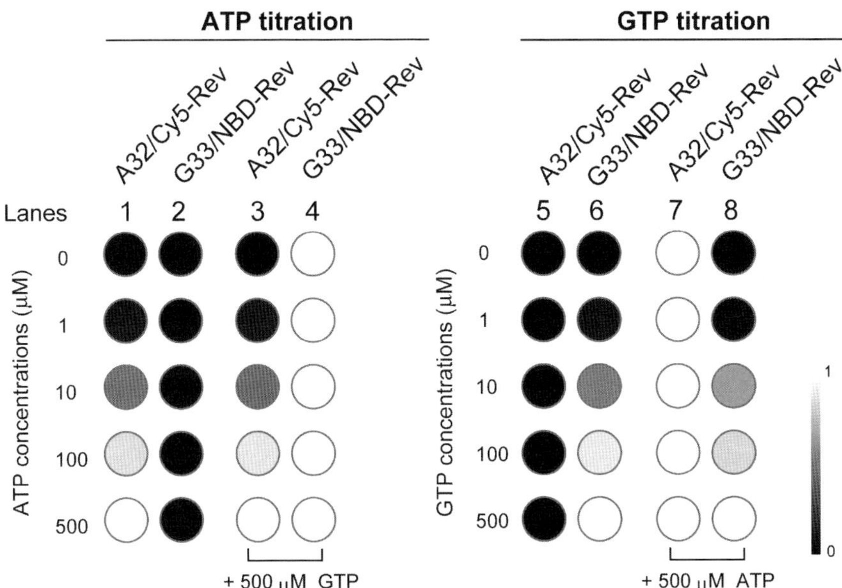

Fig. 10.11 Titrations of the ATP sensor A32/Cy5-Rev by addition of ATP (1, 10, 100, and 500 μM) showed increase in the fluorescence intensity both in the absence (*lane 1*) and presence (*lane 3*) of 500 μM GTP. Fluorescence intensities of the ATP sensor A32/Cy5-Rev were constant upon addition of GTP (1, 10, 100, and 500 μM) in the absence (*lane 5*) and presence (*lane 7*) of 500 μM ATP. Fluorescence intensities of the GTP sensor G33/NBD-Rev were constant by addition of ATP (1, 10, 100, and 500 μM) in the absence (*lane 2*) and presence (*lane 4*) of GTP (500 μM). Titrations of the GTP sensor G33/NBD-Rev by addition of GTP (1, 10, 100, and 500 μM) showed increase in the fluorescence intensity both in the absence (*lane 6*) and presence (*lane 8*) of 500 μM ATP. Fluorescence intensities were monitored at 670 nm and 535 nm for A32/Cy5-Rev and G33/NBD-Rev, respectively

3 Fluorescent RNP Sensors for Biologically Important Targets

3.1 *Phosphotyrosine*

Protein phosphorylation at the tyrosine residue controls many cellular signaling events.[43,44] Protein tyrosine kinases, protein tyrosine phosphatases, and their substrates play an important role in regulating processes such as proliferation, differentiation, motility, and immune responses, as well as pathological conditions such as cancer.[45,46] Methods for the selective recognition and sensing of protein tyrosine phosphorylation are quite important in biochemical, proteomic, and cell biological studies. However, current analytical methods of protein tyrosine phosphorylation generally need multiple laborious steps. Typical methods utilize detection by autoradiography of ^{32}P-radiolabeled proteins separated by SDS-PAGE, and immunoblotting or immunoprecipitation by anti-phosphotyrosine antibody.[47,49] Antibody detection is an especially sensitive method (fmol level), which is highly suitable for the detection of specific tyrosine phosphorylated residues in individual phosphoproteins.[47] Until now, more than 600 distinct tyrosine phosphorylation sites have been reported in mammalian proteome, and anti-phosphotyrosine antibodies are commercially available for more than 80 different tyrosine phosphorylation sites.[50,51] Mass spectral analysis[49,51–53] of peptide fragments obtained by utilizing antiphospho-tyrosine antibody has advantages of being rapid and does not require radiolabeling, which unveiled temporal interaction of many signaling proteins. Although these methods were powerful and exploited our understanding of cellular signaling, it is difficult to obtain quantitative interpretation due to the washing steps involved in the analytical procedures. There still exists a demand for direct detection methods of tyrosine-phosphorylated proteins, such as using artificial biosensors.

Several receptors for tyrosine phosphorylated peptides or *O*-phospho-L-tyrosine (pTyr) have been reported. Smith et al.[54] reported artificial pTyr sensors based on β-cyclodextrin with pendant ammonium or guanidinium groups on the primary rim of the cavity. A β-cyclodextrin derivative modified with the guanidinium group showed dissociation constant of 2.8 mM for pTyr, which are not low enough to detect protein tyrosine phosphorylation practically. Hamachi et al.[55,56] developed an artificial fluorescent chemosensor for tyrosine phosphorylated peptides based on coordination chemistry. The anthracene bis (zinc(II)-dipicolylamine) complex (Zn(II)-Dpa) bound and sensed a pTyr residue fluorescently within the Glu-rich peptide sequence with a dissociation constant of 10^{-7} M. Almost no fluorescence change was observed for the nonphosphorylated peptide up to 10 μM. This result revealed that Zn(II)-Dpa distinguished the phosphate in native tyrosine phosphorylation site. However, the binding selectivity of Zn(II)-Dpa for the phosphorylated peptides and biological pyrophosphate derivatives (ATP or ADP) is not high enough to rapidly and easily detect tyrosine-phosphorylated sites both in vitro and in vivo.

RNP-based fluorescent probes would be ideal to detect protein tyrosine phosphorylation directly in the solution. To develop a fluorescent RNP sensor for the pTyr residue, RNP receptors has been developed for the pTyr residue.[57] RNP receptors for pTyr were isolated from a pool of RNA sequences (4^{30}) as described

```
            ◀ RRE ▶ ◀              Ligand bindig region                    ▶ ◀    RRE    ▶
pY03  GGUCUGGGCGCA UCUUUUUGGGUGAACAAGCAGGACCUUGUU                             UGACGGUACAGGCC
pY09  ------------ GUCAU UGC AGGCUGUGGCUCCU          GGUAGAA G               --------------
pY12  ------------       UGC AGGUGUACUGAGCUGUACUU    GGUAGAA                 --------------
pY16  ------------   UAG UGC AGUGCGUUGGCAGU          GGUAGAA CUU             --------------
pY27  ------------    AG UGC AUUUCUUGUUGAGU          GGUAGAA UUUG            --------------
pY28  ------------     U UGC AUUGUGUUAUUGCUUU        GGUAGAA GUG             --------------
pY34  ------------       UGC AUCUUCUUGAUUUUGGGGAU    GGUAGAA                 --------------
pY43  ------------   UGU UGC AUUCGCGU                GGUAGAA GCAUGACGU       --------------
pY04  ------------       UGC AGCGGUAUCUUGGCACCGUU    GGUAGA  U               --------------
pY37  ------------ AUACU UGC CGGUGCGCCG              GGUAGAA GGUUU           --------------
pY35  ------------  AGAU GGUAGAA GCGGUUGACCGU            UGC GUCU            --------------
pY36  ------------ CUGUCU GGUAGAA GCGCGUUUU              UGC GUUGC           --------------
pY32  ------------   CUC GGUAGAA GGCAUUAUUGCU            UGC GGGAU           --------------
pY21  ------------       GGUAGAA GUCUUCGUUUCUUCUUAGAU   UGC                 --------------
pY29  ------------       GGUAGAA GAUUUCCUUUAGUGGAAGUU   UGC                 --------------
pY06  ------------ CAUUCUGAGGUGUGAAGUUGGUGGGCCCGC                            --------------
pY25  ------------ UUAUUUUUGCUCCGUUAGAUUCGGUGUGCU                            --------------
```

Fig. 10.12 Nucleotide sequences in the random region of the RNA subunit from the pool obtained after the sixth round of in vitro selection. A possible consensus sequence UGC-GGUAGA indicated *bold*

for the ATP-binding RNP receptors.[39] Analysis of the nucleotide sequences of clones revealed distinct consensus sequences (Fig. 10.12). Among the 29 clones, nine revealed the same nucleotide sequence pY03. A 10-nucleotides consensus sequence 5′-UGC-GGUAGAA-3′ was thoroughly observed for other clones. Equilibrium dissociation constant for the complex of pY03RNA and the Rev peptide (pY03RNP) to pTyr were determined from the saturation curves to be 376 μM. The RNP complex pY03RNP preferentially bound pTyr over *o*-phospho-L-serine (pSer), Gly-Tyr, or Leu-pTyr. The phosphate charge of pTyr contributed to the specific binding complex formation of pY03RNP as the nonphosphorylated dipeptide Gly-Tyr did not bind pY03RNP. Because pSer did not bind pY03RNP, the aromatic ring of pTyr is also a key determinant for the selective binding of pY03RNP. A pTyr-containing dipeptide Gly-pTyr bound pY03RNP with a similar affinity to pTyr. Interestingly, a pTyr-containing dipeptide Leu-pTyr showed almost no affinity to pY03RNP. It is likely that the steric hindrance at the *N*-terminal portion of pTyr group strongly prohibits the binding of pTyr group to the binding pocket of pY03RNP. It would be possible to isolate RNP receptors that can recognize not only a pTyr residue, but also amino acid residues surrounding the pTyr. By increasing the recognition surface, it would be possible to increase the affinity of RNP to pTyr-containing amino acid sequences.

By using the RNP-based modular strategy[39] for the fluorescent ATP sensors, it will be possible to obtain fluorescent pTyr biosensors from the pTyr-binding RNP receptors to detect a pTyr residue within a defined amino acid sequence on the protein surface.

3.2 Biogenic Amines

The biogenic amines, such as dopamine, histamine, serotonin and norepinephrine, epinephrine (Fig. 10.13), play a critical role in the function of thehypothalamic–pituitary–

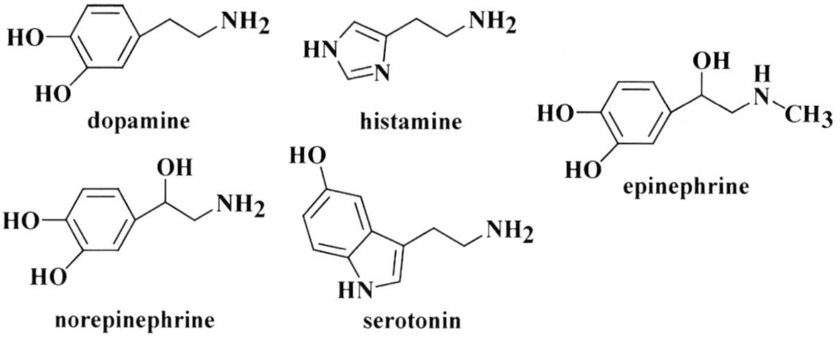

dopamine

histamine

epinephrine

norepinephrine

serotonin

Fig. 13 Biogenic amines of the central nervous system relevant to neurodegenerative disease

adrenal axis and in the integration of information in sensory, limbic, and motor systems.[58,59] The major neurodegenerative diseases in human, Parkinson's disease, have been clearly shown to correlate with low levels of dopamine in the brain.[60] However, the direct assessment of dopamine and its derived metabolites in the brain of living human is not possible and *post-mortem* samples suffer from a number of limitations. Peripheral measures of these compounds in urine, blood plasma, and cerebrospinal fluid have been carried out for the purpose of assessing the presence of a dysfunction in a given condition.[61] Specific and sensitive methods for quantification of the biogenic amines would offer the quantitative measure of the metabolite concentrations for patients. Direct detection methods of the biogenic amines would permit a quantitation with less amount of the sample without a need of complicated handling.

High-performance liquid chromatography (HPLC) with fluorimetric detection,[62–64] electrochemical detection,[65–67] or combined fluorimetric and electrochemical detection[68] has been applied for the analyses of biogenic amines. However, all these methods require pretreatment of the samples and long analysis times, making them inefficient for routine work.[69] In addition, these methods do not always present reliable selectivity and/or sensitivity for a given biogenic amine.

Substantial efforts have been devoted to the development of molecular sensors for dopamine. Raymo et al.[70] reported a two-step procedure to coat silica particles with fluorescent 2,7-diazapyrenium dications sensing toward dopamine. The analysis of the fluorescence decay with multiple-equilibria binding model revealed that the electron deficient dications and the electron-rich analytes form 1:1 and 1:2 complexes at the particle/water interface. The interfacial dissociation constants of the 1:1 complexes were 5.6 mM and 3.6 mM for dopamine and catechol, respectively. Dopamine was dominated by the interaction of its electron-rich dioxyarene fragment with the electron-deficient fluorophore in neutral aqueous environments. Ahn et al.[71] reported tripodal oxazoline-based artificial receptors, capable of providing a preorganized hydrophobic environment by rational design, which mimics a hydrophobic pocket predicted for a human D2 receptor. A moderate binding affinity, a dissociation constant of 8.2 mM was obtained by NMR titrations of tripodal oxazoline-based artificial receptor with dopamine in a phosphate buffer solution (pH 7.0). Structurally related ammonium ions, norepinephrine, 2-phenylethylamine, and tyramine indicated twofold lower affinity when compared with that for dopamine. Glass et al.[72] reported a molecular sensor that binds to catecholamines by forming an iminium ion with the amine group and a boronate ester with the catechol moiety. A boronic acid-containing coumarin aldehyde was synthesized, and examined spectrophotometrically by titration with dopamine under neutral aqueous conditions. Fitting the decrease in fluorescence to a 1:1 binding isotherm gave a dissociation constant of $294\,\mu M$. The sensor also bound norepinephrine with a similar efficiency to dopamine.

All of these molecular sensors and receptors can directly detect dopamine. A major problem arises in using these molecular tools in real biological matrixes where many interfering compounds coexist is the selectivity and sensitivity of the sensor. An attractive alternative for the detection and quantitation of biogenic amines is the biosensor, which combine an efficient recognition through a macromolecular receptor, such as proteins and nucleic acids, possessing high specificity with an optical signaling function. Among these, aptamer-based biosensors are suitable for sensing biogenic amines due to the high affinity and selectivity relied on molecular structure containing distinctive functional gruops.[30]

In fact, aptamers for biogenic amines or structurally-related biogenic amines have been reported, such as dopamine-binding aptamers,[74] histidine-binding aptamers,[75] and tryptophan-binding aptamers.[76] RNA aptamers that specifically bind dopamine were isolated by in vitro selection method.[74] One aptamer, which dominated the selected pool, was characterized and bound to the dopamine affinity column with a dissociation constant of $2.8\,\mu M$. The specificity was demonstrated by studying binding properties of a number of dopamine-related molecules. The interaction of dopamine with the RNA might be mediated by the hydroxyl group at position 3 and the proximal aliphatic chain in the dopamine molecule.

The above results indicate that it is likely to construct tailor-made biosensors for biogenic amines by the RNP-based strategy. Aptamer-based sensors or RNP-based sensors would enable selective and reliable simultaneous measurements of biogenic amines at physiological concentrations.

4   Conclusions

The combinatorial strategy using the modular RNP receptor described here enables efficient tailoring of fluorescent sensors for specific ligands, ATP, GTP, phosphotyrosine, and biogenic amines. The approach consisting of a selection followed by a screening, namely, in vitro selection of RNP receptors and an efficient screening of the combinatorial RNP sensor library, affords fluorescent biosensors with a series of emission wavelengths, high I/I_0 ratios, and/or responding ligand concentration ranges without a chemical modification of the ligand-binding RNA region and without detailed knowledge of the three-dimensional structure for the RNP receptors. By choosing the fluorescent RNP sensors with appropriate optical characteristics, sensitive detections of multiple ligands without inhibitory effects of intrinsic fluorescence in the samples such as cellular extracts would be possible. Combination of the fluorescent RNP sensors offers a fluorescent sensing system that detects a wide range of ligand concentrations. Fluorescent sensor systems with wide detection concentration ranges were previously achieved[6–10] by manipulating the ligand-binding affinity of the parent receptor by means of site-directed mutations that needed detailed knowledge of three-dimensional structural information. In contrast, in vitro selection of the RNA-based RNP library provides RNP receptors with various affinities to the target ligands. Such an inherent diversity of the ligand-binding affinity is a definitive advantage for RNP to construct a fluorescent sensor. The noncovalent modular structure of RNP provides a distinct advantage for the screening strategy of fluorescent sensors described here. Once an optimal fluorescent RNP sensor is constructed, the noncovalent complex of RNP sensor could be covalently linked by tethering between the RNA and the fluorophore-labeled peptide subunits. Such a stable RNP sensor can be immobilized by hybridization of the attached RNA sequences to complementary surface-bound DNA probes[77,78] to organize an RNP sensor microarray that is useful for analyzing biologically active molecules simultaneously.

References

1. Zhang, J.; Campbell, R. E.; Ting, A. Y.; Tsien, R. Y., Creating new fluorescent probes for cell biology, *Nat. Rev. Mol. Cell Biol.* 2002, 3, 906–918
2. Weiss, S., Fluorescence spectroscopy of single biomolecules, *Science* 1999, 283, 1676–1683
3. Pollack, S. J.; Nakayama, G. R.; Schultz, P. G., Introduction of nucleophiles and spectroscopic probes into antibody combining sites, *Science* 1988, 242, 1038–1040
4. Renard, M.; Belkadi, L.; Hugo, N.; England, P.; Altschuh, D.; Bedouelle, H. J., Knowledge-based design of reagentless fluorescent biosensors from recombinant antibodies, *J. Mol. Biol.* 2002, 318, 429–442
5. Gilardi, G.; Zhou, L. Q.; Hibbert, L.; Cass, A. E., Engineering the maltose binding protein for reagentless fluorescence sensing, *Anal. Chem.* 1994, 66, 3840–3847
6. de Lorimier, R. M.; Smith, J. J.; Dwyer, M. A.; Looger, L. L.; Sali, K. M.; Paavola, C. D.; Rizk, S. S.; Sadigov, S.; Conrad, D. W.; Loew, L.; Hellinga, H. W., Construction of a fluorescent biosensor family, *Protein Sci.* 2002, 11, 2655–2675
7. Marvin, J. S.; Corcoran, E. E.; Hattangadi, N. A.; Zhang, J. V.; Gere, S. A.; Hellinga, H. W., The rational design of allosteric interactions in a monomeric protein and its applications to the construction of biosensors, *Proc. Natl. Acad. Sci. USA* 1997, 94, 4366–4371
8. Hamachi, I.; Nagase, T.; Shinkai, S., A general semisynthetic method for fluorescent saccharide-biosensors based on a lectin, *J. Am. Chem. Soc.* 2000, 122, 12065–12066
9. Benson, D. E.; Conrad, D. W.; de Lorimier, R. M.; Trammell, S. A.; Hellinga, H. W., Design of bioelectronic interfaces by exploiting hinge-bending motions in proteins, *Science* 2001, 293, 1641–1644
10. Morii, T.; Sugimoto, K.; Makino, K.; Otsuka, M.; Imoto, K.; Mori, Y., A new fluorescent biosensor for inositol trisphosphate, *J. Am. Chem. Soc.* 2002, 124, 1138–1139
11. Jayasena, S. D., Aptamers: An emerging class of molecules that rival antibodies in diagnostics, *Clin. Chem.* 1999, 45, 1628–1650
12. Llano-Sotelo, B.; Chow, C. S., RNA-aminoglycoside antibiotic interactions: fluorescence detection of binding and conformational change, *Bioorg. Med. Chem. Lett.* 1999, 9, 213–216
13. Jhaveri, S.; Rajendran, M.; Ellington, A. D., In vitro selection of signaling aptamers, *Nat. Biotechnol.* 2000, 18, 1293–1297

14. Stojanovic, M. N.; de Prada, P.; Landry, D. W., Fluorescent sensors based on aptamer self-assembly, *J. Am. Chem. Soc.* 2000, 122, 11547–11548

15. Jhaveri, S. D.; Kirby, R.; Conrad, R.; Maglott, E. J.; Bowser, M.; Kennedy, R. T.; Glick, G.; Ellington, A. D., Designed signaling aptamers that transduce molecular recognition to changes in fluorescence intensity, *J. Am. Chem. Soc.* 2000, 122, 2469–2473

16. Stojanovic, M. N.; de Prada, P.; Landry, D. W., Aptamer-based folding fluorescent sensor for cocaine, *J. Am. Chem. Soc.* 2001, 123, 4928–4931

17. Fang, X. Cao, Z. Beck, T.; Tan, W., Molecular aptamer for real-time oncoprotein platelet-derived growth factor monitoring by fluorescence anisotropy, *Anal. Chem.* 2001, 73, 5752–5257

18. Stojanovic, M. N.; Landry, D. W., Aptamer-based colorimetric probe for cocaine, *J. Am. Chem. Soc.* 2002, 124, 9678–9679

19. Nutiu, R., Li, Y., Structure-switching signaling aptamers, *J. Am. Chem. Soc.* 2003, 125, 4771–4778

20. Stojanovic, M. N.; Green, E. G.; Semova, S.; Nikic, D. B.; Landry, D. W., Cross-reactive arrays based on three-way junctions, *J. Am. Chem. Soc.* 2003, 125, 6085–6089

21. Stojanovic, M. N.; Kolpashchikov, D. M., Modular aptameric sensors, *J. Am. Chem. Soc.* 2004, 126, 9266–9270

22. Yamana, K.; Ohtani, Y.; Nakano, H.; Saito, I., Bis-pyrene labeled DNA APTAMER as an intelligent fluorescent biosensor, *Bioorg. Med. Chem. Lett.* 2003, 13, 3429–3431

23. Ho, H. A.; Leclerc, M., Optical sensors based on hybrid aptamer/conjugated polymer complexes, *J. Am. Chem. Soc.* 2004, 126, 1384–1387

24. Kirby, R.; Cho, E. J.; Gehrke, B.; Bayer, T.; Park, Y. S.; Neikirk, D. P.; McDevitt, J. T.; Ellington, A. D., Aptamer-based sensor arrays for the detection and quantitation of proteins, *Anal. Chem.* 2004, 76, 4066–4075

25. Savran, C. A.; Knudsen, S. M.; Ellington, A. D.; Manalis, S. R., Micromechanical detection of proteins using aptamer-based receptor molecules, *Anal. Chem.* 2004, 76, 3194–3198

26. Jiang, Y.; Fang, X.; Bai, C., Signaling aptamer/protein binding by a molecular light switch complex, *Anal. Chem.* 2004, 76, 5230–5235

27. Nutiu, R.; Li, Y., In vitro selection of structure-switching signaling aptamers, *Angew. Chem. Int. Ed.* 2005, 44, 1061–1065

28. Merino, E. J.; Weeks, K. M., Facile conversion of aptamers into sensors using a 2′-ribose-linked fluorophore, *J. Am. Chem. Soc.* 2005, 127, 12766–12767

29. Mayer, B. J.; Ren, R.; Clark, K. L.; Baltimore, D., A putative modular domain present in diverse signaling proteins, *Cell* 1993, 73, 629–630

30. Haslam, R. J.; Koide, H. B.; Hemmings, B. A., Pleckstrin domain homology, *Nature* 1993, 363, 309–310

31. Lemmon, M. A.; Ferguson, K. M.; O'Brien, R.; Sigler, P. B.; Schlessinger, J., Specific and high-affinity binding of inositol phosphates to an isolated pleckstrin homology domain, *Proc. Natl Acad. Sci.* USA 1995, 92, 10472–10476

32. Hyvonen, M.; Macias, M. J.; Nilges, M.; Oschkinat, H.; Saraste, M.; Wilmanns, M., Structure of the binding site for inositol phosphates in a PH domain, *EMBO J.* 1995, 14, 4676–4685

33. Ferguson, K. M.; Lemmon, M. A.; Schlessinger, J. Sigler, P. B., Structure of the high affinity complex of inositol trisphosphate with a phospholipase C pleckstrin homology domain, *Cell* 1995, 83, 1037–1046

34. Ellington, A. D.; Szostak, J. W., In vitro selection of RNA molecules that bind specific ligands, *Nature* 1990, 346, 818–822

35. Wilson, D. S.; Szostak, J. W., In vitro selection of functional nucleic acids, *Annu. Rev. Biochem.* 1999, 68, 611–647

36. Lin, C. H.; Patel, D. J., Structural basis of DNA folding and recognition in an AMP-DNA aptamer complex: distinct architectures but common recognition motifs for DNA and RNA aptamers complexed to AMP, *Chem. Biol.* 1997, 4, 817–832

37. Jiang, F.; Kumar, R. A.; Jones, R. A.; Patel, D. J., Structural basis of RNA folding and recognition in an AMP-RNA aptamer complex, *Nature* 1996, 382, 183–186

38. Dieckmann, T.; Suzuki, E.; Nakamura, G. K.; Feigon, J., Solution structure of an ATP-binding RNA aptamer reveals a novel fold, *RNA* 1996, 2, 628–648
39. Hagihara, M.; Fukuda, M.; Hasegawa, T.; Morii, M., A modular strategy for fluoresent biosensors from ribonucleopeptide complexes, *J. Am. Chem. Soc.* 2006, 128, 12932–12940
40. Morii, T.; Hagihara, M.; Sato, S.; Makino, K., In vitro selection of ATP-binding receptors using a ribonucleopeptide complex, *J. Am. Chem. Soc.* 2002, 124, 4617–4622
41. Sato, S.; Fukuda, M.; Hagihara, M.; Tanabe; Y., Ohkubo, K.; Morii, T., Stepwise molding of a highly selective ribonucleopeptide receptor, *J. Am. Chem. Soc.* 2005, 127, 30–31
42. Battiste, J. L.; Mao, H.; Rao, N. S.; Tan, R.; Muhandiram, D. R.; Kay, L. E.; Frankel, A. D.; Williamson, J. R., Alpha helix-RNA major groove recognition in an HIV-1 rev peptide-RRE RNA complex, *Science* 1996, 273, 1547–1551
43. Johnson, L. N.; Lewis, R. J., Structural basis for control by phosphorylation, *Chem. Rev.* 2001, 101, 2209–2242
44. Chen, Z.; Gibson, T. B.; Robinson, F.; Silvestro, L.; Pearson, G.; Xu, B.; Wright, A.; Vanderbilt, C.; Cobb, M. H., MAP kinases, *Chem. Rev.* 2001, 101, 2449–2476
45. Hunter, T., Protein modification: phosphorylation on tyrosine residues, *Curr. Opin. Cell Biol.* 1989, 1, 1168–1181
46. Hunter, T., Signaling-2000 and beyond, *Cell* 2000, 100, 113–127
47. Yan, J. X.; Packer, N. H.; Gooley, A. A.; Williams, K. L., Protein phosphorylation: technologies for the identification of phosphoamino acids, *J. Chromatogr. A* 1998, 808, 23–41
48. Pestka, S.; Lin, L.; Wu, W.; Izotova, L., Introduction of protein kinase recognition sites into proteins: a review of their preparation, advantages, and applications, *Protein Expr. Purif.* 1999, 17, 203–214
49. Machida, K.; Mayer, B. J.; Nollau, P., Profiling the global tyrosine phosphorylation state, *Mol. Cell Proteomics* 2003, 2, 215–233
50. Mandell, J. W., Phosphorylation state-specific antibodies: applications in investigative and diagnostic pathology, *Am. J. Pathol.* 2003, 163, 1687–1698
51. Rush, J.; Moritz, A.; Lee, K. A.; Guo, A.; Goss, V. L.; Spek, E. J.; Zhang, H.; Zha, X. M.; Polakiewicz, R. D.; Comb, M. J., Immunoaffinity profiling of tyrosine phosphorylation in cancer cells, *Nat. Biotech.* 2005, 23, 94–101
52. Salomon, A. R.; Ficarro, S. B.; Brill, L. M.; Brinker, A.; Phung, Q. T.; Ericson, C.; Sauer, K.; Brock, A.; Horn, D. M.; Schultz, P. G.; Peters, E. C., Profiling of tyrosine phosphorylation pathways in human cells using mass spectrometry, *Proc. Natl. Acad. Sci. USA* 2003, 100, 443–448
53. Blagoev, B.; Ong, S. E.; Kratchmarova, I.; Mann, M., Temporal analysis of phosphotyrosine-dependent signaling networks by quantitative proteomics, *Nat. Biotechnol.* 2004, 22, 1139–1145
54. Cotner, E. S.; Smith, P. J., Phosphotyrosine binding by ammonium- and guanidinium-modified cyclodextrins, *J. Org. Chem.* 1998, 63, 1737–1739
55. Ojida, A.; Mito-oka, Y.; Inoue, M. A.; Hamachi, I., First artificial receptors and chemosensors toward phosphorylated peptide in aqueous solution, *J. Am. Chem. Soc.* 2002, 124, 6256–6258
56. Ojida, A.; Mito-oka, Y.; Sada, K.; Hamachi, I., Molecular recognition and fluorescence sensing of monophosphorylated peptides in aqueous solution by bis(zinc(II)-dipicolylamine)-based artificial receptors, *J. Am. Chem. Soc.* 2004, 126, 2454–2463
57. Hasegawa, T.; Ohkubo, K.; Yoshikawa, S.; Morii, T., A ribonucleopeptide receptor targets phosphotyrosine, e-J. Surf. *Sci. Nanotech.* 2005, 3, 33–37
58. Hoffman, B. J.; Hansson, S.R.; Mezey, E.; Palkovits, M., Localization and dynamic regulation of biogenic amine transporters in the mammalian central nervous system, *Front. Neuroendocrinol.* 1998, 19, 187–231
59. Bloom, F. E.; Giarman, N. J., Physiologic and pharmacologic considerations of biogenic amines in the nervous system, *Annu. Rev. Pharmacol.* 1968, 8, 229–258
60. Houghton, P. J.; Howes, M. J., Natural products and derivatives affecting neurotransmission relevant to Alzheimer's and Parkinson's disease, *Neurosignals.* 2005, 14, 6–22
61. Davis, B. A., Biogenic amines and their metabolites in body fluids of normal, psychiatric and neurological subjects, *J. Chromatogr.* 1989, 466, 89–218

62. Fotopoulou, M. A.; Ioannou, P. C., Post-column terbium complexation and sensitized fluorescence detection for the determination of norepinephrine, epinephrine and dopamine using high-performance liquid chromatography, *Anal. Chim. Acta* 2002, 462, 179–185

63. Chan, E. C.; Wee, P. Y.; Ho, P. Y.; Ho, P. C., High-performance liquid chromatographic assay for catecholamines and metanephrines using fluorimetric detection with pre-column 9-fluorenylmethyloxycarbonyl chloride derivatization, *J. Chromatogr B Biomed. Sci. Appl.* 2000, 749, 179–189

64. Panholzer, T. J.; Beyer, J.; Lichtwald, K., Coupled-column liquid chromatographic analysis of catecholamines, serotonin, and metabolites in human urine, *Clin. Chem.* 1999, 45, 262–268

65. McKenzie, J. A.; Watson, C. J.; Rostand, R. D.; German, I.; Witowski, S. R.; Kennedy, R. T., Automated capillary liquid chromatography for simultaneous determination of neuroactive amines and amino acids, *J. Chromatogr. A.* 2002, 962, 105–115

66. Raggi, M. A.; Sabbioni, C.; Casamenti, G.; Gerra, G.; Calonghi, N.; Masotti, L., Determination of catecholamines in human plasma by high-performance liquid chromatography with electrochemical detection, *J. Chromatogr B Biomed. Sci. Appl.* 1999, 730, 201–211

67. Xu, F.; Gao, M.; Shi, G.; Wang, L.; Zhang, W.; Xue, J.; Jin, L.; Jin, J., Simultaneous detection of monoamines in rat striatal microdialysate at poly(para-aminobenzoic acid) modified electrode by high-performance liquid chromatography, *Anal. Chim. Acta.* 2001, 439, 239–246

68. Slingerland, R. J.; van Kuilenburg, A. B.; Bodlaender, J. M.; Overmars, H.; Voûte, P. A.; van Gennip, A. H., High-performance liquid chromatographic analysis of biogenic amines in cells and in culture media using on-line dialysis and trace enrichment, *J. Chromatogr B Biomed. Sci. Appl.* 1998, 716, 65–75

69. Grossi, G.; Bargossi, A.; Sprovieri, G.; Bernagozzi, V.; Pasquali, R., Full automation of serotonin determination by column-switching and HPLC, *Chromatographia* 1990, 30, 61–68

70. Raymo, F. M.; Cejas, M. A., Supramolecular association of dopamine with immobilized fluorescent probes, *Org. Lett.* 2002, 4, 3183–3185

71. Kim, J.; Raman, B.; Ahn, K. H., Artificial receptors that provides a preorganized hydrophobic environment: a biomimetic approach to dopamine recognition in water, *J. Org. Chem.* 2006, 71, 38–45

72. Secor, K. E.; Glass, T. E., Selective amine recognition: development of a chemosensor for dopamine and norepinephrine, *Org. Lett.* 2004, 6, 3727–3730

73. Tuerk, C.; Gold, L., Systematic evolution of ligands by exponential enrichment: RNA ligands to bacteriophage T4 DNA polymerase, *Science* 1990, 249, 505–510

74. Mannironi, C.; Di Nardo, A.; Fruscoloni, P.; Tocchini-Valentini, G. P., In vitro selection of dopamine RNA ligands, *Biochemistry* 1997, 36, 9726–9734

75. Majerfeld, I.; Puthenvedu, D.; Yarus, M., RNA affinity for molecular L-histidine; genetic code origins, *J. Mol. Evol.* 2005, 61, 226–235

76. Famulok, M.; Szostak, J. W., Stereospecific recognition of tryptophan agarose by in vitro selected RNA, *J. Am. Chem. Soc.* 1992, 114, 3990–3991

77. Weng, S.; Gu, K.; Hammond, P. W.; Lohse, P.; Rise, C.; Wagner, R. W.; Wright, M. C.;Kuimelis, R. G., Generating addressable protein microarrays with PROfusion covalent mRNA-protein fusion technology, *Proteomics* 2002, 2, 48–57

78. Collett, J. R.; Cho, E. J.; Ellington, A. D., Production and processing of aptamer microarrays, *Methods.* 2005, 37, 4–15

Section 5
Inorganic Gas-Sensing Materials

Chapter 11
Impedometric Screening of Gas-Sensitive Inorganic Materials

Maike Siemons and Ulrich Simon

Abstract This chapter presents a setup for high throughput impedance spectroscopy (HT-IS) on gas sensing materials at different temperatures and in various gas atmospheres. Time consuming steps could be parallelized by using multielectrode substrate plates for 64 samples. Screening results for a surface doped $CoTiO_3/La$ and $LnFeO_3$ sample plates are shown to illustrate the relevance of HT-IS in the search for new gas sensing materials.

1 Introduction

Resistive gas sensors are used in a wide range of applications like air quality monitoring, fire detection, process control, or medical diagnostics. Hence, the mass market for gas sensors is still growing.[1] However, the majority of research activities on the development of fast responding, sensitive, and especially highly selective gas sensor materials is often restricted to improvement of known systems like SnO_2 or ZnO, and only rarely being directed toward the search for alternative sensor materials.

With rapidly growing demands for better functional materials, complex materials are receiving more and more attention in the materials research community.[2] In sensor research, materials are typically discovered and optimized by an empirical trial and error process that is both time consuming and costly. Obviously, the discovery process becomes even more difficult and time-consuming when the materials include more chemical elements.[3] Compared with the possible combinations of ternary or quaternary systems that one might reasonably make and which could exhibit several phases, it becomes clear that large numbers of compositions have

M. Siemons and U. Simon (✉)
Institute of Inorganic Chemistry, RWTH Aachen University, D-52056 Aachen, Germany
e-mail: ulrich.simon@ac.rwth-aachen.de

R.A. Potyrailo and V.M. Mirsky (eds.), *Combinatorial Methods for Chemical and Biological Sensors*, DOI: 10.1007/978-0-387-73713-3_11,
© Springer Science + Business Media, LLC 2009

yet to be examined for their gas sensing properties. Furthermore, the limited number of samples, which are tested in conventional sensor screenings, usually does not allow to find important correlations between structure and composition on the one hand, and the response toward different gases over a broad temperature range on the other hand. In recent years, much research effort was paid on understanding the mechanism of gas sensing and the relationships between structure and properties, which cannot be predicted precisely resulting from the complex interplay of different parameters like synthesis, calcination temperature, film preparation, etc.[4]

A possible way to overcome these problems is the use of high throughput and combinatorial approaches. The use of high throughput experimentation (HTE) techniques accelerates material synthesis and characterisation, and thus enables an investigation of a multitude of different materials. In addition, HTE allows the application of combinatorial strategies, in which algorithms define new material combinations based on previous characterisations. These processes are successfully used in pharmaceutical and biotechnological industries. In these fields, automation of the fabrication of multivariate specimen arrays, screening, and analysis techniques force the development of important new drugs and drug variants.[5,6]

The first experiments using high throughput experimentation in materials science were published in the 1960s. Experimentation in these years was limited by the analysis and data reduction equipment, which did not have the speed, automation, and/or resolution required. Since the 1990s, HTE and combinatorial strategies were established in material sciences, e.g. in the development of superconducting oxide compounds,[7,8] improved heterogeneous catalysts,[9–12] phosphors,[13] magnetore-sisitive materials,[14] photochemical materials,[15] dielectrics,[16] microwave dielectrics,[17] and electronic noses.[18] These high throughput research efforts are also extensively discussed in previous review papers.[11,19,20] However, up to now examples on resistive gas sensing materials research like in the work presented here are still very rare.

The materials in HTE approaches are typically combined to create a library with common features, like a base material, gaining diversity by addition of dopants or additives. High throughput characterization (screening) can be achieved by using sample plates containing several materials under test resulting in parallel processing of time-consuming steps in the characterisation workflow. HTE methods should allow to rapidly screen a large number of solid-state materials, and therefore should have a dramatic impact on the search for new classes of gas sensing materials with enhanced properties. In addition, HTE provides numerous data available for statistical investigations, which may help to map structure–property relationships by means of data mining.[5]

Our approach in search of novel and improved gas sensor materials is focussed on semiconducting nanoscaled metal oxides derived from the polyol method, using a high throughput impedance spectroscopy (HT-IS) system. The starting point of this work was a collaborative research project between industrial and academic research groups, focussing on the development of electrical and optical sensors by use of high throughput processes.[21] In this approach, large collections of different materials are rapidly processed with help of laboratory robotic systems and screened for specific properties of interest.

2   High Throughput Setup

The next sections describe the realization of the high throughput workflow and its corresponding steps within.

2.1   Multielectrode Array

For the high throughput screening, a multielectrode array substrate has been developed, on which 64 different materials can be electrically characterized. The sample plate is made of aluminum oxide (side length of 106 mm). This material is a suitable substrate because of its chemical resistance and good long-term stability against mechanical stress and thermal treatment. Additionally, the high electrical resistivity of the material does not interfere with the sample measurements. The multielectrode array consists of screen-printed platinum leads forming 64 interdigital capacitors (IDCs) as shown in Fig. 11.1. The IDCs are arranged in an 8 × 8 array in the center of

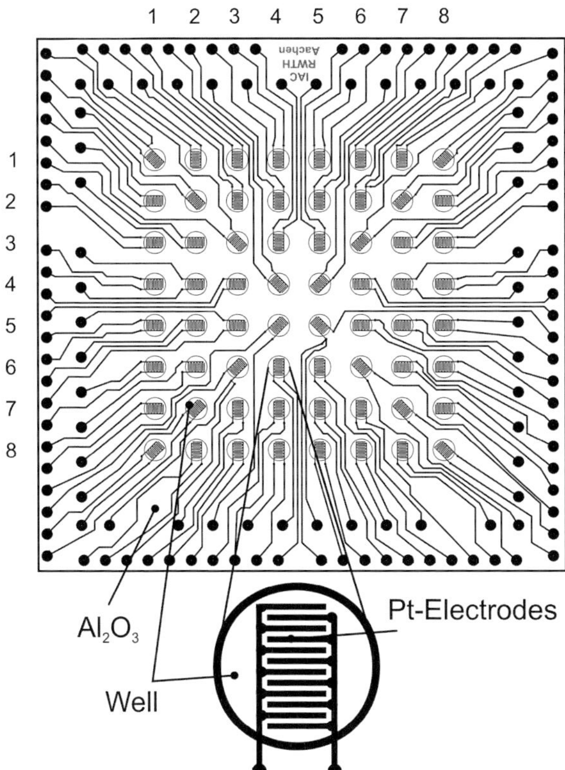

Fig. 11.1 Layout of the multielectrode substrate plate

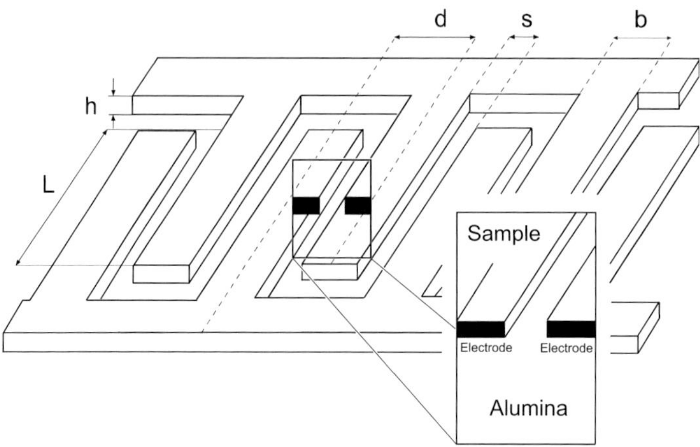

Fig. 11.2 IDC structure with a unit cell (adapted from[22])

the substrate plate. Positions on the substrate plates are labeled with numbers: each vertical column is labeled from top to bottom, while each horizontal row is labeled from left to right with numbers 1–8. In position descriptions, the row number is followed by the column number. For example, the position on the top left corner is termed 1–1 and the position in the bottom right corner is termed 8–8.

To realize the contact to the IDCs, a conductor path for each IDC leads to a contact pad outside the array. Figure 11.2 shows schematically the arrangement of the electrodes of the IDC.

One IDC consists of N fingers with length L, width b, distance s, and height h. Structural parameters of the electrodes could be optimized with focus on measurability of high sample resistivities and small changes in sample capacitances. The ideal structure width of the IDC is predicted by the permittivity of the substrate material and the thickness of the sample layer. The electrical field of the electrodes decays within the sample layer if the thickness of the layer is in the same dimension as the distance between the electrodes.[23] Layer thickness varies generally from 50 to 200 µm depending on synthesis and sintering conditions. In the examples described in this chapter, a distance of $s = 150$ µm between the electrodes was chosen.

In addition, the width and number of electrodes was optimized with regard to the preparation accuracy. The grain size of the platinum paste, the sinter process, and the substrate preparation has limited accuracy, resulting in variations in width and distance between the electrodes. These variations should have the smallest possible effect on the distribution of the IDC's capacity. Calculations showed that assuming a constant electrode distance of $s = 150$ µm, the average error will be sufficiently small, if the electrode width is >100 µm. Based on these calculations, b was taken as 125 µm.

The compact setup causes differences in length and distance of conductor pathways to each IDC. Thus, a broad distribution of the measured capacities can be found. Figure 11.3 shows an IDC structure with parallel supply leads and the corresponding circuit equivalent.

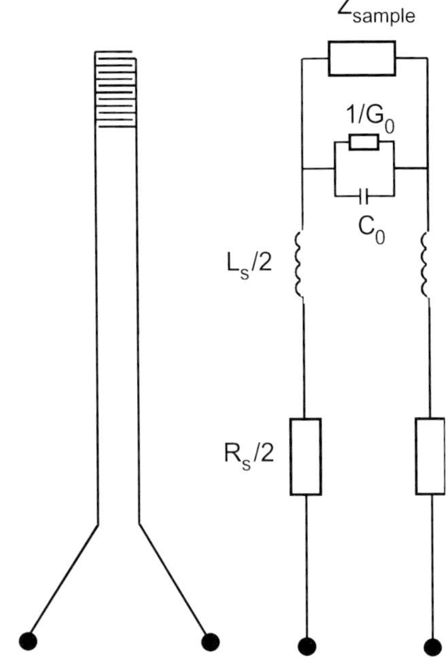

Fig. 11.3 IDC structure with parallel supply leads (*left*) and corresponding equivalent circuit (*right*) (adapted from[22])

The circuit equivalent taking into account parasitic effects consists of a contact resistance (R_s), a lead inductance L_s in series as well as a stray capacitance C_0 and the resistance of the substrate material between the leads $1/G_0$ parallel to the impedance of the sample. These parasitic effects, resulting from the compact arrangement of the electrodes can be eliminated by an offset adjustment. Capacitive parasitic effects are acquired by compensation measurements. The electrode array comprises some IDCs, which have almost negligible parasitic pathway effects, like position 5–5 (Fig. 11.1), where the conductor pathways are perpendicular to each other.[22] These positions show the smallest capacities, which result only from the IDC and can be taken to determine the parasitic capacitive values of the other positions.[24] Inductive parasitic effects L_s are acquired by data fitting and elimination from the sample impedance. Resistances of the conductor path (R_s generally $<10\,\mathrm{m\Omega}$) and conductivity of the substrate material ($1/G_0$, $>20\,\mathrm{m\Omega}$, exceeds measurement limit of the impedance analyser) are unaccounted.

The array presented is applicable for resistive as well as for capacitive measurements. In addition, the design allows efficient and automated pipetting, robot assisted sample preparation and coating.

2.2 High Throughput Impedance Spectroscopy Setup

Electrical impedance measurements are extremely valuable in many applications including measurements of superconductivity, magnetoresistivity, ferroelectric, and predominantly of gas sensing properties. The use of impedance spectroscopy over a wide frequency range allows screening of electrical properties, which are determined by the microstructure of the material, such as grain boundary conductance, interfacial polarization, or polarization of the electrodes. In the temperature range studied, ionic as well as electron conducting materials can be characterized.

A contact probe consisting of a matrix of spring-loaded contact pins was fabricated and used. This system is capable of measuring 2-probe conductivity of 64 samples in one experiment. The multisample measurement is accomplished by a computer-controlled multi-channel switching (high frequency capable multiplexer) and data acquisition system. Temperature, gas control, and data acquisition are fully automated by computer software. Figure 11.4 schematically shows the setup. Time consuming process steps such as heating up or sample conditioning can be proceeded in parallel. Sample plates are placed in a sample rack in contact to the mobile measuring head (5), which is separately shown on scheme and photograph in Fig. 11.5. The measuring head allows thermal decoupling and electrical contact to the electrode array via Al_2O_3 covered platinum wires. Constant contact pressure is assured by the spring-loaded contact tips, which are electrically connected by coaxial plugs (SMA). Materials were chosen because of their thermal stability. Load-bearing elements were made of high-temperature stable steel. Guidance for the insulated wires and sample holder were made of marcor®, a high temperature, stress-tolerant, and pore-free ceramic.

The measuring head connects the sample plate via high frequency capable relay matrices (multiplexers, model KRE-2450-TFCU, MTS-Systemtechnik, 3) to the measuring instruments (1, 2). The setup is usable for complex impedance spectroscopy

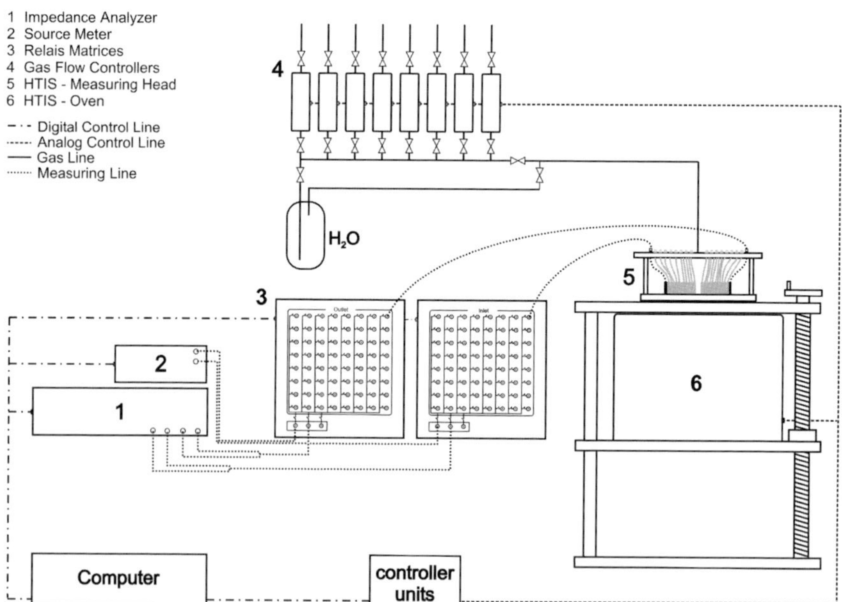

Fig. 11.4 The HT-IS Setup. For clarity only two measuring lines are displayed (adapted from[25])

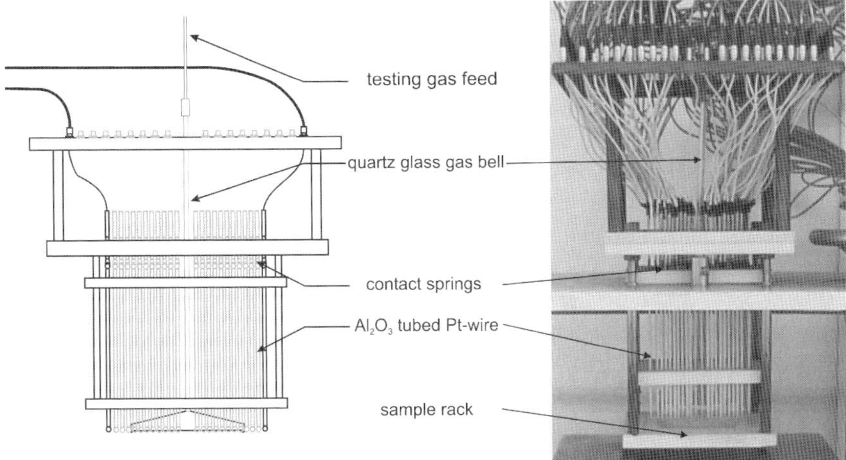

Fig. 11.5 The measuring head

measurements (Agilent 4192A, f = 10–10^7 Hz, 1) and dc measurements (Keithley 2400, 2). Connection between SMA plugs and the measuring equipment is achieved by using coaxial cables. For clarity only two of them are shown in the scheme. The measuring head is inserted into a furnace, which can heat samples up to 800°C (6). Temperature is achieved by four heating elements in the furnace associated with a thermo-controller (Eurotherm 2416) equipped with a Ni–Cr/Ni thermocouple (Thermocoax) placed below the sample rack. A set of gas flow controllers (model 1179/2179, MKS Instruments, 4) is used to compose the different test gases. Up to eight test gases can be mixed by variation of the flow rate. In addition, the test gases can be humidified by bubbling the carrier gas through a water reservoir at variable temperature. Stainless steel inlet pipes (Swagelok, 3 mm in diameter) transfer the test gases to the measuring head. Test gases are fed to the sample plate by a quartz glass tube whose lower end forms a quartz glass bell that covers all sample positions on the plate. A diffuser inlay helps to evenly feed the bell with test gases and a quartz glass plate with four drilled holes helps to achieve evenly gas flow on the sample positions. Steady gas and temperature distribution on the sample substrate is crucial for reproducible results.

The combined control and measuring software manages the process flow via interfaces (GPIB and RS232, respectively) in all devices. The program supports a basic script language for flexible and nonrestrictive processing of batches.

2.3 Flexible Data Handling

High throughput research generates large amounts of data, which need to be organised. For data evaluation, a flexible database was required, which was adjusted to the complex workflow of the sensor preparation and screening presented here. A common database concept was used to combine datasets resulting from different work steps and to store data for and from all project partners.[26] Flexibility was desired to adjust to requirements of new projects. In addition, the database should allow searching for correlations of properties and structure of the materials. For data mining, the most important parameter was the chemical composition to allow efficient comparison of different data sets produced in the project. The base material and surface or bulk dopants are storable with the respective concentrations. Additionally, the synthesis method and all electrical characterization data could be added. This means for the described impedance spectroscopy, values of resistance and capacitance reduced from the frequency dependence of the real part and the imaginary part of the complex impedance and the fit data are inserted in the data base.

3   Experimental Section

In the following section, we present exemplifying results for the use of HT-IS. Further examples for the successful establishment of this HT-IS setup can be found in.[25,27–30] The polyol method[31–35] was applied for the synthesis of nanocrystalline $CoTiO_3$/La and a series of $LnFeO_3$ perovskites (Ln = La, Pr, Nd, Sm-Lu).

$CoTiO_3$ is a p-type semiconductor and was introduced by Chu et al. for the detection of ethanol.[36] Lanthanum volume doped $CoTiO_3$ showed higher conductivity and better ethanol sensitivity of the material as shown in Siemons.[29] The use of HT-IS enabled the analysis of surface doping with different amounts of Au, Ce, Pd, Pt, Rh, and Ru in order to find out the highest selectivity and sensitivity of the respective material composition. Measurements of $CoTiO_3$/La samples at different temperatures and under different atmospheres will be shown in order to illustrate the high reproducibility of the impedance measurements and sensitivities of identical samples on different positions.

The $LnFeO_3$ shows a wide range of application, i.e., as catalysts,[37] electrodes of fuel cells,[38] and magneto-optics.[39] A number of perovskite oxides have been proposed previously as gas sensor materials for different test gases such as CH_3SH, NO_2,[40] H_2,[41] ethanol, and CO[42] because of their stability in thermally and chemically aggressive atmospheres. With help of the HT-IS system, it was possible to screen the $LnFeO_3$ materials under the same conditions, and to find a correlation between sensitivity and binding energy of the materials.

3.1   Sample Preparation

The preparation of $CoTiO_3$/La and the $LnFeO_3$ perovskite compounds is described in detail elsewhere.[28,30] Suitable amounts of precursors were dispersed in diethylene glycol. The mixture was heated up to T_1 (80–140°C) under vigorous stirring until a

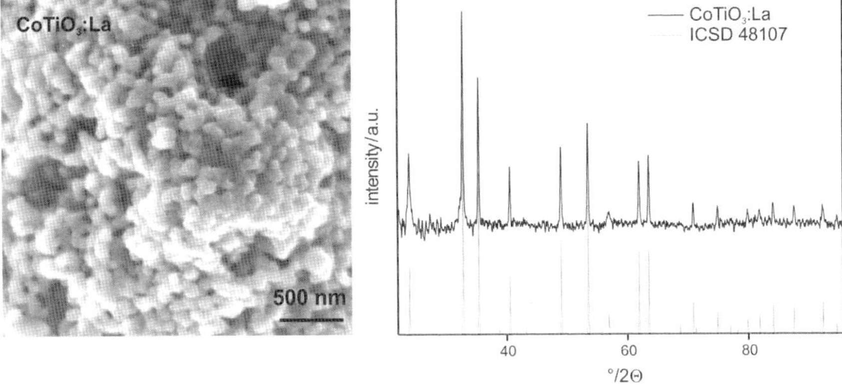

Fig. 11.6 SEM image and X-ray diffraction pattern ($\lambda = 1.5418\,\text{Å}$) of $CoTiO_3$/La powder after annealing for 12 h at $T = 700°C$

clear solution was obtained. If necessary, a sufficient surplus of hydrolysis agent was added. The emerging suspension was heated for several hours up to T_2 (160–190°C) and then cooled to room temperature.

To obtain crystalline materials, the suspensions were first dried and then annealed at T_3 (700–900°C) for 2–12 h. For each individual synthesis, reaction parameters needed to be optimized with respect to particle size, yield, and purity.

The initial characterization of the products was carried out by powder XRD measurements on thin films in transmission. Powder morphology was examined by SEM analysis. As an example in Fig. 11.6 the powder X-ray diffraction pattern of $CoTiO_3$/La displays the formation of single phase compound with $CoTiO_3$-structure (literature data ICSD 48107). The SEM image of the annealed material shows particles with a diameter of 30–100 nm. All annealed materials appeared as open porous networks of interconnected particles, which should allow good interaction between gas and surface.

3.2 Thick Film Preparation

For thick film deposition, the powder material was dispersed by mixing in a mortar with a solution of polyethylene imine in water. The suspension was applied directly to the electrode structure by deposition into a stack of individual cells that form an 8 × 8 array (Fig. 11.7). The design is similar to the sol–gel reactor set-up described in Frantzen et al.[43] In brief, the assembly consisted of a teflon block (1) 17 mm in thickness with 4 mm diameter holes positioned according to the substrate plate design. Each well had a viton fitting (O-ring, 2) for sealing against the substrate. The block was affixed to the electrode substrate (3), forming up to 64 wells (volume per well 214 µL). The base (4) and top plate (5) were made of metal.

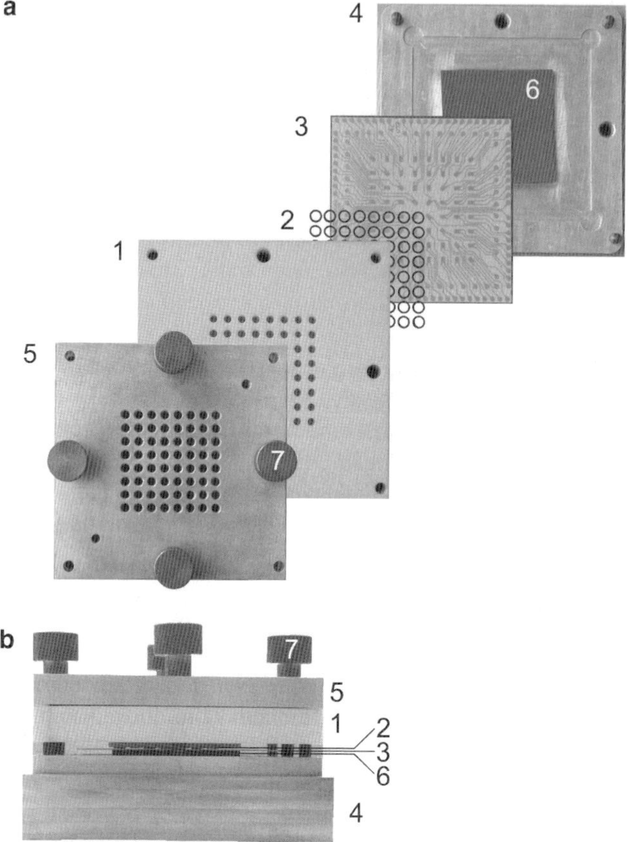

Fig. 11.7 Reactor set-up for thick film deposition, (a) *top view*, (b) *side view*

A viton plate (6) was placed between the substrate plate and the metal base to evenly distribute the pressure from the screws (7).

This design is well-suited for automated synthesis through pipetting robots, in our approach, a Lissy, Zinsser Analytic GmbH was used. Different aqueous metal salt solutions with concentrations from 0.2 to 0.6%atm were used for surface doping. The impregnation was also performed by use of the laboratory robotic system. The applied doping elements were Au (as $HAuCl_4 \cdot 3H_2O$), Ce (as $(NH_4)_2Ce(NO_3)_6$), Ir (as $Ir(C_5H_7O_2)_3$), Pd (as $Pd(NO_3)_2 \cdot 2H_2O$), Pt (as $Pt(NH_3)_4(NO_3)_2$), Rh (as $Rh(NO_3)_3 \cdot 2H_2O$), and Ru (as $Ru(NO)(O_2C_2H_3)_3$). Drying at room temperature and annealing for 12 h at 700°C resulted in homogenous sensing layers (like shown in Fig. 11.8). The distribution of the different material combinations are shown in Fig. 11.9. To register possible gradients in gas concentration and temperature as well as single failures due to contact defects over the substrate, three statistically selected positions were equipped with the same material composition.

Fig. 11.8 Scheme of thick film deposition and photograph of thick film

3.3 *Impedance Screening*

Impedance measurements were performed in a temperature range between 400 and 300°C in 25°-steps for CoTiO$_3$/La and between 500 and 200°C in 25°-steps for the LnFeO$_3$ materials. The measuring gas sequence was H$_2$ (25 ppm), air, CO (50 ppm), air, NO (5 ppm), air, NO$_2$ (5 ppm), air, ethanol (40 ppm), air, propylene (25 ppm), air. Reference measurements were performed in synthetic air (air), which was also used as carrier gas for the test gases H$_2$, CO, NO$_2$, ethanol, and propylene; for NO we choose N$_2$ as the carrier gas to avoid formation of NO$_2$. For the gas sensing experiments, the test gases were mixed with synthetic air to reach a consistent volume flow of 100 sccm. Conditioning of the materials was carried out at 400°C for 240 min under air and at the following temperatures for 90 min. Relative humidity of the gases was 45% at room temperature to perform the measurements in a typical humidity range representative for ambient conditions. For sample conditioning, a preliminary gas flow was applied over 30 min to reach adsorption equilibrium.

Figure 11.10 plots the imaginary –Z″ vs. the real part Z″ of the complex impedance (Argand plot) exemplary for undoped CoTiO$_3$/La at 400°C under synthetic air. All plots showed semicircles and could be described with the impedance function of a parallel RC circuit equivalent.

3.4 *Data Fitting and Evaluation*

An automated data fitting software was developed allowing to describe the impedance spectra as the impedance function of a circuit equivalent, e.g., consisting of a resistor and a capacitor in parallel. Also other elements, such as inductances or constant phase elements (CPE), could be implemented. The admittance of the parallel RC circuit is just the sum of the admittances of the two elements which give the impedance[44]:

$$Z^* = \frac{R}{1 + Ri\omega C}.$$

(1)

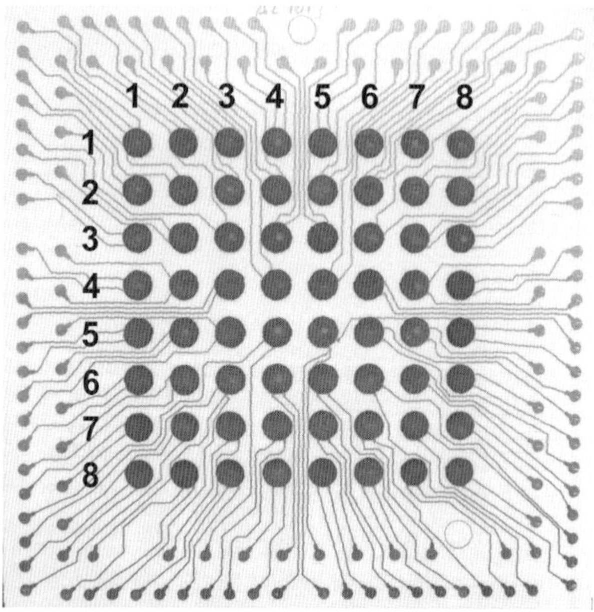

	1	2	3	4	5	6	7	8
1	Ru 0.6 at%	Pd 0.4 at%	Ce 0.2 at%	Ru 0.2 at%	Ce 0.6 at%	Pt 0.2 at%	Pd 0.2 at%	Ru 0.4 at%
2	Rh 0.2 at%	Pt 0.2 at%	Rh 0.4 at%	Pt 0.4 at%	un-doped	Pt 0.6 at%	Rh 0.6 at%	Ru 0.2 at%
3	Rh 0.6 at%	Pt 0.6 at%	Ir 0.2 at%	Ir 0.4 at%	Ir 0.6 at%	Ce 0.2 at%	Pt 0.4 at%	Ir 0.2 at%
4	Pd 0.2 at%	Ce 0.4 at%	Ce 0.6 at%	Au 0.2 at%	Au 0.4 at%	Au 0.6 at%	Ce 0.4 at%	Ir 0.4 at%
5	Pd 0.4 at%	Rh 0.4 at%	Ru 0.2 at%	Ru 0.4 at%	Ru 0.6 at%	Rh 0.2 at%	Rh 0.4 at%	Ir 0.6 at%
6	Au 0.2 at%	un-doped	Rh 0.6 at%	Pd 0.2 at%	Pd 0.4 at%	Pt 0.2 at%	Pt 0.6 at%	Au 0.4 at%
7	Au 0.6 at%	Ce 0.2 at%	Pt 0.6 at%	Pt 0.4 at%	un-doped	Ce 0.4 at%	Ir 0.2 at%	Ce 0.6 at%
8	Au 0.4 at%	Ir 0.6 at%	Ru 0.6 at%	Rh 0.2 at%	Au 0.6 at%	Ru 0.4 at%	Ir 0.4 at%	Au 0.2 at%

Fig. 11.9 The prepared sample plate and distribution of the doping elements on the base material $CoTiO_3$/La

From this $|Z|$ can be calculated as:

$$|Z| = \sqrt{\left(\frac{1}{R}\right)^2 + (\omega C)^2}^{\,-1} . \tag{2}$$

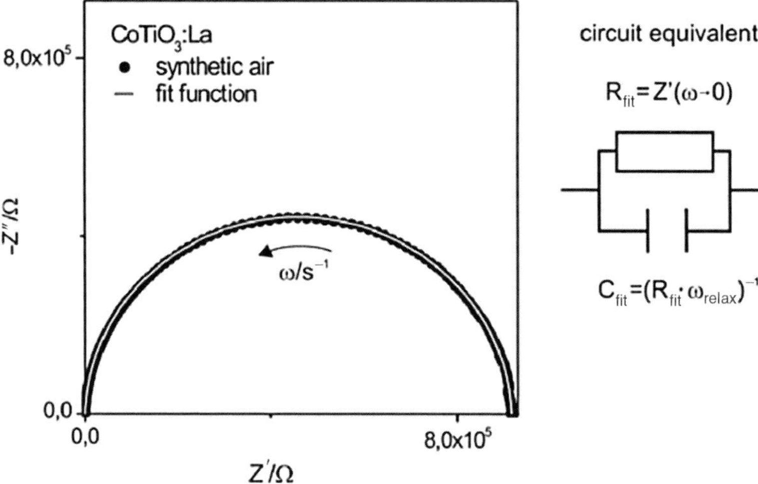

Fig. 11.10 Argand plot of CoTiO₃/La in synthetic air at $T = 400°C$ and the corresponding circuit equivalent

With the phase $\varphi = \arctan(R\omega C)$, the impedance function used for the real part Z' and the imaginary part Z'' is given by:

$$Z' = \cos[\arctan(\omega RC)] . \sqrt{\left(\frac{1}{R}\right)^2 + (\omega C)^2}^{\,-1} , \tag{3}$$

$$Z'' = \sin[\arctan(\omega RC)] . \sqrt{\left(\frac{1}{R}\right)^2 + (\omega C)^2}^{\,-1} . \tag{4}$$

In this case R is the extrapolated dc-resistance of the material, C the capacitance, and ω the angular frequency. The use of these fit functions reduces the complex impedance function describing the electrical properties of each material to one value for the resistance and the capacitance, respectively. Although the capacitance can be assigned to the geometric capacitance of the IDC, the resistance values under the respective measuring conditions are taken for the determination of the material sensitivity.

Samples with depressed semicircles have been adjusted with a CPE. The CPE is a nonintuitive circuit element. In case of a resistance R and a CPE in parallel, we observe the arc of a circle with the center some distance below the x-axis in the Argand plot. The corresponding circuit equivalent is called a ZARC-element.[44]

$$Z^* = \frac{R}{1 + RQ(i\omega)^n}, \tag{5}$$

where R is the low frequency real-axis intercept, n is related to the depression angle, i is the imaginary number, and Q has the numerical value of the admittance Y at ω.[45]

$$Y^* = Q(i\omega)^n \qquad \text{for } 0 < n < 1, \tag{6}$$

$$Y^* = i\omega C \qquad \text{for } n = 1. \tag{7}$$

The behavior of a resistor in parallel with an ideal capacitor (see above) is recovered when n is 1 ($Q = C$). When n is close to 1, the CPE resembles a capacitor, but the phase angle is not 90°. The real capacitance can be calculated from Q and n. When n is zero, only a resistive influence is found. For all impedance spectra shown in this work, fitting with a single RC circuit was found to be sufficient, i.e., n was in all cases larger than 0.9. Figure 11.10 shows that a good accordance of measuring data and fit function is evident.

Figure 11.11 shows the temperature dependence of the sample resistance achieved from fitting for two identical samples. A narrow distribution of the two position resistances can be observed. This gives evidence that film deposition, measuring method and fitting shows reproducible results.

3.4 Gas Sensing Properties

From the resistivity values obtained, the sensitivity of the measured materials in different test gas atmospheres could be calculated. Sensitivity is defined as the ratio of the resistance of a sensing material in the presence of an analyte to the baseline resistance measured in air.

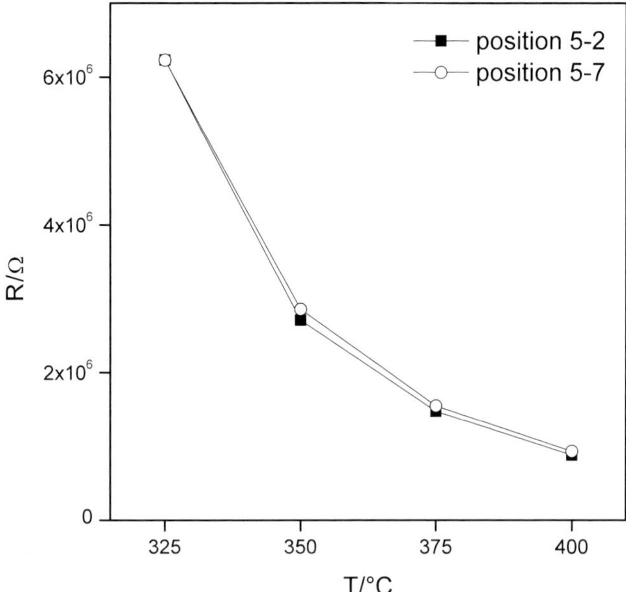

Fig. 11.11 Comparison of two different positions with same sample layer (undoped CoTiO$_3$/La) in air

Hence, the absolute sensitivity S of a material to a test gas is defined by:

$$S = \frac{R_g}{R_r}; \quad R_g > R_r,\tag{8}$$

$$S = \frac{R_r}{R_g}; \quad R_g < R_r,\tag{9}$$

where R_g is the resistance obtained from the measurement under test gas conditions, and R_r is the reference resistance value for a measurement under synthetic air.

We describe the response of a material to a test gas by the relative sensitivity S_Δ, which is defined as

$$S_\Delta = -\frac{R_g - R_r}{R_g}; \quad R_g > R_r,\tag{10}$$

$$S_\Delta = +\frac{R_r - R_g}{R_r}; \quad R_g < R_r,\tag{11}$$

where values for S_Δ range from 1 to -1. Decrease of resistance while applying a test gas gives values between 0 and 1, rise of resistance results correspondingly in values for S_Δ between -1 and 0. Using S_Δ instead of the commonly used sensitivity S including algebraic signs turned out to be beneficial for the data mining process in our high throughput workflow. Furthermore, it allowed an immediate differentiation between n and p-type semiconducting materials and/or the differentiation between oxidizing and reducing test gases, respectively.

Figure 11.12 shows the fingerprints for the entire sample plate at 400°C. Positions 1–6, 1–7, 2–4, 2–6, 3–5, 5–4, 6–6, 6–8, and 8–3 showed solely capacitive behavior. This may occur due to insufficient contact between sample and IDC or defect conductor paths on the microelectrode array. For these positions, no sensitivities were achieved. Equally occupied positions show good correspondence in the sensitivities. All materials are highly sensitive toward hydrocarbons. For the majority of the surface dopants, no significant influence on the sensitivity could be observed. The most pronounced effect on the sensitivity can be found for Au and Pd-doped materials. Here we will focus on the gold doped samples. Almost no dependence on the concentration of the surface dopant was found. Figure 11.13 shows the temperature dependence of the relative sensitivity toward the tested gases for undoped and Au-doped $CoTiO_3$/La. The undoped material is highly sensitive and selective towards hydrocarbons at higher temperatures. At lower temperatures, the cross sensitivity toward the other gases increases.

Au doping causes an increased sensitivity towards CO over the whole temperature range having its maximum at 325°C. At the same time, the sensitivity toward hydrogen is increased, which is most pronounced at lower temperatures.

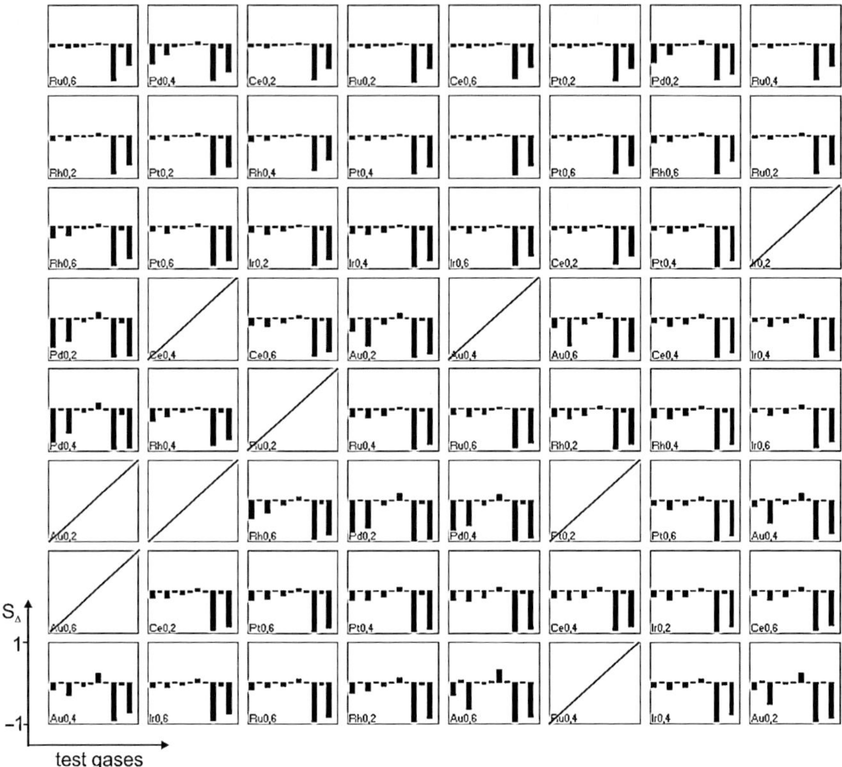

Fig. 11.12 Fingerprint of the sample plate AC1013 at 400°C. Many surface dopants show the same sensitivties as the undoped material. Besides the nine fields, which reflect defects in the electrical contacts of the respective materials, the measurements show good reproducibility

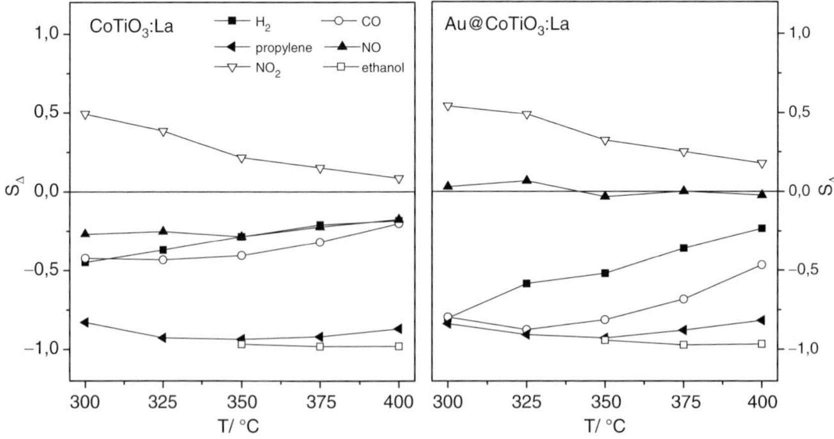

Fig. 11.13 Relative sensitivity of CoTiO$_3$/La vs. operating temperature (300–500°C) for various gases

Remarkable changes do also appear in the sensitivity toward NO_x. Although the undoped material shows an almost temperature independent S_Δ between -0.2 and -0.29 for NO, the sensitivity toward NO_2 decreases with increasing temperature from 0.5 to 0.1 between 300 and 400°C. The Au-doped material shows a slightly increased sensitivity toward NO_2, whereas the sensitivity toward NO appears to be suppressed.

When a doping element is added to the surface, the equilibrium state and/or velocity of the surface reactions is modified. Au is well known as active catalysts for e.g. CO oxidation,[46,47] hydrogen peroxide synthesis from H_2 and O_2,[48] or hydrocarbon oxidation. The doping element supports the catalytic conversion of the reducing gas into its respective oxidation product. This may be due to spill-over of activated fragments to the semiconductor surface to react with the adsorbed oxygen.[49] A catalytic effect inducing a located temperature change and a change in resistance may cause the sensitivity change of the surface doped material. These models may describe the increased sensitivity of the Au-doped material toward H_2, NO_2, and CO observed here. The reaction toward the hydrocarbons originates from the metal oxide and therefore not highly influenced by the surface dopant.

Most interesting for practical use is the almost fully suppressed sensitivity towards NO with a simultaneously pronounced sensitivity toward NO_2 observed for the Au-doped material. The NO-tolerance was also observed in Filippini et al. and Sun et al.[50,51]

The sensitivities of the $LnFeO_3$ materials are described in detail elsewhere.[30] Here, we present a structure property relation that was revealed by the HT-IS measurements.

The bond strength of the oxygen at the surface near sites is an important factor for the sensing properties that has already been proposed by Arakawa et al. for methanol sensing activity of different rare-earth perovskites.[52] The binding energy $\Delta H(M–O)$ is calculated by:

$$\Delta H(\mathrm{Ln} - O) = \frac{1}{2CN}\left(H_f - H_s.2 - \frac{3}{2}D_o \right), \qquad (12)$$

where H_f is the formation energy of one mole of oxide Ln_2O_3, H_s is the sublimation energy of the metal, D_O is the dissociation energy of O_2, and CN is the coordination number of metal ions. As a result we can show that the sensitivity toward ethanol, propylene, and nitrogen dioxide is correlated to the binding energy, as shown in Fig. 11.14. In the $LnFeO_3$ series, increasing sensitivity is connected with a decreasing absolute value of the binding energy. Considering that $\Delta H(\mathrm{Ln}–O)$ reflects the bond strength between lanthanide cation on the surface and adsorbed oxygen, a weak metal-to-oxygen bond corresponds to an easier reaction of the adsorbed oxygen in the presence of testing gas species. Only $LuFeO_3$ showed a higher sensitivity toward ethanol ($S = 13$) than it was expected from this model. HT-IS allowed to find this correlation because of the possibility to screen all 13 materials under same conditions.

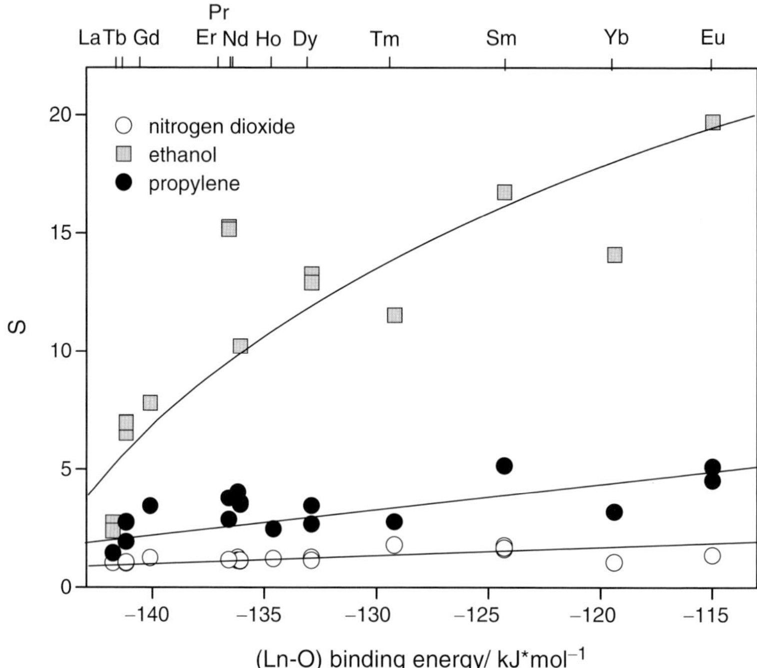

Fig. 11.14 Sensitivity S vs. (Ln–O) binding energy at 300°C (without LuFeO$_3$)

4   Conclusion

This chapter presents a detailed description of a high throughput impedance spectroscopy measuring setup. HT-IS allows high throughput electrical characterisation of metal oxides in a wide range of temperatures and in different atmospheres. Time reduction is achieved by the use of multielectrode arrays supporting fast and parallel processing, assisted by complete automation of the measurement work flow.

Measurements on surface doped CoTiO$_3$/La showed that measurements on different positions with equally deposition are reproducible and with low deviations. With help of HT-IS a CO sensitive material with also NO-tolerant NO$_2$-sensing properties at 350°C was found. In addition, the screening of a series of LnFeO$_3$ materials was able to show a correlation between sensitivity and (Ln–O) binding energy of the materials. Thus, the methodology described will be a powerful tool for our future research on gas sensing materials.

Acknowledgments This work was financially supported by BMBF under contract No.03 C 0305 A. M. S. gratefully acknowledges generous support by the Studienstiftung des Deutschen Volkes.

References

1. Yamazoe, N., Towards innovations of gas sensor technology, *Sens. Actuators B: Chem.* **2005**, 108(1–2), 2–14
2. Schüth, F., Hochdurchsatzuntersuchungen. In: *Chemische Technik: Prozesse und Produkte, Winnacker/Küchler, Band 2: Neue Technologien*, Dittmeyer, R.; Keim, W.; Kreysa, G.; Oberholz, A., (Eds.) Wiley-VCH, Weinheim **2004**
3. Maier, W. F., Kombinatorische Chemie-Herausforderung und Chance für die Entwicklung neuer Katalysatoren und Materialien, *Angew. Chem.* **1999**, 111(9), 1295–1296
4. Franke, M. E.; Koplin, T. J.; Simon, U., Metal and metal oxide nanoparticles in chemiresistors: Does the nanoscale matter? *Small* **2006**, 2(1), 36–50
5. Amis, E. J.; Xiang, X.-D.; Zhao, J.-C., Combinatorial Materials Science: What's New Since Edison? *MRS Bull.* **2002**, 4, 295–297
6. Amis, E. J., Combinatorial materials science: Reaching beyond discovery, *Nat. Mater.* **2004**, 3, 83–84
7. Xiang, X.-D.; Sun, X.; Briceño, G.; Lou, Y.; Wang, K. A.; Chang, H.; Wallace-Freedman, W. G.; Chen, S.-W.; Schultz, P. G., A combinatorial approach to material discovery, *Science* **1995**, 268, 1738–1740
8. Xiang, X.-D.; Schultz, P. G., The combinatorial synthesis and evaluation of functional materials, *Physica C* **1997**, 282–287, 428–430
9. Reichenbach, H. M.; McGinn, P. J., Combinatorial synthesis of oxide powders, *J. Mater. Res.* **2001**, 16(4), 967–974
10. Moates, F. C.; Somani, M.; Annamalai, J.; Richardson, J. T.; Luss, D.; Wilson, R. C., Infrared Thermographic screening of combinatorial libraries of heterogenous catalysts, *Ind. Eng. And Chem. Res.* **1996**, 35, 4801–4803
11. Jandeleit, B.; Schaefer, D. J.; Powers, T. S.; Turner, T. W.; Weinberg, W. H., Combinatorial materials science and catalysis, *Angew. Chem.* 111, 2648–2689; *Angew. Chem. Int. Ed.* **1999**, 38, 2494–2532
12. *High-Throughput Screening in Chemical Catalysis*, Hagemeyer, A.; Stasser, P.; Volpe, Jr. A. F. (Eds.) Wiley-VCH, Weinheim, **2004**
13. Danielson, E.; Devenney, M.; Giaquinta, D. M.; Golden, J. H.; Haushalter, R. C.; McFarland, E. W.; Poojary, D. M.; Reaves, C. M.; Weinberg, W. H.; Wu, X. D., A rare-earth phosophor containing one-dimensional chains identified through combinatorial methods, *Science* **1998**, 279, 837–839
14. Briceño, G.; Chang, H.; Sun, X.; Schultz, P. G.; Xiang, X.-D., A class of cobalt oxide magnetoresistance materials discovered with combinatorial synthesis, *Science* **1995**, 270, 273–275
15. Baeck, S. H.; Jaramillo, T. F.; Brändli, C.; McFarland, E. W., Combinatorial electrochemical synthesis and characterization of tungsten-based mixed-metal oxides, *J. Comb. Chem.* **2002**, 4, 563–568
16. Van Dover, R. B.; Schneemeyer, R. F.; Fleming, R. M., Discovery of a useful thin-film dielectric using a composition-spread approach, *Nature* **1998**, 392, 162
17. Chang, H.; Gao, C.; Takeuchi, L.; Yoo, Y.; Wang, J.; Schultz, P. G.; Xiang, D.; Sharma, P. R.; Downes, M.; Venkatesan, T., Combinatorial synthesis and high throughput evaluation of ferroelectric/dielectric thin-film libraries for microwave applications, *Appl. Phys. Lett.* **1998**, 72, 2185–2187
18. Aronova, M. A.; Chang, K. S.; Takeuchi, I.; Jabs, H.; Westerheim, D.; Gonzalez-Martin, A.; Kim, J.; Lewis, B., Combinatorial libraries of semiconductor gas sensors as inorganic electronic noses, *Appl. Phys. Lett.* **2003**, 83(6), 1255–1257
19. Sekan, S., Combinatorial heterogeneous catalysis – a new path in an old field, *Angew. Chem. Int. Ed.* **2001**, 40(2), 312–329

20. Yoo, Y. K.; Xiang, X.-D., Combinatorial material preparation, *J. Phys.: Condens. Matter* **2002**, 14, R49–R78

21. Brinz, T., Maier, W. F.; Wolfbeis, O.; Simon, U., Gassensoren durch High-Throuput-Methoden, *Nachrichten aus der Chemie* **2004**, 52, 247–251

22. Simon, U.; Sanders, D.; Jockel, J.; Heppel, C.; Brinz, T., Design strategies for multielectrode arrays applicable for high-throughput impedance spectroscopy on novel gas sensor materials, *J. Comb. Chem.* **2002**, 4, 511–515

23. Scheibe, C.; Obermeier, E.; Maunz, W.; Plog, C., Development of a high-temperature basic device for chemical sensors based on an IDC with on-chip heating, *Sens. Acturators B* **1995**, 25, 584–587

24. Sanders, D., Entwicklung von Gassensoren auf Indiumoxid-Basis mittels Hochdurchsatz-Impedanzspektroskopie, PhD thesis, RWTH Aachen, **2004**

25. Simon, U.; Sanders, D.; Jockel, J.; Brinz, T., Setup for high-throughput impedance screening of gas-sensing materials, *J. Comb. Chem.* **2005**, 7(5), 682–687

26. Frantzen, A.; Sanders, D.; Scheidtmann, J.; Simon, U.; Maier, W. F., A flexible database for combinatorial and high-throughput materials science, *QSAR Comb. Sci.* **2005**, 24, 22–28

27. Sanders, D.; Siemons, M.; Koplin, T.; Simon, U., Development of a high-throughput impedance spectroscopy screening system (HT-IS) for characterization of novel nanoscaled gas sensing materials, *Mater. Res. Soc. Symp. Proc.* **2005**, 876E, R6.1.1–R6.1.6

28. Koplin, T. J.; Siemons, M.; Océn-Valéntin, C.; Sanders, D.; Simon, U., Workflow for high-throughput screening of gas sensing materials, *Sensors* **2006**, 6, 298–307

29. Siemons, M.; Simon, U., Preparation and gas sensing properties of nanocrystalline La-doped $CoTiO_3$, *Sens. Actuators B* **2007**, 120(1), 110–118

30. Siemons, M.; Leifert, A.; Simon, U., Preparation and gas sensing characteristics of nanoparticulate p-type semiconducting $LnFeO_3$ and $LnCrO_3$ (Ln = La, Pr, Nd, Sm, Eu, Gd, Tb, Dy, Ho, Er, Tm, Yb, Lu), *Adv. Funct. Mater.* **2007**, 17(13), 2189–2197

31. Feldmann, C., Polyol-mediated synthesis of nanoscale functional materials, *Adv. Funct. Mater.* **2003**, 13(2), 101–107

32. Feldmann, C., Darstellung und Charakterisierung der nanoskaligen Vb-Metalloxide M_2O_5 (M = V, Nb, Ta), *Z. Anorg. Allg. Chem.* **2004**, 630, 2473–2477

33. Fiévet, F., Polyol Process in T. Sugimoto, *Fine Particles: Synthesis, Characterization, and Mechanisms of Growth*, Marcel Dekker, NY, pp. 460–496, **2000**

34. Poul, L.; Jouini, N.; Fievet, F., Layerd Hydoxide Metal Acetates (Metal = Zinc, Cobalt and Nickel): Elaboration via hydrolysis in polyol medium and comparative study, *Chem. Mater.* **2000**, 12, 3123–3132

35. Siemons, M.; Weirich, Th.; Mayer, J.; Simon, U., Preparation of nanosized perovskite-type oxides via polyol method, *Z. Anorg. Allg. Chem.* **2004**, 630, 2083–2089

36. Chu, X.; Liu, X.; Wang, G.; Meng, G., Peparation and gas sensing properties of nano-$CoTiO_3$, *Mat. Res. Bull.* **1999**, 34(10/11), 1789–1795

37. Peña, M. A.; Fierro, J. L. G., Chemical structures and performance of perovskite oxides, *Chem. Rev.* **2001**, 101, 1981–2017

38. Yokokawa, H.; Sakai, N.; Horita, T., Yamaji, K., Recent developments in solid oxide fuel cell materials, *Fuel Cells* **2001**, 1(2), 117–131

39. Keller, N.; Mistrik, J.; Visnovsky, S.; Schmool, D. S.; Dumont, Y.; Renaudin, P.; Guyot, M.; Krishnan, R., Magneto-optical Faraday and Kerr effect of orthoferrite thin films at high temperatures, *Eur. Phys. J. B* **2001**, 21(1), 67–73

40. Aono, H.; Traversa, E.; Sakamoto, M.; Sadaoka, Y., Crystallographic characterization and NO_2 gas sensing property of $LnFeO_3$ prepared by thermal decomposition of Ln–Fe hexacyanocomplexes, $Ln[Fe(CN)_6] \cdot nH_2O$, Ln = La, Nd, Sm, Gd, and Dy, *Sens. Actuators B* **2003**, 94, 132–139

41. Niu, X.; Du, W.; Du, W., Preparation, characterization and gas-sensing properties of rare earth mixed oxides, *Sens. Actuators B*, **2004**, 99, 399–404

42. Martinelli, G.; Carotta, M. C.; Ferroni, M.; Sadaoka, Y.; Traversa, E., Screen-printed perovskite-type thick films as gas sensors for environmental monitoring, *Sens. Actuators B* **1999**, 55, 99–110

43. Frantzen, A.; Sanders, D.; Jockel, J.; Scheidtmann, J.; Frenzer, G.; Maier, W. F.; Brinz, Th.; Simon, U., High-throughput method for the impedance spectroscopic characterization of resistive gas sensors, *Angew. Chem.*, 116, 770–773; *Angew. Chem. Int. Ed.* **2004**, 43, 752–754

44. Macdonald, J. R. *Impedance Spectroscopy*, Wiley, New York, **1987**

45. Orazem, M. E.; Shukla, P.; Membrino, M. A., Extension of the measurement model approach for deconvolution of the underlying distributions for impedance measurements, *Electrochimica Acta* **2002**, 47, 2027–2034

46. Janssens, T. V.; Carlsson, A.; Puig-Molina, A.; Clausen, B. S., Relation between nanoscale Au particle structure and activity for CO oxidation on supported gold catalysts, *J. Catal.* **2006**, 240(2), 108–113

47. Haruta, M., Size-and support-dependency on the catalysis of gold, *Catal. Today* **1997**, 36, 153–166

48. Solsona, B. E.; Edwards, J. K.; Landon, P.; Carley, A. F.; Herzing, C., Direct synthesis of hydrogen peroxide from H_2 and O_2 using Al_2O_3 supported Au–Pd catalysts, *Chem. Mater.* **2006**, 18(11), 2689–2695

49. Korotcenkov, G., Gas response control through structural and chemical modification of metal oxide films: state of the art and approaches, *Sens. Actuators B* **2005**, 107, 209–232

50. Filippini, D.; Fraigi, L.; Aragón, R.; Weimar, U., Thick film Au-gate field-effect devices sensitive to NO_2, *Sens. Actuators. B* **2002**, 81(2–3), 296–300

51. Sun, H.-T.; Cantalini, C.; Faccio, M.; Pelino, M., NO_2 gas sensitivity of sol–gel-derived α-Fe_2O_3 thin films, *Thin Solid Films* **1995**, 269, 97–101

52. Arakawa, T.; Kurachi, H.; Shiokawa, J., Physicochemical properties of rare earth perovskite oxides used as gas sensor material, *J. Mater. Sci.* **1985**, 20, 1207–1210

Chapter 12
Design of Selective Gas Sensors Using Combinatorial Solution Deposition of Oxide Semiconductor Films

Jong-Heun Lee, Sun-Jung Kim, and Pyeong-Seok Cho

Abstract The study examined sensing behavior of multicompositional gas sensing materials prepared through combinatorial deposition of SnO_2, ZnO, and WO_3 sols. Selective detection of C_2H_5OH and CH_3COCH_3 in the presence of CO, C_3H_8, H_2 and NO_2 was accomplished by combinatorial manipulation of the gas sensor composition. A further tuning of the gas-sensing materials and gas-sensing temperature allowed discrimination between C_2H_5OH and CH_3COCH_3, which is a challenging issue due to their similar chemical nature. The discrimination of similar gases and selective gas detection are discussed with respect to the gas sensing mechanism. Combinatorial approach is very convenient and useful for determining an optimal composition for selective-gas detection.

1 Introduction

Since the discovery of oxide semiconductor gas sensors in the 1960s, many sensing materials, such as SnO_2, ZnO, WO_3, Fe_2O_3, and TiO_2, have been explored to achieve the requirements of (1) high sensitivity, (2) selective detection of a specific gas, (3) rapid response time, (4) long-term stability, (5) cost efficiency.[1–5]

The operation principle of oxide semiconductor type gas sensors provides the directions for accomplishing the above requirements. At 200–400°C, ambient oxygen is adsorbed on the surface of a *n*-type semiconductor with a negative charge, which forms an electrical core-shell structure: an electron depleted layer near the surface and a semiconducting layer at the core region. If a trace concentration of reducing gases, such as CO, C_3H_8, C_2H_5OH, CH_4, and C_2H_6, is present in the atmosphere, they are oxidized to CO_2 and H_2O gases through a reaction with

J.-H. Lee (✉), S.-J. Kim, and P.-S. Cho
Department of Materials Science and Engineering, Korea University, Seoul 136-713, Korea
jongheun@korea.ac.kr

R.A. Potyrailo and V.M. Mirsky (eds.), *Combinatorial Methods for Chemical and Biological Sensors*, DOI: 10.1007/978-0-387-73713-3_12,

negatively charged surface oxygen (O_{surf}^-), and the remnant electrons return to the core n-type semiconductor. This increases the sensor conductivity proportionally to the concentration of the reducing gas.

The main approach to reach higher sensor sensitivity is based on maximization of the electron depletion layer compared with the semiconducting core; this can be realized by decreasing the particle size down to the scale of the depletion layer thickness.[6–8] Oxide semiconductors in the form of nanocrystalline powders,[9–11] nanorods,[12] nanowires,[13–15] nanotubes,[16,17] and nanobelts[18] with a high surface area/volume ratio have been studied intensively as highly sensitive materials.

The device structure of an oxide semiconductor gas sensor is relatively simple compared with that of an electrochemical-type sensor; this enables cost-effective fabrication. However, the simple sensing algorithm hinders the selective detection of a specific gas to a great extent. For example, the SnO_2 gas sensor shows an increase in conductivity when exposed to various different reducing gases. Therefore, selective gas sensing using an oxide semiconductor is a challenging issue.

Various methods have been used to achieve selective gas detection. These include a modulation of the sensing temperature,[19,20] an addition of a noble metal[21] or oxide catalyst,[22] a deposition of a coating layer for the selective filtering of a gas,[23,24] design of multicompositional sensing materials,[25,26] surface modification,[27,28] manipulation of the nanostructure,[29] and an application of a neural network algorithm.[30] Almost all these approaches except manipulation of the sensing temperature and nanostructure can be best optimized by using of combinatorial routes.

In this contribution, combinatorial approaches to fabricate oxide semiconductor gas sensors are reviewed, and the design of multicompositional gas sensing materials for selective gas detection is demonstrated; it was realized by combinatorial deposition of SnO_2, ZnO, and WO_3 sols. We focused mainly on the design of gas sensor composition for the selective detection of C_2H_5OH and CH_3COCH_3 in the presence of CO, C_3H_8, H_2, NO_2, and C_2H_5OH, and on the discrimination between C_2H_5OH and CH_3COCH_3.

2 Combinatorial Approaches in Oxide Semiconductor Gas Sensors

Combinatorial chemistry is a rapid, convenient, and effective tool for discovering or searching for multicompositional materials with the best performance from a large compositional landscape.[31–35] The development of combinatorial method was originally motivated by requirements of drug discovery; later it was applied for the development of catalysts.[31,36–38] Then it was extended to the discovery of new materials, this approach is now referred as "combinatorial materials science"[31–35]. The combinatorial methodology was then expanded to the high-throughput analysis of the physico-chemical properties of materials.[39–44] So far, combinatorial materials science has been applied to the discovery and design of luminescent materials,[45] transparent conducting materials,[46] ferromagnetic materials,[47] piezoelectric materials,[48] and shape memory alloys.[49,50]

In design of gas sensors, combinatorial approaches have two main advantages. The first one is the easiness in discovering a new gas sensor composition or surface-doping element, using high-throughput analysis. High sensitivity and selectivity are the main targets. The second motivation is the establishing of a wide scope of materials library for gas sensing. Selective detection is aimed on the sensing of a specific gas. However, if a target gas is not a singular chemical quantity but the complex mixture containing multiple chemical quantities, a selective detection of a specific gas may be difficult. In this case, a pattern recognition using the sensor arrays provides an effective way for selective detection of different gases. The groups[48–50] of Maier and Simon fabricated an 8×8 sensor array on the Al_2O_3 plates and tested the gas sensing characteristics using high-throughput impedance spectroscopy[51,52] and DC-resistance measurement with the programmed switching.[53] From the different signal fingerprints of the WO_3 and In_2O_3 sensors doped with various surface doping elements, they successfully demonstrated the validity of their combinatorial gas sensor design. Aronova et al.[54] deposited 16 different SnO_2-based gas sensors using combinatorial pulsed-laser deposition with a computer controlled two-dimensional physical shadow mask, and tested the gas sensing behaviors using a pattern recognition mechanism.

Figure 12.1 shows the representative approaches for selective gas detection and the potential use of the combinatorial technology for design of sensing materials. When the sensing temperature is varied, different forms of adsorbed oxygen (O_2^-, O^-, O^{2-}) are formed on the surface of the oxide semiconductor. These species interact differently with different gases thus providing selective gas sensing (Fig. 12.1a).

The second approach is based on addition of a precious metal or oxide catalyst (Fig. 12.1b). The design of a catalyst usually requires a large number of experiments because various catalyst materials are available and their concentrations should be optimized. In this respect, the combinatorial method is advantageous. Indeed, there are reports on the design of catalysts or surface doping elements that promote the selective gas sensing reaction.[48–50]

Selective filtering layer can be fabricated through the coating of oxide or catalyst-added oxide layers[23,24] (Fig. 12.1c). The filtering layer promotes the selective detection of the target gas through oxidation of interfering gases into less reactive (neither oxidizing nor reducing) ones such as CO_2 or H_2O. The oxide and catalyst materials need to be designed very carefully considering the chemical nature of the target and interfering gases. For this, combinatorial materials design will be also advantageous. It should be noted that the sensor response times might become longer when the filtering layer is dense and thick. Therefore, the compensation between the in-diffusion of the target gas through the filtering layer and the selective oxidation of interfering gases is indispensable.

Multicompositional films can be produced (Fig. 12.1d). The catalytic materials in semiconductor-type gas sensors are usually added after forming a sensing layer through impregnation,[55] sputtering,[56] and vapor deposition,[57] even though it can be

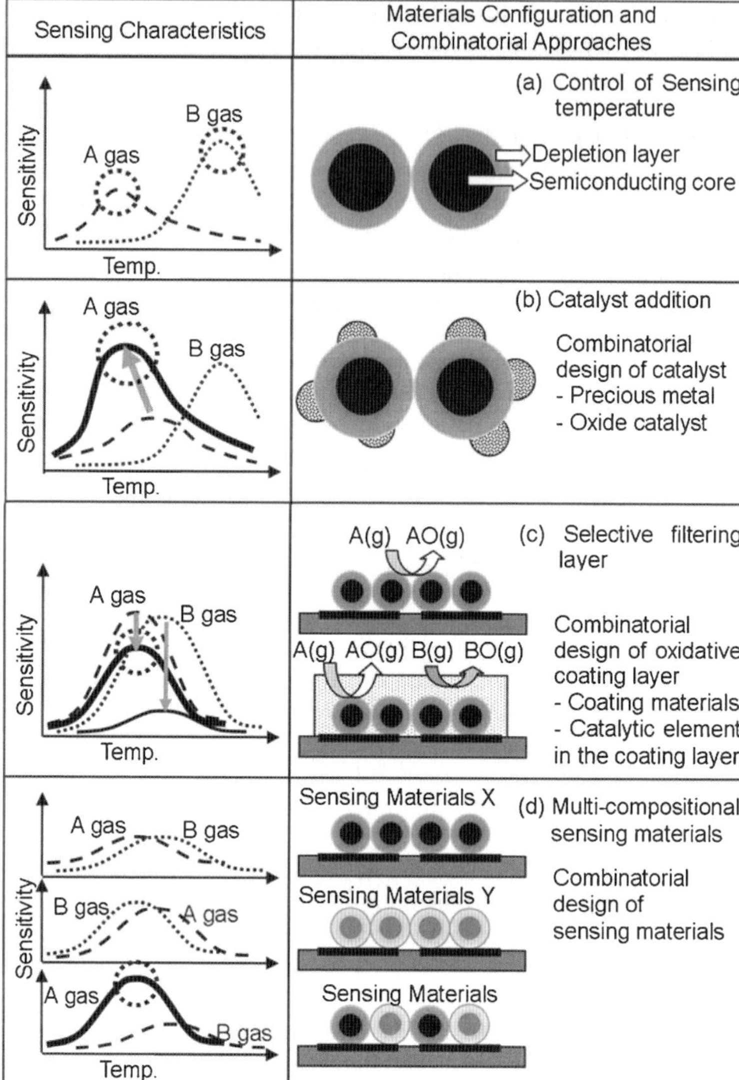

Fig. 12.1 Main approaches for selective detection of gases and possibilities for applications of combinatorial methods

also included during the fabrication of sensor materials. In contrast, the design of multicompositional films requires intensive efforts to prepare sensing materials of various compositions. In this respect, high-throughput methods for fabrication of new multicompositional sensing materials are promising approach to discover selective gas sensing materials.

3 Design of Selective Gas Sensing Materials by Combinatorial Solution Deposition

3.1 Formation of Oxide Sols

The precursor sols of SnO_2 and ZnO were prepared according to literature[58,59]. Figure 12.2 shows a schematic diagram of the methodology used for preparation of SnO_2, ZnO, and WO_3 sols. 200 ml of zinc acetate in 2-methoxyethanol (0.35 M) was added dropwise to 200 ml of a monoethanolamine (MEA) solution 2-methoxyethanol (0.35 M) under constant stirring; a white precipitate was formed. The mixture was heated to 60°C and stirred for 2 h to obtain a clear homogeneous liquid.

Hydrated tin oxide precipitates were prepared by mixing of aqueous solutions of ammonium bicarbonate and tin chloride pentahydrate. The precipitate was washed in distilled water using a centrifuge and suspended in aqueous ammonia solution (pH 10.5). The suspension was transferred to a Teflon-lined stainless steel autoclave and treated hydrothermally at 200°C for 3 h. This resulted in a clear and homogeneous SnO_2 sol (SnO_2 sol). The concentration of the SnO_2 sol was adjusted to 0.065 M by evaporating at ~70°C with constant stirring.

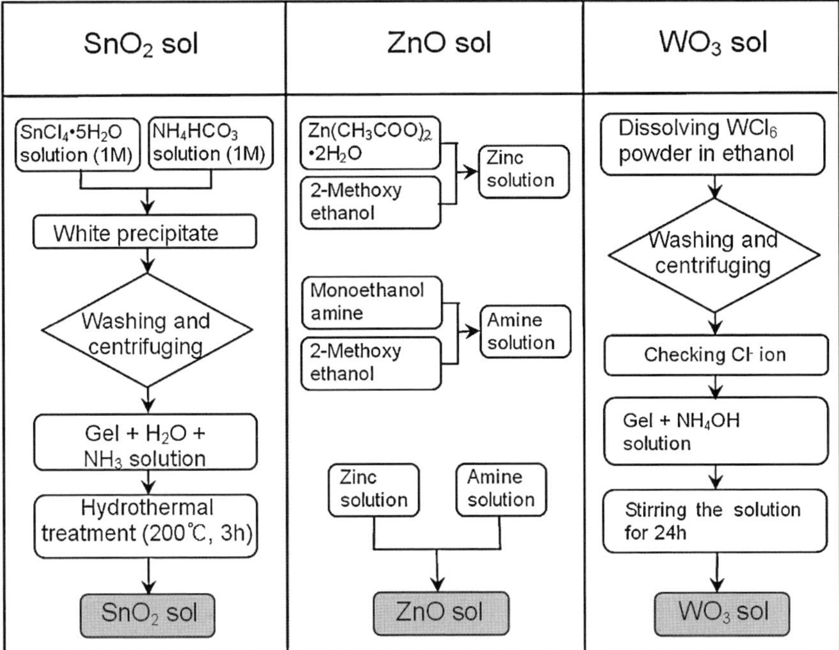

Fig. 12.2 Preparation of SnO_2, ZnO, and WO_3 sols

WCl$_6$ was dissolved in ethanol. A blue gel precipitated after addition of distilled water. Then it was washed with distilled water using a centrifuge, added 30% NH$_4$OH solution and stirred for 24 h; this resulted in a clear liquid; the concentration was 0.27 M. For simplicity, the above three stock liquids will be referred further as SnO$_2$, ZnO, and WO$_3$ sols.

3.2 *Combinatorial Solution Deposition of Gas Sensing Films*

Figure 12.3 shows a preparation of thin film gas sensors. The sensors were fabricated by combinatorial dropping of the source SnO$_2$, ZnO, and WO$_3$ sols onto alumina substrate ($1.5 \times 1.5\,\text{mm}^2$) with two Au electrodes separated by 0.1 mm gap. A precise micropipette was used for deposition. Six thin-film sensors of different compositions were fabricated by alternate dropping of SnO$_2$, ZnO, and WO$_3$ sols (1) S (20 droplets of SnO$_2$ sol, 1 droplet = ~0.87 µl), (2) SZ (10 droplets of SnO$_2$ sol + 10 droplets of ZnO sol), (3) Z (20 droplets of ZnO sol), (4) ZW (10 droplets of ZnO sol + 10 droplets of WO$_3$ sol), (5) W (20 droplets of WO$_3$ sol), and (6) SW (10 droplets of SnO$_2$ sol + 10 droplets of WO$_3$ sol). For further analysis of the correlation between the gas sensor composition and the gas selectivity, (7) S$_{1/4}$Z$_{3/4}$ (5 droplets of SnO$_2$ sol + 15 droplets of ZnO sol) and (8) S$_{3/4}$Z$_{1/4}$ (15 droplets of SnO$_2$

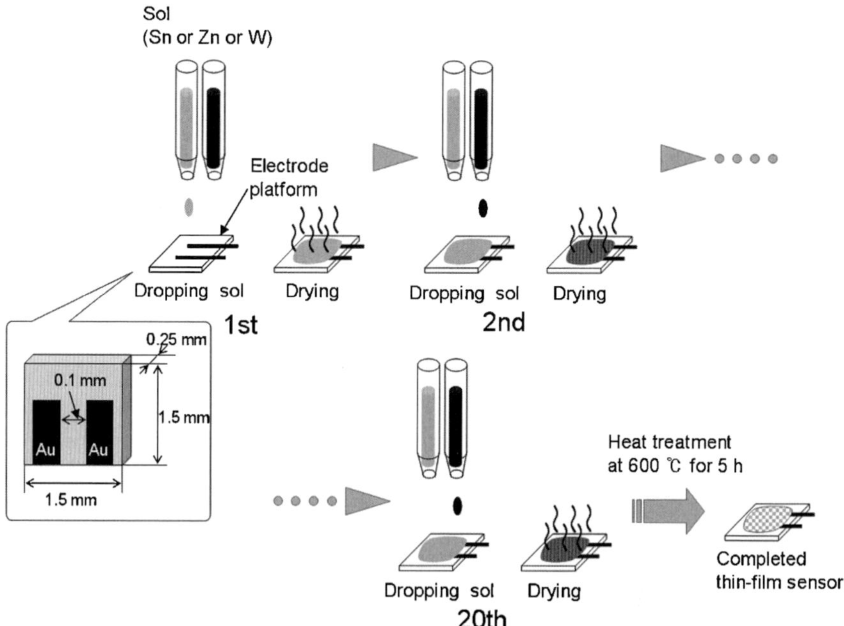

Fig. 12.3 The sensor fabrication procedure using the combinatorial solution deposition method

sol + 5 droplets of ZnO sol) sensors were also fabricated. After each deposition step, the substrate was dried at ~70°C. Special attention was paid to the dropping sequence in order to attain a homogeneous film composition. For example, a SZ film was prepared by alternative dropping of ZnO and SnO_2 sols. To decompose organic substances and to prevent the growth of nanoparticles at sensing temperature (300–400°C), the sensor elements were heated at 600°C for 5 h in air.

3.3   Phase and Microstructure

Figures 12.4 and 12.5[60] show the X-ray diffraction patterns and scanning electron micrographs of sensing films after treatment at 600°C. The phases of the S, Z, and W sensors were identified as SnO_2(tetragonal), ZnO(hexagonal), and WO_3(monoclinic), respectively. The SZ and SW sensors showed a mixed phase configuration. Note that in the ZW sensor, a $ZnWO_4$ phase was formed with a minor WO_3 component. The S sensor contained very fine particles (10–15 nm). The average sizes of the primary particles of the SZ, Z, ZW, SW, and W sensors are approximately 30, 40, 70, 500 and 200 nm, respectively. The ZW sensor showed a uniform and porous microstructure. In contrast, the secondary particles of the W sensor showed a cubic morphology.

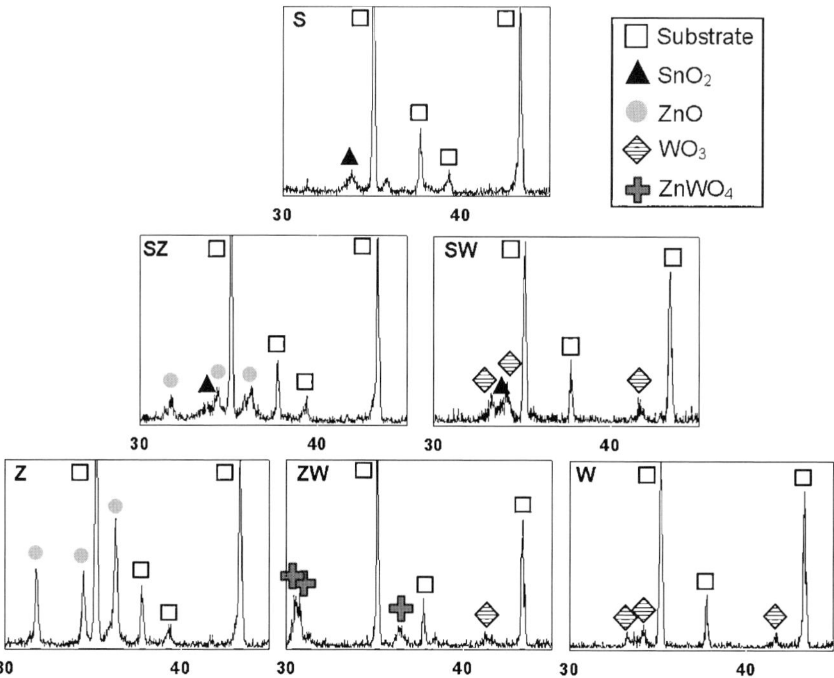

Fig. 12.4 X-ray diffraction patterns of the S, SZ, Z, ZW, W, and SW sensing films after treatment at 600°C for 1 h in air

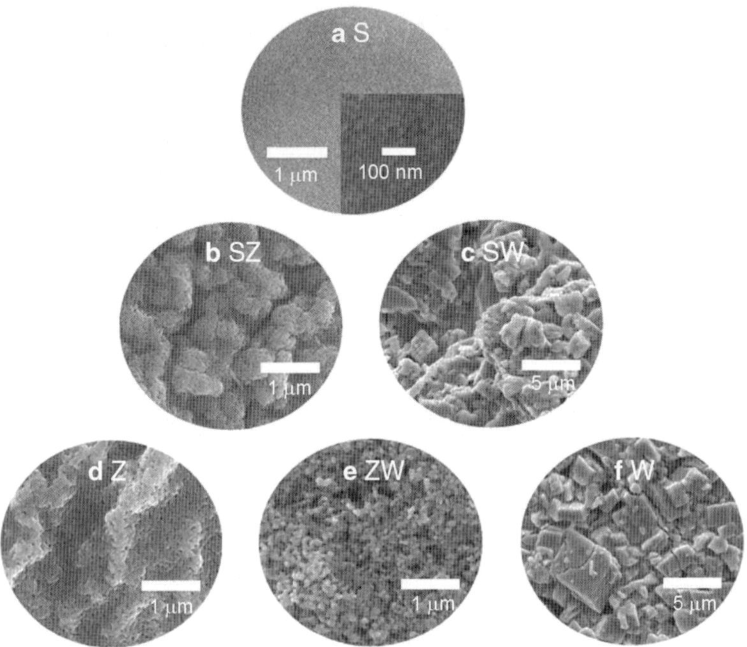

Fig. 12.5 Scanning electron micrographs of (**a**) S, (**b**) SZ, (**c**) SW, (**d**) Z, (**e**) ZW, and (**f**) W sensor films after heat treatment at 600°C for 5 h in air, according to Kim et al.[60]

3.4 Gas Sensing Characteristics

The sensor was installed in a quartz tube and the temperature of the furnace was stabilized at a constant sensing temperature. The gas concentration was controlled by changing the mixing ratio of the parent gases (200 ppm C_2H_5OH, 200 ppm CH_3COCH_3, 200 ppm C_3H_8, 100 ppm CO, 200 ppm H_2, and 5 ppm NO_2, all in an air balance) and dry synthetic air. A flow-through technique with a constant flow rate (500 ml/min) was used. At 300–400°C, the gas response was measured by comparing the dc 2 probe resistance of the sensor in high-purity air (R_a) and those in the target gases (R_g). The gas responses for reducing gases such as the C_2H_5OH, CH_3COCH_3, C_3H_8, CO, and H_2 gases was defined as R_a/R_g, while that for the oxidizing NO_2 gas was defined as R_g/R_a.

Figure 12.6 shows the sensing characteristics of the S, SZ, Z, ZW, W, and SW sensors at 300°C. The Z and SW sensor shows relatively low sensitivities to all gases. In contrast, the S, SZ, ZW, and W sensors showed high sensitivity to C_2H_5OH and CH_3COCH_3 compared with C_3H_8, CO, H_2, and NO_2. These results show that the selective detection of C_2H_5OH and CH_3COCH_3 in the presence of C_3H_8, CO, H_2, and NO_2 is possible using S, SZ, ZW, and W sensors.

The sensitivity to C_2H_5OH and CH_3COCH_3 (S_{et} and S_{ac}) were also significantly higher than those to C_3H_8, CO, H_2, and NO_2 at 400°C. Figure 12.7 shows the S_{et} and

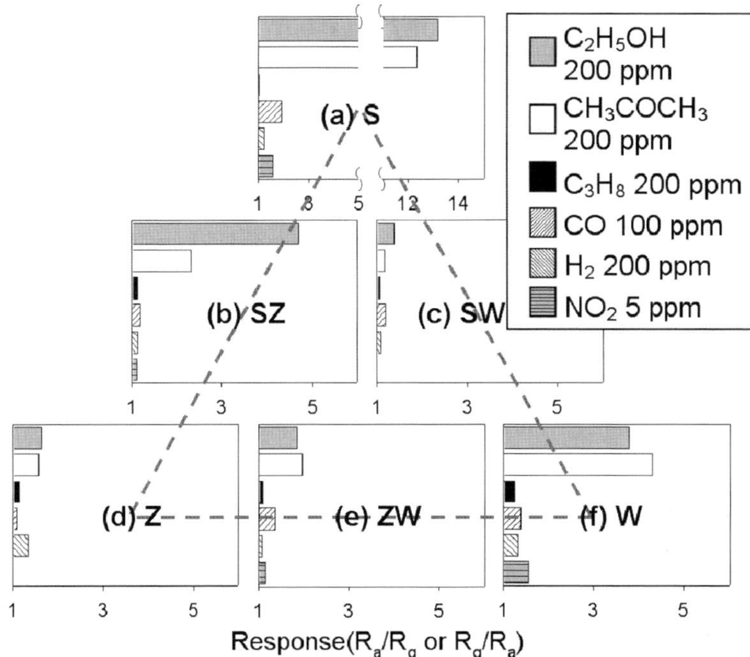

Fig. 12.6 The gas responses of (**a**) S, (**b**) SZ, (**c**) SW, (**d**) Z, (**e**) ZW, and (**f**) W sensors at 300°C. The responses to C_2H_5OH, CH_3COCH_3, C_3H_8, H_2, and CO are R_a/R_g and that to NO_2 is R_g/R_a

S_{ac} values at 300 and 400°C. At 300°C, the S_{et} values of the S and W sensors were similar to the S_{ac} value. In contrast, the S_{et} value of the SZ sensor was approximately twice that of the S_{ac} value. At 400°C, the S_{et} value was substantially higher than the S_{ac} value in the ZW sensor, while the S_{et} and S_{ac} values of the SZ, Z, SW sensors were similar. Note that the S_{ac} value in the S and W sensors at 400°C is higher than the S_{et} value.

The oxide-semiconductor-based ethanol sensor is being used to screen intoxicated drivers. In the test condition on the road, the ambient concentrations of CO and NO_2 can be up to 100 and 10 ppm due to the emissions from gasoline and diesel engines, respectively.[61] The results shown in Fig. 12.6 suggest that the present sensors may be applied for selective detection of ethanol. Acetone is a very rare component in an ordinary ambient atmosphere. However, the expiration of a diabetes patient can contain acetone.[62] Acetone concentration in breath air can reach up to 300 ppm in the case of an aceto-acidotic coma related to diabetes mellitus.[63,64] This might interfere the ethanol sensor. A high selectivity to ethanol is required for such applications. The SZ sensor at 300°C and the ZW sensor at 400°C can satisfy these requirements. On the contrary, to examine the health condition of a diabetes patient, selective detection of acetone without the interference with alcohol is desirable. In this case, the W sensor at 400°C will be of advantageous.

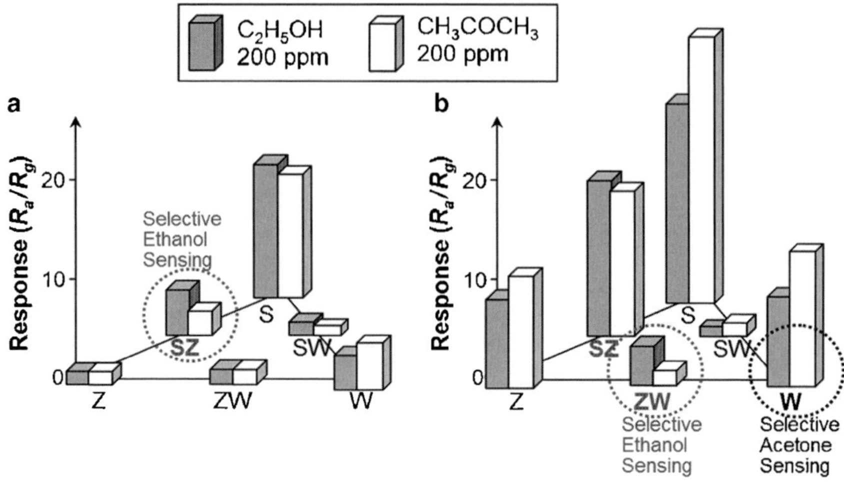

Fig. 12.7 The responses of the sensors to C$_2$H$_5$OH 200 ppm and CH$_3$COCH$_3$ 200 ppm (**a**) at 300 and (**b**) 400°C, according to Kim et al.[60]

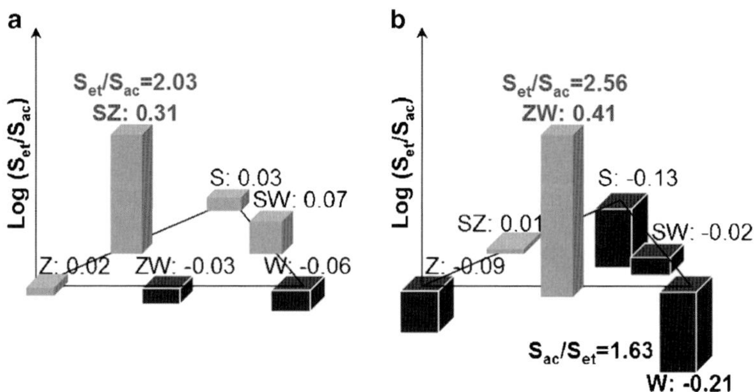

Fig. 12.8 The ratio between the sensitivities to C$_2$H$_5$OH 200 ppm and CH$_3$COCH$_3$ 200 ppm (log (S_{et}/S_{ac})) (**a**) at 300 and (**b**) 400°C

To quantify the selectivity toward ethanol and acetone, the log(S_{et}/S_{ac}) values were calculated and plotted in Fig. 12.8. The positive log(S_{et}/S_{ac}) value means selective detection of ethanol in the presence of acetone, while the negative log(S_{et}/S_{ac}) one means selective detection of acetone in the presence of ethanol. Substantial positive log(S_{et}/S_{ac}) values were found on the SZ and ZW sensors at 300 and 400°C, respectively. In contrast, the W sensor at 400°C showed a negative log(S_{et}/S_{ac}) value.

3.5 Discussion

There are many reports on various oxide semiconductor materials to detect ethanol: SnO_2[65–68], ZnO[69], TiO_2[70], WO_3[71], γ-Fe_2O_3[72,73], $ZnFe_2O_4$[74], SnO_2–ZnO[75,76], $ZnSnO_3$[77], Fe_2O_3–SnO_2[78], Fe_2O_3–In_2O_3[79], TiO_2–WO_3[80], and CeO_2–Fe_2O_3[81]. However, in most case, the cross-sensitivity to acetone was not checked,[65,66,70–74] or observed to be very similar.[67–69,75–81] The similar sensitivities to ethanol and acetone can be understood in relation to their oxidation reaction at the surface of the oxide semiconductor. Jinakawa et al.[82] suggested that the decomposition of ethanol dependens on the acid–base properties of the oxide catalyst used:

$$C_2H_5OH(g) \rightarrow CH_3CHO\ (g) + H_2\ (g)\ \text{(for basic oxide)} \tag{1}$$

$$C_2H_5OH(g) \rightarrow C_2H_4\ (g) + H_2O\ (g)\ \text{(for acidic oxide)} \tag{2}$$

The $CH_3CHO(g) + H_2(g)$ rather than $C_2H_4(g) + H_2O(g)$ induces a higher increase in conductivity in a n-type semiconductor-based ethanol sensor. This suggests that the ethanol sensitivity of basic oxide is higher than that of the acidic oxide. Indeed, Jinakawa et al.[82] reported that the addition of basic oxides such as Cs_2O, La_2O_3, Sm_2O_3, and Gd_2O_3 to SnO_2 increase the response to 1,000 ppm C_2H_5OH by up to 10–40 times. The oxidation of acetone is known to produce the following intermediate phases.[83]

$$CH_3COCH_3 \rightarrow CH_2COCH_2^{\ *} \rightarrow (CH_3CHO,\ CH_3)$$
$$\rightarrow H_2C_2O \rightarrow H_2CO \rightarrow CO \rightarrow CO_2 \tag{3}$$

Similar sensing behaviors to ethanol and acetone can be attributed to the similar intermediate phase (CH_3CHO) and oxidation reactions. The similar values of S_{et} and S_{ac} for most sensors can be understood in this viewpoint. The electro-negativities of W^{6+}, Sn^{4+}, and Zn^{2+} are 2.36, 2.0, and 1.7, respectively. In the sole viewpoint of electronegativity, the acidic WO_3 promotes reaction (2), which can decrease the S_{et}. The selective detection of acetone using the W sensor at 400°C can be understood in this way. However, the electronegative does not show any significant relationship with the S_{et}/S_{ac} values in the S and Z sensors. It is interesting that, at 300°C, the $\log(S_{et}/S_{ac})$ values of the S and Z sensors are close to zero (0.03 and 0.02, respectively) while the $\log(S_{et}/S_{ac})$ value of the SZ sensor is 0.31 ($S_{et}/S_{ac} = 2.03$). Moreover, the $\log(S_{et}/S_{ac})$ value of the SZ sensor becomes very small (0.01) at 400°C. This cannot be explained in the viewpoint of electronegativity.

To understand the corresponding reason, the S_{et}/S_{ac} values were measured with variations of the gas sensing materials (S, $S_{3/4}Z_{1/4}$, SZ, $S_{1/4}Z_{3/4}$, and Z sensors at 300°C) and the gas sensing temperatures (250, 300, 350, 400°C for SZ sensor). Figure 12.9 shows the result.[84] The S_{et}/S_{ac} ratio was higher than 2 at 250°C but decreased gradually with increasing sensing temperature, finally becoming ~1 at 400°C. This suggests that etha-

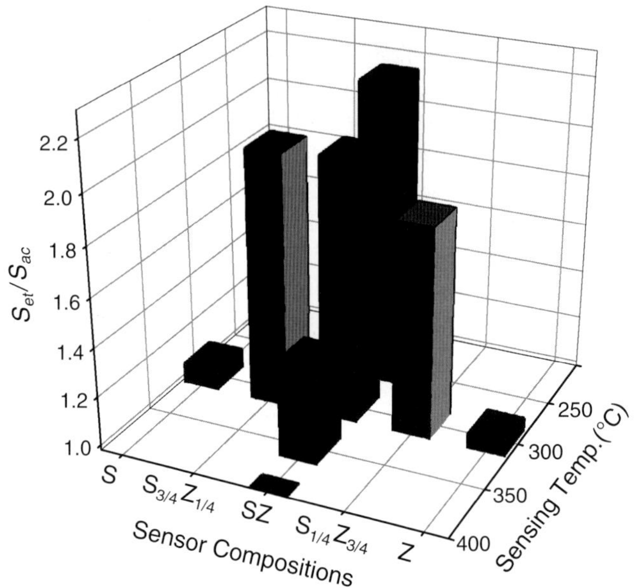

Fig. 12.9 The ratio between the sensitivities to C_2H_5OH 200 ppm and CH_3COCH_3 200 ppm Log (S_{et}/S_{ac}) at various sensor compositions and sensing temperatures, according to Kim et al.[84]

nol and acetone are oxidized into a different intermediate phase at <300°C, and that the sensing temperature should be tuned carefully to achieve selective gas sensing.

At 300°C, all three composite sensors ($S_{3/4}Z_{1/4}$, SZ, and $S_{1/4}Z_{3/4}$) showed high S_{et}/S_{ac} values (1.8–2.1), while the S and Z sensors did not. Ivanovskaya et al.[25] reported that a γ-Fe_2O_3/In_2O_3 heterostructure that was prepared through the alternative deposition of two different oxide layers showed 4–6 times higher S_{et} values compared with single layer γ-Fe_2O_3 or In_2O_3 films. They attributed the high ethanol sensitivity of the hetero-structured film to the promotion of adsorption and oxidation of ethanol due to the presence of the two types of centers with different reductive-oxidative and acid–base properties. The composite sensors in this study ($S_{3/4}Z_{1/4}$, SZ, and $S_{1/4}Z_{3/4}$) were prepared by alternative solution deposition and were identified not as a single $SnZnO_3$ phase but as a mixed phase between SnO_2 and ZnO. Therefore, the S_{et}/S_{ac} value, which is significantly higher than 1 in the present $S_{3/4}Z_{1/4}$, SZ, and $S_{1/4}Z_{3/4}$ sensors, can be attributed to the hetero-structure effect.

4 Combinatorial Solution Deposition: Materials and Processing Issue and Future Outlook

Figure 12.10 summarizes applications of combinatorial solution deposition to development of gas sensing materials. First, uniform sol solutions are needed to guarantee precise and uniform chemical composition throughout the sensing films.

Additionally, the amount of a single droplet should be reproducible. Both alternative dropping of different sol solutions and dropping of a single mixed solution can be used. The former is of advantageous providing an achievement of a "hetero-structure effect," while the latter is more effective for preparation of uniform films.

The Si substrate provides a well-established semiconductor-processing and is compatible with MEMS (micro-electromechanical system) technology, both factors are very useful for fabrication of electrodes and heater patterns, manipulation the

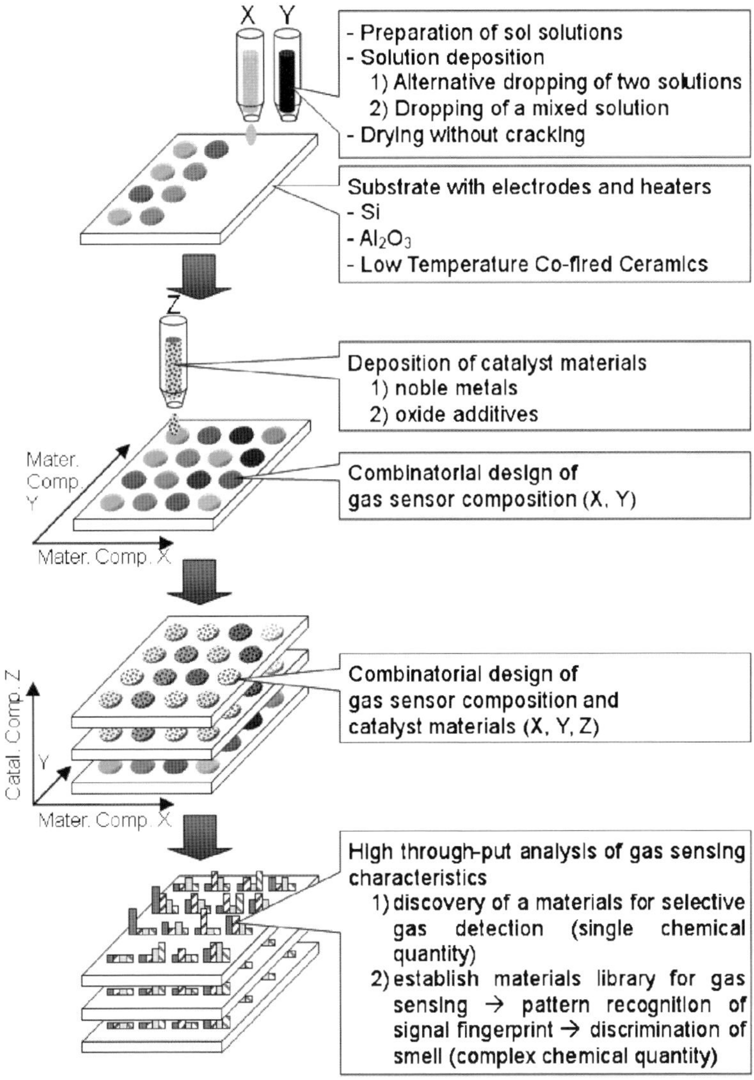

Fig. 12.10 The application of the combinatorial solution deposition to gas sensor research and their related materials and processing issues

device structure, and integration into the sensor array. However, on the one hand, heat treatment at >600°C is not easy due to the materials endurance. On the other hand, an Al_2O_3 substrate shows a low thermal expansion coefficient, high thermal conductivity, chemical stability, and high resistance. High-temperature heat treatment is possible, even though there is less freedom to pattern the electrode and heater patterns compared with the Si-based substrate. Finally, the sensor array can be made using LTCC (Low Temperature Co-fired Ceramics) technology. The LTCC substrate was originally developed to realize an integrated multilayer circuit with Ag or Au electrodes and the typical temperature for co-firing the laminated green sheets is ~850°C. Considering that the typical heat-treatment temperatures of oxide semiconductor gas sensors are 600–700°C, the multilayer substrates with exposed electrodes, buried heaters, and the connections to the package prepared using LTCC technology provide a promising platform for the fabrication of a sensor array. Indeed, we demonstrated[85] that the combination between a combinatorial solution deposition and the LTCC-based substrates is effective and convenient approach for the development of a gas sensor array.

5 Conclusions

Design of a multicompositional gas sensing materials is important not only for selective gas detection of a single chemical quantity and but also for development of libraries of sensor materials for pattern recognition. The multicompositional gas sensors were fabricated by combinatorial solution deposition of SnO_2, ZnO, and WO_3 sols, and their gas sensing behaviors were characterized. The combinatorial modulation of the gas sensor composition enabled the selective detection of C_2H_5OH and CH_3COCH_3 in the presence of interference gases such as CO, C_3H_8, H_2, and NO_2, and the discrimination between C_2H_5OH and CH_3COCH_3. The results demonstrated high efficiency of combinatorial methodology for development of gas sensing materials.

Acknowledgements This work was supported by KOSEF NRL program grant funded by the Korean government (MEST) (No.R0A-2008-000-20032-0) and by a grant from the Core Technology Development Program funded by the Ministry of Commerce, Industry and Energy (MOCIE), Republic of Korea.

References

1. Yamazoe, N., Review. Toward innovation of gas sensor technology, *Sens. Actuators B* **2005**, *108*, 2–14
2. Park, C. O.; Akbar, S. A., Ceramics for chemical sensing, *J. Mater. Sci.* **2003**, *38*, 4611–4637
3. Göpel, W.; Schierbaum, K., SnO_2 sensors: current status and future prospects, *Sens. Actuators B* **1995**, *26–27*, 1–12
4. Shimizu, Y.; Egashira, M., Basic aspects and challenges of semiconductor gas sensors, *MRS Bull.* **1999**, *24*, 18–24

5. Sberveglieri, G., Recent developments in semiconducting thin-film gas sensors, *Sens. Actuators B* **1995**, *23*, 103–109

6. Yamazoe, N., New approaches for improving semiconductive gas sensors, *Sens. Actutators B* **1991**, *5*, 7–19

7. Comini, E., Review. Metal oxide nano-crystals for gas sensing, *Anal. Chem. Acta.* **2006**, *568*, 28–40

8. Xia, Y.; Yang, P.; Sun, Y.; Wu, Y.; Mayers, B.; Gates, B.; Yin, Y.; Kim, F.; Yan, H., One-dimensional nanostructures: Synthesis, characterization, and applications, *Adv. Mater.* **2003**, *15*, 353–389

9. Shimizu, Y.; Jono, A.; Hyodo, T.; Egashira, M., Preparation of large mesoporous SnO_2 powder for gas sensor application, *Sens. Actuators B* **2005**, *108*, 56–61

10. Carotta, M. C.; Guidi, V.; Malagù, C.; Vendemiati, B.; Zanni, A.; Martinelli, G.; Sacerdoti, M.; Licoccia, S.; Vona, M. L. D.; Traversa, E., Vanadium and tantalum-doped titanium oxide (TiTaV): a novel material for gas sensing, *Sens. Actuators B* **2000**, *108*, 89–96

11. Savage, N. O.; Akbar, S. A.; Dutta, P. K., Titanium dioxide based high temperature carbon monoxide selective sensor, *Sens. Actuators B* **2001**, *72*, 239–248

12. Cho, P. -S.; Kim, K. -W.; Lee, J. -H., NO_2 sensing characteristics of ZnO nanorods prepared by hydrothermal method, *J. Electroceram.* **2006**, *17*, 975–978

13. Wang, Y.; Jiang, X.; Xia, Y., A solution-phase, precursor route to polycrystalline SnO_2 nanowires that can be used for gas sensing under ambient conditions, *J. Am. Chem. Soc.* **2003**, *125*, 16176–16177

14. Kolmakov, A.; Zhang, Y.; Cheng, G.; Moskovits, M., Detection of CO and O_2 using tin oxide nanowire sensors, *Adv. Mater.* **2003**, *15*, 997–1000

15. Li, C., Zhang; D., Liu, X.; Han, S.; Tang, T.; Han, J.; Shou, C., In_2O_3 nanowires as chemical sensors, *Appl. Phys. Lett.* **2003**, *82*, 1613

16. Kong, J.; Franklin, N. R.; Zhou, C.; Chapline, M. G.; Peng, S.; Cho, K.; Dai, H., Nanotube molecular wires as chemical sensors, *Science* **2000**, *287*, 622–624

17. Varghese, O. K.; Gong, D.; Paulose, M.; Ong, K. G.; Dickey, E. C.; Grimes, C. A., Extreme changes in the electrical resistance of titania nanotubes with hydrogen exposure, *Adv. Mater.* **2003**, *15*, 624–627

18. Comini, E.; Faglia, G.; Sberveglieri, G.; Pan, Z.; Wang, Z. L., Stable and highly sensitive gas sensors based on semiconducting oxide nanobelts, *Appl. Phys. Lett.* **2002**, *81*, 1869–1871

19. Chiorino, A.; Ghiotti, G.; Prinetto, F.; Carotta, M. C.; Gnani, D.; Marinelli, G., Preparation and characterization of SnO_2 and MoO_x-SnO_2 nanosized powders for thick film gas sensors, *Sens. Actuators B* **1999**, *58*, 338–349

20. Chakraborty, S.; Sen, A.; Maiti, H. S., Selective detection of methane and butane by temperature modulation in iron doped tin oxide sensors, *Sen. Actuators B* **2006**, *115*, 610–613

21. Choi, U. -S.; Sakai, G.; Shimanoe, K.; Yamazoe, N., Sensing properties of Au-loaded SnO_2-Co_3O_4 composites to CO and H_2, *Sens. Actuators B* **2005**, *107*, 397–401

22. Tamaki, J.; Shimanoe, K.; Yamada, Y.; Yamamoto, Y.; Miura, N.; Yamazoe, N., Dilute hydrogen sulfide sensing properties of CuO-SnO_2 thin film prepared by low-pressure evaporation method, *Sens. Actuators B* **1998**, *49*, 121–125

23. Cabot, A.; Arbiol, J.; Cornet, A.; Morante, J. R.; Chen, F.; Liu, M., Mesoporous catalytic filters for semiconducting gas sensors, *Thin Solid Films* **2003**, *436*, 64–69

24. Tulliani, J. M.; Moggi, P., Development of a porous layer catalytically activated for improving gas sensor performances, *Ceram. Int.* **2007**, *22*, 1199–1203

25. Ivanovskaya, M.; Kotsikau, D.; Faglia, G.; Nelli, P.; Irkaev, S., Gas-sensitive properties of thin film heterojunction structures based on Fe_2O_3-In_2O_3 nanocomposties, *Sens. Actuators B* **2003**, *93*, 422–430

26. Costello, B. P. J. de L.; Ewen, R. J.; Ratcliffe, N. M.; Sivenand, P. S., Thick film organic vapour sensors based on binary mixtures, *Sens. Actuators B* **2003**, *92*, 159–166

27. Hyodo, T.; Abe, A.; Shimizu, Y.; Egashira, M., Gas-sensing properties of ordered mesoporous SnO_2 and effect of coatings thereof, *Sens. Actuators B* **2003**, *93*, 590–600

28. Shimizu, Y.; Bartolomeo, E. D.; Traversa, E.; Gusmano, G.; Hyodo, T.; Wada, K.; Egashira, M., Effect of surface modification on NO_2 sensing properties of SnO_2 varistor-type sensors, *Sens. Actuators B* **1999**, *60*, 118–124

29. Zhang, G. -Y.; Guo, B.; Chen, J., MCo$_2$O$_4$ (M = Ni, Cu, Zn) nanotubes: Template synthesis and application in gas sensors, *Sens. Actuators B* **2006**, *114*, 402–409
30. Huang, J. R.; Li, G. Y.; Huang, Z. Y.; Huang, X. J.; Liu, J. H., Temperature modulation and artificial neural network evaluation for improving the CO selectivity of SnO$_2$ gas sensor, *Sens. Actuators B* **2006**, *114*, 1059–1063
31. Jandeleit, B.; Schaefer, D. J.; Powers, T. S.; Turner, H. W.; Weinberg, W. H., Combinatorial materials science and catalysis, *Angew. Chem. Int. Ed.* **1999**, *38*, 2494–2532
32. Koinuma, H.; Takeuchi, I., Combinatorial solid state chemistry of inorganic materials, *Nat. Mater.* **2004**, *3*, 429–438
33. Takeuchi, I.; Lauterbach J.; Fasolka, M. J., Combinatorial materials synthesis, *Mater. Today* **2005**, *8*, 18–22
34. Schultz, P. G.; Xiang, X. -D., Combinatorial approaches to materials science, *Curr. Opin. Solid State Mater. Sci.* **1998**, *3*, 153–158
35. Amis, E. J., Combinatorial materials science: Reaching beyond discovery, *Nat. Mater.* **2004**, *3*, 83–85
36. Greeley, J.; Jaramillo, T. F.; Bonde, J.; Chorkendorff, I.; Nørskov, J. K., Computational high-throughput screening of electrocatalytic materials for hydrogen evolution, *Nat. Mater.* **2006**, *5*, 909–913
37. Ng, H. T.; Chen, B.; Koehne, J. E.; Cassell, A. M.; Li, J.; Han, J; Meyyappan, M., Growth of carbon nanotubes: a combinatorial method to study the effects of catalysts and underlayers, *J. Phys. Chem. B.* **2003**, *107*, 8484–8489
38. Kim, D. K.; Maier, W. F., Combinatorial discovery of new autoreduction catalysts for the CO$_2$ reforming of methane, *J. Catal.* **2006**, *238*, 142–152
39. Potyrailo, R. A.; Lemmon, J. P.; Terry, K. L., High-throughput screening of selectivity of melt polymerization catalysts using fluorescence spectroscopy and two-wavelength fluorescence imaging, *Anal. Chem.* **2003**, *75*, 4676–4681
40. Yanase, I.; Ohtaki, T.; Watanabe, M., Combinatorial study on nano-particle mixture prepared by robot system, *Appl. Surf. Sci.* **2002**, *18*, 292–299
41. Wroczynski, R. J.; Rubinsztajn, M.; Potyrailo, R. A., Evaluation of process degradation of polymer formulation utilizing high-throughput preparation and analysis methods, *Macromol. Rapid Commun.* **2004**, *25*, 264–269
42. Potyrailo, R. A.; Wroczynski, R. J.; Lemmon, J. P.; Flanagan, W. P.; Siclovan, O. P., Fluroescence spectroscopy and multivariate spectral descriptor analysis for high-throughput multiparameter optimization of polymerization conditions of combinatorial 96-microreactor arrays, *J. Comb. Chem.* **2003**, *5*, 8–17
43. Danielson, E.; Devenney, M.; Giaquinta, D. M.; Golden, J. H.; Haushalter, R. C.; McFarland, E. W.; Poojary, D. M.; Reaves, C. M.; Weinberg, W. H.; Wu, X. D., A rare-Earth phosphor Containing one-dimensional chains identified through combinatorial methods, *Science* **1998**, *279*, 837–839
44. Jiang, R.; Rong, C.; Chu, D., Combinatorial approach toward high-throughput analysis of direct methanol fuel cells, *J. Comb. Chem.* **2005**, *7*, 272–278
45. Danielson, E.; Golden, J. H.; McFarland, E. W.; Reaves, C. M.; Weinberg, W. H.; Wu, X. D., A combinatorial approach to the discovery and optimization of luminescent materials, *Nature* **1997**, *389*, 944–948
46. Matsumoto, Y.; Murakami, M.; Hasegawa, T.; Fukumura, T.; Kawasaki, M.; Ahmet, P.; Nakajima, K.; Chikyow, T.; Koinuma, H., Structural control and combinatorial doping of titanium dioxide thin films by laser molecular beam epitaxy, *Appl. Surf. Sci.* **2002**, *189*, 344–348
47. Murakami, M.; Matsumoto, Y.; Nagano, M.; Hasegawa, T.; Kawasaki, T.; Koinuma, H., Combinatorial fabrication and characterization of ferromagnetic Ti–Co–O system, *Appl. Surf. Sci.* **2004**, *223*, 245–248
48. Rende, D.; Schwarz, K.; Rabe, U.; Maier, W. F.; Arnold, W., Combinatorial synthesis of thin mixed oxide films and automated study of their piezoelectric properties, *Prog. Solid State Chem.* **2004**, *223*, 245–248

49. Cui, J.; Chu, Y. S.; Famodu O. O.; Furuya, Y.; Hattrick-Simpers, J.; James, R. D.; Ludwig, A.; Thienhaus, S.; Wuttig, M.; Zhang, Z.; Takeuchi, I, Combinatorial search of thermoelastic shape-memory alloys with extremely small hysteresis width, *Nat. Mater.* **2006**, *5*, 286–290

50. Takeuchi, I.; Famodu, O. O.; Read, J. C.; Aronova, M. A.; Chang, K. –S.; Craciunescu, C.; Lofland, S. E.; Wuttig, M.; Wellstood, F. C.; Knauss, L.; Orozco, A., Identification of novel composition of ferromagnetic shape-memory alloys using composition spreads, *Nat. Mater.* **2003**, *2*, 180184

51. Frantzen, A.; Scheidtmann, J.; Frenzer, G.; Maier, W. F.; Brinz, T.; Sanders, D.; Simon, U., High-throughput method for the impedance spectroscopic characterization of resistive gas sensors, *Angew. Chem. Int. Ed.* **2004**, *43*, 752–754

52. Sanders, D.; Simon, U., High-throughput gas sensing screening of surface-doped In_2O_3, *J. Comb. Chem.* **2007**, *9*, 53–61

53. Scheidtmann, J.; Frantzen, A.; Frenzer, G.; Maier, W. F., A combinatorial technique for the search of solid state gas sensor materials, *Mater. Sci. Technol.* **2005**, 16, 119–127

54. Aronova, M. A.; Chang, K. S.; Takeuchi, I.; Jabs, H.; Westerheim, D.; Gonzalez-Martin, A.; Kim, J.; Lewis, B., Combinatorial libraries of semiconducting gas sensors as inorganic electronic nose, *Appl. Phys. Lett.* **2003**, *83*, 1255–1257

55. Mitra, P.; Maiti, H. S., A wet-chemical process to form palladium oxide sensitiser layer on thin film zinc oxide based LPG sensor, *Sens. Actuators B* **2004**, *97*, 49–58

56. Lee, H. -J.; Song, J. -H.; Yoon, Y. -S.; Kim, T. -S.; Kim, K. -J.; Choi, W. -K., Enhancement of CO sensitivity of indium oxide-based semiconductor gas sensor through ultra-thin cobalt adsorption, *Sens. Actuators B* **2001**, *79*, 200–205

57. Kolmakov, A.; Klenov, D. O.; Lilach, Y.; Stemmer S.; Moskovits, M., Enhanced gas sensing by individual SnO_2 nanowires and nanobelts functionalized with Pd catalyst particles, *Nano Lett.* **2005**, *5*, 667–673

58. Baik, N. S.; Sakai, G.; Miura, N.; Yamazoe, N., Hydrothermally treated sol solution of tin oxide for thin film gas sensor, *Sens. Actuators B* **2000**, *63*, 74–79

59. Li, H.; Wang, J.; Liu, H.; Zhang, H.; Li, X., Zinc oxide films prepared by sol–gel method, *J. Cryst. Growth* **2005**, *275*, e943–e946

60. Kim, S. -J.; Cho, P. -S.; Lee, J -H.; Kang, C. -Y.; Kim, J. -S.; Yoon, S. -J., Preparation of multi-compositional gas sensing films by combinatorial solution deposition, *Ceram. Int.* **2008**, 34, 827-831

61. Nakagawa, H.; Okazaki, S.; Asakura, S.; Fukuda, K.; Akimoto, H.; Takahashi, S.; Shigemori, S., An automated car ventilation system, *Sens. Actuators B* **2000**, *65*, 133–137

62. Ryabsev, S. V.; Shaposhnick, A. V.; Lukin, A. N.; Domashevskaya, E. P., Application of semiconductor gas sensors for medical diagnostics, *Sens. Actuators B* **1999**, *59*, 26–29

63. Fleischer, M.; Simon, E.; Rumpel, E.; Ulmer, H.; Harbeck, M.; Wandel, M.; Fietzek, C.; Weimar, U.; Meixner, H., Detection of volatile compounds correlated to human diseases trough breath analysis with chemical sensors, *Sens. Actuators B* **2002**, *83*, 245–249

64. Yu, J. -B.; Byun, H. -G.; So, M. -S.; Huh, J. -S., Analysis of diabetic patient's breath with conducting polymer sensor array, *Sens. Actuators B* **2005**, *108*, 305–308

65. Gong, H.; Wang, Y. J.; Teo, S. C.; Huang, L., Interaction between thin-film tin oxide gas sensor and five organic vapors, *Sens. Actuators B* **1999**, *54*, 232–235

66. Jie, Z.; Li-Hua, H.; Shan, G.; Hui, Z.; Jing-Gui, Z., Alcohols and acetone sensing properties of SnO_2 thin films deposited by dip-coating, *Sens. Actuators B* **2006**, *115*, 460–464

67. Liu, Y.; Koep, E.; Liu, M., A highly sensitive and fast-responding SnO_2 sensor fabricated by combustion chemical vapor deposition, *Chem. Mater.* **2005**, *17*, 3997–4000

68. Chen, Y. J.; Nie, L.; Xue, X. Y.; Wang, Y. G.; Wang, T. H., Linear ethanol sensing of SnO_2 nanorods with extremely high sensitivity, *Appl. Phys. Lett.* **2006**, *88*, 083105

69. Wan, Q.; Li, Q. H.; Chen, Y. J.; Wang, T. H.; He, X. L.; Li, J. P.; Lin, C. L., Fabrication and ethanol sensing characteristics of ZnO nanowire gas sensors, *Appl. Phys. Lett.* **2004**, *84*, 3654–3656

70. Zhu, B. L.; Xie, C. S.; Wang, W. Y.; Huang, K. J.; Hu, J. H., Improvement in gas sensitivity of ZnO thick film to volatile organic compounds (VOCs) by adding TiO_2, *Mater. Lett.* **2004**, *58*, 624–629

71. Li, X.; Zhang, G.; Cheng, F.; Guo, B.; Chen, J., Synthesis, characterization, and gas-sensor application of WO_3 nanocuboids, *J. Electrochem. Soc.* **2006**, *153*, H133–H137

72. Jing, Z.; Wang, Y.; Wu, S., Preparation and gas sensing properties of pure and doped γ-Fe_2O_3 by an anhydrous solvent method, *Sens. Actuators B* **2006**, *113*, 177–181

73. Jing, Z.; Wu, S., Synthesis, characterization and gas sensing properties of undoped and Co-doped γ-Fe_2O_3-based gas sensors, *Mater. Lett.* **2006**, *60*, 952–956

74. Jiang, Y.; Song, W.; Xie, C.; Wang, A.; Zeng, D.; Hu, M., Electrical conductivity and gas sensitivity to VOCs of V-doped $ZnFe_2O_4$ nanoparticles, *Mater. Lett.* **2006**, *60*, 1374–1378

75. Ho, J. -J.; Fang, Y. K.; Wu, K. H.; Hsieh, W. T.; Chen, C. H.; Chen, G. S.; Ju, M. S.; Lin, J. -J.; Hwang, S. B., High sensitivity and ethanol gas sensor integrated with a solid-state heater and thermal isolation improvement structure of legal drink-drive limit detecting, *Sens. Actuators B* **1998**, *50*, 227–233

76. Costello, B. P. J. de L.; Ewen, R. J.; Guernion, N.; Ratcliffe, N. M., Highly sensitive mixed oxide sensors for detection of ethanol, *Sens. Actuators B* **2002**, *87*, 207–210

77. Xue, X. Y.; Chen, Y. J.; Wang, Y. G.; Wang, T. H., Synthesis and ethanol sensing properties of $ZnSnO_3$ nanowires, *Appl. Phys. Lett.* **2005**, *86*, 233101–233103

78. Reddy, C. V. G.; Cao, W.; Tan, O. K.; Zhu, W., Preparation of $Fe_2O_{3(0.9)}$–$SnO_{2(0.1)}$ by hydrazine method: application as an alcohol sensor, *Sens. Actuators B* **2002**, *81*, 170–175

79. Ivanovskaya, M.; Kotsikau, D.; Faglia, G.; Nelli, P., Influence of chemical composition and structural factors of Fe_2O_3/In_2O_3 sensors on their selectivity and sensitivity to ethanol, *Sens. Actuators B* **2003**, *96*, 498–503

80. Reddy, C. V. G.; Cao, W.; Tan, O. K.; Zhu, W.; Akbar, S. A., Selective detection ethanol vapor using $xTiO_2$-$(1-x)WO_3$ based sensor, *Sens. Actuators B* **2003**, *94*, 99–102

81. Neri, G.; Bonavita, A.; Rizzo, G.; Galvagno, S.; Capone, S.; Siciliano, P., A study of the catalytic activity and sensitivity to different alcohols of CeO_2–Fe_2O_3 thin films, *Sens. Actuators B* **2005**, *111–112*, 78–83

82. Jinakawa, T.; Sakai, G.; Tamaki, J.; Miura, N.; Yamazoe, N., Relationship between ethanol gas sensitivity and surface catalytic property of tin oxide sensor modified with acidic or basic oxides, *J. Mol. Catal. Chem.* **2000**, *155*, 193–2000

83. Tsuboi, T.; Ishii, K.; Tamura, S., Themal oxidation of acetone behind reflected shock wave, in Proceedings of the 17th International Colloquium on the Synamics of Explosions and Reactive Systems, Heidelberg, Germany, July 25–30, 1990

84. Kim, K. -W.; Cho, P. -S.; Kim, S. -J.; Lee, J -H.; Kang, C. —Y.; Kim, J. -S.; Yoon, S. -J., The selective detection of C_2H_5OH using SnO_2–ZnO thin film gas sensors, *Sens. Actuators B* **2007**, *123*, 318–324

85. Kang, C. -Y.; Yoon, S. -J.; Kim, J. -S.; Lee, J -H.; Kim, K. -W.; Cho, P. -S., *Multi-functional Olfactory Sensor Using LTCC and Method of Making*, Korean Patent KR 10-0655367-0000: **2005**

Section 6
Electrochemical Synthesis of Sensing Materials

Chapter 13
Combinatorial Development of Chemosensitive Conductive Polymers

Vladimir M. Mirsky

Abstract Conductive polymers are established materials for development of chemical and biological sensors. Properties of these polymers are influenced by a number of different physical and chemical factors. Application of combinatorial and high-throughput techniques to development and optimization of chemo and biosensors is reviewed. Methods for addressable synthesis of conductive polymers and protocols for comprehensive description of chemosensitive properties are discussed.

1 Functions of Conductive Polymers in Chemo and Biosensors

The group of polymers with electrical conductivity is formed by polymers with a conjugated backbone. To this group belong polyaniline, polypyrrole, polythiophene, polyindole, polyazine, polyacetylene, polyphenylene, and numerous derivatives of these compounds.[1,2] Like other polymeric sensor materials, conductive polymers can be used as recognition elements.[3,4] Their binding properties can be provided by chemical structure of the main polymer backbone (for example, binding of HCl or ammonia to polyaniline,[3,5,6] protonation/deprotonation of polyaniline and polypyrrole,[7,8] binding of oppositely-charged polyelectrolytes, etc.), by introduction of receptor groups into conductive polymers or by copolymerization of conductive polymers with monomers possessing receptor groups (for example, incorporation of biological receptors[9,10] phenylboronic acid,[11,12] crown ethers,[13–15] calixarenes,[16,17] tetraphenylborate,[18] cyclodextrines) or by molecularly-imprinted polymerization.*

V.M. Mirsky
Lausitz University of Applied Sciences,
Department of Nanobiotechnology, 01968 Senftenberg, Germany
vmirsky@fh-lausitz.de

---

*This technology is presented by Subrahmanyam and Piletsky and by Sellergren et al. in the Sect. 3 (eds.).

R.A. Potyrailo and V.M. Mirsky (eds.), *Combinatorial Methods for Chemical and Biological Sensors*, DOI: 10.1007/978-0-387-73713-3_13,

Molecularly imprinted conductive polymers were used for synthesis of artificial receptors for theophylline[19] or for enantioselective detection of L- and D-glutamic acid[20] (overoxidized polypyrrole was used).

In contrast to other materials for chemical sensors and biosensors that have only receptor function, conductive polymers may have a number of very different functions in chemical sensors. Conductive polymers can be used as transducers of binding events into optical or electrical signals. For example, polyaniline can convert its HCl or ammonia binding to changes of its optical properties,[5,21] and binding of ascorbic acid to poly-*o*-methoxyaniline leads to changes of its electrical conductivity.[22] Protonation/deprotonation leads to changes of optical or electrical properties of polyaniline, polypyrrole, polyindole and their derivatives. Some conductive polymers have electrocatalytical activity. For example, nanostructured polyaniline catalyses nitrate reduction,[23] poly-*o*-methoxyaniline catalyses oxidation of ascorbic acid,[24] poly-1-naphthylamine catalyses reduction of hydrogen peroxide.[25] Each of these electrocatalytically active materials can be also used as a chemosensitive material: in combination with simple electrochemical techniques (voltammetric, chronoamperometry, or others), these materials allow one to get a generation of electrical signals depending on the substrate (analyte) concentration.

Other applications of conductive polymers in chemical and biological sensors include a building of electrical connection between different elements of these devices[6] or electrical connection and mediators of electron transfer between enzyme and electrode[26,27] or between electrocatalytically active functional groups or incorporated nanoparticles and electrode,[28] as a filter,[29] as matrix for immobilization of enzymes or even as a combination of many of these functions.[26] The requirements for materials for these applications are very specific and will not be analyzed here.

The combination of transducing and sensing functions leads to extension of general requirements for sensing materials, such as an appropriate binding constant, binding selectivity, stability (chemical, long-time, and thermal), etc. The new requirements include high sensitivity (in IUPAC definition, i.e., a concentration derivative of sensor signal), response reproducibility, response reversibility, response time, recovery time.

2   Synthesis of Conductive Polymers

2.1   Chemical Synthesis and Electropolymerization

Synthesis of conductive polymers can be realized either by addition of oxidizing agents or by electrochemical oxidation at anodic potentials. To distinguish from electrochemical polymerization, polymerization by external oxidizer is often marked in literature as a "chemical polymerization." Many types of conductive polymers formed by this way have a strong trend to adsorb on the surfaces (probably the synthesis is catalyzed by surfaces), therefore this technology can be also used for deposition of polymer coatings on solid supports, electrodes, optical surfaces. This approach is flexible and general; some conductive polymers can be obtained only by this way; however, it does not provide a possibility to control thickness of the polymer layer.

Electrochemical polymerization of organic molecules on electrodes was developed more than 50 years ago and since 1960s is used in industry for deposition of corrosion protective coatings. An intensive investigation of conductive polymers in 1970s and 1980s was additional motivation for further development of electropolymerization. It has been shown that most of the conductive polymers can be synthesized by electrochemical oxidation of corresponding monomers. In many cases, a synthesized polymer (or essential amount of it) remains to be adsorbed on the electrode surface, and this phenomenon led to a simple technology to form thin polymer layers on conductive surfaces. Moreover, many polymers (polyanilines, polypyrroles) demonstrate much faster (more than two order of magnitude) growth along the surface. The reason for this catalytic effect of surface on the polymerization is still unclear; probably it is caused by surface activity of monomers or short polymeric oligomers. If the polymer is conductive, electrochemical synthesis can lead to formation of polymer layers with a thickness over 1 mm. If the polymer is poor conductive, a gradual increase of electrical resistance of the organic layer suppressed electropolymerization. This leads to the principal limitation of the layer thickness below ~100 nm. The mechanisms of electropolymerization are reviewed.[30–32] Polymerization in the presence of several monomers can lead to formation of copolymers[33–35] and block copolymers.[36,37] Oxidized conductive polymers can be electrochemically (or chemically) reduced, and it is usually accompanied by the loss of counter-ions and by dramatic changes in optical and electrical properties.

2.2 Variable Parameters of Conductive Polymers

Properties of conductive polymers as sensing (and transducing) material depend on many factors (Fig. 13.1). Polymers can be synthesized from one or more different monomers. Chemosensitive properties (as well as many physical properties that are crucial for sensor design) depend strongly on the counterions[38] used for polymer synthesis and on the presence and type of additional dopants. The type of polymerization (chemical or electrochemical) and physical conditions of polymerization (for example, temperature, mode of electrochemical polymerization, the last applied potential during polymerization) as well as the solvent also effect on the material properties. Mechanical parameters, such as the thickness and porosity of polymers are also important. Some sensor configurations exploit not bulk polymer properties, but the properties of the metal/polymer contacts. An optimization of sensing material, which is sensitive to so large number of parameters, can be hardly performed without combinatorial technology.

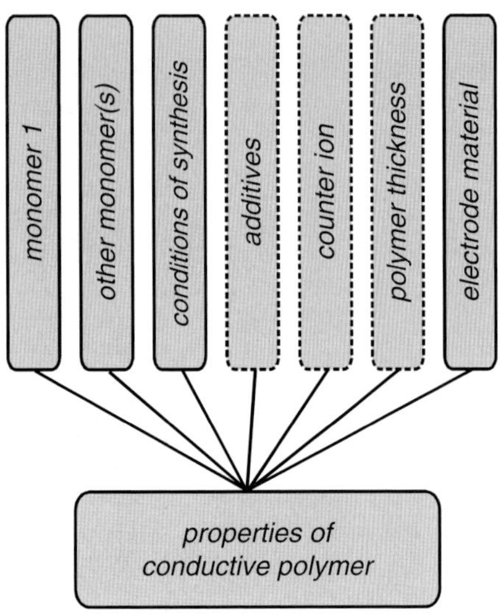

Fig. 13.1 Properties of conductive polymers depend on a number of different parameters, which can be varied

3   Combinatorial Synthesis of Conductive Polymers

Experimental approaches to combinatorial synthesis of conductive polymers are summarized in the Fig. 13.2. Combinatorial polymerization driven by an introduced chemical oxidizer can be performed with a usual liquid handling robot or with a microfluidic system. Surprisingly, much more sophisticated techniques were usually used (Fig. 13.2).

A combination of a mixing of reagents by means microfluidics and subsequent electropolymerization was reported[39] (Fig. 13.2). Instead of microfluidic delivery of reagents, a mechanical robot immersing electrodes into the wells with solution of monomers was suggested.[40,41] Another approach is based on electrically addressed polymerization, and this technology does not include either mechanical robotics or microfluidics[42–44] (Fig. 13.2).

The combinatorial synthesis based on the combination of microfluidics and electropolymerization was realized with a microchip consisting of two areas: microfluidic channels for generation of gradient of two substances and a parallel electrochemical reactor with platinum electrodes (Fig. 13.3). The system was tested for synthesis of polyaniline in the presence of polysulfonic acid. The reagent ratio providing the best efficiency of the polymer synthesis was evaluated.

A general purpose set-up for combinatorial electrochemistry was developed by a combination of a mechanical robot and electrochemical system.[40,41]* The set-up can operate with one electrode set (consisting of classical three-electrode configuration) moving it between different cells or with 8-electrode set providing

---

*This technology is presented by S. Borgmann and W. Schuhmann in the Chap. 14 (eds.).

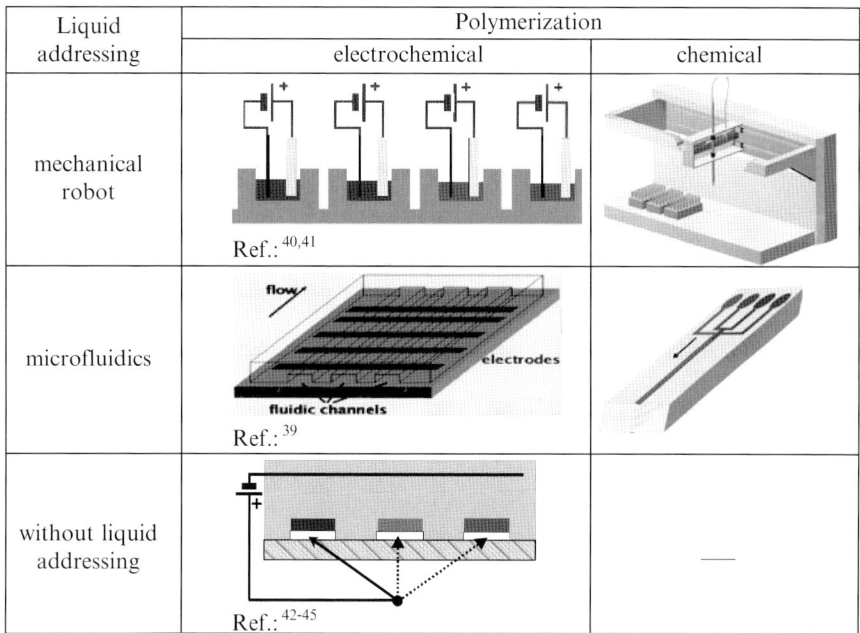

Fig. 13.2 Experimental approaches for combinatorial synthesis of conductive polymers. The figure includes a parts of illustration reprinted with permissions from Xiang and LaVan[39] (copyright 2006 The Institute of Electrical and Electronics Engineers, Inc.)

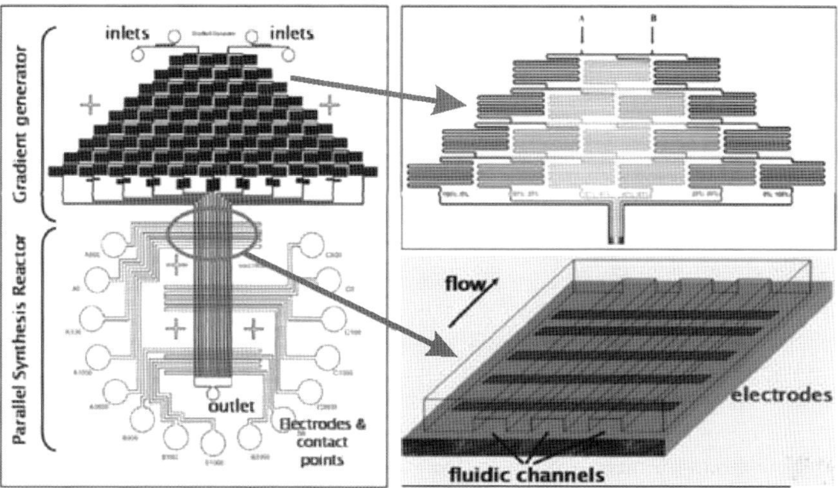

Fig. 13.3 Design of microfluidic system for combinatorial electrochemical synthesis of conductive polymers. The device is composed of a gradient generator and a parallel electrochemical synthesis reactor. Two different species A and B were injected into the system from the top. Reprinted with permission from Xiang and LaVan[39] (copyright 2006 The Institute of Electrical and Electronics Engineers, Inc.)

simultaneous 8-channel measurements. The electrodes geometry was adapted for the work with microtiterplates. The system was used for automated characterization of combinatorially synthesized π-conjugated thiophenes.[45]

Mechanical addressing can be also used for combinatorial postsynthetical modifications of conductive polymers. Postsynthetical modification was applied to formations of a number of different derivates of polyaniline (Fig. 13.4), the modification was performed by nucleophilic addition (Fig. 13.4), coupling with diazonium salts and by electrophilic aromatic substitution.[46]

A system for combinatorial polymerization of conductive polymers based on electrical addressation was reported.[43,47] This same approach was used also for preparation of DNA and protein-arrays.[10,48] In the case of combinatorial applications,

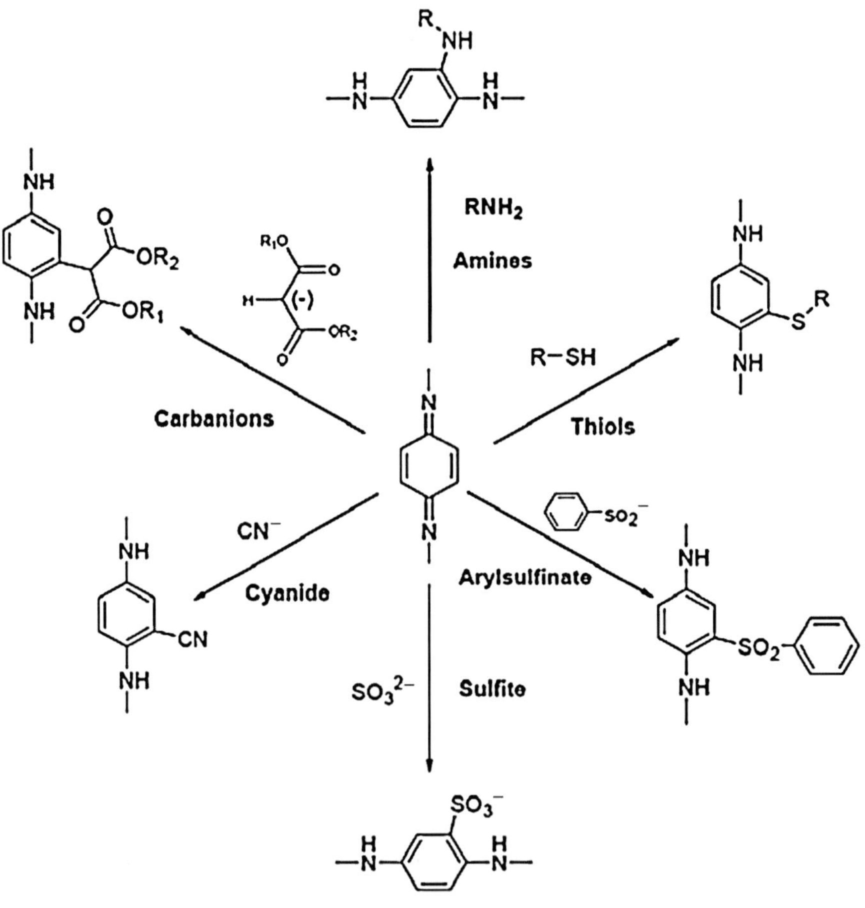

Fig. 13.4 Postsynthetic modifications of polyaniline by nucleophilic additions. Reprinted with permission from[46] (copyright 2004, Elsevier)

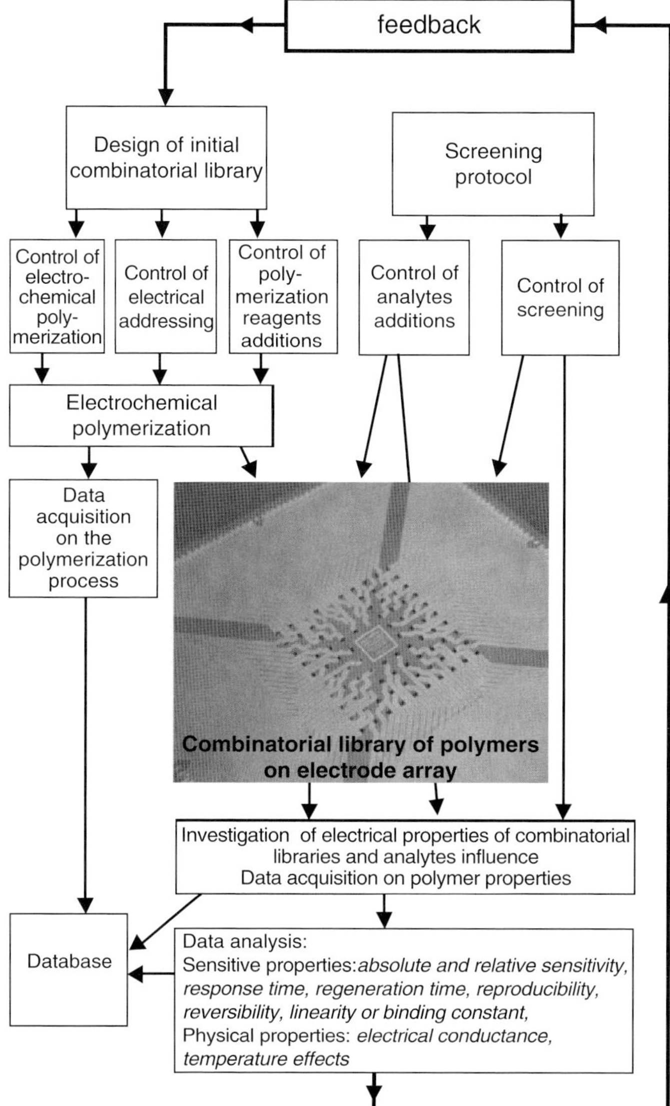

Fig. 13.5 The complete system for combinatorial electro-polymerization and high-throughput characterization. The synthesized polymers are observed as dark point in the center of combinatorial library[49]

it includes also modules for automated screening of chemosensitive properties of synthesized materials. Information flow in the whole system is shown in the Fig. 13.5. Control of electrical potential of single electrode groups consisting of four electrodes (designed for 4-point measurements) on array containing 96 such electrode groups (each group of $400 \times 400\,\mu m$) enables addressable electrochemical polymer synthesis on the defined electrodes. The polymerization is performed until defined polymerization charge is passed. During the electropolymerization, an electrical potential causing this electrochemical reaction is applied to the definite electrode group of the electrode array while all other electrode groups are hold at the potential below polymerization potential. Then the polymerization solution is changed, and the electro-polymerization cell is rinsed. This procedure is repeated for each electrode group of the electrode array. Afterwards, electrical characteristics of polymers and influence of potential analytes on these characteristic are measured. Finally, an automated data analysis is performed.

The completely automated set-up for combinatorial electropolymerization includes an automated dosing station, an electronic multiplexer and a socket board with electro polymerization cell. The photograph of the system is shown in Fig. 13.6.

The system was first applied for development of chemosensors for gaseous hydrogen chloride. Polyaniline, and its copolymers with different derivates of aniline were used. Then a similar approach was tested in the author's group for optimization of amperometric biosensors for glucose based on electrocatalytical detection of hydrogen peroxide. A pigment Prussian blue was used as an electrocatalyst for decomposition of this product of enzymatic oxidation of

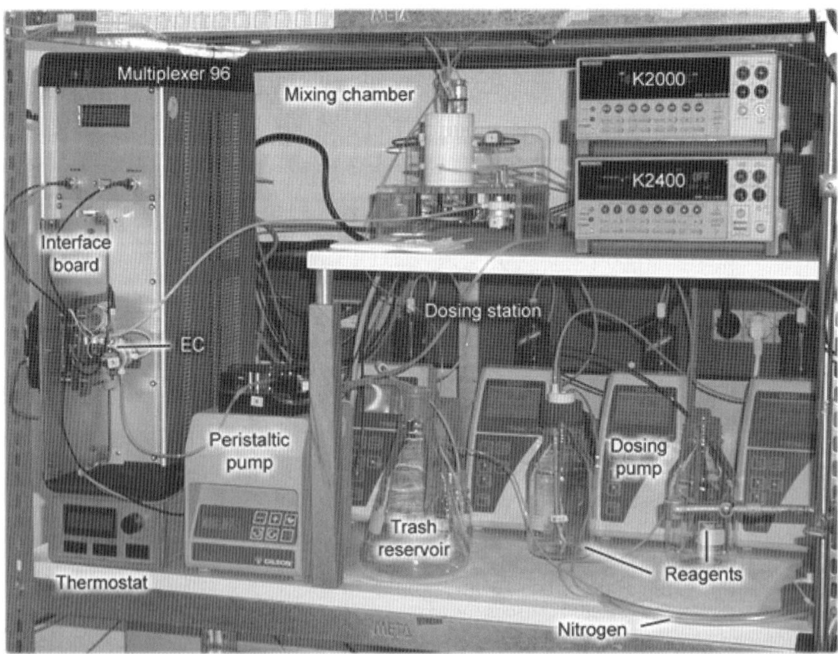

Fig. 13.6 The main part of the system for combinatorial electropolymerization and high-through-put characterization of the synthesized polymers

glucose. The sensor was prepared in two steps. First, Prussian blue was synthesized electrochemically on the electrode array. Then an electrochemical synthesis of polypyrrole in the presence of glucose oxidase was performed. The thicknesses of the dye layer and of the layer of polypyrrole with glucoseoxidse were varied (Fig. 13.7). The formation of blue pigment on electrodes with systematically varied color intensity in the first step of the library preparation was detected also by visual control. Subsequent deposition of polypyrrole led to very dark (almost black) spots on the electrodes.

A possibility to make chemical polymerization of conductive polymers can be used for formation of gradient libraries of polymers on solid supports by the same way as it has been established for other materials.[50] Investigation of electrical parameters of such films can be performed directly, with in situ four-contact technique[51] or by simultaneous two and four-point technique.[45,52] In the case of highly inhomogeneous distribution of conductivity in this combinatorial library, an influence of neighboring layer on measurement conductivity can be minimized by stamping out, for example using the technique of the intermediate layer lithography[53].

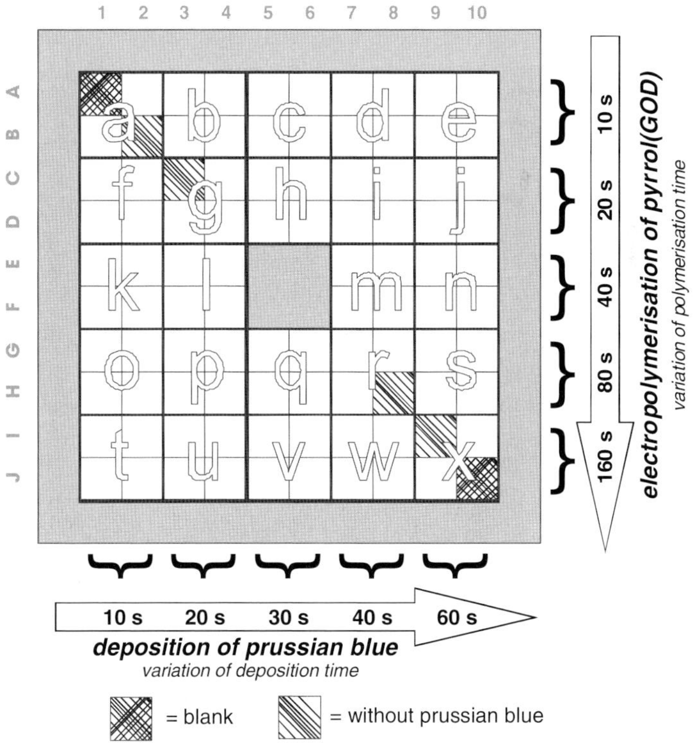

Fig. 13.7 Design of two-dimensional array for optimization of amperometric biosensors

4 Multiparameter Characterization of Chemosensitive Properties of Conductive Polymers

Functional high-throughput multifunctional screening of chemosensitive properties of conductive polymers was described recently.[44] The authors developed a minimal test procedure consisting of two concentration pulses of an analyte at the same concentration and a sequence of concentration pulses at increased concentrations. Automated analysis of the materials responses has given the following information:

- Absolute analytical sensitivity $\left(\text{as} \dfrac{d[\text{absolute signal value}]}{d[\text{analyte concentration}]} \right)$

- Relative analytical sensitivity $\left(\text{as} \dfrac{d[\text{relative signal value}]}{d[\text{analyte concentration}]} \right)$

- Response rate; it was defined as relative changes of the signal during fixed time of the analyte addition)
- Recovery rate (if the recovery relative occurred as a desorption at zero analyte concentration) or recovery efficiency (if an external influence for the signal recovery is required); it was defined as signal changes during the fixed desorption time or during another defined recovery procedure (i.e., a temperature pulse)
- Reversibility (was defined as the ratio of the signal values before analyte adsorption and after recovery or for the recovery after the first and after the second analyte concentration pulses)
- Reproducibility (was defined as the ratio of the signal changes for the first and second analyte addition)
- Binding constant (for sensor materials that obey Langmuir adsorption isotherm) or response linearity (for sensor material that obey Henry adsorption isotherm)

This analysis technology was applied to screening of chemosensitive properties of polyanilines and copolymers of anilines and aniline derivatives. Combinatorial libraries were prepared by electrically addressed polymerization.[42,43] Gaseous hydrogen chloride was selected as an analyte. The measured signal was electrical resistance detected simultaneously by 2 and 4-point techniques (s24 technique[44,52]), and the ratio of these values provides information on the contribution of the polymer/contact resistance into the total resistance measured by 2-point configuration.

As an example, the Fig. 13.8 demonstrates screen-shots of data analysis for absolute (upper panel) and relative (lower panel) sensitivities. Polymerization charge was varied. In the bottom part of the screen, an operator can select the analyte concentration range for the data analysis, data for either 2 or 4-point resistance measurement mode, abscissa-axis and information on the design of the combinatorial library. In the left part of the screen, an operator can select one of the analyzed parameters: initial conductance, changes of conductivity, relative changes in conductivity, resistivity, changes of resistivity, relative changes of resistivity, kinetics

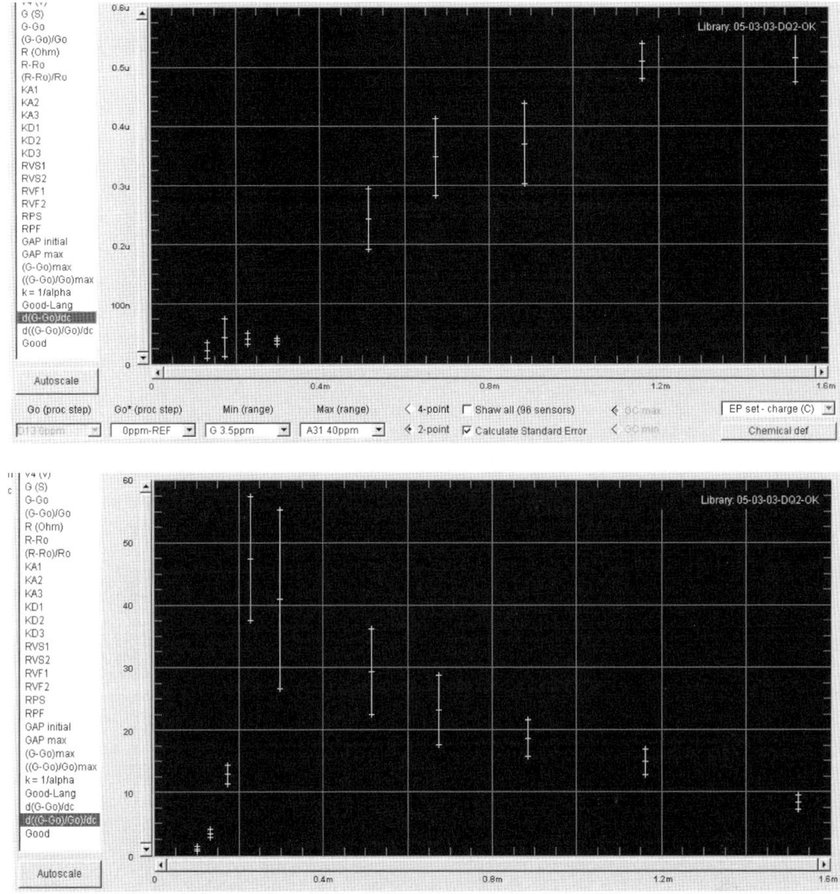

Fig. 13.8 An example of screen-shot for analysis of absolute (up) and relative (down) sensitivity vs. polymer thickness. The values are normalized to the relative sensitivity of pure polyaniline (ANI). Polymer: polyaniline. Polymer thickness was varied as the polymerization charge, the value in abscissa corresponds the charge in coulombs, the electrode size $400 \times 400\,\mu$m. Analyte: 3.5 ppm HCl. Reprinted from[50] with permission (copyright 2008 American Chemical Society)

of the sensor response during first and second time intervals, trend (second derivative) of the sensor response, desorption efficiencies for the first and second time intervals, trend of the desorption efficiency, reversibility for long or short analyte pulses, reproducibility for long or short analyte pulses, adsorption constant (for Langmuir adsorption isotherm), saturation value (for Langmuir adsorption isotherm), absolute sensitivity, relative sensitivity, correlation coefficient.

The results (Fig. 13.9) demonstrate that while the absolute sensitivity increases almost linearly, the relative sensitivity reaches its maximum at polymerization charge of ~0.2–0.4 mC. This value was selected for further screening of sensing

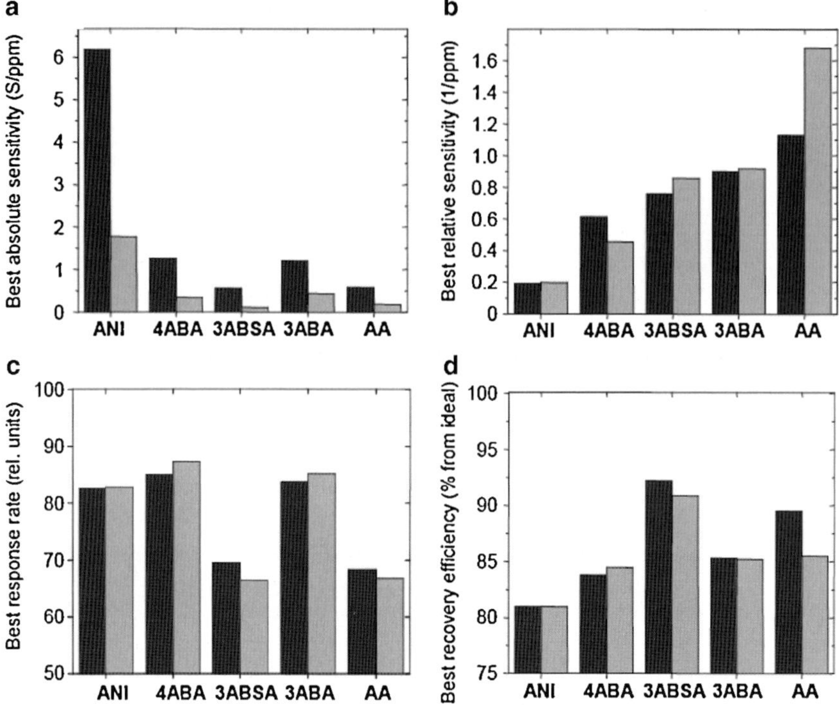

Fig. 13.9 The best absolute and relative sensitivity, response rate and recovery efficiency reached for different binary copolymers. *ANI* aniline, *4ABA* 4-aminobenzoic acid, *AA* anthranilic acid, *3ABA* 3-aminobenzoic acid, *3ABSA* 3-aminobenzenosulfonic acid. Reprinted from[50] with permission (copyright 2008 American Chemical Society)

materials. Concentration dependence of conductivity changes on HCl concentration was linear in the whole studied range (3–40 ppm). Also an introduction of aniline derivates did not lead to saturation of the concentration dependence. Therefore, the reciprocal binding constant remains to be relatively low (less than reciprocal maximal concentration). However, strong conductivity changes provide very high analytical sensitivity (over 1,000% per ppm) and therefore low detection limit (below 1 ppm). An introduction of aniline derivatives into the polymerization mixture results in increasing of relative analytical sensitivity by a factor of two. Some screening results are presented in the Fig. 13.10. It is to note that optimization of different analytical parameters, as for example, sensitivity, adsorption rate, reversibility, etc. leads to different results. The authors reported[49] ~100 times increase of the throughput, ~4–5 times decrease of toxic waste and ~100–10,000 times about decrease possible exposure of labor personnel to hazardous substances.

A very similar protocol was developed for screening of combinatorically synthesized electrocatalysts or whole biosensors. Electrocatalytical activity of Preussian blue was detected as changes of electrical current on addition of hydrogen peroxide;

the signal of enzymatic biosensor based on Prussian blue and glucose oxidase was registered as changes of electrical current on addition of glucose.

5   Outlook

The results demonstrate several approaches to combinatorial synthesis of conductive polymers and an example of automated high-throughput screening of their chemosensitive properties. Convenient format of combinatorial libraries in the form of small amounts of polymers (few µg) deposited on silicon substrate with electrodes allows one not only simple characterization, but also simple storage and organization of banks of materials with electrical and optical access for further investigations.

The applications of high-throughput experimentations with conductive polymers for development of gas sensors, optimization of electrocatalytic layer and amperometric enzymatic biosensors are just a few examples of this approach. A possibility to use electrochemical polymerization for preparation of multilayer structures[6] indicates an application potential for combinatorial preparation and high-throughput analysis of complicated monolayer devices. The same approach can be certainly used for development of the based on electropolymerizable compounds molecularly imprinted polymers[19,20,54] or ion-selective electrodes.[55] Additionally to combinatorial electropolymerization, an electrochemical deposition of metals or a combination of electrical addressing with other techniques (for example, ink-jet deposition, photopolymerizytion, photolithography) may be applied. This way can be used for in-situ synthesis of combinatorial libraries of different devices for analytical applications: chemosensitive diodes, Schottki-contacts and transistors, ion-selective electrodes, optrodes, selective filters, microsystems for electrophoresis, orchromatography containing integrated detectors.

Acknowledgments The author is grateful to R.A. Potyrailo for interesting and motivating discussions on applications of high-throughput methodologies in material science and for reading and critical comments of the present manuscript. The described technique of combinatorial electropolymerization with electrical addressing and its applications were realized in the frame of the projects Kombisense (supported by the German Ministry for Science and Education) and short-term fellowships of the German Science Foundation for V. Kulikov and T. Delaney. The main contribution into practical realization of the concept has been done by V. Kulikov during his Ph.D work. An assistance of Q. Hao, T. Delaney, C. Swart, and Th. Hirsch in the performing of particular tasks of the work and fruitful advices of O. S. Wolfbeis, partners of the Kombisense-project, as well as N. Roznyatovskaya and U. Lange are acknowledged.

References

1. Wallace, G.; Kane-Maguire, L. Conductive polymers. Encyclopedia Biomater. Biomed. Eng. **2004**, 1, 374–383

2. Hansen, G. What we dreamt as children-how conductive polymers are bringing our dreams to reality. J. Adv. Mater. **2006**, 38, 68–74

3. Bai, H.; Shi, G. Gas sensors based on conducting polymers. Sensors **2007**, 7, 267–307

4. Malhotra, B. D.; Chaubey, A.; Singh, S. P. Prospects of conducting polymers in biosensors. Anal. Chimica Acta **2006**, 578, 59–74

5. Samoylov, A. V.; Mirsky, V. M.; Hao, Q.; Swart, C.; Shirshov, Y. M.; Wolfbeis, O. S. Nanometer-thick SPR sensor for gaseous HCl. Sens. Actuators. **2005**, B106, 369–372

6. Hao, Q.; Wang, X.; Lu, L.; Yang, X.; Mirsky, V. M. Electropolymerized multilayer conducting polymers with response to gaseous hydrogen chloride. Macromolecul. Rapid Commun. **2005**, 26, 1099–1103

7. Pringsheim, E.; Zimin, D.; Wolfbeis, O. S. Fluorescent beads coated with polyaniline. A novel nanomaterial for optical sensing of pH. Adv. Mater. **2001**, 13, 819–822

8. Krondak, M.; Broncova, G.; Anikin, S.; Merz, A.; Mirsky, V. M. Chemosensitive properties of poly-4,4′-dialkoxy-2,2′-bipyrroles. J. Solid State Electrochem. **2006**, 10, 185–191

9. Lassalle, N.; Mailley, P.; Vieil, E.; Livache, T.; Roget, A.; Correia, J. P.; Abrantes, L. M. Electronically conductive polymer grafted with oligonucleotides as electrosensors of DNA Preliminary study of real time monitoring by in situ techniques. J. Electroanal. Chem. **2001**, 509, 48–57

10. Grosjean, L.; Cherif, B.; Mercey, E.; Roget, A.; Levy, Y.; Marche, P. N.; Villiers, M. B.; Livache, T. A polypyrrole protein microarray for antibody-antigen interaction studies using a label-free detection process. Analyt. Biochem. **2005**, 347, 193–200

11. Shoji, E.; Freund, M. S. Potentiometric saccharide detection based on the pKa changes of poly(aniline boronic acid). J. Am. Chem. Soc. **2002**, 124, 12486–12493

12. Pringsheim, E.; Terpetschnig, E.; Piletsky, S. A.; Wolfbeis, O. S. A polyaniline with near-infrared optical response to saccharides. Adv. Mater. **1999**, 11, 865–868

13. Marsella, M. J.; Swager, T. M. Designing conducting polymer-based sensors: selective ionochromic response in crown ether-containing polythiophenes. J. Am. Chem. Soc. **1993**, 115, 12214–12215

14. Fabre, B.; Simonet, J. Electroactive polymers containing crown ether or polyether ligands as cation-responsive materials. Coord. Chem. Rev. **1998**, 178–180 (Pt. 2), 1211–1250

15. Pandey, P. C.; Prakash, R. Polyindole modified potassium ion-sensor using dibenzo-18-crown-6 mediated PVC matrix membrane. Sens. Actuators. **1998**, B46, 61–65

16. Kaneto, K.; Bidan, G. Electrochemical recognition and immobilization of uranyl ions by polypyrrole film doped with calix[6]arene. Thin Solid Films. **1998**, 331, 272–278

17. Marsella, M. J.; Newland, R. J.; Carroll, P. J.; Swager, T. M. Ionoresistivity as a highly sensitive sensory probe: investigations of polythiophenes functionalized with calix[4]arene-based ion receptors. J. Am. Chem. Soc. **1995**, 117, 9842–9848

18. Pandey, P. C.; Singh, G.; Srivastava, P. K. Electrochemical synthesis of tetraphenylborate doped polypyrrole and its applications in designing a novel zinc and potassium ion sensor. Electroanalysis. **2002**, 14, 427–432

19. Ulyanova, Y., V; Blackwell, A. E.; Minteer, S. D. Poly(methylene green) employed as molecularly imprinted polymer matrix for electrochemical sensing. Analyst. **2006**, 131, 257–261

20. Deore, B.; Chen, Z.; Nagaoka, T. Potential-induced enantioselective uptake of amino acid into molecularly imprinted overoxidized polypyrrole. Anal. Chem. **2000**, 72, 3989–3994

21. Nicho, M. E.; Trejo, M.; Garcia-Valenzuela, A.; Saniger, J. M.; Palacios, J.; Hu, H. Polyaniline composite coatings interrogated by a nulling optical-transmittance bridge for sensing low concentrations of ammonia gas. Sens. Actuators. **2001**, B76, 18–24

22. Ivanov, S.; Tsakova, V.; Mirsky, V. M. Conductometric transducing in electrocatalytical sensors: Detection of ascorbic acid. Electrochem. Commun. **2006**, 8, 643–646

23. Luo, X.; Killard, A. J.; Smyth, M. R. Nanocomposite and nano-porous polyaniline conducting polymers exhibit enhanced catalysis of nitrite reduction. Chemistry. **2007**, 13, 2138–2143

24. Komsiyska, L.; Tsakova, V. Ascorbic acid oxidation at nonmodified and copper-modified polyaniline and poly-ortho-methoxyaniline coated electrodes. Electroanalysis. **2006**, 18, 807–813

25. Li, X.; Ta, N.; Sun, C. Electrochemical polymerization of 1-naphthylamine and properties of poly-1-naphthylamine. Bull. Electrochem. **2005**, 21, 173–177

26. Kranz, C.; Wohlschlaeger, H.; Schmidt, H. L.; Schuhmann, W. Controlled electrochemical preparation of amperometric biosensors based on conducting polymer multilayers. Electroanalysis. **1998**, 10, 546–552

27. Habermuller, K.; Mosbach, M.; Schuhmann, W. Electron-transfer mechanisms in amperometric biosensors. Fresenius J. Anal. Chem. **2000**, 366, 560–568

28. Galal, A. Electrocatalytic oxidation of some biologically important compounds at conducting polymer electrodes modified by metal complexes. J. Solid State Electrochem. **1998**, 2, 7–15

29. Guerrieri, A.; De Benedetto, G. E.; Palmisano, F.; Zambonin, P. G. Electrosynthesized non-conducting polymers as permselective membranes in amperometric enzyme electrodes: a glucose biosensor based on a co-crosslinked glucose oxidase/overoxidized polypyrrole bilayer. Biosens. Bioelectron. **1998**, 13, 103–112

30. Waltman, R. J.; Bargon, J. Electrically conducting polymers: a review of the electropolymerization reaction, of the effects of chemical structure on polymer film properties, and of applications towards technology. Canad. J. Chem. **1986**, 64, 76–95

31. Yano, J. Electro-oxidative polymerization mechanism of polyaniline. Curr. Trends Polym. Sci. **1998**, 3, 131–143

32. Sabouraud, G.; Sadki, S.; Brodie, N. The mechanisms of pyrrole electropolymerization. Chem. Soc. Rev. **2000**, 29, 283–293

33. Kan, J.; Pan, X.; Zhou, W. Electrochemical copolymerization of aniline and o-chloroaniline. Bull. Electrochem. **2005**, 21, 65–70

34. Xu, J.; Nie, G.; Zhang, S.; Han, X.; Hou, J.; Pu, S. Electrochemical copolymerization of indole and 3,4-ethylenedioxythiophene. J. Mater. Sci. **2005**, 40, 2867–2873

35. Nie, G.; Xu, J.; Zhang, S.; Cai, T.; Han, X. Electrochemical copolymerization of carbazole and 3-methylthiophene. J. Appl. Polym. Sci. **2006**, 102, 1877–1885

36. Yilmaz, F.; Sel, O.; Guner, Y.; Toppare, L.; Hepuzer, Y.; Yagci, Y. Controlled synthesis of block copolymers containing side chain thiophene units and their use in electrocopolymerization with thiophene and pyrrole. J. Macromol. Sci., Pure Appl. Chem. **2004**, A41, 401–418

37. Asawapirom, U.; Guentner, R.; Forster, M.; Scherf, U. Semiconducting block copolymers-synthesis and nanostructure formation. Thin Solid Films. **2005**, 477, 48–52

38. Hao, Q.; Rahm, M.; Weiss, D.; Mirsky, V. M. Morphology of electropolymerized poly (N-methylaniline) films. Microchim. Acta. **2003**, 143, 147–153

39. Xiang, Y.; LaVan, D. Parallel microfluidic synthesis of conductive biopolymers. Proceedings of the 2nd IEEE/ASME International Conference on Mechatronic and Embedded Systems and Applications; The Institute of Electrical and Electronic Engineers, Inc.: Piscataway, 2006; p.1

40. Ryabova, V.; Schulte, A.; Erichsen, T.; Schuhmann, W. Robotic sequential analysis of a library of metalloporphyrins as electrocatalysts for voltammetric nitric oxide sensors. Analyst. **2005**, 130, 1245–1252.

41. Erichsen, T.; Reiter, S.; Ryabova, V.; Bonsen, E. M.; Schuhmann, W.; Markle, W.; Tittel, C.; Jung, G.; Speiser, B. Combinatorial microelectrochemistry: Development and evaluation of an electrochemical robotic system. Rev. Sci. Instr. **2005**, 76, 1–11

42. Mirsky, V. M.; Kulikov, V. Combinatorial electropolymerization: Concept, equipment, and applications. In: High-Throughput Analysis: A Tool for Combinatorial Material Science; Potyrailo R. A., Amis E. J., Eds. Kluwer/Plenum: New York, NY, 2003; 431–446

43. Mirsky, V. M.; Kulikov, V.; Hao, Q.; Wolfbeis, O. S. Multiparameter high throughput characterization of combinatorial chemical microarrays of chemosensitive polymers. Macromol. Rap. Commun. **2004**, 25, 253–258

44. Kulikov, V.; Mirsky, V. M.; Delaney, T. L.; Donoval, D.; Koch, A. W.; Wolfbeis, O. S. High-throughput analysis of bulk and contact conductance of polymer layers on electrodes. Meas. Sci. Technol. **2005**, 16, 95–99

45. Briehn, C. A.; Schiedel, M. S.; Bonsen, E. M.; Schuhmann, W.; Bauerle, P. Single-compound libraries of organic materials: from the combinatorial synthesis of conjugated oligomers to structure-property relationships. Angew. Chemie, Intern. Ed. **2001**, 40, 4680–4683

46. Barbero, C.; Salavagione, H. J.; Acevedo, D. F.; Grumelli, D. E.; Garay, F.; Planes, G. A.; Morales, G. M.; Miras, M. C. Novel synthetic methods to produce functionalized conducting polymers I. Polyanilines. Electrochim. Acta. **2004**, 49, 3671–3686.
47. Kulikov, V.; Mirsky, V. M. Equipment for combinatorial electrochemical polymerization and high-throughput investigation of electrical properties of the synthesized polymers. Meas. Sci. Technol. **2004**, 15, 49–54
48. Bidan, G.; Billon, M.; Livache, T.; Mailley, P.; Roget, A. Molecular engineering and electro-chemistry for the implementation of DNA chips. Act. Chimiq. **2003**, 11/12, 39–46
49. Mirsky, V. M.; Kulikov, V.; Wolfbeis, O. S. Complete system for combinatorial synthesis and functional investigation of conductive polymers. PMSE Preprints. **2005**, 93, 1053
50. Potyrailo, R. A.; Mirsky, V. M. Combinatorial and high-throughput development of sensing materials: The first ten years. Chem. Rev. **2008**, 108, 770-813
51. Wang, Q. High-throughput conductivity measurements of thin films. In: In: High-Throughput Analysis: A Tool for Combinatorial Material Science; Potyrailo R. A., Amis E. J., Eds. Kluwer/Plenum: New York, NY, **2003**; 395–414.
52. Hao, Q.; Kulikov, V.; Mirsky, V. M. Investigation of contact and bulk resistance of conducting polymers by simultaneous two- and four-point technique. Sens. Actuators. **2003**, B94, 352–357
53. Chakraborty, A.; Liu, X.; Parthasarathi, G.; Luo, C. An intermediate-layer lithography method for generating multiple microstructures made of different conducting polymers. Microsys. Technol. **2007**, 13, 1175–1184
54. Panasyuk, T. L.; Mirsky, V. M.; Piletsky, S. A.; Wolfbeis, O. S. Electropolymerized molecularly imprinted polymers as receptor layers in capacitive chemical sensors. Anal. Chem. **1999**, 71 (20), 4609–4613
55. Bobacka, J.; Ivaska, A.; Lewenstam, A. Potentiometric ion sensors based on conducting polymers. Electroanalysis. **2003**, 15, 366–374

Chapter 14
Robotic Systems for Combinatorial Electrochemistry

Sabine Borgmann and Wolfgang Schuhmann

Abstract The scope of this chapter is to present the current state of the art in the field of combinatorial electrochemistry with the main focus on plate-based technologies. In particular, it is focused on the development and use of electrochemical robotic systems.

1   Motivation for the Electrochemical Robotic System

The concept of combinatorial synthesis and screening was introduced by Hanak in 1970 in the field of material science (Fig. 14.1).[1] Revival of this idea found application in modern biology and pharmacology in concert with high-throughput technologies.[2,3] Combinatorial chemistry[4] is the chemistry of a large number of different but structurally related compounds that are called compound libraries or collections.[5]

Synthesizing large libraries in short time is of interest as well as creating libraries that show actual activity in screening for selected properties. Technical approaches dealing with synthesis of the library components and their screening are mainly driven by the large number of library elements involved. Therefore, specific high-throughput compatible technologies were and are still developed to meet these requirements.

Several reaction procedures that are employed in classical chemistry – comprising also redox chemistry[6] – can also be used in combinatorial chemistry, but in a miniaturized fashion and at a considerably higher throughput. This normally results

S. Borgmann (✉)
Technische Universität Dortmund, Fachbereich Chemie - Biologisch- Chemische Mikrostrukturtechnik, Otto-Hahn Strasse 6, D-44227 Dortmund, Germany & ISAS - Institute for Analytical Sciences, Bunsen-Kirchhoff-Str. 11, D-44139 Dortmund, Germany
e-mail: borgmann@isas.de

W. Schuhmann
Analytische Chemie - Elektroanalytik & Sensorik, Ruhr-Universität Bochum, D-44780 Bochum, Germany
e-mail: wolfgang.schuhmann@rub.de

R.A. Potyrailo and V.M. Mirsky (eds.), *Combinatorial Methods for Chemical and Biological Sensors*, DOI: 10.1007/978-0-387-73713-3_14,

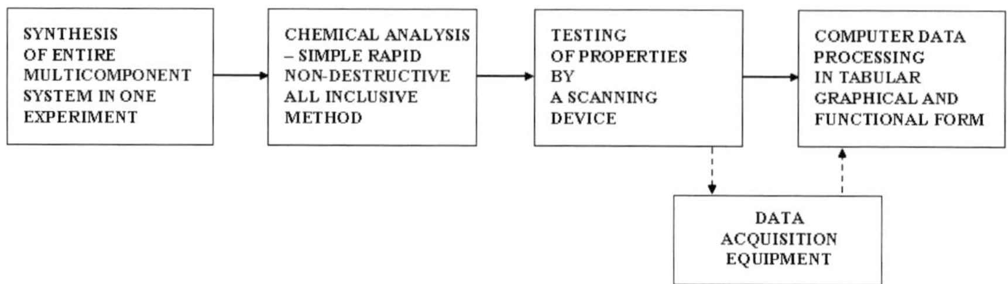

Fig. 14.1 Hanak's representation of combinatorial synthesis and screening (Copyright (1970) Chapman and Hall Ltd reprinted with kind permission of Springer Science and Business Media[1])

in higher investment costs (instrumentation) and more complex data management, but saves space, reagents, human resources, and produces less waste. Advances in automation technology have led to dramatically increased use of combinatorial chemistry in research and industry, mainly in the fields of material and life science.[7] Screening of large compound libraries is often done in a plate-based format in fully automated process routines, documentation, barcoding, and data processing. Commercially available microtiter plates containing 24, 96, 384, or 1,536 wells are used because they are compatible with most of the current analytical techniques that are available for high-throughput analysis. Integrated synthetic and analytic methodology takes advantage of solid phase or polymer supported strategies. One could claim that the roots of combinatorial chemistry were already planted in 1963, when Robert B. Merrifield introduced solid phase peptide synthesis.[8]

Electrochemistry integrates analytical technique (determination of concentrations, reaction mechanisms, or properties[9]) and synthetic methods such as electrolysis.[10] Electrons needed for redox reactions are provided by an electric current supplied through electrodes in a highly controlled and selective manner. Products can be isolated easier. It is well known that electrochemical redox reactions may result in reactive intermediates under mild conditions.[11] Electrochemistry is a clean and convenient methodology even on the preparative scale.

Electrochemical experiments can easily be miniaturized.[12,13] Thus, technological developments in electrochemistry are currently driven by miniaturization. However, implementing the idea of combinatorial chemistry and high-throughput screening (HTS) in field of electrochemical research and application has been quite recent compared with other chemistry branches. Despite the fact that electro-organic reactions are proven tools for synthesis and have been applied even in large-scale industrial processes,[10,14,15] they were not applied in combinatorial chemistry until the early 2000s. Merging the concepts (Fig. 14.2) of combinatorial chemistry and HTS with traditional electrochemical synthesis, analysis and present miniaturization approaches required a certain shift in thinking.

To date, combinatorial libraries are screened mainly by optical detection using for example UV/Vis absorbance or fluorescence as readout. Nevertheless, the obvious advantages of electrochemical over classical redox reactions only slowly found use in combinatorial chemistry, which is illustrated by the fact that the term "combinatorial electrochemistry" was introduced only in 1998 by Reddington et al.[16]. They first used

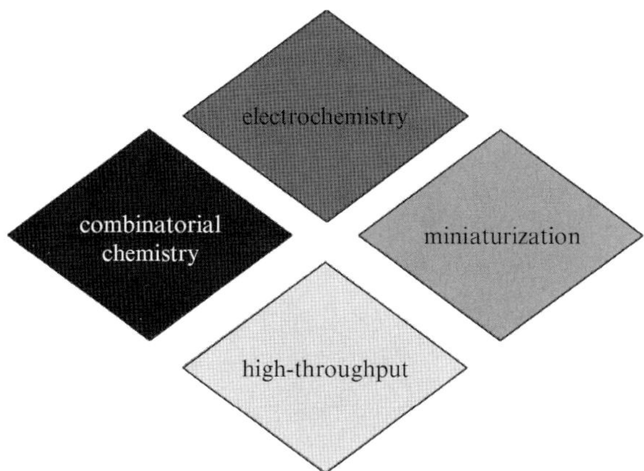

Fig. 14.2 Merging of electrochemistry with the concepts of combinatorial chemistry and high-throughput screening employing miniaturization strategies

this wording in the context of generating and screening a library of metal alloy electrocatalysts for methanol oxidation. Interestingly, neither the synthesis nor the analysis – an optical readout was chosen – were electrochemical. Subsequently, so-called "combinatorial electrode arrays" were introduced that were produced by various (including electrochemical) microscopic techniques. Potential catalytic[17] or sensing materials[18] were placed on miniaturized electrode arrays. Screening for activity was performed either directly by current measurements[19] or indirectly by fluorescence.[20] Advantages of electrochemical techniques applied in combinatorial chemistry were clearly revealed by this work in the early 2000s. Mallouk and Smotkin summarized combinatorial methods for the discovery of electrocatalysts.[21] However, a solid basis for synthetic application of combinatorial electrochemistry was not established. Even today, the use of electrochemical synthesis of organic compounds in combinatorial approaches is not frequent, but publication numbers are slowly growing.

Yudin and coworkers promoted the use of a 16-well spatially addressable electrolysis platform (SAEP) starting in 2000.[22–25] The SAEP uses a rectangular array of 16 electrochemical cells each using a tubular stainless-steel cathode and a graphite rod anode (Fig. 14.3a). The connection between the individual electrodes and the galvanostat was manually switched. The SAEP platform was used for anodic oxidation of carbonates, amides, and sulfonamides,[22] copolymerization of 2,2′-bithiophene and pyrrole-bound 2,2,6,6-tetramethylpiperidin-1-yloxy (TEMPO) as well as parallel screening of the resulting polymers[26] with respect to their catalytic activity toward oxidation of primary alcohols. Good yields were obtained by this technique (Fig. 14.3b). However, using a galvanostat for electrochemical synthesis does not offer the selectivity and control that a potentiostatic approach would provide.

In addition, other electrochemical strategies for synthetic combinatorial chemistry with considerable potential were introduced. Flow production of carbocations and their coupling products was presented by Suga et al.[27] Localized electropolymerization was shown by Kulikov et al. in 2004.[28] Pilard et al. introduced a novel solid-phase synthesis process by immobilization of electrochemically cleavable protecting reagents on a conducting polymer support.[29,30] Nad et al. reported the first use polymer-bead supported indirect electrosynthesis on a solid phase via a mediator (Fig. 14.4).[31] They have chosen electrochemical oxidation of furans as a model reaction. This approach seems to be applicable for large-scale synthesis of combinatorial libraries and provide a general and practical strategy for solid-phase synthesis.

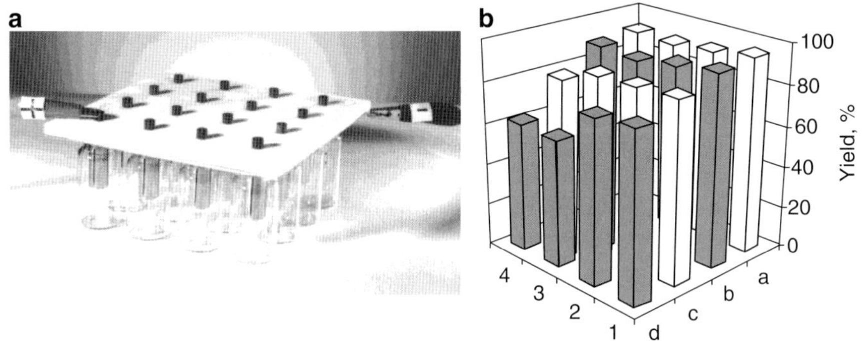

Fig. 14.3 (**a**) Photograph of the spatially addressable electrolysis platform (SAEP); (**b**) Yields of parallel electrolysis of substrates 1–4 in alcohols a–d (Reprinted with permission from Siu et al.[22] Copyright (2000) American Chemical Society)

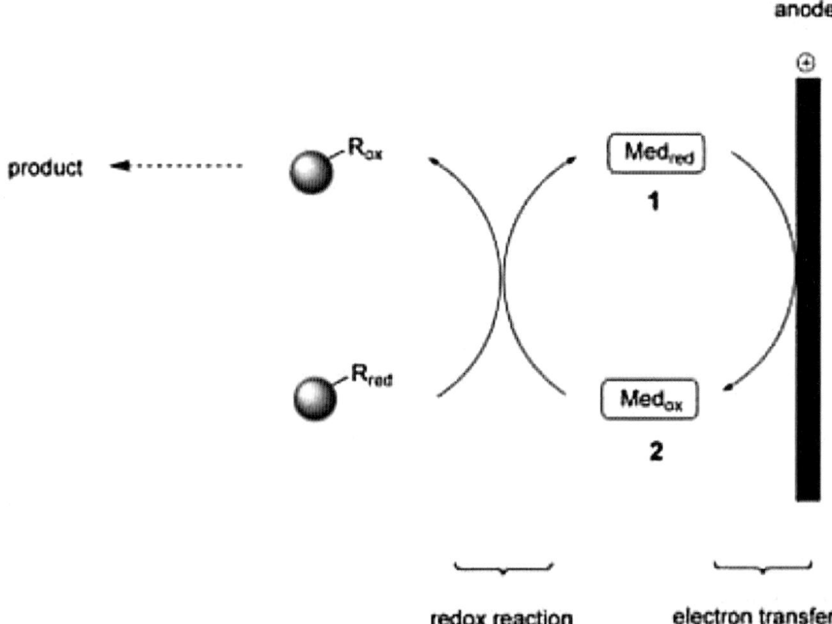

Fig. 14.4 Principle of redox-catalyst-mediated electro-organic synthesis on a solid phase.[31] Reproduced by permission of the author and Wiley-VCH Verlag GmbH & Co KGaA)

Microchip-assisted generation of addressable libraries was presented by Tesfu, Moeller, and coworkers.[32–35] Two of the introduced reactions are highlighted in Fig. 14.5. The Heck reaction using a Pd(0) reagent was realized as explained in Fig. 14.5a.[33] A chip-based amination reaction to couple fluorescent dyes (red and green) is schematically depicted in Fig. 14.5b.[34]

As more and more electrochemical approaches for arrays and microchips were introduced, the use of electrochemical detection for gene expression analysis became more vital. The current method of choice to readout microarrays is optical detection such as fluorescence.[36,37] Microarrays are of crucial interest for in vitro diagnostic applications.[38] Here, we would like to mention just a few of the increasing number of publications in this area addressing electrochemical approaches. More detailed reviews for further reading can be found in literature[39–44]. There has been a significant progress in electrochemically addressed arrays for DNA hybridization,[45,46] bacterial RNA detection,[47] and in electrochemical label-free detection approaches.[48,49] As an example, we present here the recent work by Ghindilis and coworkers from CombiMatrix Corporation that introduces an oligonucleotide microarray platform consisting of 12,544 individually addressable microelectrodes in a semiconductor matrix.[50] The electrochemical detection scheme is depicted in Fig. 14.6. The authors compared electrochemical and fluorescence detection schemes for their chips and observed an improved detection limit for the electrochemical detection.

In addition, CombiMatrix Corporation provides instrumentation for electrochemical reading of their chips, the ElectraSense™ Reader (Fig. 14.7). The main advantages of employing electrochemical instead of optical detection technologies are (1) lower investment costs in instrumentation, (2) easier data management because no special imaging software is required, (3) extended dynamic range, (4) high potential for further automation in conjunction with other assays on the chip (using e.g. microfluidics or robotics).

There has been a trend toward electrochemical reactions in lab-on-a-chip devices in the last few years.[51,52] This is mainly because miniaturized electrodes can be fabricated using microfabrication methods and solutions can be transferred by microfluidics approaches.[53,54] Flow injection analysis and sequential injection analysis techniques were also employed for electrochemical enantioselective high-throughput screening of drugs.[55]

Guerin et al. introduced a novel high-throughput technique that allows investigation of the impact of metal particles and their support on electrocatalytic activity in an array format (Fig. 14.8).[56] The electrode arrays were made by high-throughput physical vapor deposition.[57,58] Oxygen reduction on gold nanoparticles supported on titanium dioxide and carbon was analyzed on arrays containing gradients of particle sizes ranging from 1.4 to 6.3 nm. Catalytic activity decreased for particle sizes below 3.0 nm. This high-throughput approach was successfully used to link redox surface behavior of the supported gold particles with their particle size.

Despite using array-based technologies in combinatorial chemistry, most of the current synthesis methods in research laboratories or industry are based on highly automated plate-based technologies. The concept of parallel synthesis in combinatorial chemistry involves the generation of discrete compounds in spatial separated reaction compartments. The typical employed format is the 96-well microtiter plate.

Fig. 14.5 Generation of addressable libraries on chips (**a**) Scheme of Heck reaction under employment of an transition metal catalyst (Reprinted with permission from Tian et al.[33] Copyright (2005) American Chemical Society); (**b**) Scheme of reductive amination reactions (Reprinted with permission from Tesfu et al.[34] Copyright (2006) American Chemical Society)

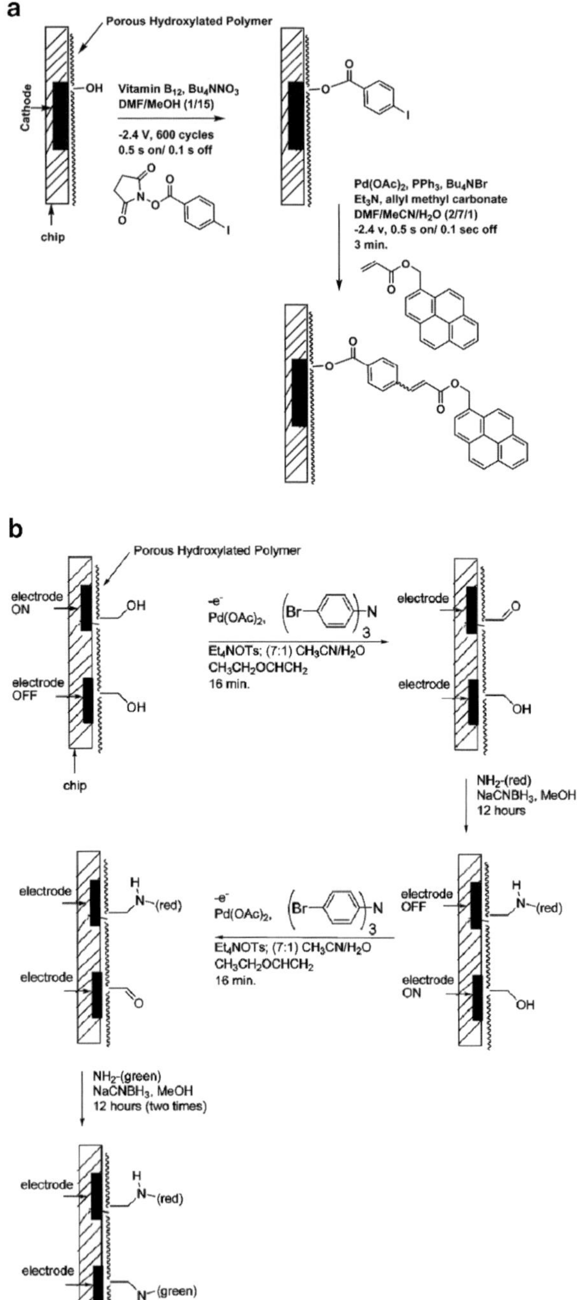

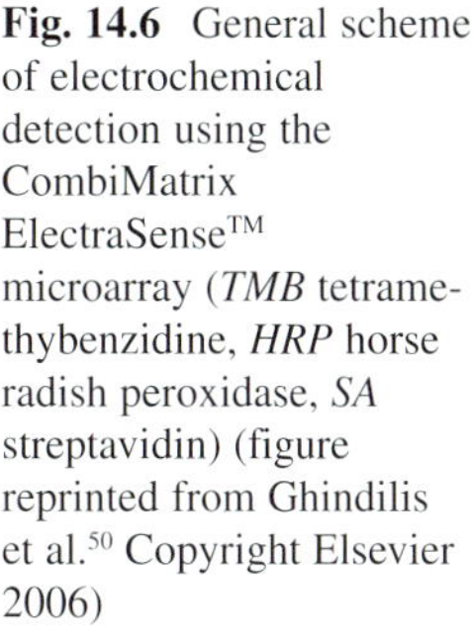

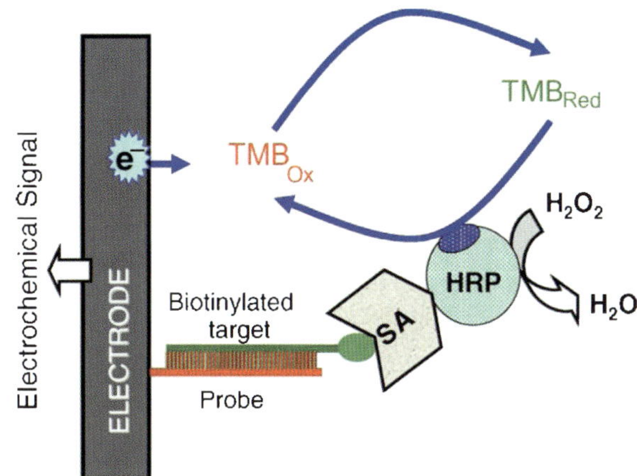

Fig. 14.6 General scheme of electrochemical detection using the CombiMatrix ElectraSense™ microarray (*TMB* tetramethybenzidine, *HRP* horse radish peroxidase, *SA* streptavidin) (figure reprinted from Ghindilis et al.[50] Copyright Elsevier 2006)

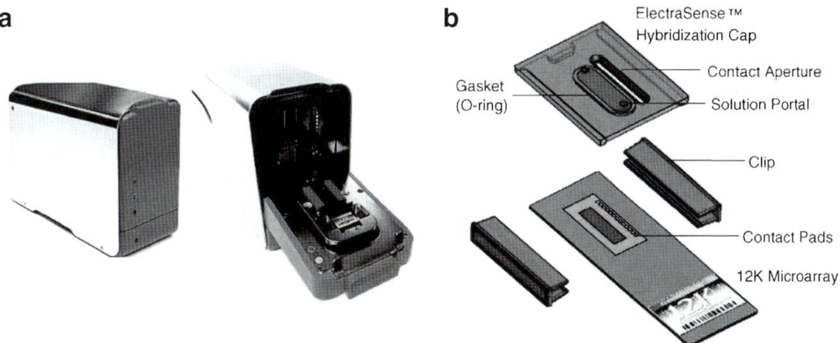

Fig. 14.7 ElectraSense™ Reader (**a**) and ElectraSense™ 12 K microarray with hybridization cap (**b**) (figure reprinted from Ghindilis et al.[50] Copyright Elsevier 2006)

Using commercial available microtiter plates as reaction compartments for electrochemical reactions was presented in 2001 for the first time by Reiter et al.[59] The driving force of this approach was to facilitate the development of a technology that allows integration in current plate-based high-throughput screening (HTS) schemes. The presented multielectrode sequential analyzer (MESA; Fig. 14.9) records up to 16 cyclic voltammograms (CVs) or differential pulse voltammograms (DPVs) by sequentially relay switching the working electrodes of the 16 individual three-electrode bundles placed in 16 wells of a 96-well microtiter plate.

Aim of the study was to evaluate coordinative labeling of redox proteins with Ru-complexes. The ligand exchange of the [Ru(bpy)$_2$CO$_3$] complex with imidazole, serving as a model compound for histidine residues in proteins, was visualized by recording DPVs over time as shown in Fig. 14.10.

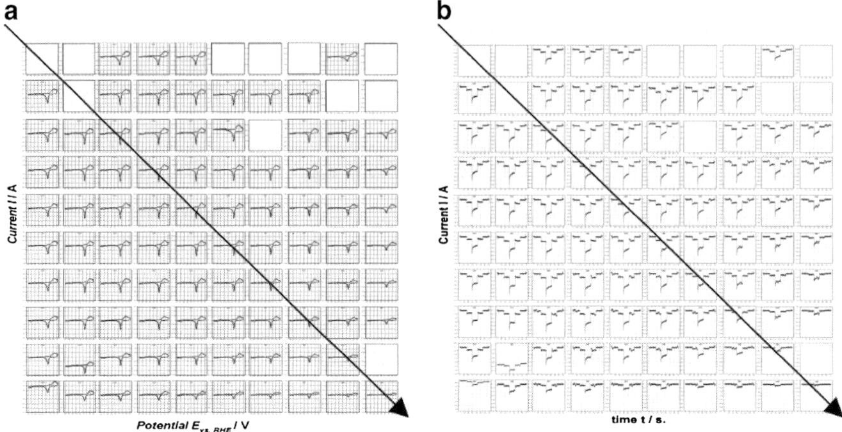

Fig. 14.8 (**a**) Cyclic voltammetry ($v = 100\,\text{mV s}^{-1}$, between 0.00 and +1.63 V) on a 100 electrode array of carbon-supported gold particles in argon-purged solution. (**b**) Oxygen reduction at the same array electrode as determined by potential step measurements from 0.80 V (baseline, 60 s) to 0.50 V (90 s), 0.40 V (90 s), and 0.3 V (90 s) in oxygen-saturated solution. All measurements were made in 0.5 M $HClO_4$ electrolyte at room temperature. The *arrows* indicate the direction of the graduated flux; in this array the *upper left-hand* corner corresponds to a nominal gold thickness of 1.8 nm and the lower right-hand corner to a nominal thickness of 0.21 nm. (Reprinted with permission from Guerin et al.[56] Copyright (2006) American Chemical Society)

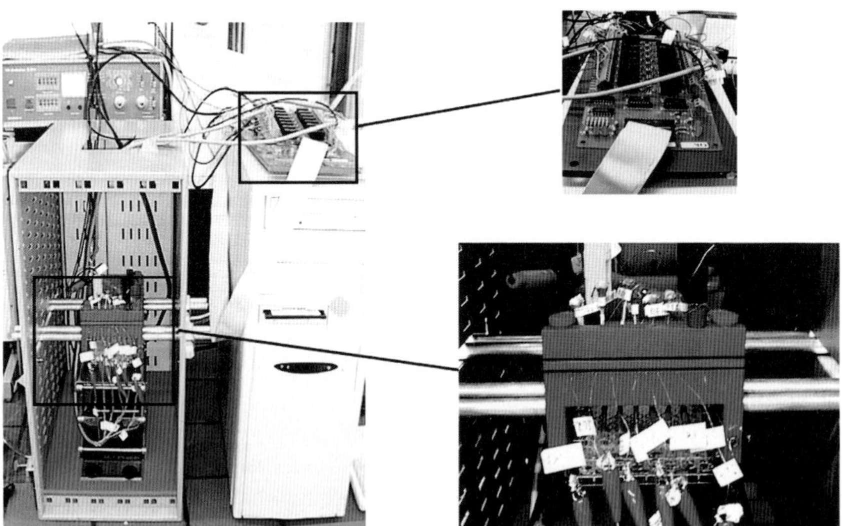

Fig. 14.9 The sequential-parallel electroanalytical set-up showing the Faraday cage with the microtiter-plate compatible electrode system, relais board, personal computer, and potentiostat. The relais board and the electrode holder are magnified on the *right*.[59] (Reproduced by permission of The Royal Society of Chemistry)

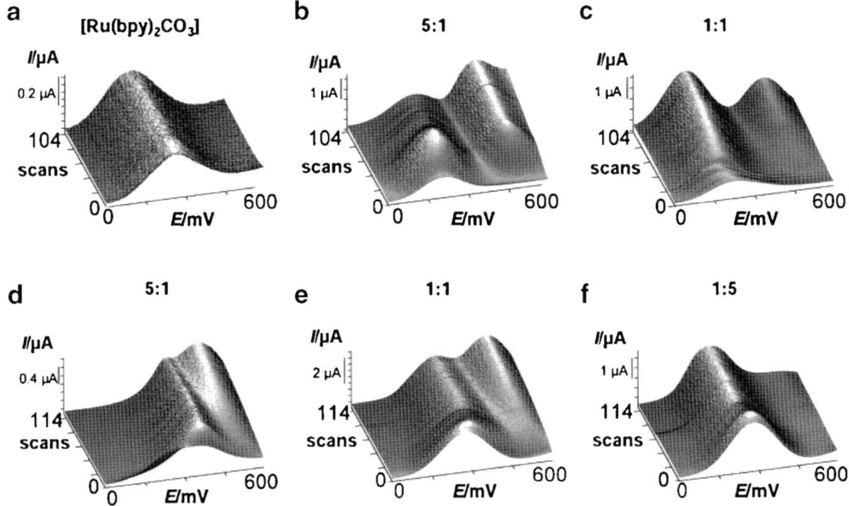

Fig. 14.10 Sequential-parallel DPV investigation of the course of the ligand exchange reaction of [Ru(bpy)$_2$CO$_3$] with imidazole. In 50 mM Na$_2$CO$_3$–NaHCO$_3$ buffer, pH 9.6: (**a**) [Ru(bpy)$_2$CO$_3$] as control; (**b**) [Ru(bpy)$_2$CO$_3$]:imidazole = 5:1; (**c**) [Ru(bpy)$_2$CO$_3$]:imidazole = 1:1; in 50 mM HEPES buffer, pH 7: (**d**) [Ru(bpy)$_2$CO$_3$]:imidazole = 5:1; (**e**) [Ru(bpy)$_2$CO$_3$]:imidazole = 1:1; (**f**) [Ru(bpy)$_2$CO$_3$]:imidazole = 1:5. (DPV from 0 to 600 mV vs. Ag/AgCl; 20 mV s^{-1}, pulse width 60 ms, pulse height 150 mV).[59] (Reproduced by permission of The Royal Society of Chemistry)

Tang et al. introduced a multichannel electrochemical detector (MED) consisting of eight sets of electrodes inserted into a row of eight wells of a microtiter plate as depicted in Fig. 14.11.[60] The device was successfully used to perform parallel amperometric immunoassays.

Jiang and Chu developed an electrolyte-probe screening method that allows half-cell and full-cell electrochemical experiments.[19] The electrochemical oxidation of methanol was reproducibly performed in an electrode array for various methanol-air cells. The effects of catalyst loading and methanol concentration on the discharge performance for the methanol-air cells were investigated by fast screening of the anode electrodes in the array. The same group also presented a combinatorial approach toward high-throughput analysis of direct methanol fuel cells in a 40-member array.[61] In 2005, Jian and Chu introduced an electrode probe for high-throughput screening of electrochemical libraries (Fig. 14.12).[62] They employed a pen-shaped O$_2$ electrode probe that is made from a large-area O$_2$ electrode and a cylindrical electrolyte sponge with a short cone tip. Screening of an electrochemical library of 128 micro zinc/air batteries was performed by moving the tip of the O$_2$-electrode probe at constant potential (1.0 V) while recording the current.

Andreescu et al. introduced a 96-electrode well-type device enabling oxygen sensing for monitoring respiratory activity of biological cells.[63–65] The principal set-up of the multichannel dissolved oxygen sensor system (DOX-96) is schematically in shown in Fig. 14.13. This highly parallelized approach shows successful

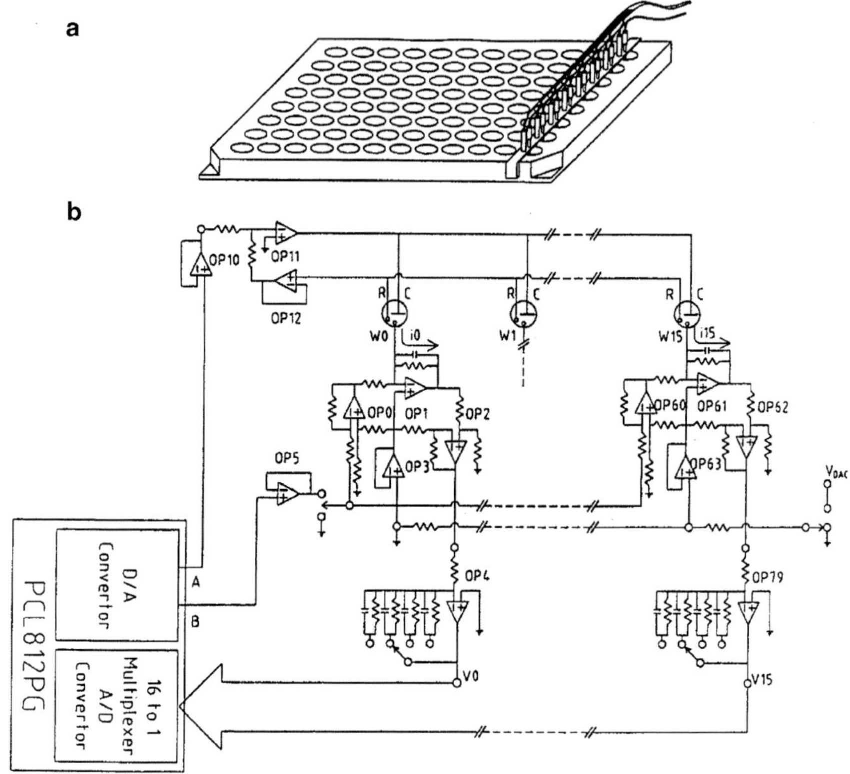

Fig. 14.11 (**a**) The electrode assembly consisting of eight sets of Pt electrodes fitted with a row of microtiter wells. (**b**) Circuitry of the software-controlled 16-channel potentiostat (Reprinted with permission from Tang et al.[60]. Copyright (2002) American Chemical Society)

real-time monitoring of cell cultures. This might be a viable alternative to current cytotoxicity tests.

Although the approaches discussed are valuable contributions to the area of combinatorial electrochemistry and high-throughput screening, most of them do not offer sufficient throughput, flexibility to conduct the whole palette of electrochemical techniques on the same technical platform, and easy interfacing with a variety of different reaction and analysis schemes. Screening for novel drugs or lead structures by combinatorial approaches sets high demands on high-throughput analysis technologies. Table 14.1 summarizes the ideal properties of a HTS instrument, in general. The design, flexibility in using different test/assay formats, information content of the data collected, throughput and costs are important factors for establishing new high-throughput methods. Thus, main focus in this compilation is on the detailed description of the conception, development, features, and applications of a recently developed electrochemical robotic system[66] that seeks to meet several of the requirements mentioned in Table 14.1.

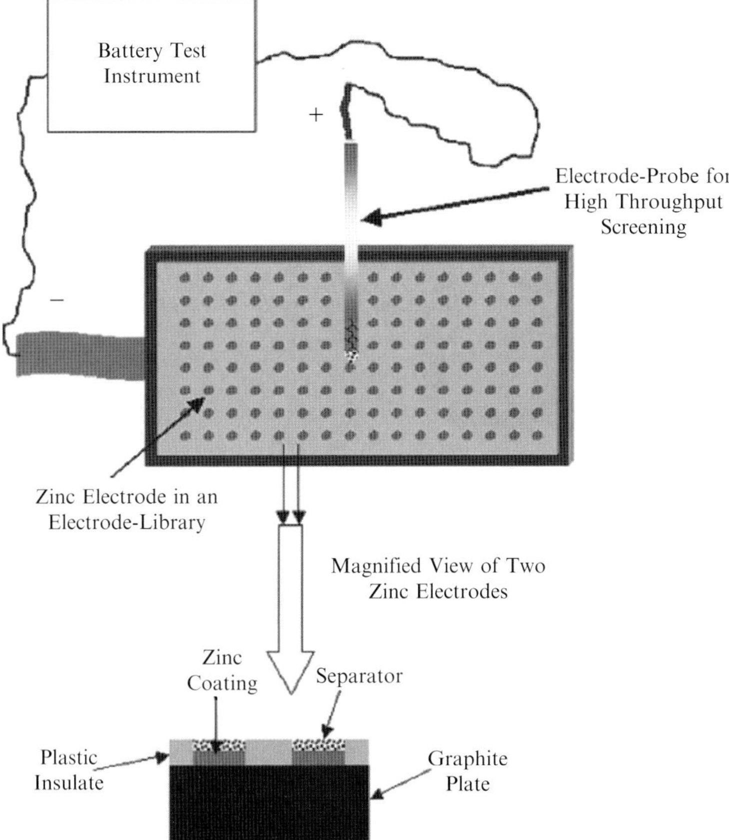

Fig. 14.12 A material spot library and the electric connections to it with an O_2 electrode probe and a battery test instrument. The material spot library contains 128 zinc spots spatially separated on a graphite plate. The surface of the graphite is coated with a plastic layer on the non-spot areas to keep ionic insulation between these material spots. The area of each spot is $0.196\,cm^2$. (Reused with permission from Rongzhong Jiang and Deryn Chu.[62] Copyright 2005, American Institute of Physics.)

1.1 Conception and Evaluation of the Electrochemical Robotic System

For the conception of an automated instrument for electrochemical synthesis, analysis and screening several aspects have to be considered:

(1) *Compatibility to established high-throughput technologies:* To date, most of the applied high-throughput systems use microtiter plates with 24, 96, 384, or 1,536 wells as platforms. Microtiter plates are available in various materials such as glass, Teflon, polyethylene, polystyrene, qualities, and additional features such as special coatings, transparent bottoms, combined wells, deep

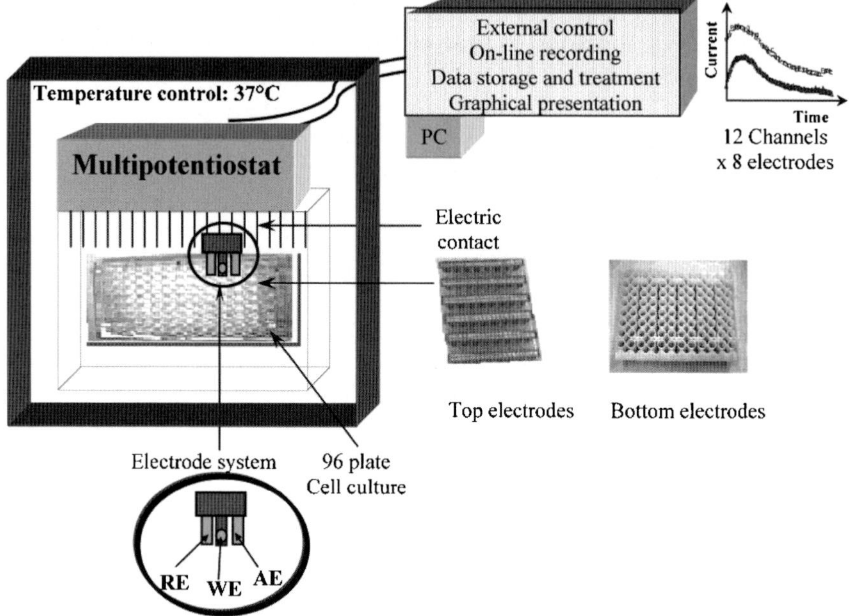

Fig. 14.13 The multichannel DOX oxygen sensor system used for measuring cytotoxicity. Two configurations of the 96 electrodes are available disposed in the top or bottom of the well plates. Each sensor consists of three electrodes: reference (RE), auxiliary (CE), and working (WE) lectrode. (Reprinted with permission from Andreescu et al.[63] Copyright (2004) American Chemical Society)

Table 14.1 Characteristics of an "ideal" high throughput screening tool

Easy-to-use
Minimal or no sample preparation (at best automated sample preparation)
Usage of low volumes/amounts of samples, materials, and additional solutions
Fast and efficient duty cycle that allows rapid serial or parallel analysis of multiple samples leading to a quick overall throughput
Additional features possible such as performing combinatorial synthesis or other reactions in the same device or a coupled instrument
On-line and real-time analysis (before, during, and after reaction or assay)
Performing analysis of one or multiple parameters
High precision, accuracy and reproducibility of measurements
Automated instrument operation, communication with other instruments, data acquisition, processing, documentation, and management
Compatible to other high-throughput technologies or processes
Flexibility of the users to adapt the instrument to their specific needs
Potential for further miniaturization and automation
Reasonable costs (investment and running costs)

wells. Therefore, the proposed electrochemical robotic system should be a plate-based device, and use commercial available microtiter plates as reaction vessels for assays and combinatorial synthesis. The system should be user-friendly and easy to interface with other high-throughput devices.

(2) *Miniaturization of the electrochemical cells:* The dimensions of the wells of the microtiter plates require the use of miniaturized working, counter and reference electrodes. At the same time, reagent consumption will be significantly reduced when compared with conventional electrochemical experiments.

(3) *Automation of all necessary procedures:* The electrochemical robotic system should enable efficient sequential and parallel electroanalysis, flexible selection and variation of parameters, and easy data management. Additional features such as automatic transfer and dispensing of reagents with suitable pumps, integration of a microtiter plate shaker, and control of the temperature within the wells need to be implemented.

(4) *Facilitating a broad spectrum of possible assays and applications:* The developed electrochemical robotic system should be open for diverse applications such as electrosynthesis, electroanalysis, chemical and biological assays. Adaptations of the device to future requirements should be easily possible.

(5) *Standardization of developed assays:* The developed electrochemical assays should be easily standardized, or have a potential for later standardization.

On the basis of these initial considerations, a plate-based electrochemical robotic system was conceived (Fig. 14.14). The system was described in detail by Erichsen et al. in 2005.[66] However, the first prototype was introduced already in 2001.[67] Standard microtiter plates are used as reaction wells in which the miniaturized electrodes are immersed in an automated fashion to perform electrochemical experiments. Accurate positioning of the miniaturized electrodes is achieved by

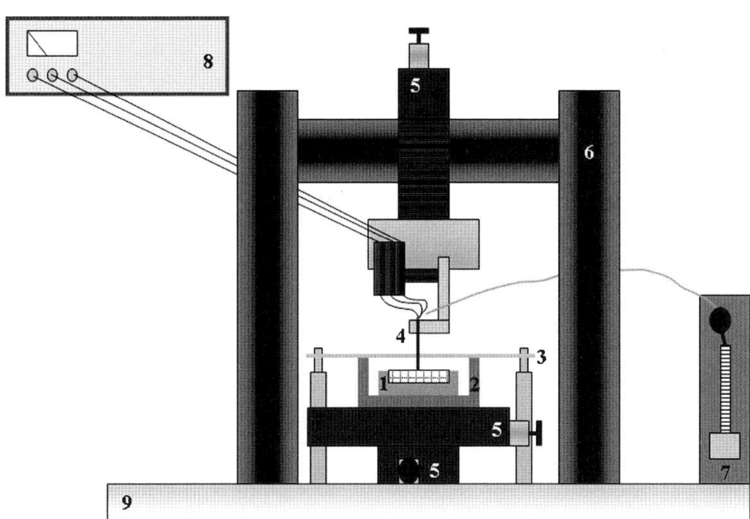

Fig. 14.14 The electrochemical robotic system: (1) microtiter plate, (2) chamber, (3) glass plate, (4) equipped with a single microelectrode bundle consisting of the working, counter and reference electrode, (5) step motor-driven stages, (6) H-shaped column, (7) syringe pump, (8) potentiostat

means of a z-axis step motor-driven stage. The electrode assembly varies in design in accordance to the specific experimental needs.

The microtiter plate is placed into a specially designed chamber mounted on step motor-driven x- and y-positioning stages. The step motor-driven stages exhibit a nominal accuracy in positioning of 30 nm in the micro-step mode, a maximal speed of 8 mm s^{-1}, and accuracy in repositioning better than 2 μm. To minimize evaporation of the electrolyte, the chamber is covered with a glass plate having up to eight holes for insertion of the electrodes. A Teflon lid is mounted on the chamber for sealing.

Figure 14.15 shows a photograph of commonly used microelectrodes and miniaturized reference and counter electrode. Note that even a dual microelectrode (double-barrel electrodes[68]) can be used in the robotic system.[69]

Its design allows for the flooding of the chamber with inert gas that may be saturated with water or solvent. The system is compatible to aqueous and organic solvents. All reaction or assay sequences including positioning of electrodes and microtiter plate, performing electroanalytical techniques, mixing and washing steps, addition, transfer or removal of reagents by means of a syringe pump can be performed automatically. This is ensured by a modular hardware and software structure schematically displayed in Fig. 14.16.

A script language based on simple text files facilitated the user to predefining complex experimental sequences to be performed by the electrochemical robotic system. This allows executing of complex experimental sequences without actually knowing the programming code. Available action modules consist of a word built by the x,y-position, the name of the procedure, and all necessary parameters that are used by the procedure. A list of action modules, which can be easily extended

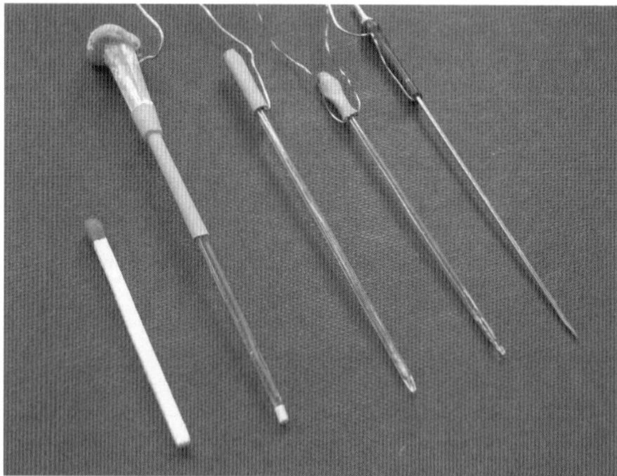

Fig. 14.15 Miniaturized electrodes for usage in the electrochemical robotic system. From *left* to *right*: 3 M Ag/AgCl reference electrode, platinum microelectrode, dual platinum microelectrode, stainless steel cannula that serves as pump outlet and counter electrode[44]

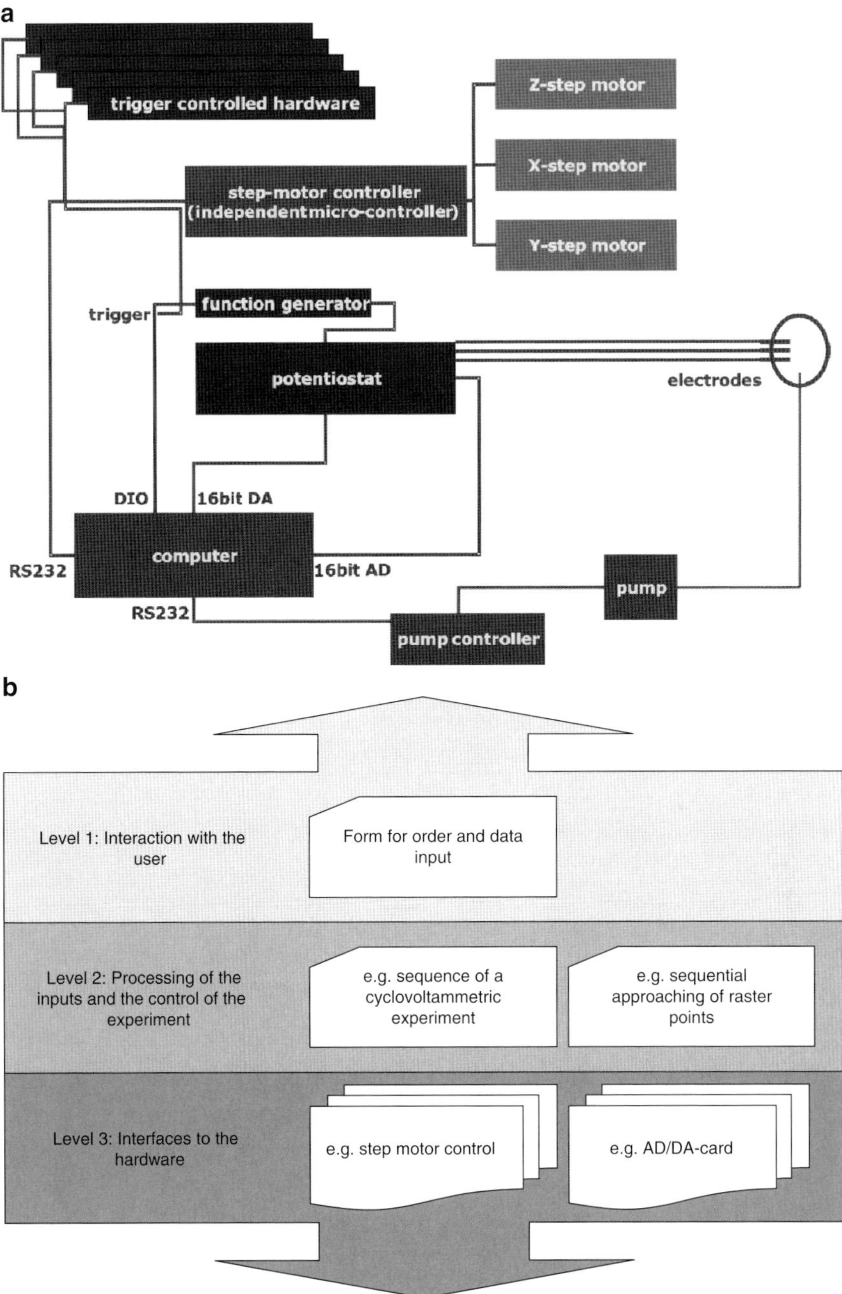

Fig. 14.16 The modular hardware (*top*) and layered software structure for controlling the electrochemical robotic system (*bottom*) (Reused with permission from Erichsen et al.[66] Copyright 2005, American Institute of Physics.)

on demand to meet specific needs of an experimental sequence, are given in Table 14.2. The system allows performance of potentiostatic and galvanostatic electrosynthesis, high-throughput electroanalysis, and electrochemical assays even under inert-gas atmosphere.

The first generation of the electrochemical robotic system was tested for a variety of applications to demonstrate its feasibility for automatic electrochemical synthesis and analyses with increased throughput. Developing automated synthesis procedures and screening formats required the implementation of additional routines such as quality control of electrode surfaces by means of cyclic voltammetry, washing steps, chemical and electrochemical cleaning of electrode surfaces, transfer of reagents by syringe pump(s) and data file management.

Summarizing, the electrochemical robotic system enables the combination of electrosynthesis and electroanalysis in a single experimental set-up using galvanostatic and potentiostatic methods depending on the particular problem. The system is based on the conventional microtiter-plate format but may be miniaturized in the future. To date, the 24, 96, and 384-well format have been used in the electrochemical robotic system (Fig. 14.17). It introduces a high degree of automation enabling complex synthetic and analytical protocols based on a highly flexible hardware and software concept. It is possible to use up to eight electrode bundles in parallel by means of suitable electrode holders and multi-potentiostats (Fig. 14.17).

Table 14.2 Action modules and list of parameters which are transferred from the script file to the individual program modules for the different experimental sequences

Action modules	Transferred parameters
Amperometry	Position, electrochemical technique (A), measuring time (s), applied potential (mV)
	Optional: speed of pump (μl/s), pumped volume (μl) (up to sequentially added eight volumes), delay time between pumping events (s)
z-approach (amperometric SECM feedback control)	Position, command (approach), applied potential (mV), stop case (% change of bulk current signal), maximum driving way (μm), driving speed (μm/s)
Cyclic voltammetry	Position, electrochemical technique (cv), number of cycles, scan rate (mV/s), start (vertex 1, vertex 2 and end) potential (mV)
Digital out	Position, command (digout), value (0,1,2)
Differential pulse amperometry	Position, electrochemical technique (dpa), measuring time (s), applied base potential (mV), pulse height (mV), pulse time (ms), time between pulses (ms)
	Optional: speed of pump (μl/s), pumped volume (μl) (up to sequentially added eight volumes), delay time between pumping events (s)
Differential pulse voltammetry	Position, electrochemical technique (dpv), measuring time (s), applied base potential (mV), pulse height (mV), pulse time (ms), time between pulses (ms), scan rate (mV/s), start (vertex 1, vertex 2 and end) potential (mV)

(continued)

Table 14.2 (continued)

Action modules	Transferred parameters
Electrolysis with electrode movement	Position, electrochemical technique (ekl), measuring time (s), applied potential (mV), driving way (μm), DA channel (0,1,2)
Mixing (moving electrodes up and down)	Position, command (mix), number of times, driving way (μm)
Potentiostatic pulses (max, eight pulses)	Position, electrochemical technique (pulse), delay time before pulses (s), number of cycles, potential 1 (mV), pulse time 1 (ms),...
	Optional:..., potential 2 (mV), pulse time 2 (ms),..., potential 8 (mV), pulse time 8 (ms)
Pumping	Position, command (pump), pumped volume (μl), speed of pump (μl/s), pump direction (1,0), value number (0,1)
Driving a square	Position command (square), x-way (μm), y-way (μm), driving speed (μm/s)
Stop	Position, command (stop)

Reused with permission from Erichsen et al.[66] Copyright 2005, American Institute of Physics

2   Applications of the Electrochemical Robotic System

Several feasibility and application studies have been performed using the developed electrochemical robotic system aiming on the exploration of the instrumental limits in automation of electrochemical synthesis and analysis. Parallelization of processes and miniaturization (Fig. 14.17) for higher throughput was a major objective.

2.1   *Electrochemical Characterization of Compound Libraries*

One problem of HTS is the occurrence of false positive hits during the screening process, which may result in high expenses for industry in areas such as drug discovery. Therefore, there is a need for complementary analytical methods to fluorescent-based assays that allow broadening of the spectrum of compound classes screened and that will provide an increase in knowledge about compound libraries. This may promote more efficient product development in industry or academic research. Electrochemical and electroanalytical methods are ideal for synthesis and screening of compound libraries because of their high sensitivity and selectivity obtained by exact control of the applied electrode potentials.

The electrochemical robotic system was first applied for characterization of a combinatorial substance library.[67] 256 quarter (3-arylthiophenes) were screened by means of cyclic voltammetry to reveal their redox properties. The first oxidation potentials leading to the formation of the radical cation of the investigated species could be correlated with the electronic effects of the individual substituents (Fig. 14.18).

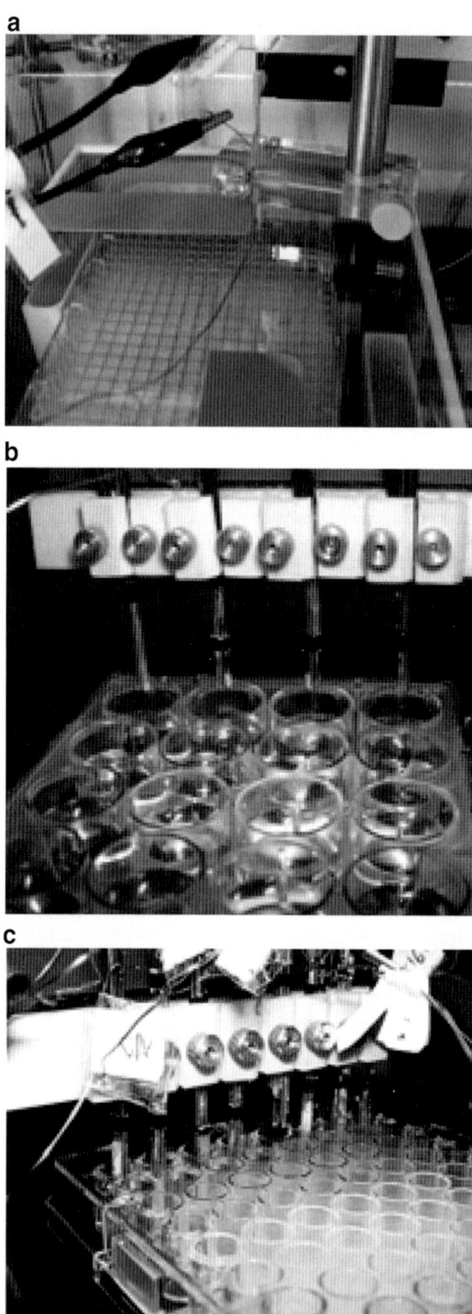

Fig. 14.17 Measurements in the electrochemical robotic system[44] (**a**) Using a 384-well plate for redox screening of redox mediators with one electrode bundle; (**b**) Using a 24-well plate for biosensor characterization with four electrodes in parallel; (**c**) Using a 96-well plate for a set-up with eight electrode bundles in parallel

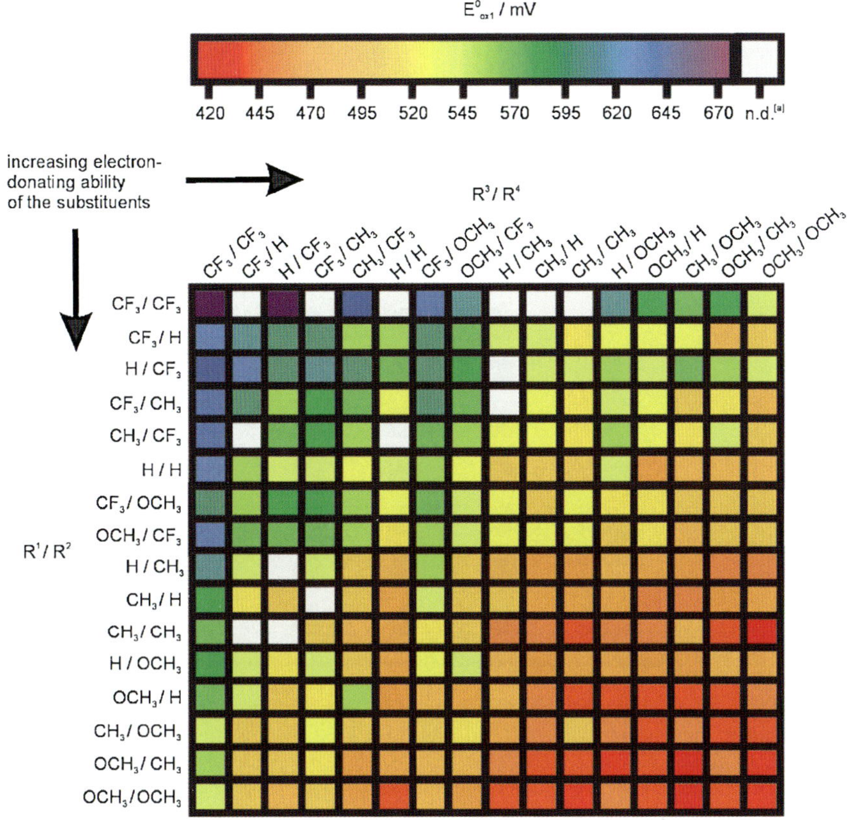

Fig. 14.18 First oxidation potentials of the 256-membered quater(3-arylthiophene) library. Potentials are color-coded according to the scale at the top. The arylthiophene units are given along the axis. [a] not determined.[70] Reproduced by permission of The Royal Society of Chemistry, original data published first in[67])

The performance of the robotic system was further demonstrated by redox screening of a Ru-complex library that serves as the basis for homogeneous catalysts for hydrogenation.[71] Figure 14.19a shows the loading of the 96-well microtiter plate with Ru(II) complexes (sample wells in black; structures shown in Fig. 14.19b). Washing wells (open circles) were filled with plain electrolyte only. To avoid any long-term potential shift caused by the use of a chloridized silver wire as pseudo-reference electrode, ferrocene (Fc) was added to the reference wells to keep track of the actual potential at distinct time intervals during the screening (refer also to Fig. 14.20 row 1). Figure 14.20 displays the results of the redox screening of a library of Ru(II) complexes.

Shape and position of the redox waves in the cyclic voltammograms were correlated to the catalytic activity of the compounds. Compared with standard classical ways for evaluating electroactive compound collections by chemical reaction and product analysis, the robotic strategy enables a very fast alternative at considerably reduced effort.

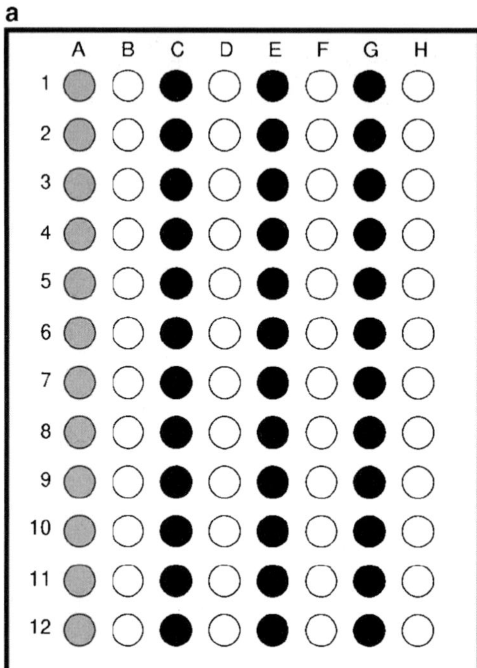

Fig. 14.19 (**a**) 96-well microtiter plate used for cyclic voltammetric redox screening of Ru(II) complexes; *light grey*: reference wells (filled with 0.5 mM Fc solution); *open circles*: wash wells (filled with 0.1 M NBu$_4$PF$_6$/CH$_3$CN electrolyte and no redox species); *black*: sample wells with Ru complexes; (**b**) Schematic of used Ru(II) complexes in this study; (Figure reprinted from Lindner et al.[71] Copyright Elsevier B.V. (2005).)

2.2   *Electrochemical Synthesis*

Combinatorial electrosynthesis was accomplished in the electrochemical robotic system. Transferring classical electrosynthesis into the wells of a microtiter plate allows for working with an electrolyte volume of only ~250 µl. The significantly decreased volume makes the approach potentially applicable for combinatorial applications. Moreover, the electrochemical robotic system is able to additionally perform a voltammetric in situ analysis for quality control.

Libraries were generated from iminoquinol ether and [1,2,4]triazolo[4,3-*a*] pyridinium perchlorate collections in the wells of a microtiter plate by potentiostatic microelectrolysis.[72] The progress of electrolysis was monitored by microelectrode steady-state voltammetry, and product formation was screened by semi-quantitative HPLC/MS. A schematic of the employed four-electrode assembly and a photograph of that bundle in action in the robotic system are shown in Fig. 14.21.

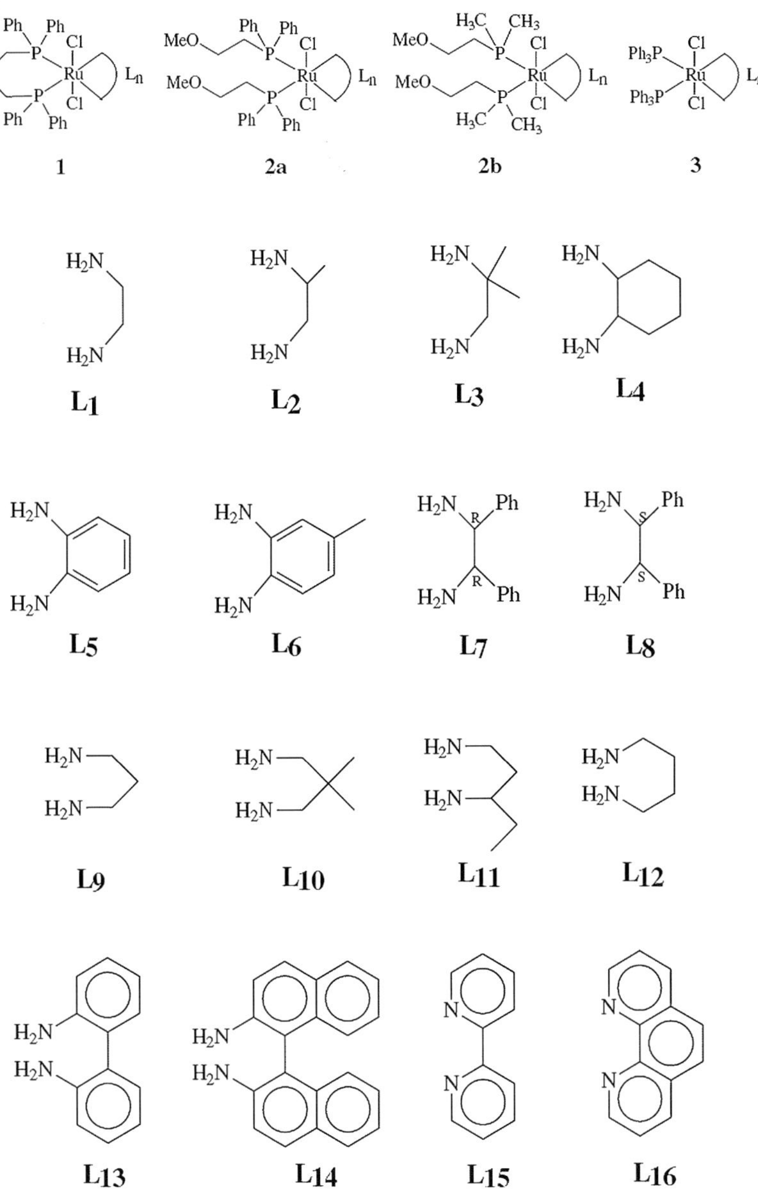

Fig. 14.19 (continued)

As example, the reaction sequence for oxidizing aniline derivatives in the presence of alcoholic nucleophiles is illustrated in Fig. 14.22. Aniline derivatives **1a,b** were electrolyzed in $CH_3CN/0.1\,M\;NBu_4PF_6$ in the presence of lutidine and various alcohols R_2OH.

Fig. 14.20 Cyclic voltammetric array (potential axes range from −0.1 to 1.1 V vs. Ag/AgCl, current axes ranges from −80 to 180 nA) of samples in a microtiter plate as shown in Fig. 14.19; potential scan rate 0.5 V s^{-1}, all wells with 0.1 M NBu$_4$PF$_6$/CH$_3$CN electrolyte; reference wells with 0.5 mM Fc solution; sample wells with ~ 0.5 mM complex solution, wash wells (WW): electrolyte without redox active species; 200 μm Pt disc electrode (Figure reprinted from Lindner et al.[71] Copyright Elsevier B.V. (2005))

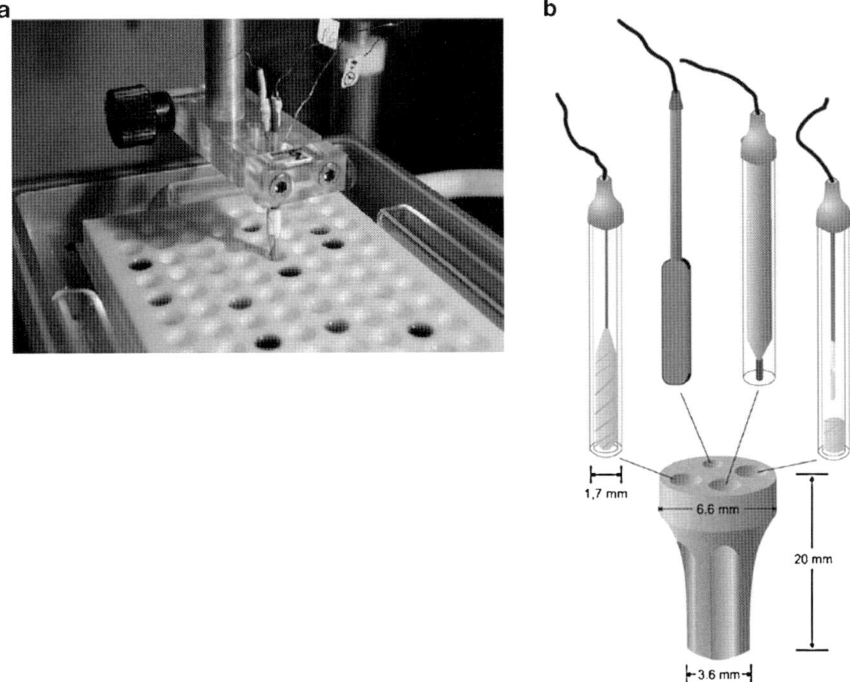

Fig. 14.21 Electrochemical robotic system for microelectrosynthesis (**a**) Detailed view of an electrode bundle inserted into well E4 of the microtiter plate (covering glass plate removed to enable view into wells; containers for solvents are placed left and right of the microtiter plate and keep the atmosphere solvent saturated); (**b**) Assembly of the electrode bundle using a PTFE holder (*bottom*) and individual electrodes from left to right: counter, electrolysis working, CV microdisk, reference electrode; while the electrolysis working electrode is inserted into the holder from below, the other three electrodes are inserted from the top. (Figure reprinted from Markle et al.[72] Copyright Elsevier Ltd. (2005).)

During the course of electrolysis, the aniline concentration was monitored by means of steady-state voltammetry at the microdisk electrode (Fig. 14.23)a for the example of **1b** and methanol). A decrease in aniline concentration was observed concomitantly with a decreasing electrolysis current (Fig. 14.23b). The electrolysis is complete after 2–3 h. 2.15 electrons were transferred per molecule aniline. The value higher than the expected 2.0 is due to redox recycling of 1b in the uncompartmented electrolysis cell. The reaction can be accelerated by mixing the electrolyte in the well by convection (Fig. 14.23c, compare curves 2 and 3). Similar behavior was observed for other alcohols and for aniline **1a**. To our knowledge, the work from Märkle et al.[72] shows for the first time a successful application of potentiostatic electrolysis in microtiter plates. Later, the same authors examined in detail the potential of using microelectrode steady-state voltammetry to follow the progress of preparative electrolyses on-line in the microtiter plate.[73] This approach helps to monitor the quality of the electrolysis process by keeping track of the starting

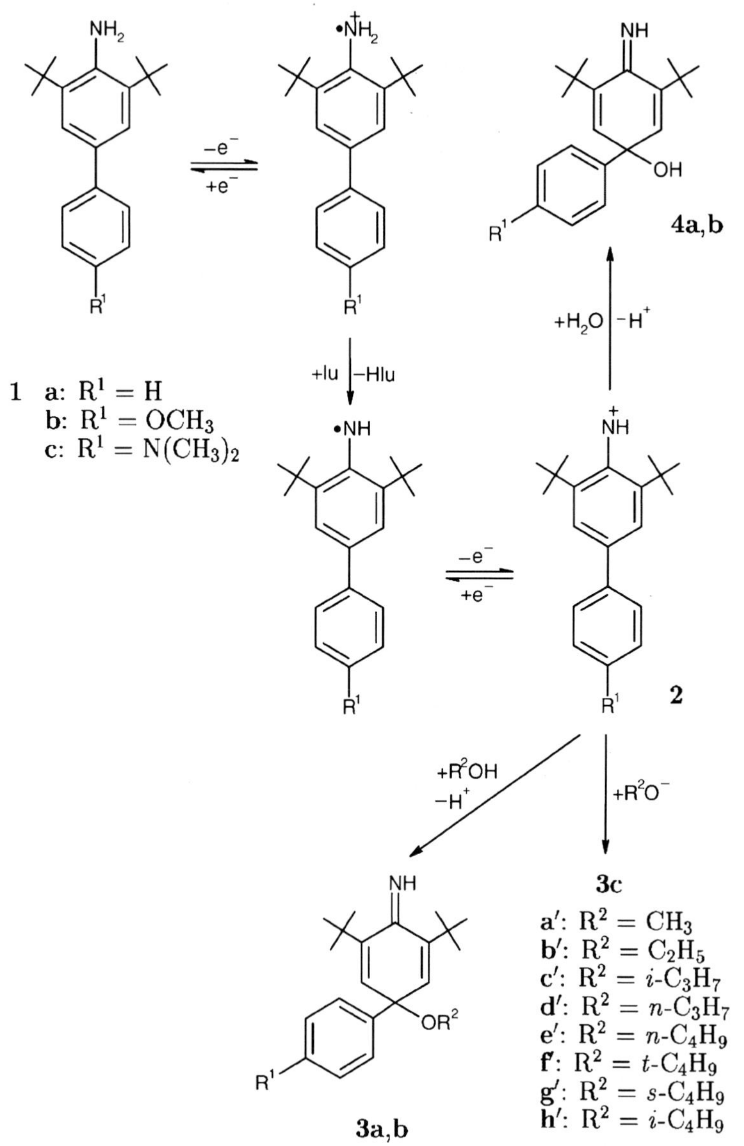

Fig. 14.22 Anodic oxidation of sterically hindered anilines 1 in the presence of alcoholic nucleophiles (lu = lutidine) (figure reprinted from Markle et al.[72] Copyright Elsevier Ltd. (2005).)

materials and product concentrations. Endpoints can be detected, number of electrons transferred per molecule can be calculated, and diffusion coefficient ratios for redox couples can be determined.

The main advantage of employing microelectrodes for analytical purposes in electrolysis is that CV monitoring can be done without introducing errors because

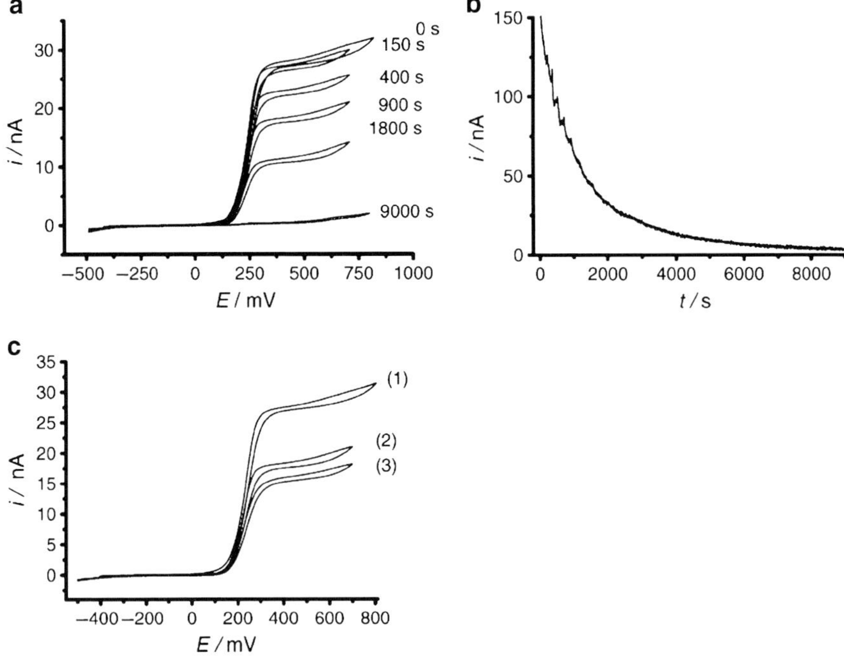

Fig. 14.23 Microtiter plate anodic electrooxidation from 1b to form 3ba' in the presence of CH_3OH, c(1b) = 4 mM, c(lu) = 50 mM, c(CH_3OH) = 2 M, electrolysis potential E = +0.4 V vs. fc/fc+ (**a**) steady-state microdisk electrode (d = 25 μm) cyclic voltammetry during electrolysis, v = 0.02 V s^{-1}, times after start of electrolysis indicated, (**b**) current development during electrolysis, (**c**) steady-state voltammograms before (1) and after 900 s of electrolysis with (3) and without (2) mixing by convection. (Figure reprinted from Markle et al.[72]). Copyright Elsevier Ltd. (2005)

the steady-state current is reached at the microelectrode within a very short time after application of a potential. Other advantages are more generally related to electro-organic synthesis. (1) High degree of control over the selectivity of the reaction by controlling the redox potential. Contrary to conventional chemical redox reagents potentials can be varied. (2) Current density controls reaction speed. (3) Reaction turnover can be determined by transferred charge. (4) Selectivity and speed of reaction can be fine-tuned by selecting suitable electrode materials and electrolytes. (5) Electro-organic synthesis is advantageous due to mild reaction conditions. The impact of temperature on electron transfer is low. Therefore, a suitable temperature may be chosen to reduce undesired side reactions. (6) Electron transfer occurs at an electrode. Thus, there is no need of later separation of redox reagents. Disadvantages of electrochemical synthesis are (1) the need of special equipment, (2) large electrode surfaces for efficient reactions, (3) and disturbance of later purification steps during product separation.

There is a clear potential in increasing the throughput of microelectrolysis in microtiter plates by further parallelization of the robotic approach, for example, using eight electrode bundles to perform eight electrolysis procedures within eight wells simultaneously. As a matter of fact, this would require expensive multiple potentiostats. However, potential solutions to this problem are known. Multiplexing and serial read-out of electrode arrays using a simple potentiostat and integrated circuits was reported[74] and parallel electrochemical generation of libraries in the chip format was presented.[32–35]

2.3 Development and Evaluation of Chemical Sensors and Biosensors

Considering the increasing number of compounds made available by combinatorial chemistry, their evaluation as sensing materials needs to be done in an automated manner. Therefore, it was proposed to explore the use of the robotic system as platform to develop and optimize chemical sensors and biosensors.

A library of 83 metalloporphyrins was generated by parallel synthesis.[75] The substitution pattern was varied at the *meso* position of the porphyrines, and the central metal ions were altered (Table 14.3). To investigate the use of the members of the library as electrocatalytic material for sensitive sensors for nitric oxide (NO), a complex assay sequence was fulfilled by means of the electrochemical robotic system (Fig. 14.24). In first step, residual water was removed

Table 14.3 A comprehensive list of the NO peak potentials (mV)/oxidation currents (μA) determined for all individual member of the metalloporphyrin library (*left*: general chemical structure) as assessed by means of automated differential pulse voltammetry (*right*: Ryabova et al.[75] -Reproduced by permission of The Royal Society of Chemistry)

R	Fe	Co	Ni	Cu	Zn
H		538/2.55	529/1.59	534/4.36	531/2.43
Me	525/0.77	528/1.98	520/0.71		500/0.83
^{i}Pr	509/0.85	541/1.38	556/1.58	465/1.39	489/1.29
2.4.6-Me		516/2.43			522/1.50
2.3.4.5.6-Me					
OH		527/1.58			
OMe		531/0.87	561/0.84	462/1.52	517/2.50
OEt	524/1.05	525/0.70	534/0.88	488/1.29	500/0.85
OBu	487/014		502/1.74	547/0.92	536/1.39
Ph	505/0.62	515/0.49	449/0.19	503/1.11	519/0.90
NMe$_2$					
NEt$_2$					
SMe	522/4.70	548/6.04	553/4.34	462/1.92	541/4.47
CF$_3$	580/5.93	579/3.30	587/3.43	581/5.60	579/3.19
OCF$_3$	517/1.61	509/0.27	561/3.46	537/1.84	494/2.39
CN	529/1.42	534/4.57	523/4.25	521/0.86	483/1.92
NO$_2$	546/1.97	532/2.36	520/12.8	528/3.94	541/1.29
F	495/0.50	558/1.54	498/1.99	497/0.62	520/2.66
Cl	501/1.97		520/2.56	454/1.55	472/1.98
Br		509/2.86	525/2.47	497/3.04	529/4.44
M–T			508/1.80		

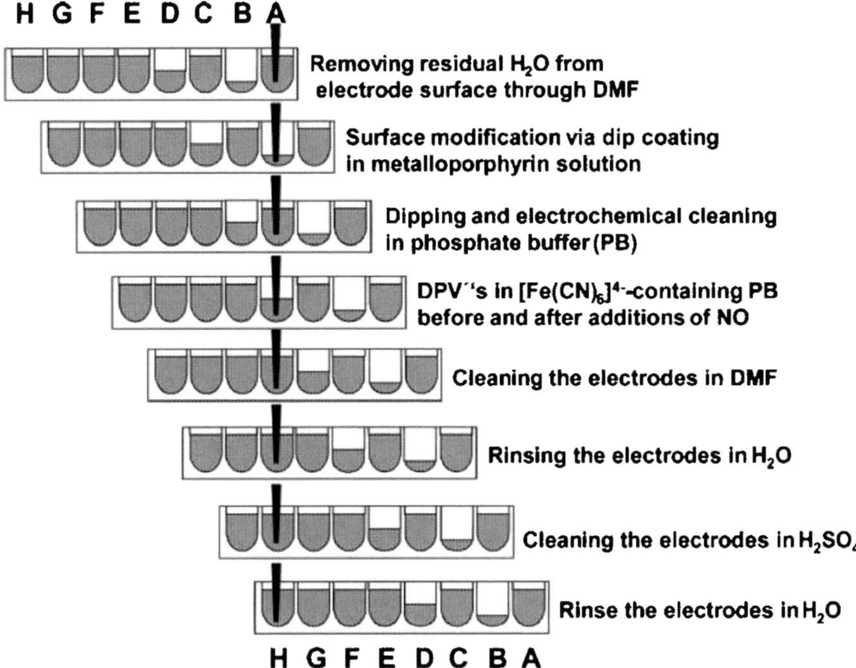

Fig. 14.24 Typical standardized sequence of steps performed within an automated test trial for fabricating and screening metalloporphyrin-based nitric oxide sensors in a microtiter plate.[75] (Reproduced by permission of The Royal Society of Chemistry)

by dipping the microelectrode into solvent. The sensor was produced by dip-coating with the metalloporphyrin solution. A cleaning step followed, and the sensor performance was characterized by differential pulse voltammetry (DPV) before and after addition of nitric oxide stock solution in the presence of $[Fe(CN)_6]^{4-}$ serving as an internal potential reference. Several cleaning steps were necessary to remove the metalloporphyrin layer from the electrode surface before starting the evaluation of the next metalloporphyrin. Screening of the entire compound collection helped in identifying the quality of the potential candidates for NO oxidation by revealing NO peak potentials and NO oxidation currents (Table 14.3).

The obtained knowledge supported initial considerations concerning the influence of varied functionalities of the metalloporphyrins on their electrocatalytic properties based on parameters such as Hammett constants.

Besides screening, the robotic system allows detailed investigation of sensor parameters and quality control measurements. A Ni-porphyrin of the library was electrochemically deposited on a platinum disc electrode ($d = 250\,\mu m$), dip coated with Nafion, and evaluated in the robotic system. Each well can be used for calibration

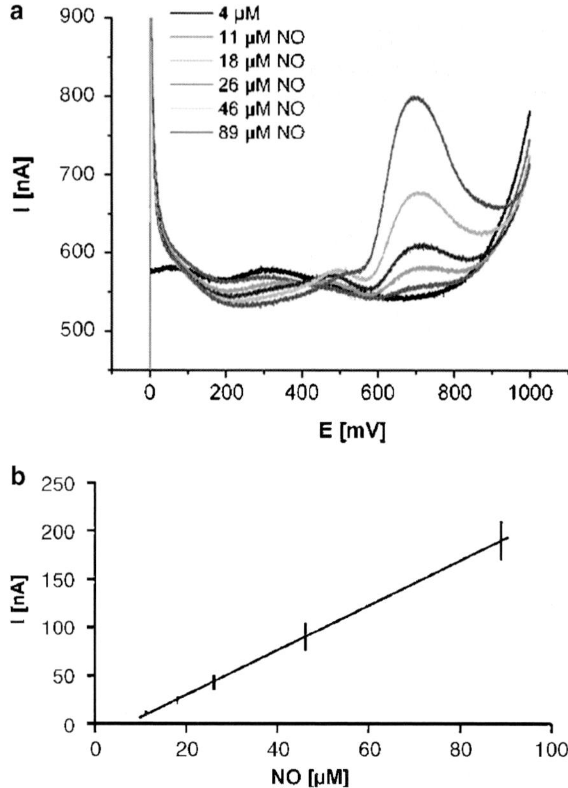

Fig. 14.25 Automated evaluation of a Ni porphyrin[44]: (a) Differential pulse voltammograms (DPVs) recorded for different NO concentrations; (b) Peak current measured from DPVs for five NO concentrations recorded in 22 sequential measurements in a microtiter plate; (c) Resulting calibration plot ($n = 22$)

of the NO sensor using amperometry, differential pulse amperometry (DPA) or as in this case differential pulse voltammetry (DPV) (Fig. 14.25a). The standard solution containing dissolved NO was stored at constant temperature under inert gas, and was added automatically into the wells by means of a syringe pump using oxygen-impermeable tubing. Long-term stability of the sensor and electrode fouling processes can be observed by repeated measurements over a selected time period. The use of the robotic system facilitates to increase the number of experiments when compared with manual experiments. Therefore, calibration plots with high statistical significance can be obtained with reasonable effort. The fully automated measuring routines ensure a high grade of reproducibility and reduce manual error sources. This is especially advantageous for measurements that involve highly labile compounds such as NO.

Nonmanual production of biosensors is well known to improve reproducibility of the manufacturing process and sensor performance.[44,77] A combinatorial library

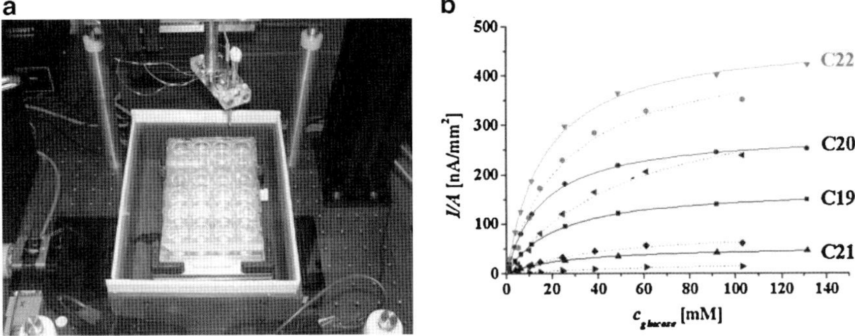

Fig. 14.26 Electrochemical robotic system applied for combinatorial biosensor development and evaluation (**a**) Photograph of the instrument applied using a 24-well microtiter plate and a microtiter-plate shaker. (**b**) Glucose calibration graphs from biosensors obtained by the robotic approach (*dotted line*; 250 μm Pt working electrode) and a manual standard procedure (*solid line*; 1 mm Pt working electrode). For comparison the current values are normalized by the geometric surface area of the working electrode. Polymers employed for biosensor preparation were C19 (*dark gray*), C20 (*gray*), C21 (black), and C22 (*light gray*).[76] Reproduced by permission of the author and Wiley-VCH Verlag GmbH & Co KGaA)

of electrodeposition paints was used to generate glucose biosensors in a conventional approach employing macroelectrodes. In parallel, the experiment was done in the robotic system using microelectrodes.[76] Figure 14.26a shows a photograph of the set-up. Here, a 24-well plate (up to ~ 2 ml electrolyte volume) and a microtiter-plate shaker were employed in addition to the standard device. For simplicity, we use a serial coding of the polymers used in this study (C19, C20, C21, C22). Details can be found in Reiter et al.[76] Figure 14.26b shows the comparison of the ampero-metric calibration graphs obtained in the classical and robotic approach. Similar trends were observed.

2.4 *Voltammetric Metal Ion Determination*

Adsorptive stripping voltammetry (AdSV) for sensitive Ni^{2+} ion detection was automated with the electrochemical robotic system.[78] Ni^{2+} ion concentrations were reproducibly determined by means of standard addition method using Bismuth film-modified glassy carbon electrodes as working electrodes (Fig. 14.27a; limit of detection: 2.1 nM Ni^{2+}, $n = 16$). This approach was used to analyze Ni^{2+} ion release from corroding NiTi shape memory alloys (Fig. 14.27b) in a 24-well microtiter-plate format. The assay strategy allowed for automatic stripping off and replating of fresh Bi films in between measurements. In general, this method employing microtiter plates for trace analysis significantly differs from earlier automation attempts where sequential flow injections systems using electrochemical flow-through cells[53,55] were used.

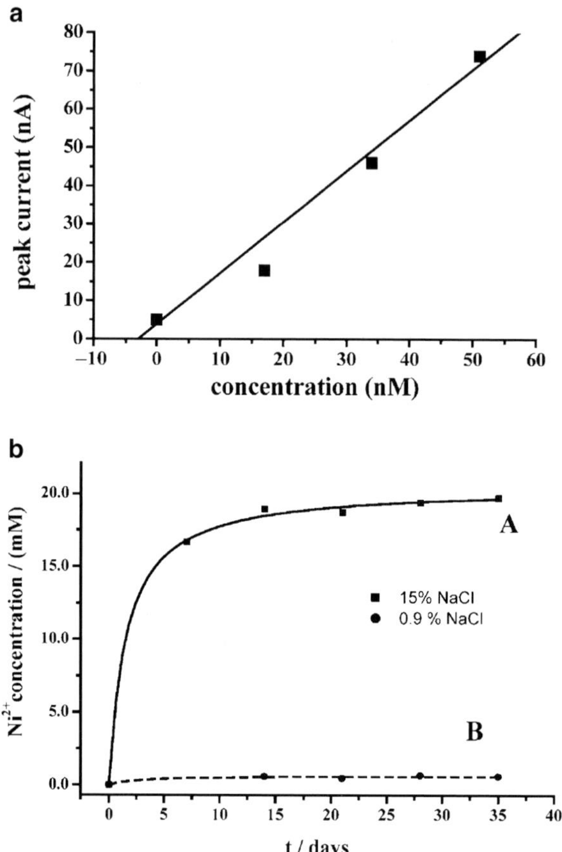

Fig. 14.27 Automated adsorptive stripping voltammetry (AdSV) of Ni^{2+} ions: (**a**) calibration plot derived from the peak current of DPVs of automated AdSV using the standard addition method; (**b**) Long-term Ni^{2+} ions release from surfaces of electropolished NiTi shape memory alloys kept in NaCl solution. NaCl concentrations were either 15 wt.% (*solid line*) or 0.9 wt.% (*dotted line*).[78] (Reproduced by permission of Wiley-VCH Verlag GmbH & Co KGaA)

2.5 SECM Applications of the Electrochemical Robotic System

The conception of the electrochemical robotic system was based on earlier instrumental developments aiming at the design of scanning electrochemical microscopes (SECM),[79–81] however using a rather coarse positioning system and hence exhibiting limited scanning resolution.

For exploring SECM features of the electrochemical robotic system, two electrodes were controlled by the robotic system to electrochemically deposit spots of an anodic electrodeposition paint[82] on a surface. The deposition of the negatively charged polymer is realized by anodic water oxidation under generating of protons. The process is schematically illustrated in Fig. 14.28. The pH shift in the vicinity of the anode surface results in protonation of the polymer and a concomitant change of its solubility. Thus, the polymer precipitates on the electrode surface (Fig. 14.28b). This technology can be used to locally generate biosensors by entrapping suitable enzymes on macro and microelectrodes (Fig. 14.28c, d).[76,82,83]

A potentiostatic pulse profile to generate H^+ was applied to two microelectrodes in the electrochemical robotic system causing the precipitation of the polymer on a suitable substrate (Fig. 14.29a). Parallel localized deposition of enzyme spots was achieved (Fig. 14.29b). This approach was a first step toward systematic improvement of biosensor microstructures within the electrochemical robotic system.

A phenomenon called "negative feedback mode" is well known in SECM.[84,85] When a microelectrode measuring a current from a redox species is approaching

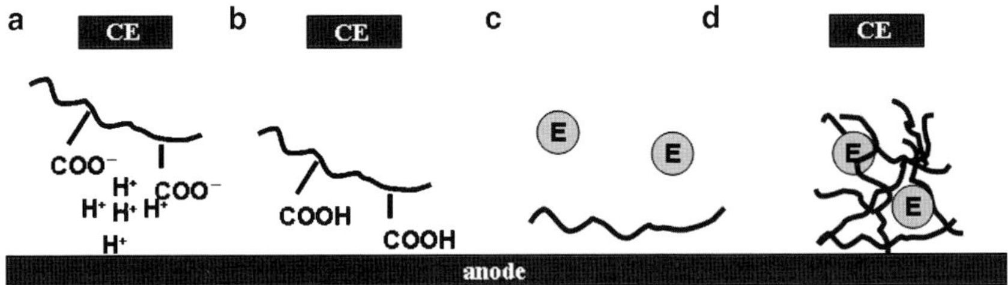

Fig. 14.28 Biosensor fabrication employing electrodeposition paints

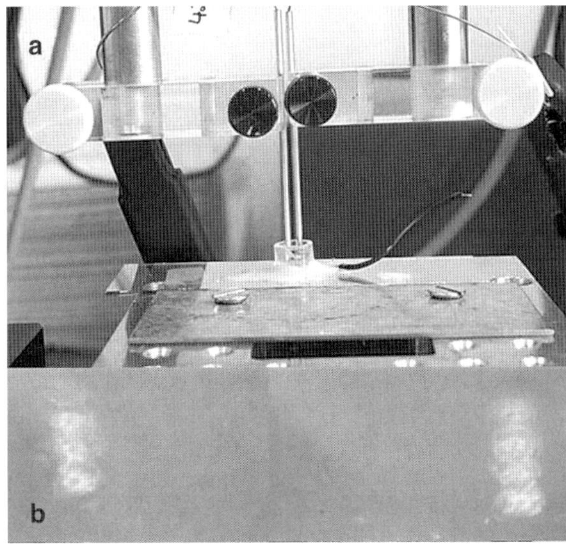

Fig. 14.29 Simultaneous deposition of enzyme-containing polymer spots using two micro-counter electrodes (**a**) photograph of the set-up, (**b**) photograph of parallel generated enzyme structures

toward a nonconducting surface during the course of an SECM experiment, diffusion of that redox species is blocked in the near-field interaction region of the surface. Consequently, the current response decreases the closer the tip of the microelectrode comes to the surface. The "feedback mode" is used for positioning in SECM. This feature was applied for a measuring situation in the electrochemical robotic system. The main objective of the study was to detect of nitric oxide (NO) released from biological cells.[69] To do this accurately, a variety of aspects have to be considered for the assay design. Among them was the fact that a gradient of NO is building up over the cell layer while NO is generated within these cells and diffusing through their membranes. The measuring situation in a well of a microtiter plate is schematically visualized in Fig. 14.30. A precise determination of NO within the gradient needs a known z-distance of the disk-shaped NO-sensor relative to the cells. This was achieved by using a dual-barrel microelectrode consisting of the NO sensor and a distance sensor (Fig. 14.30b). This approach was shown to be feasible for SECM applications by Isik et al.[68] and was later adapted for the electrochemical robotic approach by Borgmann et al.[69]

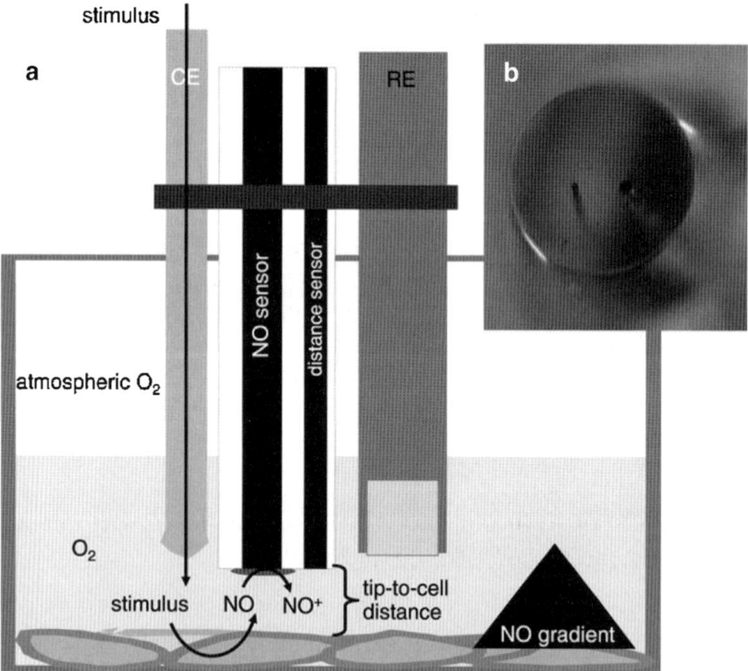

Fig. 14.30 (**a**) The electrode assembly, with the integrated distance-control sensor positioned within the wells of a microtiter plate, and the measuring situation, which indicates the complex NO gradient above the cell monolayer. (**b**) A double-barrel electrode. CE = counter electrode, RE = reference electrode.[69] Reproduced by permission of Wiley-VCH Verlag GmbH & Co KGaA)

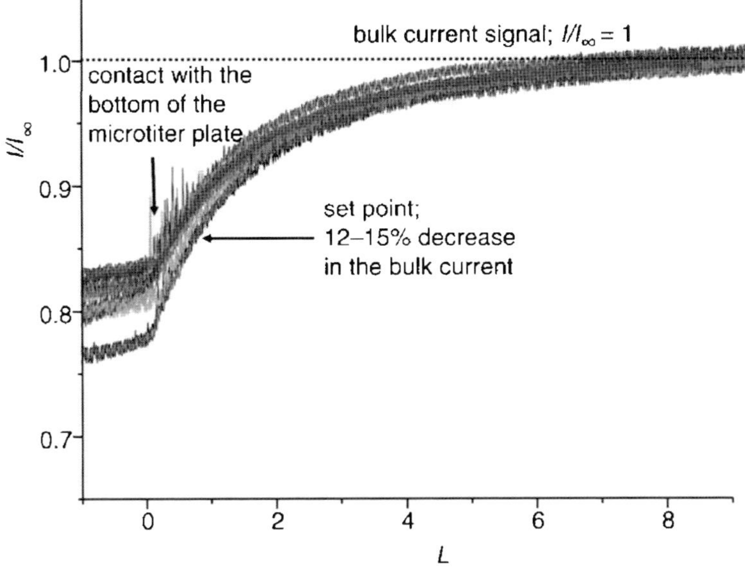

Fig. 14.31 16 normalized SECM approach curves recorded at −600 mV vs. Ag/AgCl with the distance sensor ($d = 25\,\mu m$) of a dual microelectrode within the wells of a 96-well microtiter plate filled with phosphate-buffered saline (PBS; $100\,\mu l$). The normalized distance, L, is expressed as the tip-to-sample distance divided by the electrode's radius.[69] Reproduced by permission of Wiley-VCH Verlag GmbH & Co KGaA)

Automated approach curves monitoring the reduction of any present oxygen in the buffer solution were used to control the distance of the NO sensor to the cells in the robotic system. Normalized SECM approach curves are shown in Fig. 14.31.

The use of dual microelectrodes and the implementation of an automated distance control in the electrochemical robotic system drastically improved the readout of the NO assays and made electrochemical high-content screening of released NO from endothelial cells possible.

3   Critical Evaluation of the Concept of an Electrochemical Robotic System and Conclusions

The throughput attained so far using parallel approaches and electrochemical robotics is significantly higher than in classical electrochemical experiments. However, the experiments are still rather slow compared with other HTS technologies. Much can be learned from other high-throughput approaches, but not everything is readily applicable for electrochemical robotics. There are limitations that are intrinsic to electrochemistry. Assay times are restricted by scan rates and time for calibration of sensors or standards for potential references. Smaller dimensions result in a lot of beneficial effects such as using microelectrodes,[86] but moving from a macro to a microelectrode results in a change in the diffusional behavior that may make comparisons between manual studies (mainly macroelectrodes) and automated studies (microelectrodes) more complex than expected. A much higher degree of parallelization (up to 96 three electrode systems) would improve the throughput significantly, but is challenging due to precise production of suitable electrode systems, electrode connection and handling, as well as obtaining a "real" 96 multi-potentiostat avoiding multiplexing.

Aside from technological solutions for the device itself, data processing and interpretation are challenging bottlenecks with respect to further increase of the throughput. Electrochemical data contain a high degree of information such as redox potentials and real-time data on processes monitored by redox species. To condense this data, information about redox potentials needs to be extracted from CVs and DPVs and calibration plots generated from amperometric current-time recordings in an automated fashion. Current responses may be normalized to the actual electroactive surface by using reference CVs. Data need to be processed and displayed during the screening process and later averaged for statistical purposes. There is a need for more sophisticated data processing, analysis, and interpretation software. The implementation of electrochemical simulation software would be a valuable tool in this respect.

One driving force for the design of the electrochemical robotic system was to create an "all purpose" device. The instrumental platform can readily be modified to serve different experimental needs because of its modular design. However, the second generation of electrochemical robotic systems may be focused to more specialized instruments for dedicated applications as different as combinatorial chemistry and biological assays, for example, in order to increase throughput and content of knowledge obtained by these devices.

The electrochemical robotic system provides a broad range of applications that reaches from electrochemical synthesis of combinatorial libraries, over analysis of materials and sensor optimization to automated functional assays of biological cells. The flexibility in the instrument design and its throughput allows for performing the necessary control experiments to validate assay results and avoid experimental bias. For example, using a platform like the electrochemical robotic system should make rational biosensor and chemical sensor design and optimization feasible using the benefits of sensing chemistries provided by combinatorial chemistry. For example, using a polymer film as immobilization in a specific biosensor architectures requires investigation of the properties of the deposited polymer film such as density, layer thickness, hydrophilic or hydrophobic characteristics, swelling ability, pore size, etc. as well as the polymer properties themselves because they highly influence the final sensor characteristics. To actually address the variety of parameters (Fig. 14.32) that needs to be taken into account during sensor development, a huge number of experiments have to be performed during their optimization.

This is the main reason why combinatorial libraries have not been extensively used as basis for biosensor optimization in the past. Classically, biosensor optimization

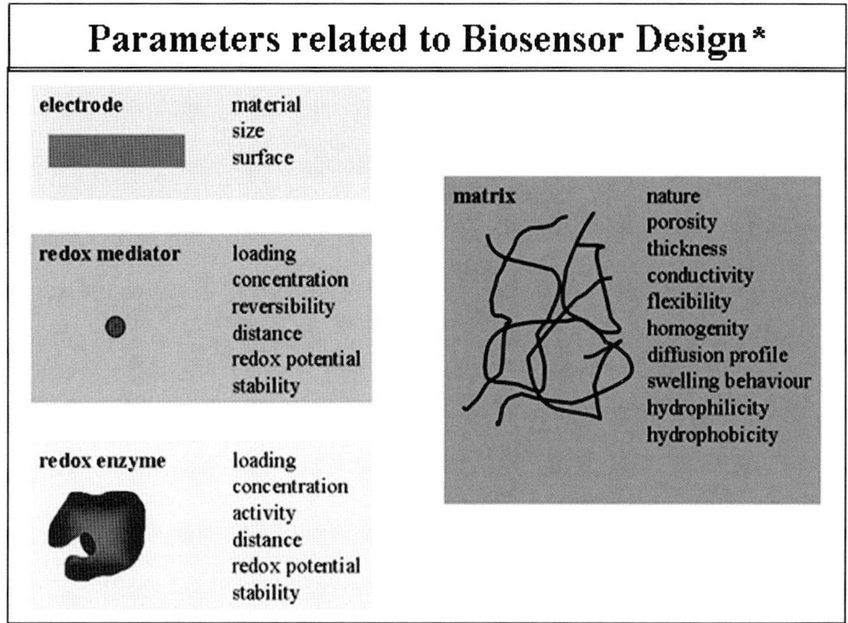

Fig. 14.32 Potential key parameters for biosensor optimization. Reprinted from Borgmann[44]. Copyright Elsevier B.V. (2005)

is done by optimizing one (assumed to be linearly independent) parameter assuming that the others are constant. But actually this assumption is not true for complex biosensor architectures. The electrochemical robotic system could come into play in providing a flexible platform that enables performing rational screening strategies for optimizing biosensor structures as an example illustrates in Fig. 14.33. In general, the main aim of biosensor development is to enable a fast electron transfer cascade from the biological recognition element to the transducer surface. Achieving this requires optimizing each individual component of the sensor architecture for a specific application.

We like emphasize that "blind" use of combinatorial approaches is not likely to provide desired results. Even though combinatorial chemistry was extensively applied in pharmaceutical industry, the number of generated and approved drug

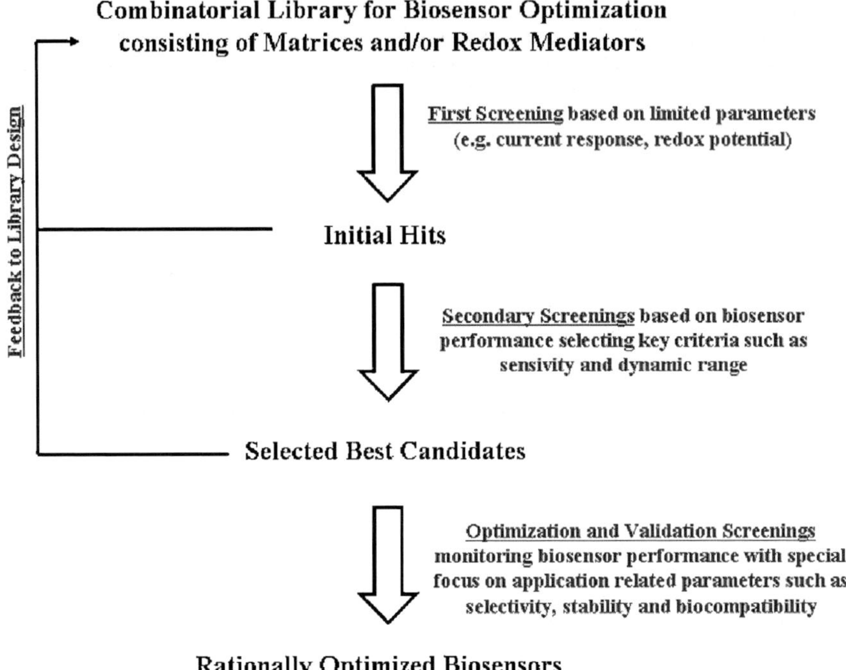

Fig. 14.33 Rational strategy to optimize biosensors in an automated approach using e.g. the electrochemical robotic system

candidates is rather limited compared with the efforts made. But we believe that a rational use (hypothesis-driven, for example) of the approaches and technologies available and still to be developed will provide very vital tools to work on and solve numerous problems. The quality of knowledge that we can gain from combinatorial approaches and other high-throughput methodologies will be only as precise and correct as our "posed questions", namely the particular screens, are. Therefore, we believe that electrochemical combinatorial approaches have a future because of the high content of information generated. The electrochemical robotic system as instrument platform thus shows great potential to be implemented in standard applications of combinatorial chemistry in research and industry.

Acknowledgments Part of the work was financially supported by the DFG (Schu 929/4-1, Ju 103/11-1,Sp265/17-1)intheframeworkofthecooperativeproject"KombinatorischeMikroelektrochemie". The authors appreciate in particular the cooperation with the research groups of Prof. Dr. Bernd Speiser and Prof. Dr. G. Jung in connection with this project. Part of the described work was financially supported by the European Commission (QLK3-2001-00244; HPRN-CT-2002-00186), the BMBF (FKZ-0312006D; FKZ-13N8607), the DFG (SFB459A5; Schu929/6-1). We are especially grateful for the conceptual contributions to the electrochemical robotic system from Dr. Thomas Erichsen (Sensolytics, GmbH). We highly appreciate the contributions from the mechanical workshop of the faculty of chemistry at the Ruhr-Universität Bochum. We thank the following students and staff for their input into projects related to the electrochemical robotic systems:

Sandra Janiak, Kathrin Eckhard, Sebastian Neugebauer, Dirk Ruhlig, Aiste Vilkanauskyte, Sonnur Isik, Ayodele O. Okunola, Christian Leson, Halyna Shkil, Eva M. Bonsen (Ruhr-Universität Bochum) and Wolfgang Märkle, Carsten Tittel (Universität Tübingen). The authors appreciate the careful reading of the manuscript by Dr. Melissa A. Jones.

References

1. Hanak, J. J., Multiple-sample-concept in materials research – synthesis, compositional analysis and testing of entire multicomponent systems, *J. Mater. Sci.* **1970**, *5*, 964–971
2. Sundberg, S. A., High-throughput and ultra-high-throughput screening: Solution- and cell-based approaches, *Curr. Opin. Biotechnol.* **2000**, *11*, 47–53
3. Bevan, C. D.; Lloyd, R. S., A high-throughput screening method for the determination of aqueous drug solubility using laser nephelometry in microtiter plates, *Anal. Chem.* **2000**, *72*, 1781–1787
4. Jung, G. *Combinatorial chemistry*; Wiley-VCH: Weinheim, **1999**
5. Hoffmann, R., Not a library, *Angew. Chem.-Int. Edit.* **2001**, *40*, 3337–3340
6. Stieber, F.; Grether, U.; Waldmann, H., An oxidation-labile traceless linker for solid-phase synthesis, *Angew. Chem.-Int. Edit.* **1999**, *38*, 1073–1077
7. Coates, W. J.; Hunter, D. J.; MacLachlan, W. S., Successful implementation of automation in medicinal chemistry, *Drug Discov. Today* **2000**, *5*, 521–527
8. Merrifield, R. B., Solid phase peptide synthesis. 1. Synthesis of a tetrapeptide, *J. Am. Chem. Soc.* **1963**, *85*, 2149–2154
9. Speiser, B., From cyclic voltammetry to scanning electrochemical microscopy: Modern electroanalytical methods to study organic compounds, materials, and reactions, *Curr. Org. Chem.* **1999**, *3*, 171–191
10. Lund, H.; Hammerich, O. *Organic electrochemistry*; 4th ed.; Marcel Dekker: New York, NY, 2001
11. Steckhan, E., *Laboratory techniques in electroanalytical chemistry*; Marcel Dekker: New York, **1996**, pp 641–681
12. Schultze, J. W.; Bressel, A., Principles of electrochemical micro- and nano-system technologies, *Electrochim. Acta* **2001**, *47*, 3–21
13. Evans, U.; Colavita, P. E.; Doescher, M. S.; Schiza, M.; Myrick, M. L., Construction and characterization of a nanowell electrode array, *Nano Lett.* **2002**, *2*, 641–645
14. Shono, T. *Electroorganic chemistry as a new tool in organic synthesis* Springer: New York, 1984
15. Weinberg, N. L. *Technique of electroorganic synthesis*; Wiley: New York, NY, **1974–1982**; Vol. 1–3
16. Reddington, E.; Sapienza, A.; Gurau, B.; Viswanathan, R.; Sarangapani, S.; Smotkin, E. S.; Mallouk, T. E., Combinatorial electrochemistry: A highly parallel, optical screening method for discovery of better electrocatalysts, *Science* **1998**, *280*, 1735–1737
17. Sullivan, M. G.; Utomo, H.; Fagan, P. J.; Ward, M. D., Automated electrochemical analysis with combinatorial electrode arrays, *Anal. Chem.* **1999**, *71*, 4369–4375
18. Feeney, R.; Kounaves, S. P., Microfabricated ultramicroelectrode arrays: Developments, advances, and applications in environmental analysis, *Electroanalysis* **2000**, *12*, 677–684
19. Jiang, R. Z.; Chu, D., A combinatorial approach toward electrochemical analysis, *J. Electroanal. Chem.* **2002**, *527*, 137–142
20. Sun, Y. P.; Buck, H.; Mallouk, T. E., Combinatorial discovery of alloy electrocatalysts for amperometric glucose sensors, *Anal. Chem.* **2001**, *73*, 1599–1604
21. Mallouk, T. E.; Smotkin, E. S., Combinatorial catalyst development methods (chapter 23), In *Handbook of fuel cells – fundamentals, technology and application*; W. Vielstich; A. Lamm and H. A. Gasteiger, Ed.; Wiley, New York, NY, 2003; Vol. Volume 2: Electrocatalysis

22. Siu, T.; Li, W.; Yudin, A. K., Parallel electrosynthesis of alpha-alkoxycarbamates alpha-alkoxyamides, and alpha-alkoxysulfonamides using the spatially addressable electrolysis platform (saep), *J. Comb. Chem.* **2000**, *2*, 545–549

23. Siu, T.; Li, W.; Yudin, A. K., Parallel electrosynthesis of 1,2-diamines, *J. Comb. Chem.* **2001**, *3*, 554–558

24. Yudin, A. K.; Siu, T., Combinatorial electrochemistry, *Curr. Opin. Chem. Biol.* **2001**, *5*, 269–272

25. Siu, T.; Yudin, A. K., Practical olefin aziridination with a broad substrate scope, *J. Am. Chem. Soc.* **2002**, *124*, 530–531

26. Siu, T.; Yekta, S.; Yudin, A. K., New approach to rapid generation and screening of diverse catalytic materials on electrode surfaces, *J. Am. Chem. Soc.* **2000**, *122*, 11787–11790

27. Suga, S.; Okajima, M.; Fujiwara, K.; Yoshida, J., "Cation flow" method: a new approach to electrochemical conventional and combinatorial organic syntheses using electrochemical microflow systems, *J. Am. Chem. Soc.* **2001**, *123*, 7941–7942

28. Kulikov, V.; Mirsky, V. M., Equipment for combinatorial electrochemical polymerization and high-throughput investigation of electrical properties of the synthesized polymers, *Meas. Sci. Technol.* **2004**, *15*, 49–54

29. Pilard, J. F.; Marchand, G.; Simonet, J., Chemical synthesis at solid interfaces. On the use of conducting polythiophenes equipped of adequate linkers allowing a facile and highly selective cathodic s-n bond scission with a fully regenerating resin process, *Tetrahedron* **1998**, *54*, 9401–9414

30. Marchand, G.; Pilard, J. F.; Simonet, J., Solid phase chemistry at a modified cathode surface. First synthesis of a polyamine precursor, *Tetrahedron Lett.* **2000**, *41*, 883–885

31. Nad, S.; Breinbauer, R., Electroorganic synthesis on the solid phase using polymer beads as supports, *Angew. Chem.-Int. Edit.* **2004**, *43*, 2297–2299

32. Tesfu, E.; Maurer, K.; Ragsdale, S. R.; Moeller, K. D., Building addressable libraries: The use of electrochemistry for generating reactive pd(ii) reagents at preselected sites on a chip, *J. Am. Chem. Soc.* **2004**, *126*, 6212–6213

33. Tian, J.; Maurer, K.; Tesfu, E.; Moeller, K. D., Building addressable libraries: The use of electrochemistry for spatially isolating a heck reaction on a chip, *J. Am. Chem. Soc.* **2005**, *127*, 1392–1393

34. Tesfu, E.; Maurer, K.; Moeller, K. D., Building addressable libraries: Spatially isolated, chip-based reductive amination reactions, *J. Am. Chem. Soc.* **2006**, *128*, 70–71

35. Tesfu, E.; Roth, K.; Maurer, K.; Moeller, K. D., Buillding addressable libraries: Site selective coumarin synthesis and the "Real-time" Signaling of antiibody-coumarin binding, *Org. Lett.* **2006**, *8*, 709–712

36. Nagl, S.; Schaeferling, M.; Wolfbeis, O. S., Fluorescence analysis in microarray technology, *Microchim. Acta* **2005**, *151*, 1–21

37. Kuswandi, B.; Tombelli, S.; Marazza, G.; Mascini, M., Recent advances in optical DNA biosensors technology, *Chimia* **2005**, *59*, 236–242

38. Petrik, J., Diagnostic applications of microarrays, *Transfus. Med.* 2006, *16*, 233–247

39. Vercoutere, W.; Akeson, M., Biosensors for DNA sequence detection, *Curr. Opin. Chem. Biol.* 2002, *6*, 816–822

40. Wang, J., Portable electrochemical systems, *Trends Anal. Chem.* **2002**, *21*, 226–232

41. Albers, J.; Grunwald, T.; Nebling, E.; Piechotta, G.; Hintsche, R., Electrical biochip technology – a tool for microarrays and continuous monitoring, *Anal. Bioanal. Chem.* 2003, *377*, 521–527

42. Kricka, L. J.; Park, J. Y.; Li, S. F. Y.; Fortina, P., Miniaturized detection technology in molecular diagnostics, *Expert Rev. Mol. Diagn.* **2005**, *5*, 549–559

43. Lagally, E. T.; Soh, H. T., Integrated genetic analysis microsystems, *Crit. Rev. Solid State Mat. Sci.* **2005**, *30*, 207–233

44. Borgmann, S. PhD thesis Thesis, Ruhr-Universität Bochum, 2005

45. Campas, M.; Katakis, I., Electrochemically arrayed and addressed DNA multi-sensor platforms, *Sens. Actuator B-Chem.* **2006**, *114*, 897–902

46. Andreu, A.; Merkert, J. W.; Lecaros, L. A.; Broglin, B. L.; Brazell, J. T.; El-Kouedi, M., Detection of DNA oligonucleotides on nanowire array electrodes using chronocoulometry, *Sens. Actuator B-Chem.* **2006**, *114*, 1116–1120

47. Elsholz, B.; Worl, R.; Blohm, L.; Albers, J.; Feucht, H.; Grunwald, T.; Jurgen, B.; Schweder, T.; Hintsche, R., Automated detection and quantitation of bacterial RNA by using electrical microarrays, *Anal. Chem.* **2006**, *78*, 4794–4802

48. Turcu, F.; Schulte, A.; Hartwich, G.; Schuhmann, W., Label-free electrochemical recognition of DNA hybridization by means of modulation of the feedback current in SECM, *Angew. Chem.-Int. Edit.* **2004**, *43*, 3482–3485

49. Mir, M.; Katakis, I., Towards a fast-responding, label-free electrochemical DNA biosensor, *Anal. Bioanal. Chem.* **2005**, *381*, 1033–1035

50. Ghindilis, A. L.; Smith, M. W.; Schwarzkopf, K. R.; Roth, K. M.; Peyvan, K.; Munro, S. B.; Lodes, M. J.; Stöver, A. G.; Bernards, K.; Dill, K.; McShea, A., Combimatrix oligonucleotide arrays: Genotyping and gene expression assays employing electrochemical detection, *Biosens. Bioelectron.* **2007**, 22, 1853–1860

51. Baldwin, R. P., Recent advances in electrochemical detection in capillary electrophoresis, *Electrophoresis* **2000**, *21*, 4017–4028

52. Yi, C. Q.; Zhang, Q.; Li, C. W.; Yang, J.; Zhao, J. L.; Yang, M. S., Optical and electrochemical detection techniques for cell-based microfluidic systems, *Anal. Bioanal. Chem.* **2006**, *384*, 1259–1268

53. Ruzicka, J., From beaker chemistry to programmable microfluidics, *Collect. Czech. Chem. Commun.* **2005**, *70*, 1737–1755

54. Le Gac, S.; Rolando, C., Microsystems in chemistry, *Actual Chim.* **2002**, *2*, 21–32

55. Stefan, R. I.; van Staden, J. K. F.; Aboul-Enein, H. Y., Design and use of electrochemical sensors in enantioselective high throughput screening of drugs. A minireview, *Comb. Chem. High Throughput Screen* **2000**, *3*, 445–454

56. Guerin, S.; Hayden, B. E.; Pletcher, D.; Rendall, M. E.; Suchsland, J. P., A combinatorial approach to the study of particle size effects on supported electrocatalysts: Oxygen reduction on gold, *J. Comb. Chem.* **2006**, *8*, 679–686

57. Guerin, S.; Hayden, B. E., Physical vapor deposition method for the high-throughput synthesis of solid-state material libraries, *J. Comb. Chem.* **2006**, *8*, 66–73

58. Guerin, S.; Hayden, B. E.; Pletcher, D.; Rendall, M. E.; Suchsland, J. P.; Williams, L. J., Combinatorial approach to the study of particle size effects in electrocatalysis: Synthesis of supported gold nanoparticles, *J. Comb. Chem.* **2006**, *8*, 791–798

59. Reiter, S.; Eckhard, K.; Blochl, A.; Schuhmann, W., Redox modification of proteins using sequential-parallel electrochemistry in microtiter plates, *Analyst* **2001**, *126*, 1912–1918

60. Tang, T. C.; Deng, A. P.; Huang, H. J., Immunoassay with a microtiter plate incorporated multichannel electrochemical detection system, *Anal. Chem.* **2002**, *74*, 2617–2621

61. Jiang, R. Z.; Rong, C.; Chu, D., Combinatorial approach toward high-throughput analysis of direct methanol fuel cells, *J. Comb. Chem.* **2005**, *7*, 272–278

62. Jiang, R. Z.; Chu, D. R., An electrode probe for high-throughput screening of electrochemical libraries, *Rev. Sci. Instrum.* **2005**, *76*, Art. No. 062213

63. Andreescu, S.; Sadik, O. A.; McGee, D. W., Autonomous multielectrode system for monitoring the interactions of isoflavonoids with lung cancer cells, *Anal. Chem.* **2004**, *76*, 2321–2330

64. Andreescu, S.; Sadik, O. A., Advanced electrochemical sensors for cell cancer monitoring, *Methods* **2005**, *37*, 84–93

65. Karasinski, J.; Andreescu, S.; Sadik, O. A.; Lavine, B.; Vora, M. N., Multiarray sensors with pattern recognition for the detection, classification, and differentiation of bacteria at subspecies and strain levels, *Anal. Chem.* **2005**, *77*, 7941–7949

66. Erichsen, T.; Reiter, S.; Ryabova, V.; Bonsen, E. M.; Schuhmann, W.; Markle, W.; Tittel, C.; Jung, G.; Speiser, B., Combinatorial microelectrochemistry: Development and evaluation of an electrochemical robotic system, *Rev. Sci. Instrum.* **2005**, *76*, Art. No. 062204

67. Briehn, C. A.; Schiedel, M. S.; Bonsen, E. M.; Schuhmann, W.; Bauerle, P., Single-compound libraries of organic materials: From the combinatorial synthesis of conjugated oligomers to structure-property relationships, *Angew. Chem.-Int. Edit.* **2001**, *40*, 4680–4683

68. Isik, S.; Etienne, M.; Oni, J.; Blochl, A.; Reiter, S.; Schuhmann, W., Dual microelectrodes for distance control and detection of nitric oxide from endothelial cells by means of scanning electrochemical microscope, *Anal. Chem.* **2004**, *76*, 6389–6394

69. Borgmann, S.; Radtke, I.; Erichsen, T.; Blochl, A.; Heumann, R.; Schuhmann, W., Electrochemical high-content screening of nitric oxide release from endothelial cells, *ChemBioChem* **2006**, *7*, 662–668

70. Briehn, C. A.; Bauerle, P., From solid-phase synthesis of pi-conjugated oligomers to combinatorial library construction and screening, *Chem. Commun.* **2002**, 1015–1023

71. Lindner, E.; Lu, Z. L.; Mayer, H. A.; Speiser, B.; Tittel, C.; Warad, I., Combinatorial micro electrochemistry. Part 4: Cyclic voltammetric redox screening of homogeneous ruthenium(ii) hydrogenation catalysts, *Electrochem. Commun.* **2005**, *7*, 1013–1020

72. Markle, W.; Speiser, B.; Tittel, C.; Vollmer, M., Combinatorial micro electrochemistry – part 1. Automated micro electrosynthesis of iminoquinol ether and [1,2,4]triazolo[4,3-alpha] pyridinium perchlorate collections in the wells of microtiter plates, *Electrochim. Acta* **2005**, *50*, 2753–2762

73. Markle, W.; Speiser, B., Combinatorial microelectrochemistry – part 3. On-line monitoring of electrolyses by steady-state cyclic voltammetry at microelectrodes, *Electrochim. Acta* **2005**, *50*, 4916–4925

74. Hintsche, R.; Albers, J.; Bernt, H.; Eder, A. E., Multiplexing of microelectrode arrays in voltammetric measurements, *Electroanalysis* **2000**, *12*, 660–665

75. Ryabova, V.; Schulte, A.; Erichsen, T.; Schuhmann, W., Robotic sequential analysis of a library of metalloporphyrins as electrocatalysts for voltammetric nitric oxide sensors, *Analyst* **2005**, *130*, 1245–1252

76. Reiter, S.; Ruhlig, D.; Ngounou, B.; Neugebauer, S.; Janiak, S.; Vilkanauskyte, A.; Erichsen, T.; Schuhmann, W., An electrochemical robotic system for the optimization of amperometric glucose biosensors based on a library of cathodic electrodeposition paints, *Macromol. Rapid Commun.* **2004**, *25*, 348–354

77. Schuhmann, W., Conducting polymer based amperometric enzyme electrodes, *Mikrochim. Acta* **1995**, *121*, 1–29

78. Ruhlig, D.; Schulte, A.; Schuhmann, W., An electrochemical robotic system for routine cathodic adsorptive stripping analysis of ni2+ ion release from corroding niti shape memory alloys, *Electroanalysis* **2006**, *18*, 53–58

79. Bard, A. J.; Fan, F. R. F.; Kwak, J.; Lev, O., Scanning electrochemical microscopy – introduction and principles, *Anal. Chem.* **1989**, *61*, 132–138

80. Mirkin, M. V.; Horrocks, B. R., Electroanalytical measurements using the scanning electrochemical microscope, *Anal. Chim. Acta* **2000**, *406*, 119–146

81. Barker, A. L.; Gonsalves, M.; Macpherson, J. V.; Slevin, C. J.; Unwin, P. R., Scanning electrochemical microscopy: Beyond the solid/liquid interface, *Anal. Chim. Acta* **1999**, *385*, 223–240

82. Kurzawa, C.; Hengstenberg, A.; Schuhmann, W., Immobilization method for the preparation of biosensors based on ph shift-induced deposition of biomolecule-containing polymer films, *Anal. Chem.* **2002**, *74*, 355–361

83. Smutok, O.; Ngounou, B.; Pavlishko, H.; Gayda, G.; Gonchar, M.; Schuhmann, W., A reagentless bienzyme amperometric biosensor based on alcohol oxidase/peroxidase and an os-complex modified electrodeposition paint, *Sens. Actuator B-Chem.* **2006**, *113*, 590–598

84. Kwak, J.; Bard, A. J., Scanning electrochemical microscopy – theory of the feedback mode, *Anal. Chem.* **1989**, *61*, 1221–1227

85. Engstrom, R. C.; Pharr, C. M., Scanning electrochemical microscopy, *Anal. Chem.* **1989**, *61*, 1099–1104

86. Heinze, J., Ultramicroelectrodes – a new dimension in electrochemistry, *Angew. Chem.-Int. Edit. Engl.* **1991**, *30*, 170–171

Section 7
Optical Sensing Materials

Chapter 15
Combinatorial Chemistry for Optical Sensing Applications

M.E. Díaz-García, G. Pina Luis, and I.A. Rivero-Espejel

Abstract The recent interest in combinatorial chemistry for the synthesis of selective recognition materials for optical sensing applications is presented. The preparation, screening, and applications of libraries of ligands and chemosensors against molecular species and metal ions are first considered. Included in this chapter are also the developments involving applications of combinatorial approaches to the discovery of sol–gel and acrylic-based imprinted materials for optical sensing of antibiotics and pesticides, as well as libraries of doped sol–gels for high-throughput optical sensing of oxygen. The potential of combinatorial chemistry applied to the discovery of new sensing materials is highlighted.

1 Introduction

Sensor technology is constantly increasing the range of applications for the control of technical and industrial processes, in environmental monitoring, health care, food analysis and technology, in automobile and many other industries.[1,2] In spite that diversity of materials with enhanced properties for sensing are possible, in analytical chemistry the conventional search for sensing materials is based on the optimization and/or modification of known materials or molecules. As a result, considerable time is consumed by trials for the sensing material with adequate transduction and recognition properties, for the best reaction conditions and concentrations of reactants, etc. Consequently, there is a strong need for a more rational approach to make the discovery of new sensing materials more efficient.

Combinatorial chemistry holds the potential to fill this gap, considering the fact that combinatorial technologies have led to a revolution in the design and discovery of novel molecules and materials, particularly in the pharmaceutical industry.[3,4] In comparison to the area of drug discovery and optimization where combinatorial

M.E. Díaz-García (✉), G.P. Luis, and I.A. Rivero-Espejel
University of Oviedo, Department of Physical and Analytical Chemistry, Spain
medg@uniovi.es

R.A. Potyrailo and V.M. Mirsky (eds.), *Combinatorial Methods for Chemical and Biological Sensors*, DOI: 10.1007/978-0-387-73713-3_15,

chemistry has already established itself as an invaluable tool, current combinatorial approaches to the development of sensing materials for analytical purposes are not as widespread and are less mature. The challenge is posed by the need for general and reliable screening methodologies that allow for the facile identification of responsive library members within a large library of potential sensing materials and the determination of their molecular structure. Recent advances demonstrate that combinatorial chemistry is in the course of becoming a powerful tool for sensing materials development.[5]

This chapter highlights examples of emerging applications of combinatorial approaches in analytical chemistry, the advantages, and drawbacks. As we have reviewed this general area recently,[6] to maintain a reasonable scope, we focus upon methodologies that involve development of materials with molecular recognition properties for optical sensing.

2   Combinatorial Libraries

Combinatorial synthesis and library generation may take place in a number of ways. Libraries can be roughly grouped into two main classes: pool libraries and parallel synthesis. Pool libraries are mainly produced on solid phase with the mix and split technology pioneered by Furka et al.[7,8] In this approach, the solid supported starting material is split into a number of equal portions that are each treated separately with a different reagent. After all the reactions are completed and excess of reagents have been removed, the individual samples are then pooled and mixed before splitting again into the desired number of portions for the second chemical step, each portion containing now a mixture of compounds. The whole process is repeated several times to construct the library. These libraries offer an attractive source of molecules for sensing, but the complexity of the mixtures presents to the analytical chemistry several challenges: continuous monitoring of reactions carried out on the solid support, determination of the structure of the compound on a single isolated bead from about 10^6 structural possibilities with only 300–500 pmol of material per bead and finally, identification of the sensing active members of these libraries. Because of the lack of on-support analytical quantitative methods, the synthetic intermediates and final products are not quantitatively characterized on support, and it is assumed that the excess of reagent and the prolonged reaction time will drive the reaction to completion.

In parallel synthesis, both solution and solid phase techniques are applied. The individual members of the library are prepared in predetermined separate reaction containers. The generated molecular diversity is then somewhat limited but each library member has a clearly defined location and can be synthesized in amounts large enough to allow for the analysis of its purity by conventional analytical techniques such as NMR-spectroscopy, HPLC, IR-spectroscopy, GC, etc. Besides, screening of the library members for a desired optical sensing activity may be performed individually in the separate compartments. In fact, synthesis and screening of a parallel library can be performed in an automated high-throughput format (e.g., if 96-well microtiter plates are used as containers, reactions can be monitored by fluorescence or absorbance measurements). So, it is not surprising that development of sensing materials by combinatorial techniques took advantage of this methodology.

In contrast to the static formats of combinatorial chemistry, the library can be generated from a set of reversibly interchanging components, introducing thus dynamics into the system. Dynamic combinatorial chemistry relies on reversible interchange between sets of basic components to generate continually interconnecting adducts, giving access to dynamic combinatorial libraries containing all possible combinations of the components in equilibrium.[9] In comparison to other type of libraries, dynamic libraries can be easily augmented, can be easily screened, and the resulting members identified.[9,10] To date, dynamic combinatorial libraries constitute an unexplored field for analytical applications.

3   Solid Supports

An ideal solid support for combinatorial synthesis should fulfill some requirements: chemical, thermal, and mechanical stability, facility of functionalization, good swelling properties, and low price. Three popular supports in solid phase organic synthesis are: (a) 1%-divinylbenzene (DVB)-based resins, copolymerized polystyrene resins (PS resins), (b) polystyrene-based resins extensively grafted with polyethyleneglycol linkers (PS–PEG resins), and (c) the surface-modified polyethylene pins. The degree of cross-linking greatly affects not only the ability of reactive sites to interact (loading, diffusion) but also the physical properties of the resin (e.g., swelling, rigidity).[11] In addition to the above factors, a key practical consideration in choosing a solid support is the linker. A linker is a specialized protecting group that covalently links molecules to the resin and provides the way for their chemical attachment and cleavage. The success the solid-phase synthesis depends on the stability of the linker, which must not be affected by the chemistry used. Also important are the loading capacity of the support (high loadings allow isolation of a significant amount of product from a single bead) and the bead size that affects the reaction rate (diffusion efficiency). From the optical sensing point of view, the solid support should be optically transparent to avoid spectral interferences and/or a high background signal. A review covering resins of general interest in the solid-phase synthesis of peptides and small organic molecules has been recently published.[12]

4   Library Sensing Efficiency

Characterization of combinatorial libraries (product structure determination) can be performed by direct analytical methods, while compounds are still attached to the beads. Analytical techniques such as NMR and MAS-NMR,[13,14] ESI-MS and HPLC-ESI-MS,[15–17] and FTIR spectroscopy[18–20] fulfill the analytical requirements for the identification of resin-bound substrates. Alternatively, off-bead analysis can be performed by cleavage of the compounds, using also UV–VIS spectrophotometric and fluorimetric techniques. Extension of these methodologies to an automated format would significantly enhance throughput, making the methods more attractive for scanning a large number of reactions in a parallel format.[21]

Library screening for sensing activity of synthetic receptor molecules (e.g., sensor molecules) can be performed by mixing the beads with a solution assay of the target analyte, and the binding event is distinguished by their color or luminescence changes. So, for example, Y.-T. Chang[22] synthesized a solid-phase parallel library of 240 rhodamine compounds. The diversity was addressed by using 12 different xhantone groups (containing oxygen, nitrogen, or sulfur bridges) and 33 different substituent groups (attached to the central ring). The compounds were cleaved from the resin by acidic hydrolysis, and their characterization was performed by HPLC-MS. The library was screened by contacting the fluorescent molecules not only with analytes of biological concern such as reduced glutathione (GSH) but also with living cells. In physiological conditions (pH 7.4), at least one member of the library showed a marked fluorescence increase upon addition of GSH, and it was possible to sense directly on the beads changes in the cellular GSH level in living cells.

Libraries produced via split synthesis can also be analyzed for sensing activity by cleaving the receptor ligands into solution and screening as soluble pools. Since a number of compounds may be active in a pool, deconvolution is performed by an iterative process of screening and resynthesis of smaller sublibraries of possible candidates in increasingly smaller sublibraries, in an attempt to fractionate a mixture into its most analytical useful recognition molecule(s). Deconvolution is one of the major challenges of combinatorial chemistry as the process is cumbersome and time consuming. To speed up this process, ingenious methods have been developed by preparing encoded libraries, where each bead may be recorded by the cosynthesis of a readily analyzable surrogate marker. Excellent reviews on analytical tools for screening combinatorial libraries have been recently published.[23,24]

5   Fluorescent Sensor Libraries

The development of new molecules with the ability to recognize and report the presence of a desired chemical species is a challenging problem from the basic aspects of synthesizing molecules with purposively-designed functions and analytical applications to real samples. The power of combinatorial chemistry and solid phase synthesis is now being utilized in the generation and/or optimization of molecules with sensing purposes. Its utility has been demonstrated in the preparation of combinatorial libraries of fluorescent chemosensors, not only for molecular species but also for metal ions. Conventionally, chemosensors are synthetic molecules able to bind selectively and reversibly the analyte of interest with a concomitant change in one or more properties of the system, such as absorption, luminescence, or redox potential.[25,26]

5.1 *Libraries for Sensing Small Molecules*

Taylor et al.[27] reported a carbohydrate chemosensor library of grafted polymers containing a saccharide recognition domain (boronic group) and a signaling group (anthracene as fluorophore) prepared by an iterative approach.[28] Although none of the prepared polymers interacted with glucose, addition of sialic acid or fructose resulted in an increase of the anthracene fluorescence, with some of the library members clearly distinguishing between the two carbohydrates. The results of this study demonstrated that molecular recognition of carbohydrates can be fine-tuned by screening libraries that contain a variable recognition domain.

Using the same concept but replacing the anthracene fluorophore with a dansyl unit as the signal transducer, a combinatorial library for fructose recognition, prepared by solid phase parallel synthesis, has also been explored.[29] Diversity was achieved by variations in dansyl building blocks (Fig. 15.1) and in the solid support used (Merrifield, ArgoPore and ArgoWang-Cl resins). Screening of the library members for sensing activity was performed using a dynamic system in which the beads were packed into a flow cell (Fig. 15.2). The library compounds exhibited a range of spectral diversities (excitation range from 345 to 400 nm and emission ranges from 395 to 500 nm) and results revealed that the signaling properties of the immobilized dansyl-boronic receptors did not parallel to those in solution,[30] displaying enhanced acidity upon binding of fructose. In flow-injection competitive binding experiments using the Merrifield-immobilized dansylamine-4-phenyl-boronic acid member, high selectivity toward fructose over other carbohydrates was observed.[31] In a typical experiment, the beads were saturated with a fluorescently-tagged sugar,

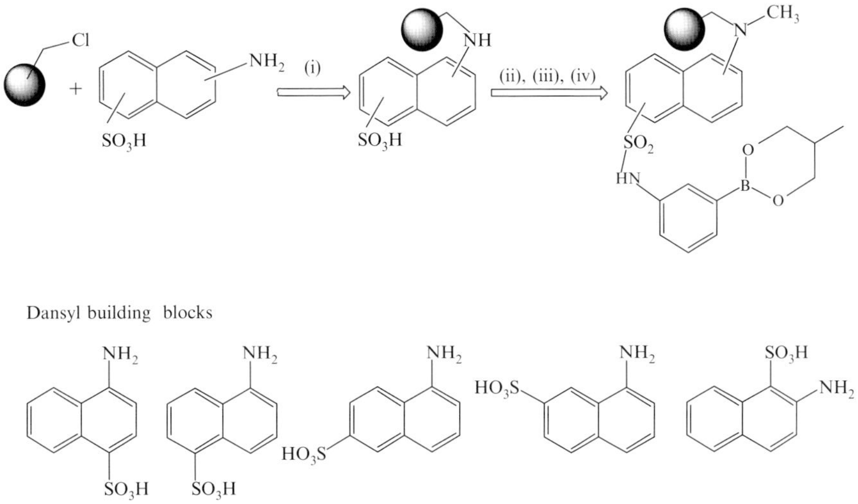

Fig. 15.1 Synthesis generating dansyl-boronic receptors for fructose and building blocks. (i) DMS/Cs$_2$CO$_3$; NH$_4$Cl/dioxane, (ii) CH$_3$I/K$_2$CO$_3$, (iii) PCl$_3$/THF, (iv) 2-methyl-1,3-propanediol-3-amine-phenyl boronate

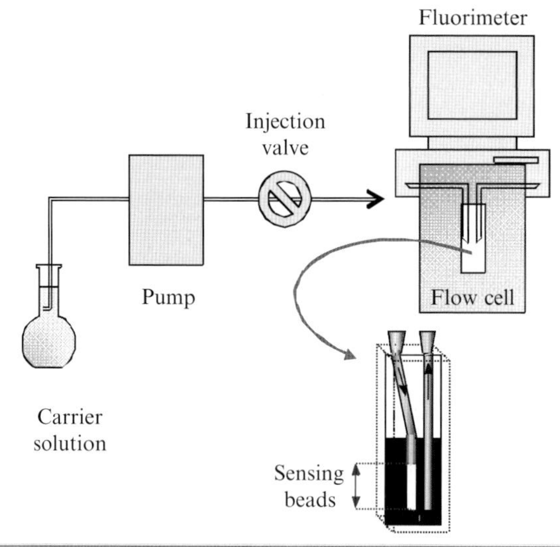

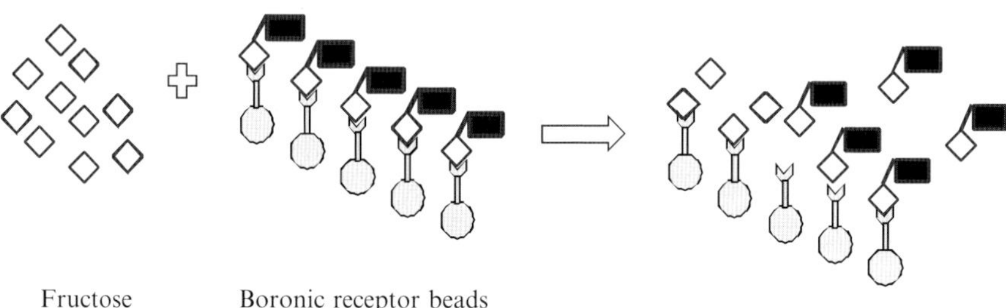

Fig. 15.2 Flow injection set-up for optical sensing and screening library members and competitive binding assay for fructose developed on-bead using fluorescent labeled glucose

4-methyl-umbellipheryl-β-D-glucose. Injection of fructose resulted in the displacement of the fluorescent-labeled sugar, thus resulting in an increase in the fluorescent conjugate concentration downstream (Fig. 15.2). A fluorescence peak was observed corresponding to the total amount of fructose injected. A detection limit of 6 mM was found for fructose, and the response was linear between the detection limit and 0.2 M fructose.[31]

More recently, Chang et al.[32] have reported the synthesis of a combinatorial benzimidazolium dye library and the discovery of the first turn-on fluorescent guanidine triphosphate (GTP) sensor reported thus far. A benzimidazolium scaffold with linker was loaded onto ethylenediamine derivatized 2-chlorotrityl polystyrene solid support. Structural diversity was introduced by different linker lengths and the coupling of aromatic aldehyde building blocks to the benzimidazolium ring on solid support. Final products were collected by acidic cleavage and screening was performed by contacting the products with GTP in 384-well microplates using a fluorescence plate reader. Of the 94-member library, two structurally related compounds showed dramatically increased fluorescence upon addition of GTP, while not responding to other nucleosides (adenosine, uridine, cytidine, guanosine) and nucleotides (XNP, where X = A, U, C, G, and N = mono, di, tri). As most of the known GTP sensors compete with adenine triphosphate, authors postulate that both the imidazolium and the 2-phenylindole moiety are important for selective GTP recognition in the discovered materials.

Peptides do not show the suitable optical properties to form luminescent or colorimetric complexes with metal ions. This problem can be overcome using an auxiliary dye to complex the metal ion. Buryak and Severin[33] described the preparation of a dynamic combinatorial library of dye complexes of Cu^{2+} and Ni^{2+} to sense the sequence isomeric tripeptides His-Gly-Gly, Gly-His-Gly, and Gly-Gly-His. The DCL of metal–dye complexes was generated by combination of three metal-binding dyes, Arsenazo I, Methylcalcein Blue, and Glycine Cresol Red with $CuCl_2$ and $NiCl_2$ in buffered aqueous solution. The addition of the peptide analytes resulted in characteristic changes in the UV–vis spectrum of the mixture, which allowed the identification of the peptide. Because of the modular nature of the sensor, it was possible to optimize the sensor by variation of the amounts and ratio of its constituent building blocks. The composition of the best sensor was found to vary substantially, depending on the sensing problem to be addressed. Although a Cu^{2+}/Ni^{2+} ratio of 1:3 gave the best differentiation for Gly-His-Gly and Gly-Gly-His, a sensor containing exclusively Cu^{2+} was found to provide the best discrimination between His-Gly-Gly and Gly-Gly-His. Authors claimed the simplicity of the approach, requiring only mixing of commercially available dyes with transition metal ions to differentiate aqueous solutions containing a variable ratio of the tripeptides Gly-His-Gly and Gly-Gly-His and to obtain quantitative information about these analytes.

5.2   *Libraries for Metal Ion Sensing*

The need of materials with the ability to sense and report in real-time and real-space the presence of metal ions is nowadays generally accepted by scientists, including environmentalists, chemists and clinical biochemists. Heavy metals contamination is recognized as a priority problem in environmental protection. Metal ions such as copper, lead, mercury, cadmium, etc. usually represent an environmental concern when present in uncontrolled and high concentrations. Calcium, magnesium, sodium, potassium are essential ions involved in biochemical processes such as regulation of cell activity, transmission of nerve impulses, or muscle contraction among others. We have reported the construction of a 17-member decapeptide sensor library for fluorescent heavy metal ions detection.[34] Diversity was achieved by variations in the decapeptide sequence (metallothionein mimics). Eu(III) ions were used as fluorescent reporters. The screening of the library for sensing heavy metal ions of environmental concern (Hg^{2+}, Pb^{2+}, Cu^{2+} and Cd^{2+}) was studied on-bead and in solution by fluorimetric techniques (confocal microscopy, fluorescence). The sensing response mechanism was based on an energy transfer process and a metal ion allosteric interaction (Fig. 15.3). Results demonstrated that combinatorial chemistry is a powerful technique for discovering chemosensor molecules, which in presence of Eu(III) ions hold promise for selective and sensitive signaling of toxic Hg(II) and Cd(II) ions. Linear calibration graphs were obtained with correlation coefficients $\geq$ 0,990 and detection limits below 300 nM for Hg(II) and Cd(II).

The metal ion receptor can also be used to provide an additional binding site or "anchor point" for the guest metal ion. So, Rivero et al.[35] have reported the parallel-synthesis of a 12-member library of materials for Cu(II) and Pb(II) determination using a phosphine sulfide unit as recognition-binding domain. Because of its strong luminescence and stability, the anthracene unit has been used as the light-emitting group and the phosphine sulfide group (recognition moiety) was attached to the 9-position on anthracene. By exploiting the selective metal-binding properties of the phosphine sulphide domain, we have sought to combine the advantageous aspects of both fluorosensing within selective metal immobilization (Fig. 15.4). The responsive materials showed quantitative retention properties, which could be improved with the use of ultrasonic bath agitation. Results demonstrated that the Wang-supported phosphine sulphides were superior for metal ion sorption than those using Merrifield, Argopore or Argogel supports. In fact, after equilibration, 89% Cu and 82% Pb were retained on the Wang-supported phosphine sulphide while ca. 40–50% retention was observed with the Merrifield resin.[36] The potential use of the materials for in situ immobilization and for fluorimetrically monitoring of toxic metals such as copper and lead was demonstrated by using a flow injection format.

Another example of immobilized chemosensors development is the automated parallel solid-phase synthesis of a 20-member library receptors for alkali-earth metal ions, recently described by our research group.[37] The library was based on the

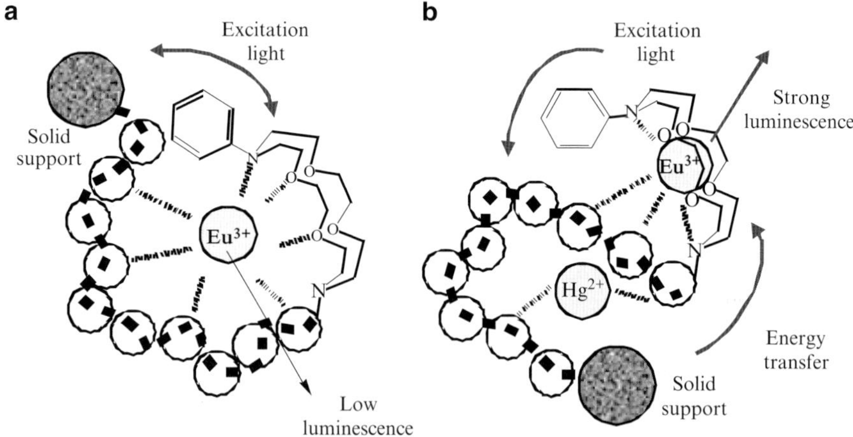

Fig. 15.3 (**a**) Intramolecular energy transfer luminescence of Eu(III) complexed with a Wang-decapeptide-Lariat ether library member; (**b**) metal ion allosteric interactions with the peptide chain induces a molecular reorganization followed by europium fluorescence enhancement. (from Rivero et al.,[34] with permission. Copyright 2006. Bentham Science Publishers)

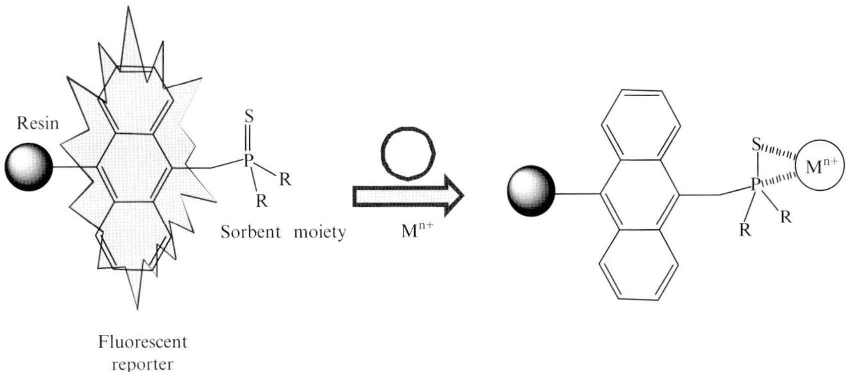

Fig. 15.4 Principle of on-bead sorption/recognition of a metal ion by a luminescent chemosensor

receptor 1,4,10,13-tetraoxa-7,15-diaza-cycloctadecane (Lariat ether) carrying a fluorescent dansyl group. The dansyl fluorophore was appropriately covalently linked so that bring it into proximity to the diazacrown receptor thus quenching its fluorescence by a photoelectron transfer (PET) process. Upon metal binding to the diazacrown, a conformational change was induced, which separated the dansyl fluorophore leading to an increase of fluorescence. The sensing properties of the library members for alkali and alkali earth metal ions were screened by packing the beads into a conventional flow-through cell in a flow injection approach. The reversibility and reproducibility of the responses were adequate enough to consider these simple systems as efficient fluorosensors for Mg^{2+}, Ca^{2+}, NH_4^+, Li^+, Na^+, and K^+ in a flow-injection mode.[36] In Fig. 15.5, the behavior of the library member 1,2,10,13-tetraoxa-7,16-diaza-cycloctadecane-dansyl supported on PS-trisamine (benzyl as the *N*-diazacrown pendant group) for different amounts of Mg^{2+} can be observed, being the calibration curve linear up to 1,000 mM Mg^{2+} ($r^2 = 0.995$).

Combinatorial chemistry will no doubt be a key part of constructing advanced receptor-luminophore systems suitable for sensing molecules of different nature and, particularly for metal ions of the *s* block and the *d* block. The large body of knowledge on the coordination chemistry of metals has been used as the starting point in the discovery of tuned receptors by combinatorial approaches not only for sensing but also for industrial and therapeutic applications.[38] In contrast to metal ions, most of the synthetic anion receptors are based on cationic polyammonium, guanidinium, quaternary ammonium or Lewis acidic receptors containing mercury, uranyl, boron, silicon and tin (anions themselves can behave as ligands toward metal ions). Combinatorial synthesis of optical receptors for single anions (e.g., chloride, fluoride, nitrate, cyanide, or phosphate) have yet to appear, taking into account that anions are also analytes of chemical, biological, or environmental concern. For example, cystic fibrosis is known to be caused by a failure in the exchange of chloride for HCO_3^- in erythrocytes[39]; thiocyanate is a detoxication product of cyanide and its content in human saliva is considered as a biomarker for identification of nonsmokers and smokers[40] and pollulants such as phosphate or nitrate from soil

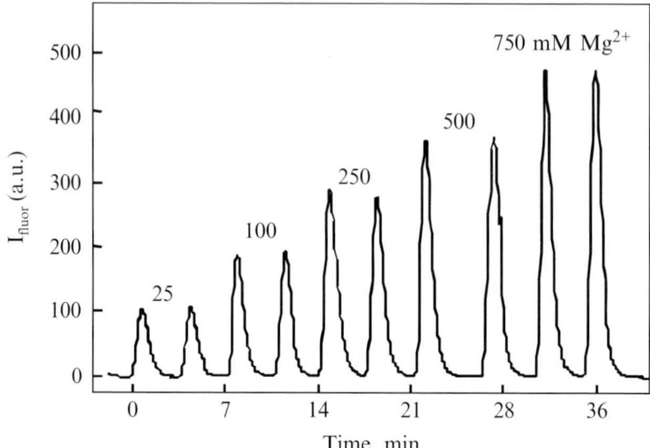

Fig. 15.5 The on-bead typical response against Mg^{2+} of a 1,2,10,13-tetraoxa-7,16-diaza-cyclocta-decane-dansyl library member in a flow-injection approach

water must be continually monitored. Design of optical-based receptor for anion sensing is a difficult task, and it is expected that combinatorial chemistry to contribute to the rational design of optically responsive receptors for anions in the future.

6 Libraries of Molecularly Imprinted Materials

Molecular imprinting is a technique for creating selective recognition sites in synthetic polymers and that requires appropriate choice of building blocks, e.g., functional monomers, cross-linkers, solvents, etc. Most of the molecularly imprinted materials prepared with analytical purposes have been designed on the basis of the chemistry of the target analyte and its possible molecular interactions with functional monomers. Optimization of all these parameters is difficult and the success of the imprinted process is not always guaranteed with the conventional one-sample-one-measurement methodology. Discovery of the lead imprinted polymer is an arduous task requiring large investments in time and chemicals. Consequently, increasing interest is being directed toward approaches that allow more rapid synthesis and screening of molecularly imprinted polymers (MIPs) to identify those with enhanced recognition properties. Recently, construction and screening of polymer libraries by combinatorial chemistry has come to the forefront of molecular imprinting.

Mosbach et al.[41] first developed a screening approach to characterize a library of 12 structurally related steroids, using polymers imprinted against either 11-α-hydrox-yprogesterone or corticosterone, in HPLC mode. It was demonstrated that imprinted polymers bound more strongly to their own print molecule than to other structurally related members of the library. The authors claimed that this approach will become specially useful in performing the initial screening of libraries against poorly char-acterized receptors and to study large libraries of compounds.

MIP libraries for determining the best formulation for the specific recognition of herbicides have also been reported.[42–44] So, Takeuchi et al.[44] made 49-member combinatorial libraries of MIPs using triazine herbicides (atrazine and ametryn) as model template molecules. The synthesis and evaluation were performed by a batch-type in situ procedure using a robot for dispensing reagents into glass vials, washing the resulting polymers, dispensing sample solutions, and injecting of the supernatant into an HPLC or a FIA system. Polymer diversity was provided by the relative proportion of two functional monomers (methacrylic acid and trifluormethacrylic acid) to the corresponding template. Once the template was removed from each of the libraries members, affinity was evaluated by incubating the polymers with a solution of the template and quantifying free atrazine or ametryn in the supernatant. Results demonstrated that combinatorial chemistry employing the batch-type in situ protocol offered a platform on which to develop important systematic correlations between the composition/recognition properties of molecularly imprinted materials, particularly when different functional monomers are simultaneously involved.

M. Kempe et al.[45] have recently reported the synthesis of a 52-imprinted polymer library against Penicillin G as well as the corresponding control library in absence of the template. The synthesis was performed by dispensing manually the reagents into screw-cap borosilicate glasses. Diversity of the library was addressed by variations in the stoichiometry and concentration of the polymers components. Once Penicillin G was removed by washings with different solvent mixtures, the rebinding of penicillin G to the imprinted and control library members was carried out by a batch-wise radioactive assay. The use of a combinatorial strategy provides significantly increased chance of discovering the imprinted polymer with the desired specifity. Even, there is a significantly increased chance of discovering imprinted polymers for molecules structurally related with the template ("dummy molecular imprinting"),[46] thus widening the spectrum of new molecular recognition materials.

Brüggemann et al.[47] have reported the preparation of a 45-member library of MIPs against R-(-)phenylbutyric acid. The corresponding control library was also generated. The authors describe a fast technique for screening the library, based on the impregnation of commercially available membrane modules with the prepolymerization mixture. Then, polymerization was performed directly on the membrane modules, after which the imprinted membranes were washed with a methanol-acetic acid mixture to remove the template. Diversity was achieved by variations in the cross-linker amount and the porogenic solvent used. The efficiency of the imprinting was evaluated by pumping a solution of the template through each MIP membrane module and determining nonbound template in the permeate by GC-MS. After optimization of the technique, it was possible to generate and evaluate the 90 materials (MIPs and control polymers) in 2 days, which was approximately 45 times faster than generating MIPs via bulk polymerization.

It is known that the advantages of imprinted ormosils (organically modified silanes) over acrylic-based MIPs are their chemical stability, the high thermal stability of sol–gel materials, synthesis in mild temperature conditions, optical transparency, photo-, electro- and chemical stability, all properties that made these materials ideal for use in optical sensing approaches, particularly in aqueous samples.[48–50] Although combinatorial chemistry is now being used in the discovery of acrylic-based imprinted polymers, the application of this methodology to imprinted sol–gels is only in its infancy. Recently, our group have introduced the concept in the discovery of new imprinted sol–gel materials against nafcillin, a β-lactamic antibiotic widely used in veterinary.[51] A 20-member imprinted sol–gel library was prepared in which diversity was addressed by variations in the molar ratio of four functional silane precursors, tetramethoxysilane, methyltrimethoxysilane, phenyltrimethoxysilane, amine-propyll-trimethoxysilane.[52] Also, diversity was introduced through different configurations: imprinted sol–gel films and bulk monoliths. HCl-catalyzed sol–gel reactions were performed by manually dispensing standard solutions of reagents into plastic vials or, alternatively, onto cleaned glass slides. After a period of curing, the bulk sol–gels were mechanically crushed and sieved in fragmented fine particles and nafcillin was removed by Soxhlet extraction with an acetic acid/methanol mixture. The corresponding control library, in the absence of the template, was also prepared. A first screening of the library for nafcillin removal was performed by checking the nafcillin concentration in the washing solutions. The second screening for rebinding of nafcillin to the imprinted sol–gels (and control library members) was carried out by continuous flow injection approach, in which the bulk imprinted sol–gels were packed in a flow-cell. In the case of the imprinted sol–gel films, plates were directly measured in a right angle surface detection configuration (Fig. 15.6). Taking advantage of the luminescent properties of nafcillin and the rigidity provided by the sol–gel microenvironment, room-temperature phosphorescence was used as the transduction mode. Results of this screening demonstrated that the hydrophobic/hydrophilic balance of the imprinted sol–gels was of paramount importance to generate materials with enhanced imprinted effects for nafcillin. From the library, two imprinted sol–gels with a high hydrophobic balance were found to be analytically useful recognition materials for nafcillin and one of them has been applied to nafcillin determination in spiked milk samples[53] using a basic flow-injection system as that shown in Fig. 15.2a.

The development of imprinted materials by combinatorial means is still in its infancy.[54,55] The rational design and high-throughput synthesis and screening of molecular imprinted materials must be the focus of research. Several features of combinatorial approaches to imprinted polymers development are clearly different from those devoted to the development of ligands and chemosensors with optical sensing activity. In particular, template extraction is difficult by high-throughput methods and, up to now, it has been performed in a discontinuous format. Also, polymerization in molecular imprinting can be performed by photochemical or thermal free-radical initiation in different solvent media. Thus, temperature, time, solvent, additives, and even catalysts (as in the case of imprinted sol–gels) are parameters that must be tested. When all these reaction parameters have to be combined with structural parameters of the template, the number of experiments becomes large, even when the size of the molecular imprinted library is small. It is expected that molecular imprinting polymerizations could possibly be performed in an automated parallel way as it has been described for other type of polymers.[56] Also, implementing the combinatorial-computational method, a virtual library of polymers could be designed containing thousands of imprinted polymers.

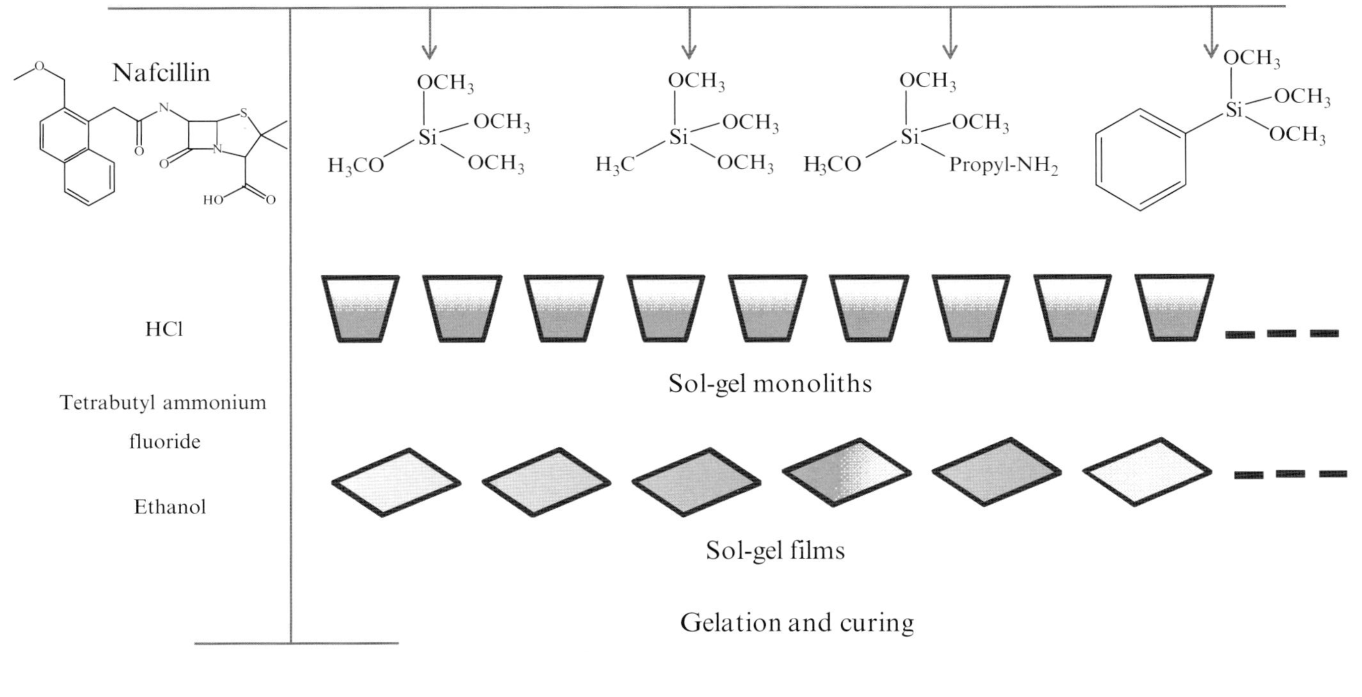

Fig. 15.6 The library synthesis of imprinted sol–gel materials against nafcillin and the screening protocol

This approach integrates synthesis, evaluation, and modeling to achieve predictions of polymer properties and performance for all members of the original polymer library. On the basis of these predictions, a small number of "lead imprinted polymers" could be selected for synthesis and detailed characterization.[57] We believe that these activities will allow a much higher level of fundamental understanding in polymer science in the future. Therefore, while many challenges remain to be solved, combinatorial chemistry is becoming an important tool for the discovery of yet unknown optically active sensing molecules and molecularly imprinted materials for applications where molecular recognition is the basis of the analytical method.

7 Libraries of Materials for Optical Sensing of Solvent Vapors and Oxygen

Combinatorial synthesis has been used by Walt et al.[58] to prepare a discrete array of sensor materials against organic vapors (hexane, methanol, benzene). Sensors were prepared by entrapping a solvatochromic fluorescent dye (Nile Red) within a polymer matrix prepared containing combinations of two monomers (methyl methacrylate and PS802, a methylsiloxane copolymer). Sensor diversity was generated by two different approaches. In the first one, four sensor materials were prepared by discrete combinations of the two monomers (sensor array), while in the second a polymer stripe was photopolymerized forming a continuous gradient between the two monomers. Results of vapor testing using an odor delivery/imaging system revealed four pairs of distinct responses, corresponding directly to the four polymer combinations used. The gradient approach was limited by the gradient to gradient reproducibility.

Wolfbeis et al.[59] have developed a library of polymeric materials for oxygen sensing. A set of 16 different polymers, 5 organic solvents, 4 plasticizers, and 4 indicators were tested in all conceivable combinations using a robotic station programmed to mix the components. Spots of dry sensing materials were screened using microtiterplates by exposing them to a carrier gas containing oxygen in various concentrations and measuring changes in decay time of indicator fluorescence. The approach enabled the fast production of 64 initial sensor materials, from which a new library of 192 compositions was produced and characterized within 1 week. Authors suggest that this methodology offers an alternative to the cumbersome search for new materials and that it holds great analytical potential as it can be easily adjusted to other target gases by changing the sensor chemistry and even, to test liquid systems.

The sol–gel processing of adequate alkoxyde precursors easily leads to a variety of mixed materials. In fact, ormosils are a versatile class of solids which controlled preparation may lead to materials with intrinsic sensing properties. Bright et al.[57] developed arrays of tuned chemical sensors by creating sol–gel-derived xerogels that were codoped with two luminophores at a range of molar ratios. Suites of quencher responsive chemical sensor elements that exhibited unique responses to oxygen were prepared by pin-printing with a 200 μm diameter solid tungsten pin into the surface of cleaned glass microscope slides. Center-to-center spacing

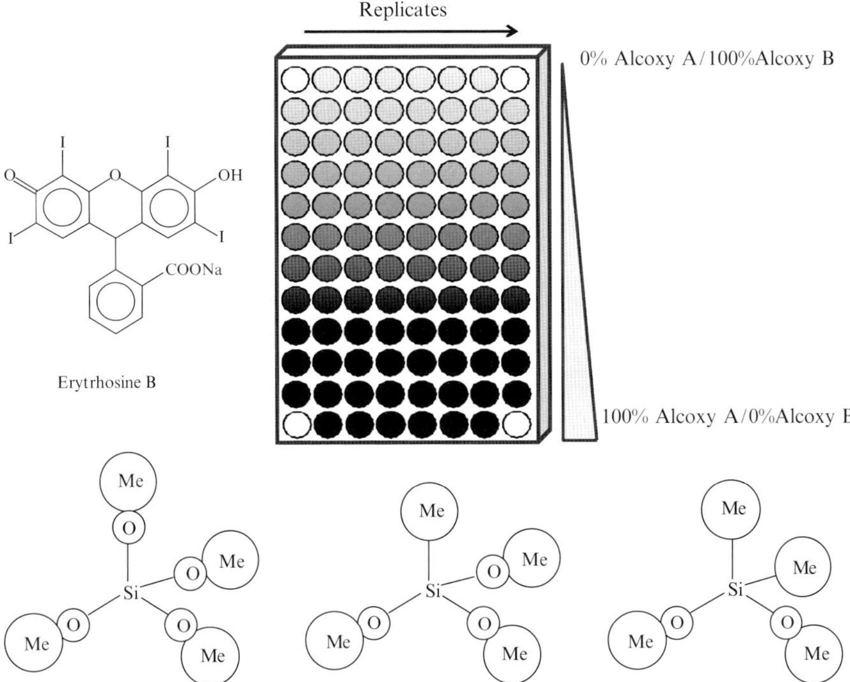

Fig. 15.7 Library of erythrosine B doped sol–gel materials for oxygen sensing. Diversity is addressed by binary mixtures of three alkoxysilanes: tetramethoxysilane, methyl-trimethoxysilane, and dimethyl-dimethoxysilane

between sensor elements was set at 350 μm. An artificial neural network (ANN) to "learn" to identify the optical outputs from these xerogel-based sensor arrays was used, thus obtaining a five to tenfold improvement in accuracy and precision for quantifying O_2 in unknown samples.

Recent work in our laboratory[60] focused, for the first time, the combinatorial synthesis of sol–gel materials doped with Erythrosine B for oxygen high-throughput screening by room temperature phosphorescence. Three-dimensional arrays (libraries) of doped sol–gels were prepared from various alkoxysilanes in which diversity was achieved by the hybrophobic/hydrophilic balance and ormosil composition. Library design shown in the Fig. 15.7 represents (8 × 10) + (6 × 2) array of sol–gels and four wells containing a solution of Erytrhosine B that served as control.

The composition of each material was based on binary mixtures of three alkoxysilanes: tetramethoxysilane, methyl-trimethoxysilane, and dimethyl-dimethoxysilane (in decreasing order of hydrophilic character). After gelation and curing of the materials screening of the library was performed by addition of a sodium sulfite solution to each well using a 6-tip micropipette. The micro plate-well was covered with an optically transparent film that avoided oxygen diffusion from the surroundings and room temperature phosphorescence was measured with a micro plate reader. As sodium sulfite is a chemical oxygen scavenger, increased room temperature phosphorescence should be observed for those doped sol–gel with enhanced oxygen sensing activity. From the 276 materials prepared, six were selected for oxygen sensing. These materials showed promise for monitoring the oxygen consumption during enzymatic reactions thus allowing substrate indirect determination (the glucose/ glucose oxidase reaction was used as proof of the principle).[60]

8 Conclusions

The systems discussed in this account provide a sample of systems that use the principles of combinatorial chemistry to discover new receptors, ligands, and materials for optical sensing applications. There is, however, substantial room for improvements in this young field, particularly in the combinatorial synthesis of new selective and highly sensitive receptors for optical sensing of anions and in the synthesis and high-throughput screening of imprinted polymer parallel libraries. Combinatorial chemistry is demonstrated to be a versatile strategy to improve the productivity of new sensing materials, and we believe that it is the way to move to the next stage from just molecular or ion recognition materials to systems with functions of naturally occurring receptors.

Acknowledgments The authors gratefully acknowledge support from Ministerio de Educación y Ciencia, Spain (MEC, Project #CTQ2006-14644-CO2-01), Consejo Nacional de Ciencia y Tecnología, México, (CONACYT, Grant #SEP-2004-C01-47835).

References

1. Optical Sensors. Industrial, Environmental and Diagnostic Applications. O.S.Wolfbeis (Series Ed.). Narayanaswamy, R., Wolfbeis, O.S. (Vol. Eds). 2004, Springer
2. Martellucci, S., Chester, A.N., Mignani, A.G. (Eds.); Optical Sensors and Microsystems – New Concepts, Materials, Technologies, 2000, Springer, Berlin
3. Koppitz, M., Eis, K.; Automated medicinal chemistry; *Drug Discov. Today*, **2006**, 11, 561–568
4. Menzella, H.G., Reeves, C.D.; Combinatorial biosynthesis for drug development; *Curr. Opin. Microbiol.*, **2007**, 10, 238–245
5. Potyrailo, R.A.; Polymeric sensor materials: toward an alliance of combinatorial and rational design tools? *Angew.Chemie Int.Ed.*, **2006**, 45, 702–723
6. Díaz–García, M.E., Pina-Luis, G., Rivero, I.A.; Combinatorial solid-phase organic synthesis for developing materials with molecular recognition properties; *Trends Anal.Chem.* **2006**, 25, 112–121
7. Sebestyen, F., Dibo, G., Kovacs, A., Furka, A.; Chemical synthesis of peptide libraries, *Bioorg. Med. Chem. Lett.*, **1993**, 3, 413–418
8. Furka, A., Sebestyen, F., Asgedom, M., Dibo, G.; General method for rapid synthesis of multicomponent peptide mixtures, *Int. J. Pep. Prot. Res.*, **1991**, 37, 487–493
9. Otto, S., Furlan, R.L.E., Sanders, J.K.M.; Recent developments in dynamic combinatorial chemistry; *Curr. Opin. Chem. Biol.*, **2002**, 6, 321–327
10. Corbett, P.T., Leclaire, J., Vial, L., West, K.R., Wietor, J.-L., Sanders, K.M., Otto, S.; Dynamic combinatorial chemistry; *Chem. Rev.* **2006**, 106, 3652–3711

11. Vaino, A.R., Janda, K.D.; Solid-phase organic synthesis: a critical understanding of the resin; *J. Comb. Chem.,* **2000**, 2, 579–596

12. Yu, Z., Bardley, M.; Solid supports for combinatorial chemistry, *Curr. Opin. Chem. Biol.,* **2002**, 6, 347–352

13. Shey, J.-Y., Sun, Ch.-M.; Liquid-phase combinatorial reaction monitoring by conventional ^{1}H NMR spectroscopy; *Tetrahedron Lett.,* **2002**, 43, 1725–1729

14. Rousselot-Pailley, P., Ede, N.J., Lippens, G.; Monitoring of solid-phase organic synthesis on macroscopic supports by high-resolution magic angle spinning NMR; *J. Comb. Chem.,* **2001**, 3 (6), 559–563

15. Triolo, A., Altamura, M., Cardinali, F., Sisto, A., Maggi, C.A.; Mass spectrometry and combinatorial chemistry: a short outline; *J. Mass Spectrom.,* **2001**, 36, 1249–1259

16. Stevens, S.M. Jr., Prokai-Tatrai, K., Prokai, L.; Screening combinatorial libraries for substrate preference by mass spectrometry; *Anal. Chem.,* **2005**, 77 (2), 698–701

17. Gerdes, J.M., Waldmann, H.; Direct mass spectrometric monitoring of solid phase organic syntheses; *J. Comb. Chem.,* **2003**, 5, 814–820

18. Yan, B., Gremlich, H.-U., Moss, S., Coppola, G.M., Sun, Q., Liu, L.; A Comparison of various FTIR and FT Raman methods: Applications in the reaction optimization stage of combinatorial chemistry; *J. Comb. Chem.,* **1999**, 1, 46–54

19. Yan, B., Yan, H.; Combination of single bead FTIR and chemometrics in combinatorial chemistry: Application of the multivariate calibration method in monitoring solid-phase organic synthesis; *J. Comb. Chem.,* **2001**, 3, 78–84

20. Schmid, D.G., Grosche, P., Bandel, H., Jung, G.; FTICR-Mass spectrometry for high-resolution analysis in combinatorial chemistry, *Biotechnol. Bioeng.,* **2001**, 71, 149–161

21. Potyrailo, R.A., Takeuchi, I.; Role of high-throughput characterization tools in combinatorial materials science, *Meas. Sci. Technol.* **2005**, 16, 1–4

22. Ahn, Y.-H., Lee, J.-S., Chang, Y.-T.; Combinatorial rosamine library and application to in vivo glutathione probe; *J. Am. Chem. Soc.* **2007**, 129, 4510–4511

23. Potyrailo, R.A.; Analytical spectroscopic tools for high-throughput screening of combinatorial materials libraries, *Trends Anal. Chem.,* **2003**, 22, 374–384

24. Kenseth, J.K., Coldiron, S.J.; High-throughput characterization and quality control of small-molecule combinatorial libraries; *Curr.Opin. Chem. Biol.,* **2004**, 8, 418–423

25. de Silva, A.P, Gunaratne, H.Q.N., Gunnlaugsson, T., Huxley, A.J.M., McCoy, C.P., Rademacher, J.T., Rice, T.E.; Signaling recognition events with fluorescent sensors and switches; *Chem. Rev.,* **1997**, 97, 1515–1565

26. Czarnik, A.W., (Ed.); Fluorescent Chemosensors for Ion and Molecule Recognition; A.C.S.Washington, 1992

27. Patterson, S., Smith, B.D., Taylor, R.E.; Tuning the affinity of a synthetic sialic acid receptor using combinatorial chemistry; *Tetrahedron Lett.,* **1998**, 39, 3111–3114

28. Gordon, E.M., Gallop, M.A., Patel, D.V.; Strategy and tactics in combinatorial organic synthesis. Applications to drug discovery; *Acc. Chem. Res.,* **1996**, 29, 144–154

29. E. Vélez-López, "Solid phase organic synthesis of chemical sensors", PhD Dissertation, Technological Institute of Tijuana, Baja California, Mexico, 2004

30. Badía, R., Pina Luis, G., Granda Valdés, M., Díaz-García, M.E.; Selective fluorescent chemosensor for fructose; *Analyst,* **1998**, 123, 155–158

31. Vélez-López, E., Pina-Luis, G., Suarez Rodríguez, J.L., Díaz-García, M.E., Rivero, I.A.; Immobilisation of a boronic receptor for fructose recognition: influence on the photoinduced electron transfer process, *Sens. Actuators, B* **2003**, 90, 256–263

32. Wang, S., Chang, Y.-T.; Combinatorial synthesis of benzimidazolium dyes and its diversity directed application toward GTP-selective fluorescent chemosensors, *J. Am. Chem. Soc.* **2006**, 128, 10380–10381

33. Buryak, A., Severin, K.; Easy to optimize: dynamic combinatorial libraries of metal-dye complexes as flexible sensors for tripeptides; *J. Comb. Chem.* **2006**, 8, 540–543

34. Rivero, I.A., Gonzalez, T., Diaz-Garcia, M.E.; Synthesis of Metallothionein-mimic decapeptides with heavy atom signaling; *Comb. Chem. High Throughput Screen.,* **2006**, 9, 535–544

35. Castillo, M., Rivero, I.A.; Combinatorial synthesis of fluorescent trialkylphosphine sulfides as sensor materials for metal ions of environmental concern; *Arkivoc*, **2003**, (xi), 193–202

36. Castillo, M., Pina-Luis, G., Díaz-García, M.E., Rivero, I.A.; Solid-phase organic synthesis of sensing sorbent materials for copper and lead recovery; *J. Braz. Chem. Soc.*, **2005**, 16, 412–417

37. Rivero, I.A., Gonzalez, T., Pina-Luis, G., Dıaz-Garcıa, M.E.; Library preparation of derivatives of 1,4,10,13-tetraoxa-7,16-diaza-cyclooctadecane and their fluorescence behavior for signaling purposes; *J. Comb. Chem.* **2005**, 7, 46–53

38. Marshall, G.R., Amruta Reddy, P., Schall, O.F., Naik, A., Beusen, D.D., Ye, Y., Slomczynska, U.; Combinatorial chemistry of metal-binding ligands. Chapter 5 in Advances in Supramolecular Chemistry Vol.8, pp. 174–243, 2002, Cerberus Press, FL

39. Hanrahan, J.W., Tabcharni, J.A., Becq, F., Matthews, C.J., Augustinas, O., Jensen, T.J., Chang, X.-B, Riordan, J.R.; In Ion Channels and Genetic Diseases, D.C.Dawson, R.A.Fritzel (Eds.), 1995, Rockefeller University Press, New York

40. Granda Valdés, M., Díaz-García, M.E.; Determination of thiocyanate within physiological fluids and environmental samples. Current practice and future trends; *Crit. Rev. Anal. Chem.*, **2004**, 34,1–13

41. Ramström, O., Ye, L., Krook, M., Mosbach, K.; Screening of a combinatorial steroid library using molecularly imprinted polymers; *Anal. Comm.*, **1998**, 35, 9–11

42. Lanza, F., Sellergren B.; Method for synthesis and screening of large groups of molecularly imprinted polymers; *Anal. Chem.* **1999**, 71, 2092–2096

43. Lanza F., Hall. A.J., Sellergren,B., Bereczki, A., Horvai, G., Bayoudh, S., Cormack, P.A.G., Sherrington, D.C.; Development of a semiautomated procedure for the synthesis and evaluation of molecularly imprinted polymers applied to the search for functional monomers for phenytoin and nifedipine, *Anal. Chim. Acta,* **2001**, 435, 91–106

44. Takeuchi, T., Fukuma, D., Matsui, J.; Combinatorial molecular imprinting: an approach to synthetic polymer receptors; *Anal. Chem.*, **1999**, 71, 285–290

45. Cederfur, J., Pei, Y., Zihui, M., Kempe, M.; Synthesis and screening of a molecularly imprinted polymer library targeted for Penicillin G; *J. Comb. Chem.*, **2003**, 5, 67–72

46. Piletsky, S., Turner, A., (Eds); Molecular Imprinting of Polymers. III Series: Biotechnology Intelligence Unit. Landes Bioscience. Georgetown, Texas, 2006

47. El-Toufaili, F.A., Visnjevski, A., Brüggemann, O.; Screening combinatorial libraries of molecularly imprinted polymer films casted on membranes in single-use membrane modules; *J. Chromatogr.*, B, **2004**, 804, 135–139

48. Díaz-García, M.E., Badía, R.; Molecular imprinting in sol–gel materials: recent developments and applications; *Microchim. Acta*, **2004**, 274,1–18

49. Marx, S., Liron, Z.; Molecular imprinting in thin films of organic–inorganic hybrid sol–gel and acrylic polymers; *Chem. Mater.*, **2001**, 13, 3624–3630

50. Cummins, W., Duggan, P., McLoughlin, P.; A comparative study of the potential of acrylic and sol–gel polymers for molecular imprinting; *Anal. Chim. Acta*, **2005**, 542, 52–60

51. The European Agency for the Evaluation of Medicinal Products; Veterinary Medicines and Information Technology. EMEA/MRL/750/00-FINAL, April, 2001

52. Guardia, L., Díaz-García, M.E.; unpublished results

53. Guardia, L., Badía Laíño, R., Díaz García, M.E.; Molecular imprinted sol–gels for nafcillin determination in milk-based products; *J. Agric. Food Chem.*, **2007**, 55, 566–570

54. Zhang, H., Hoogenboom, R., Meier, M.A.R., Schubert, U.S.; Combinatorial and high-throughput approaches in polymer science; *Meas. Sci. Technol.*, **2005**, 16, 203–211

55. Batra, D., Shea, K.J.; Combinatorial methods in molecular imprinting, *Curr. Opin. Chem. Biol.*, **2003**, 7, 434–442

56. Kohna, J., Welshb, W.J., Knight, D.; A new approach to the rationale discovery of polymeric biomaterials; *Biomaterials*, **2007**, 28, 4171–4177

57. Tang, Y., Tao, Z., Bukowski, R.M., Tehan, E.C., Karri, S., Titus, A.H., Bright, F.V.; Tailored xerogel-based sensor arrays and artificial neural networks yield improved O2 detection accuracy and precision; *Analyst*, **2006**, 131, 1129–1136

58. Dickinson, T.A., Walt, D.R., White, J., Kauer, J.S.; Generating sensor diversity through combinatorial polymer synthesis; *Anal. Chem.* **1997,** 69, 3413–3418
59. Apostolidis, A., Klimant, I., Andrzejewski, D., Wolfbeis, O.S.; A combinatorial approach for development of materials for optical sensing of gases; *J. Comb. Chem.,* **2004,** 6, 325–331
60. Samaniego Zamorano, S.; Combinatorial chemistry applied to the synthesis of luminescent sol–gel materials for high-throughput oxygen sensing; MsThesis. University of Oviedo, **2005**

Chapter 16
High Throughput Production and Screening Strategies for Creating Advanced Biomaterials and Chemical Sensors

William G. Holthoff, Loraine T. Tan, Ellen L. Holthoff, Ellen M. Cardone, and Frank V. Bright

Abstract Development of new materials is needed for numerous applications in engineering, medical, and scientific arenas. In this chapter, we describe some of our research efforts that focus on developing strategies and tools for high throughput production and screening to create advanced biomaterials and chemical sensors. Using our developed tools, we are able to produce and screen a wide array of materials in a short period of time. In several current embodiments, the system can readily produce and fully screen 100–1,000 samples/day. Our developed automated systems can provide results with minimal user input, yet with better precision and accuracy in comparison to traditional manual methods.

1 Introduction

Materials are at the heart of a number of engineering, medical, and scientific arenas. For example, materials such as high strength plastics are implemented in everyday items that impact our daily lives (e.g., automobiles and aircraft). Biodegradable polymers have emerged in the medical field as materials used in wound closure, fixation devices, and vehicles for drug delivery, to name but a few. Other polymeric materials have been used for chemical sensing applications. In the biodegradable polymer/drug delivery area, it is a challenge to develop easily loaded formulations that degrade at a desired rate(s) and that deliver the active form of the drug in a controlled manner. In chemical sensing research, engineers and scientists are faced with developing materials that will lead to favorable analytical figures of merit. Figure 16.1 presents a simplified view of a hypothetical drug-doped biodegradable

F.V. Bright (✉)
Department of Chemistry, Natural Sciences Complex, University at Buffalo
The State University of New York, Buffalo, NY 14260-3000, USA
chefvb@buffalo.edu

R.A. Potyrailo and V.M. Mirsky (eds.), *Combinatorial Methods for Chemical and Biological Sensors*, DOI: 10.1007/978-0-387-73713-3_16,

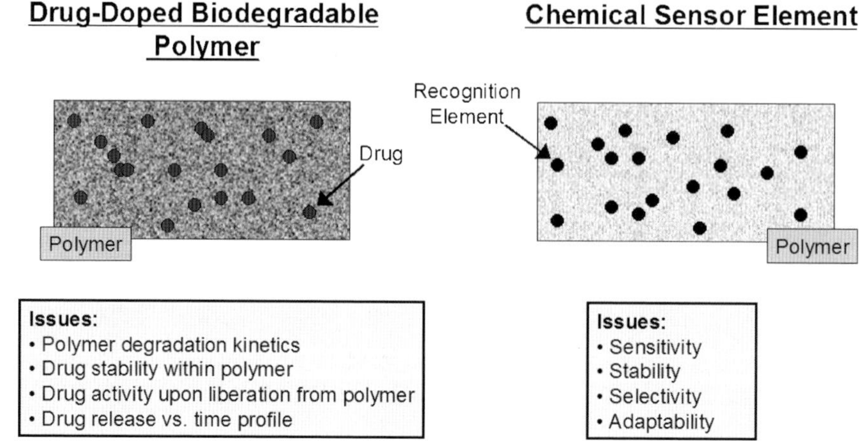

Fig. 16.1 Cross section through a hypothetical drug-doped biodegradable polymer and chemical sensing element and an associated list of important issues related to the optimization of these materials

polymer and chemical sensor element along with a list of pertinent issues that play a role in the overall material/device performance.

In this chapter, we describe some of our research efforts that focus on developing strategies and tools for high throughput production and screening to create advanced biomaterials and chemical sensors.

2 Biodegradable Polymers

Biodegradation is defined as degradation caused by biological activity, typically occurring simultaneously with abiotic degradation (e.g., photodegradation, hydrolysis), and is a two-step process.[1] The first step involves depolymerization or chain cleavage, whereby oligomers are broken down into fragments. This occurs by simple hydrolysis of the hydrolytically unstable polymer backbone. The second step is mineralization, in which enzymatic attack and polymer fragment metabolization occurs. Figure 16.2 presents the suggested mechanisms associated with the hydrolysis for poly(lactic acid) (PLA) in acidic (A) and alkaline (B) environments.[2]

Factors affecting the biological (i.e., enzymatic) degradation of biodegradable polymers (BPs) include chemical composition, chemical stability of the polymer backbone, molecular weight, morphology, presence of catalysts, additives, impurities, and/or plasticizers, as well as the geometry of the polymer.[3–5] In contrast to biological degradation, factors that affect the chemical degradation (i.e., hydrolysis) of BPs include bond type, pH, temperature, copolymer composition, and hydrophilicity.[1]

Today a vast range of BP classes are available, including polyesters, polyamides, polyurethanes, polyureas, polyethers, polyanhydrides, poly(orthoester)s, polypeptides,

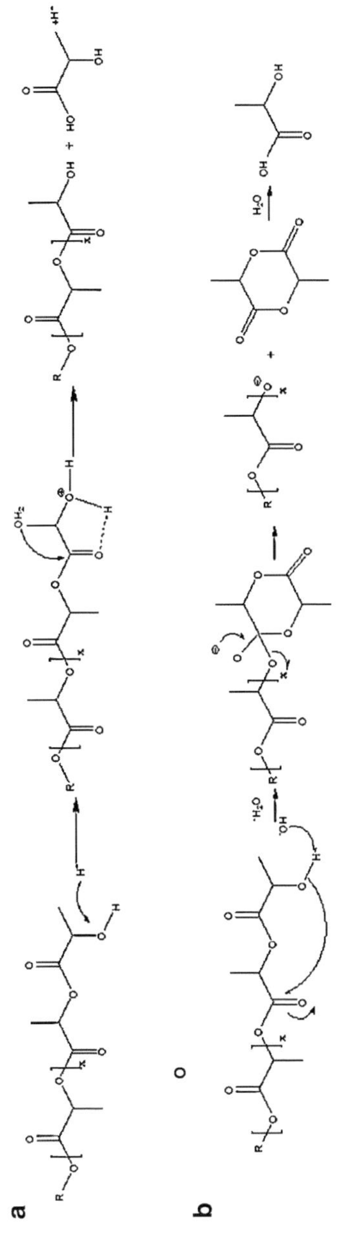

Fig. 16.2 Suggested mechanism describing PLA hydrolysis in acidic (**a**) and alkaline (**b**) environments

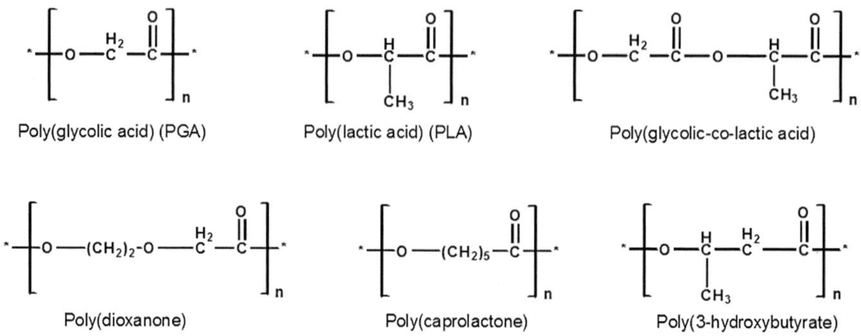

Poly(glycolic acid) (PGA)

Poly(lactic acid) (PLA)

Poly(glycolic-co-lactic acid)

Poly(dioxanone)

Poly(caprolactone)

Poly(3-hydroxybutyrate)

Fig. 16.3 Chemical structures for several common biodegradable polymers

and polysaccharides. Figure 16.3 presents the chemical structures of several frequently used BPs; the most common being synthetic and microbial polyesters.[6] BPs have found application in many important disciplines, owing to the fact that they break down safely, reliably, and relatively quickly, by biological means, into the raw materials of nature. For instance, utilization of biodegradable plastics in consumer packaging is an attractive waste management strategy given the existence of finite landfill space. BPs have also been comprehensively studied for applications in the field of medicine.[7] They are widely used in biomedical devices such as absorbable sutures and surgical implants.[3,8–11] In addition, BPs show potential for implementation in the agricultural field for the controlled release of fertilizers and pesticides, and in the automotive industry for natural fiber-reinforced polypropylene composites.[6]

While BPs can be either synthetic or natural, synthetic polymers offer greater advantages. These polymers can be tailored to give a wider range of properties and have a more predictable lot-to-lot uniformity than materials from natural sources.[3] Given this, synthetic BPs offer a superior methodology to tailor a material's mechanical properties and degradation kinetics for each specific application.[12] Many factors affect the mechanical performance of a BP, including initiator selection, material process conditions, and the presence of additives. These factors consequently influence the polymer's hydrophilicity, crystallinity, melting and glass-transition temperatures, molecular weight, molecular-weight distribution, end groups, sequence distribution (random versus blocky), as well as the presence of residual monomers or additives.[3,13] Each parameter must be carefully evaluated to properly elucidate its effect(s) on biodegradation.

Another methodology to tailor a BP material's mechanical properties and degradation kinetics for each specific application is through copolymerization and polymer blending. Copolymers are polymeric materials synthesized from more than one type of monomer; while polymer blends are a mixture of at least two macromolecular substances, polymers or copolymers, in which the ingredient content is higher than 2 wt%.[14] Although both exhibit beneficial physical and mechanical properties that cannot be matched by an individual polymer, blending is frequently used because it is more cost-effective than copolymerization.[15]

3   Sol–Gel-Derived Materials

The sol–gel process provides a direct method to fabricate glass-like or ceramic materials, under ambient conditions, where one or more reactive precursors are mixed in solution in the presence of a catalyst.[16] The solution phase first turns into a gel, followed by solidification with the loss of solvent, leading to a nano-porous, optically transparent glass.

Figure 16.4 presents an overview of the sol–gel process. Figure 16.5 illustrates the individual steps for the formation of a class I xerogel prepared from tetraalkoxysilane precursors (A) or a class II xerogel prepared from organically modified silanes (ORMOSIL)s and tetraalkoxysilanes (B). The individual steps associated with the sol–gel process are hydrolysis, condensation, and polycondensation. In the first step, hydrolysis of the metal or semi-metal alkoxide yields a hydroxylated product, or silanol, and the corresponding alcohol. This mechanism is generally acid or base catalyzed and its rate is pH dependent. In the second step, condensation between (a) an alkoxide group and a silanol produces a siloxane and alcohol, or (b) two silanols produce a siloxane and water. These constituents continue to hydrolyze and condense, eventually forming a colloidal suspension of reactive nanoscale particles called a "sol." The sol can be cast into thin films, fibers, or monoliths, while it is still in the solution phase. In the third and final step, polycondensation between the sol particles occurs, resulting in a nano-porous, glass-like, three-dimensional network that is optically transparent. During this step, aging and drying of the sol–gel processed composite occurs; this process is accompanied by the loss of solvent (alcohol, water), which increases the strength and decreases the sol–gel processed material's porosity. The method used to dry the gel can have a dramatic effect on

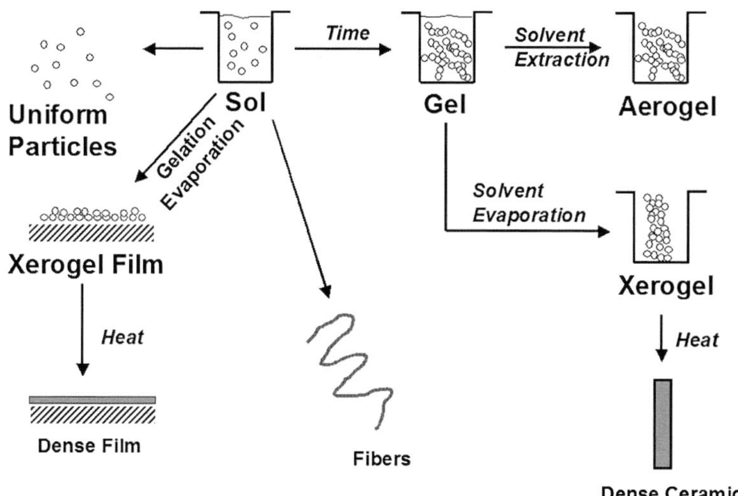

Fig. 16.4 Simplified schematic of the sol–gel process

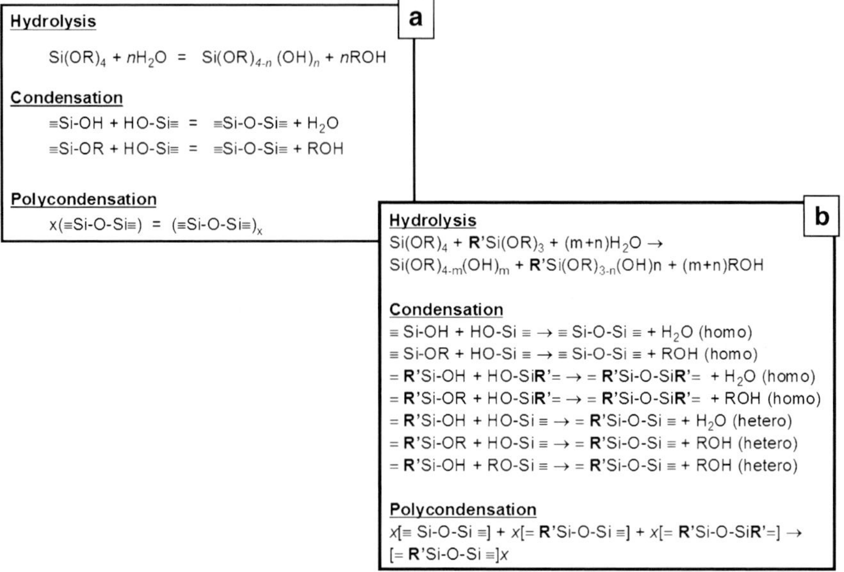

Fig. 16.5 The chemical reactions associated with the sol–gel process. (**a**) For a tetraalkoxysilane; (**b**) For the cohydrolysis and cocondensation of tetraalkoxysilane and an alkyl-trialkoxysilane to form a hybrid class II ORMOSIL

the final material's properties. When drying is performed under ambient conditions, the resulting composite is called a xerogel. An aerogel is formed when the sol is processed under supercritical conditions. Xerogels have lower surface areas, smaller pore sizes, and higher densities in comparison to aerogels.

The physicochemical properties within these sol–gel processed materials can be altered by varying the synthetic conditions (e.g., temperature, pressure, pH, nature of the catalyst (acid or base), reagent concentrations (water, alcohol, catalyst, silane precursors), and reaction times (hydrolysis and condensation times)) or the chemical composition of the precursors.[16,17] The most popular and commercially available precursors for forming silica-based xerogels are tetramethoxysilane (TMOS) and tetraethoxysilane (TEOS). However, previous research has shown that pure TMOS or TEOS-based xerogels have limitations,[18–20] which can be addressed.[21–23] The greatest advances have been realized by creating hybrid materials wherein two or more silanes (Fig. 16.6) are cohydrolyzed and cocondensed together. This allows researchers a convenient pathway to prepare xerogel-based nanocomposites with a high degree of control over the final material's properties.[24]

Sol–gel processed materials are utilized in a wide range of fields, including biotechnology, chemistry, cosmetics, electronics, mechanics, optics, and optoelectronics. They have served as a platform for materials in chromatographic supports, catalyst supports, optical coatings, nonlinear optics, photochromic and photovoltaic devices, chemical/biological sensors, protective or decorative coatings, bioreactors, controlled drug release matrices, and dielectric layers.[17,25–34]

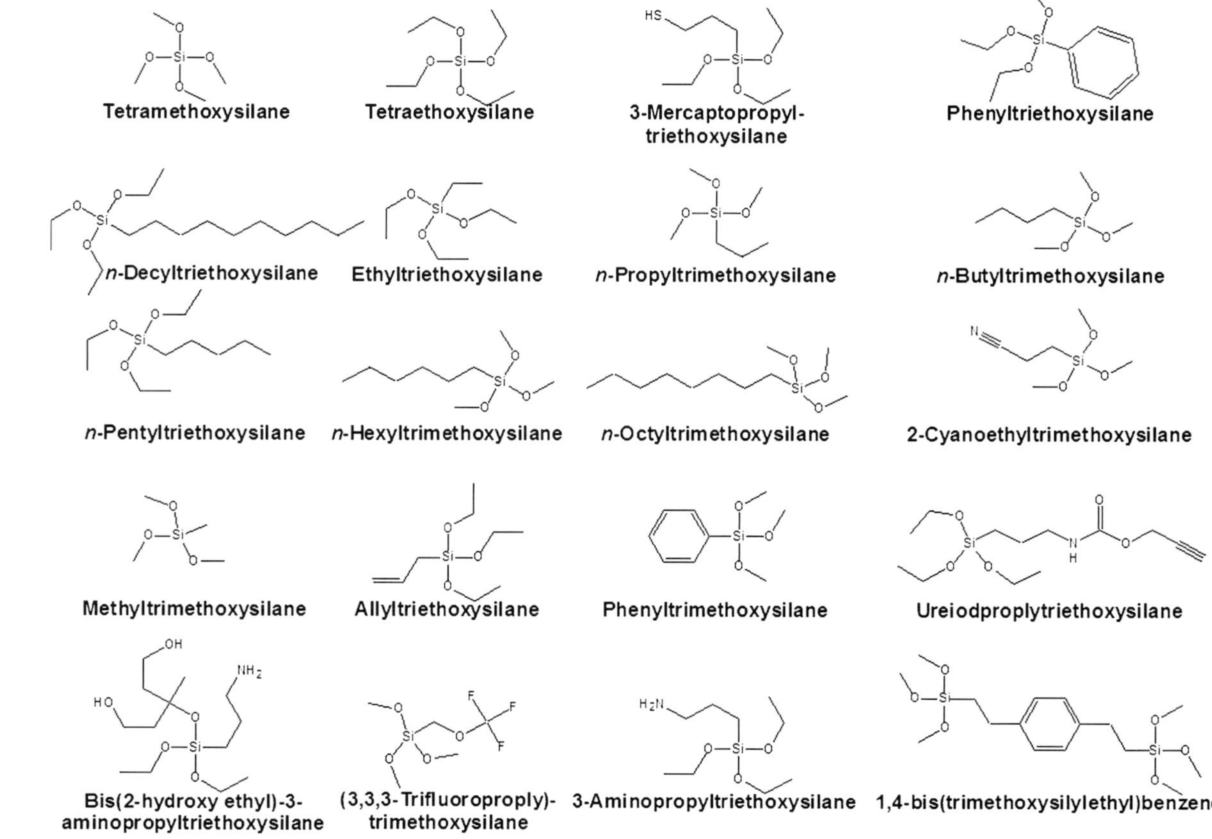

Fig. 16.6 Representative, nonexhaustive compilation of silane-based alkoxide precursors that can be used alone or together to form xerogels

Although BP and xerogel-based platforms exhibit significant promise in their respective areas, there are a plethora of materials, precursors, and additives to choose from. How does one decide on the "best" formulation for a specific need/purpose? In cases where a particular formulation or composition does not yield a material or device with adequate performance parameters, how does one get to an optimum material within a reasonable time frame? Our answer to these questions lies in the use of an automated robotic system to produce and screen large libraries of materials in a rapid manner, thus identifying optimal materials for a particular application.

4   High Throughput Production and Screening as a Pathway to Create Advanced Biomaterials and Chemical Sensing Platforms

The loading or incorporation of proteins within BP platforms for use as a drug delivery device has been extensively investigated.[35–38] In addition to factors that affect the mechanical and biodegradation properties of the BP, slight alteration of the BP platform could greatly affect "drug" release kinetics, as well as protein integrity and activity. In recent research, Chen and coworkers explored the potential of PLGA-PEG-PLGA triblock copolymers for the controlled release of a model protein (lysozyme).[39] This research focused on how copolymer concentration and the mole ratio of lactic acid segment (LA) to glycolic acid segment (GA), used to prepare the PLGA-PEG-PLGA copolymer, affected the fraction of lysozyme released from the BP. Results (Fig. 16.7) demonstrated that the percentage of released lysozyme was strongly correlated to the copolymer concentration and the LA/GA mole ratio used to prepare a given PLGA-PEG-PLGA copolymer.

Sol–gel-derived materials are popular for developing sensors.[16,40–43] The physico-chemical properties within these sol–gel-derived nanocomposites is an important factor in designing platforms for sensing applications.[18,20,22,44–53] Factors such as polarity, microviscosity, pore size, pore wall chemistry, microscopic phase separation, partition coefficient, and solubility coefficient can dramatically alter the behavior of active dopants within a nanocomposite and may lead to undesirable properties.

In previous research, the physicochemical properties within these xerogel-based materials have been probed by doping them with a fluorophore.[18,21,44,49–52,54–67] Researchers commonly assessed the "local microenvironment" surrounding the fluorophore by using steady-state and time-resolved fluorescence. There are, however, a limited number of reports that systematically map the physicochemical properties within diverse libraries of sol–gel-derived xerogels. This conceptually simple "task" becomes exceedingly complicated given the multitude of xerogel recipes, the number of possible coprecursors (Fig. 16.6) and the number of "processing" factors that govern a xerogel-based sensor's performance. For example, Tang et al.[68] showed that the sensitivity of a simple quenchometric O_2 sensor depends on how long the sol is allowed to hydrolyze and condense before forming the sensor

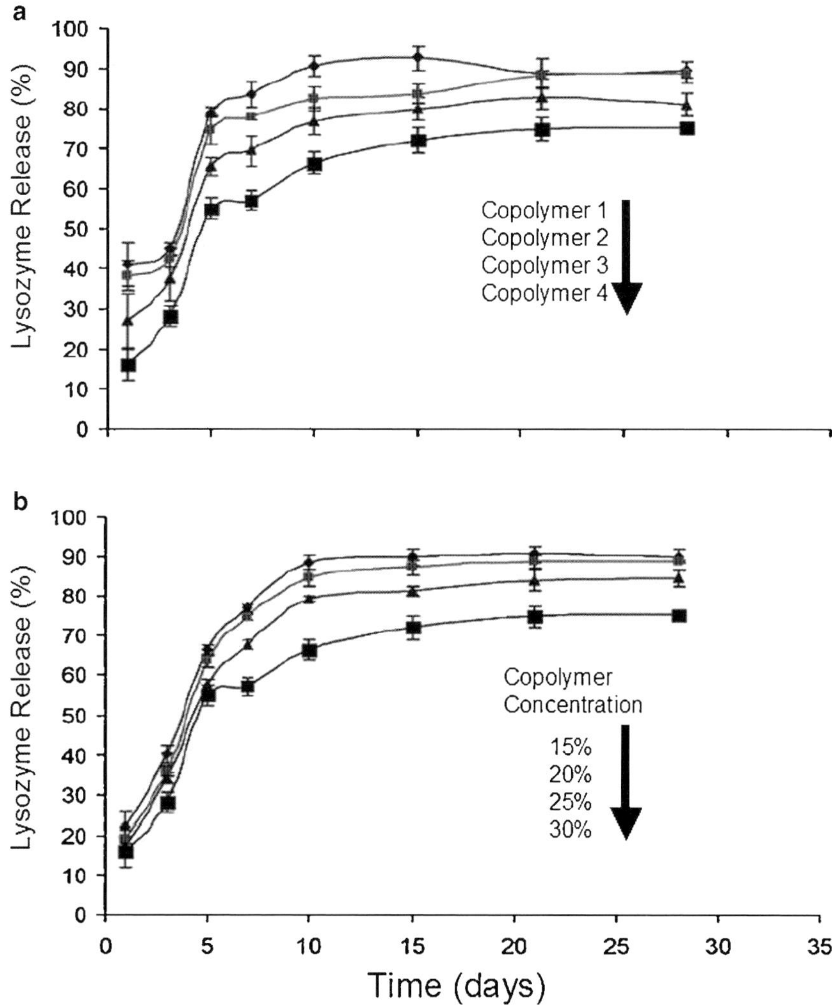

Fig. 16.7 Summary of the in-vitro release of lysozyme from a series of biodegradable copolymers. (**a**) Effect of copolymer block lengths (all 30% (w/v)); (**b**) Effect of polymer concentration (all for copolymer 4)

element. In related research, Tao et al.[52] demonstrated that one can form hybrid xerogels that are composed of TMOS or TEOS and mono-alkyltrialkoxy silanes of the form (C_nH_{2n+1}) –Si–$(OR)_3$ (n = 1–12, R = Me or Et) wherein the sensitivity was controlled by the C_n chain length (Fig. 16.8).

Given the challenges that are associated with creating BPs that can deliver active proteins and/or diversified xerogel-based sensor platforms for chemical sensor applications, we have developed automated systems that can rapidly prepare large numbers of BP or xerogel-based materials and rapidly obtain fluorescence-based spectroscopic information from these formulations. Ultimately, these laboratory

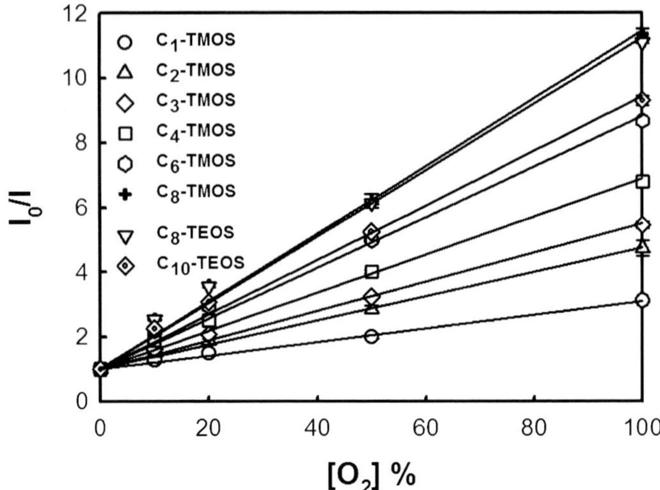

Fig. 16.8 Typical intensity-based Stern–Volmer plots for the $[Ru(dpp)_3]^{2+}$-doped pin-printed class II ORMOSIL-based sensor elements

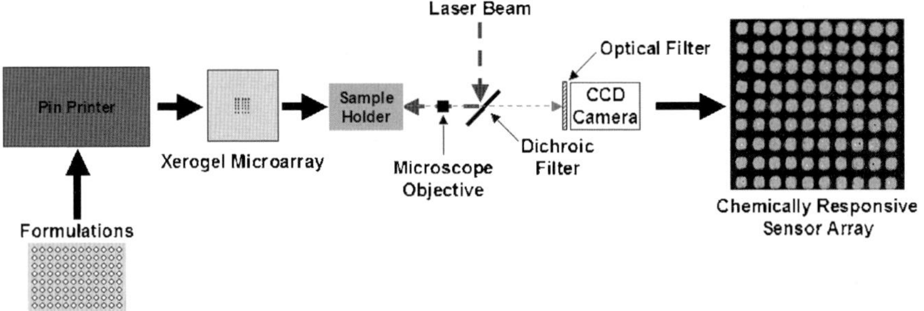

Fig. 16.9 Simplified schematic of the first generation production and screening system that was used in our laboratory

tools will be useful for constructing libraries of materials and provide comprehensive information on the physicochemical properties within these materials.

The first generation production and screening instrument (Fig. 16.9) in our laboratory was developed by Cho et al. who demonstrated its potential for rapidly producing and screening BP and xerogel-based formulations.[69,70] In the original system, formulations were made by hand with micropipettes; then introduced into a high-speed pin printer that created arrays of BP or xerogel-based formulations. Next, these arrays were introduced into an epi-fluorescence microscope and the total fluorescence, as well as the fluorescence "spectrum," for each formulation within the array was recorded and compared. Using this approach, we were able to screen more than 600 formulations per hour; enabling us to rapidly identify potentially promising BP and sol–gel-derived material formulations for a specific application.

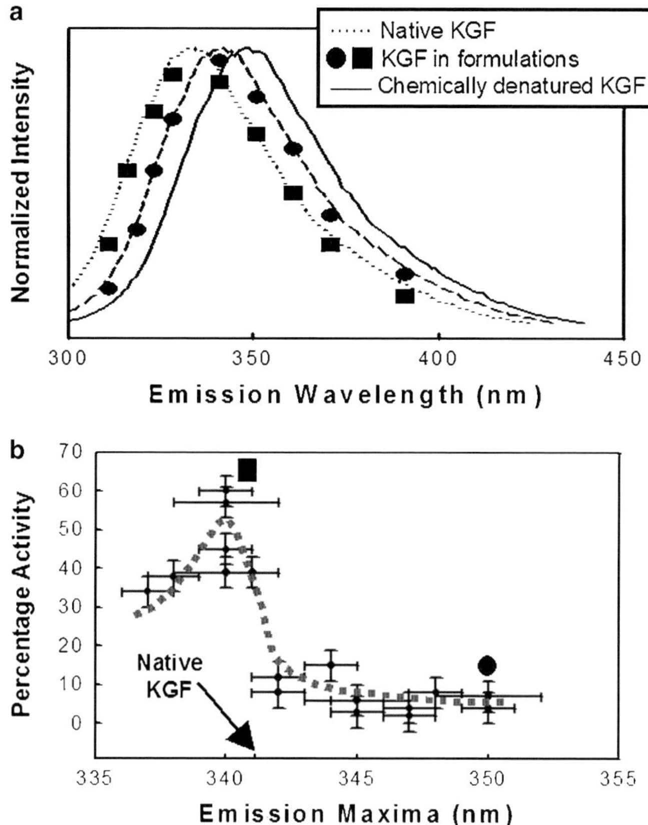

Fig. 16.10 (a) Normalized KGF emission spectra in a number of environments. KGF dissolved in buffer (*dashed line*), chemically denatured KGF dissolved in buffer (*straight line*), KGF within a somewhat denaturing biodegradable polymer formulation (*filled circle*), KGF within a nondenaturing biodegradable polymer formulation (*filled square*). As the biodegradable polymer erodes, the *filled square* formulation is more likely to release KGF in an active form in comparison to the *filled circle* formulation. All formulations are 30 days old. (b) Correlation plot of the KGF emission *within* 18 biodegradable polymer formulations and the associated KGF activity upon release from the same formulations. Note: Maximum KGF activity upon release from a formulation is seen where the KGF emission spectrum within the formulation is the most like native KGF dissolved in buffer (*arrow*). All formulations are 30 days old

Figures 16.10 and 16.11 present results whose aim was to deliver active keratinocyte growth factor (KGF) from a BP-based platform. In Fig. 16.10, we illustrate the measurement/screening strategy based on the intrinsic tryptophan fluorescence from KGF; Fig. 16.10a shows the KGF emission spectral (λ_{ex} = 257 nm) shift between native and denatured KGF. Figure 16.10b shows that there is a strong correlation between the observed KGF emission maximum and the KGF activity in vitro. Figure 16.11 illustrates the effect of BP polymers (A), PLA/PGA ratio (B), concentration of surfactant aerosol-OT (AOT) (C), molar ratio of water to AOT,

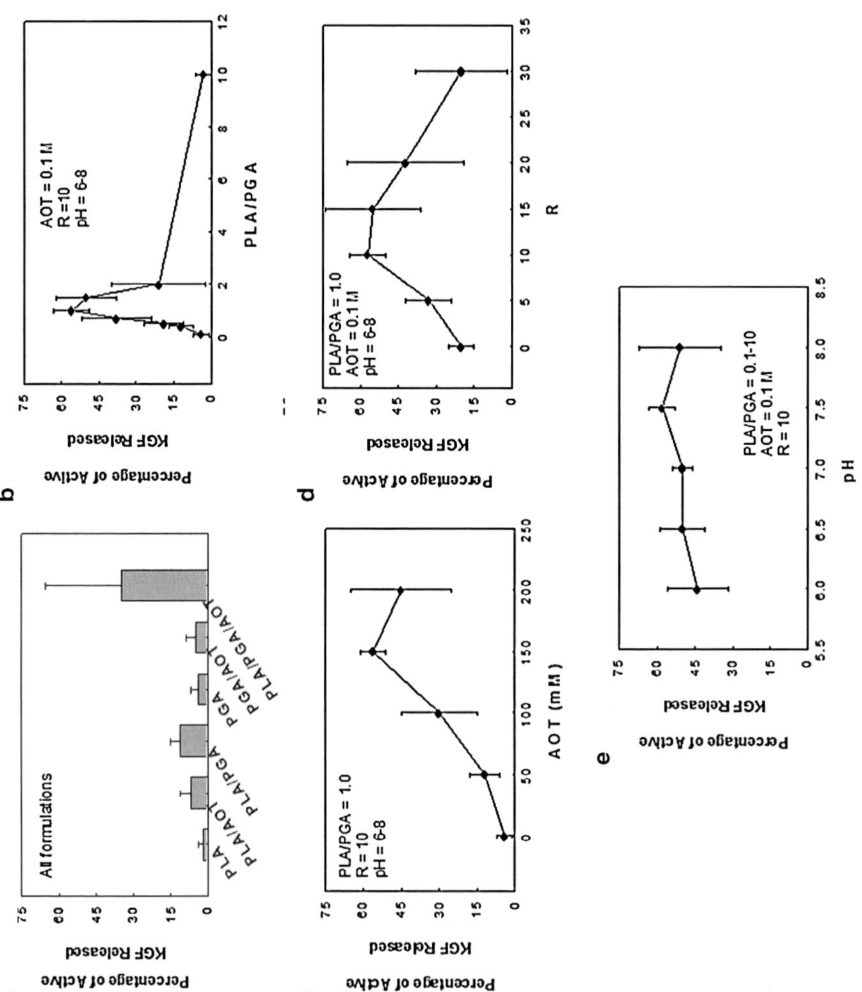

Fig. 16.11 Effects of composition (**a**), PLA/PGA weight ratio (**b**), AOT concentration (**c**), molar ratio of water to AOT, R (**d**), and water pH (**e**) on the percentage of active KGF released from a given formulation. 125 formulations were selected at random amongst the 2,500 that were printed and screened. The pure PGA formulations behave much like the pure PLA formulations

R (D), and aqueous phase pH (E) on the KGF activity upon liberation from a series of BP-based films. These results provide a roadmap that delineates activity–composition relationships over a wide range of formulation chemistry. In addition, these results revealed that the use of a binary polymer mixture in concert with AOT surfactants yielded materials that were capable of delivering up to 60% active KGF. As a benchmark, the standard approach of sequestering KGF directly within a PLA-based BP yielded ≤4% active KGF.[70]

The system presented in Fig. 16.9 exhibited several attractive features, but it relied on manually preparing a large number of formulations with micropipettes; this was a tedious and potentially error prone task. Figure 16.12 presents a simplified schematic of our current production and screening system. The system consists of a Gilson Quad Z automated liquid handling system (ALHS). The ALHS consists of a syringe pump and four probes that can independently dispense between 1 and 1,000 µL of liquid. The probes are mounted on an automated arm, which can translate in the x, y, and z directions. Solutions/reagents/precursors are manually added to Teflon® reagent reservoirs, where the probes pick up and dispense the required volume into 96-well plates or a custom assembly that holds 96 amber vials. During operation, the location of solutions/reagents/precursors and well plates is specified using a graphical interface, and the user is able to specify the order in which the solutions/reagents/precursors are to be dispensed. This system also generates a Microsoft Excel spreadsheet where the user enters the solutions/reagents/precursors volumes for each formulation/sample. Once these steps are completed, the required solutions/reagents/precursors are added to the reservoir, and the system prepares the samples without any additional input. The pin-printer and epifluorescence microscopy that we use to create and characterize these arrays of formulations is identical to the system described by Cho et al.[69,71]

Depending on the desired level of detail, this new system can perform pseudo-automated steady-state fluorescence measurements via a conventional spectrofluorometer (SLM-Aminco model 48000 MHF, SLM 8100). This can also be accomplished by using our custom built high throughput spectral acquisition system, which consists of an Ocean Optics XYZ mapping table and a S2000 fiber optic spectrophotometer. A CW laser is typically used as the excitation source whereby output is focused into the proximal end of an all-silica optical fiber mounted on a precision x, y, z translation stage. The distal end of the optical fiber is mounted within a custom bifurcated optical fiber probe, and the excitation light is directed through the optical fiber to the samples, which are housed within 1 mL amber vials. The resulting fluorescence is collected through an all-silica optical fiber whose output end is connected to the S2000 spectrophotometer; the bifurcated optical fiber probe is attached to the XYZ mapping table. Once the initial parameters are programmed, the automated spectral acquisition program records the steady-state emission spectrum for each sample/formulation.

In the current embodiment, this system can independently acquire up to 96 sample spectra before requiring operator intervention. Sampling parameters and spectral data are stored in a Microsoft Excel spreadsheet for the user. Time-resolved intensity decay measurements can be performed using a time-correlated single

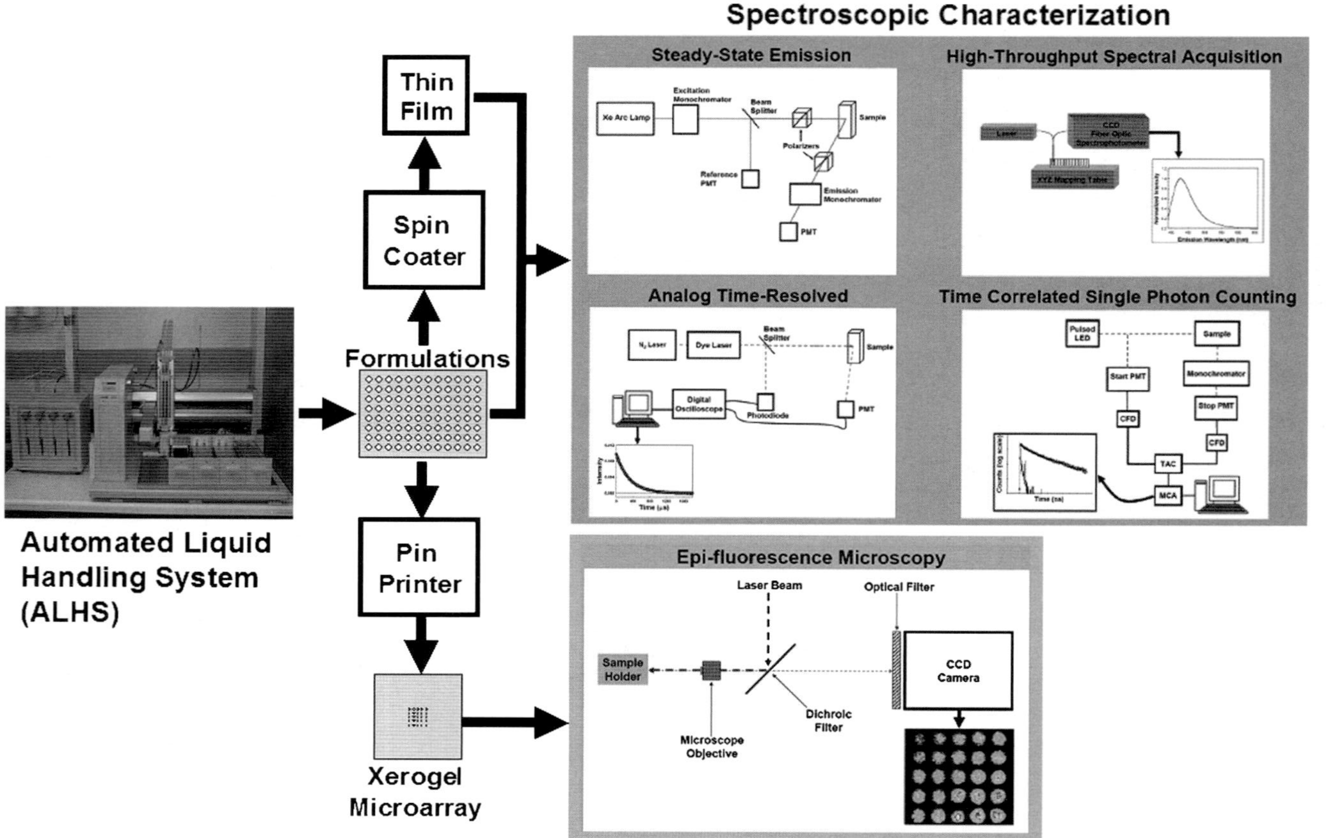

Fig. 16.12 Simplified schematic of the second generation production and screening system developed in our laboratory

photon counting (TCSPC) instrument (IBH model 5000W SAFE), or an analog time-resolved instrument described previously by Tang et al.[18]

5 Selected Results

To illustrate the capabilities of the system shown in Fig. 16.12, we present new, previously unpublished results from our laboratory. In Fig. 16.13, we illustrate results for three fluorescent probe molecules (pyrene, DCM [4-(dicyanomethylene)-2-methyl-6-(p-dimethylaminostyryl)-4H-pyran], and PRODAN [6-propionyl-2-(N,N-dimethylamino)naphthalene]) that were doped within a series of PLLA/Pluronic P104 BP blends. In the initial experiments, 16 BP formulations were prepared manually using micropipettes, while subsequent experiments utilized the ALHS to prepare 21 BP formulations. These formulations were spun cast into thin films employing quartz microscope slides as substrates. To characterize the local microenvironment surrounding each probe within a given formulation, steady-state fluorescence measurements using a conventional spectrofluorometer were performed.

Pyrene is extremely sensitive to the polarity of the microenvironment surrounding it. As the polarity of the microenvironment increases, the emission intensity of the first vibronic band (I_1) increases, while the emission intensity of the third vibronic band (I_3) decreases. Thus, I_1/I_3 is related to the dipolarity of the microenvironment surrounding the pyrene molecules; for example I_1/I_3 shifts from 0.58 in cyclohexane to 1.87 in water.[72,73] PRODAN and DCM are solvatochromic fluorophores whose fluorescence band position is extremely sensitive to the polarity of the surrounding microenvironment. For example, the fluorescence emission maximum of PRODAN shifts from 401 nm in cyclohexane to 531 nm in water.[74,75]

Figure 16.13 demonstrates that as the concentration of PLLA increases, the pyrene I_1/I_3 systematically increases, while the emission maxima of PRODAN and DCM systematically decrease. This indicates that as the concentration of PLLA increases, the microenvironment surrounding pyrene becomes more polar, while for PRODAN and DCM it becomes more nonpolar. These results argue that the dopant molecules selectively locate to specific microenevironments within these formulations, and these microenvironments are very different from each other.

These data also demonstrate the superiority of the ALHS system in comparison to manual methods. Utilization of the ALHS resulted in the production of 25% more formulations while reducing the sample preparation time. In addition, the increased accuracy and precision of the ALHS at dispensing small volumes is evident. Although the general trend in both data sets is similar, the trend is more distinct for formulations prepared using the ALHS, compared with the formulations prepared by hand.

Figure 16.14 summarizes the pyrene I_1/I_3 results from a series of hybrid class II xerogels formed from phenyl-trimethoxysilane (P-TMOS), phenyl-triethoxysilane (P-TEOS), TMOS, TEOS, and seven different monoalkyl-trialkoxysilanes (C_n-TM(E)OS). These formulations were prepared using the ALHS and screened using the high throughput spectral acquisition system; utilization of automated systems reduced

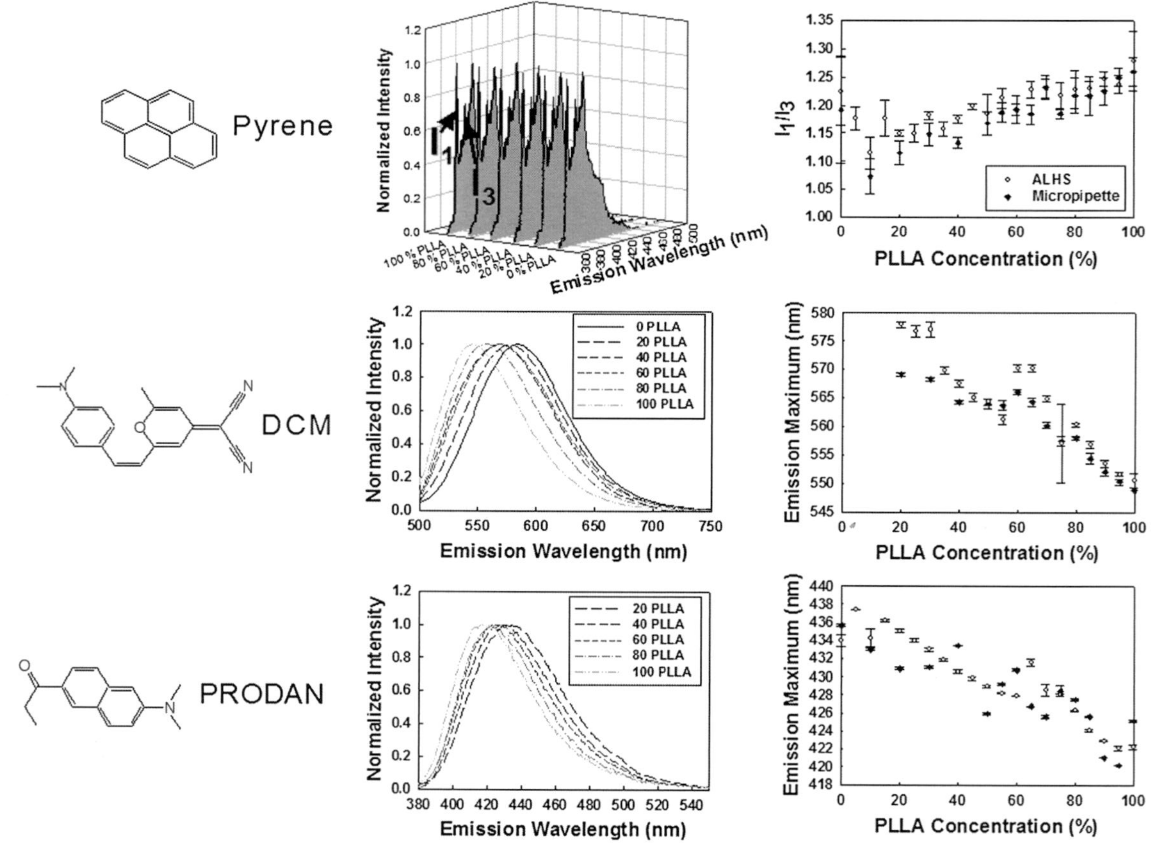

Fig. 16.13 The spectroscopy of pyrene, PRODAN and DCM. Effects of probe structure, PLLA concentration and preparation method on the local microenvironment that surrounds pyrene, PRODAN, and DCM in PLLA/P104-based BPs

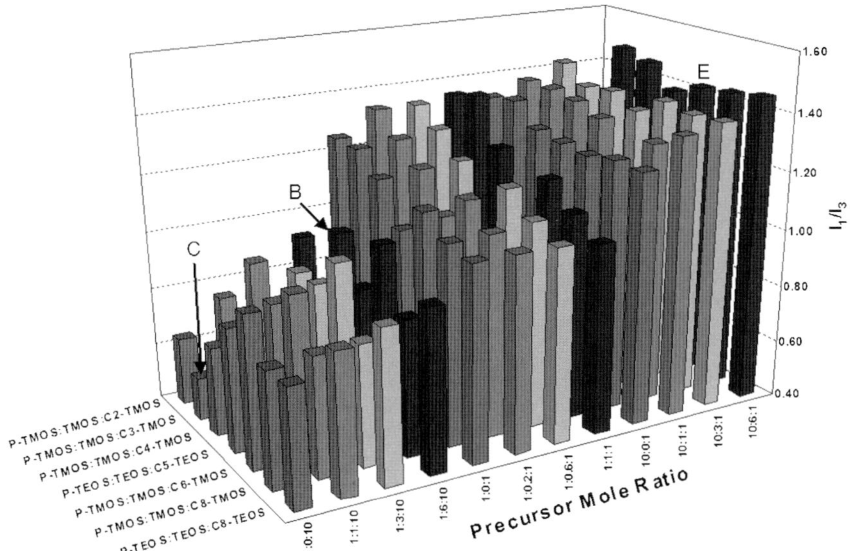

Fig. 16.14 Summary of the pyrene I_1/I_3 as a function of xerogel composition. Abbreviations: *C* cyclohexane-like, *B* benzene-like, *E* ethyl acetate-like

the sample preparation and screening time to approximately 90 min. The results illustrate that xerogels with average polarities ranging from cyclohexane-like (C), benzene-like (B), and ethyl acetate-like (E) can be readily created from these simple precursors by controlling either the precursor type or the hybrid composition. Figure 16.8 illustrated how the response of a chemical sensor derived from TMOS or TEOS and mono-alkyltrialkoxy silanes of the form (C_nH_{2n+1})–Si–$(OR)_3$ (n = 1–12, R = Me or Et) change simply with the C_n chain length. Figure 16.15 illustrates how the mole fraction of the precursors and precursor type affect the sensor response (I_0/I). In this particular case, the entire sensor library was created and screened in approximately 30 min. Recent research from our laboratory[76] has exploited these diversified response profiles to form suites of sensor elements that exhibit a continuum of response profiles and to then use these sensor arrays in concert with artificial neural networks to obtain a fivefold to tenfold improvement in accuracy and precision for quantifying O_2 in unknown samples.

The following results (Figs. 16–18) are for a series of hybrid class II xerogel formulations prepared from TEOS and 1,4-bis(trimethoxysilylethyl)benzene (BTEB). Ten formulations were prepared by varying the molar ratio of the two silane precursors in 0.5% increments from 100.0 to 96.0% TEOS, with the final formulation being 95.0% TEOS. These formulations were fabricated manually using micropipettes and with the ALHS. Both sets of formulations were doped with an O_2-responsive luminophore then spun cast and pin printed into quenchometric O_2 responsive sensor thin films and arrays.

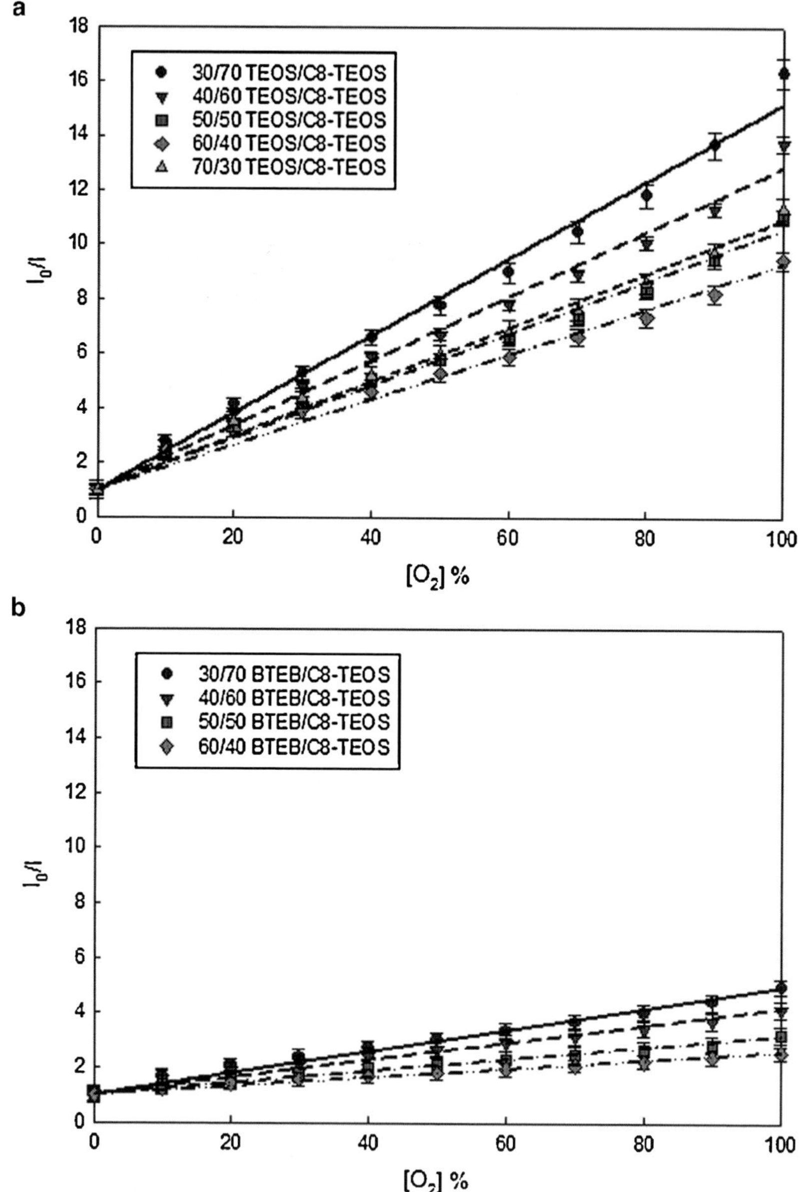

Fig. 16.15 Intensity-based Stern–Volmer plots for xerogel-based sensors formed from TEOS/C8-TEOS (**a**) and BTEB/C8-TEOS (**b**)

Figure 16.16 presents the O_2 response profiles for the hybrid class II xerogels prepared by hand using micropipettes (A), and the ALHS (B). Inspection of these results demonstrates again how critical accuracy and precision are when attempting to characterize formulations in which the molar ratio of the precursors is varied by a small

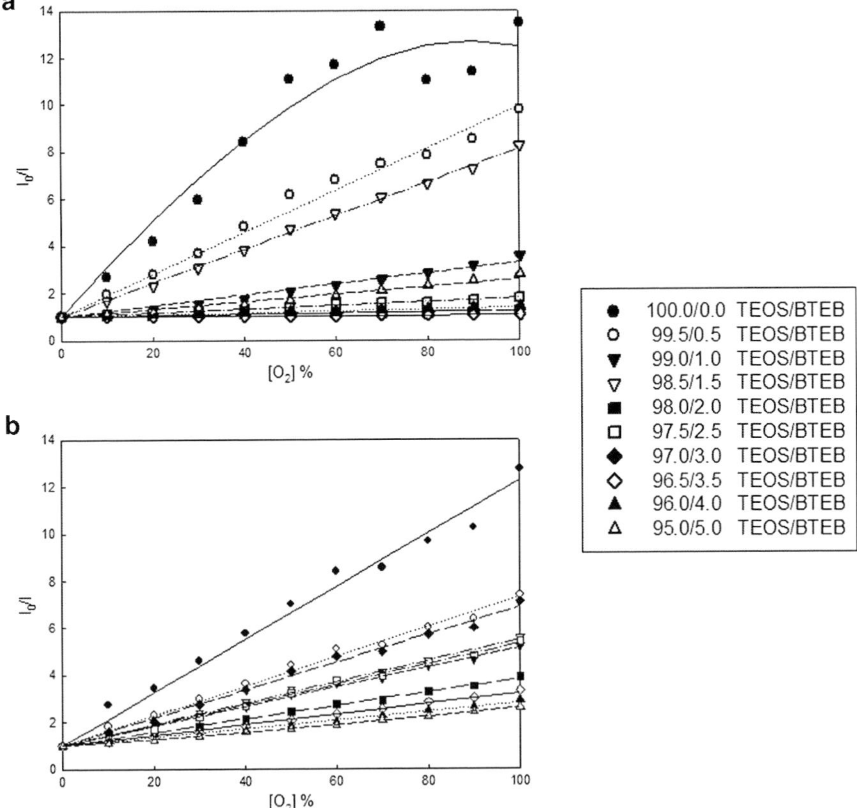

Fig. 16.16 Typical intensity-based Stern–Volmer plots for $[Ru(dpp)_3]^{2+}$-doped pin-printed class II ORMOSIL-based sensor elements (**a**) Manual prepared and (**b**) Robotically prepared

degree. The general trend in both sets of data shows that as the concentration of BTEB is increased, the sensitivity to O_2 decreases. This trend is more systematic and clearly seen for the formulations prepared using the ALHS in comparison to the formulations prepared using manual sample preparation methods.

In a quenchometric sensing scheme, the best description of luminophore quenching depends on how the luminophore molecules are distributed within the host matrix, the quencher (O_2, in this case) solubility and transport properties, and the efficiency with which the quencher molecules can interact with the luminophore molecules within the host matrix. In the simplest scenario, identical luminescent molecules are distributed within an analyte-permeable host matrix in such a way that they are each equally accessible to the quencher molecules and the microenvironment surrounding each luminophore molecule is largely similar, described as:

$$I_0/I = \tau_0 / \tau = 1 + K_{SV}[Q]. \tag{1}$$

This expression relates the ratio of the steady-state luminescence intensities or excited-state luminescence lifetimes in the absence of quencher (I_0 or τ_0, respectively) to the intensity or lifetime in the presence of quencher (I or τ, respectively) through the dynamic Stern–Volmer quenching constant, K_{SV}, and the quencher concentration, $[Q]$. For this ideal case, a plot of I_0/I or τ_0/τ vs. $[Q]$ (called the Stern–Volmer plot) will be linear with a slope equal to K_{SV} and an intercept of unity. K_{SV} depends on τ_0 and the bimolecular quenching constant (k_q) between the luminophore and the quencher molecule ($K_{SV} = \tau_0 k_q$); k_q depends on the quencher solubility coefficient in the host matrix (S) and the quencher diffusion coefficient (D) within the host matrix. In Fig. 16.17, we use the time-resolved capabilities of our system (Fig. 16.12) to determine the exact origin of the sensitivity observed in Fig. 16.16 b. In these particular materials, the observed decrease in K_{SV} as one increases the % BTEB in the xerogel arises from a concomitantly large decrease in k_q despite a τ_0 that initial drops as we progress from 0 to 1% BTEP and then rebounds between 1 and 5% BTEP.

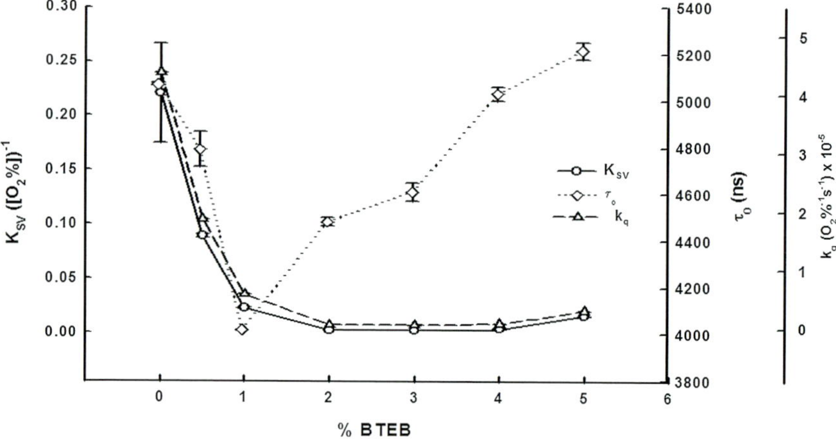

Fig. 16.17 Effects of BTEB on K_{SV}, τ_0 and k_q

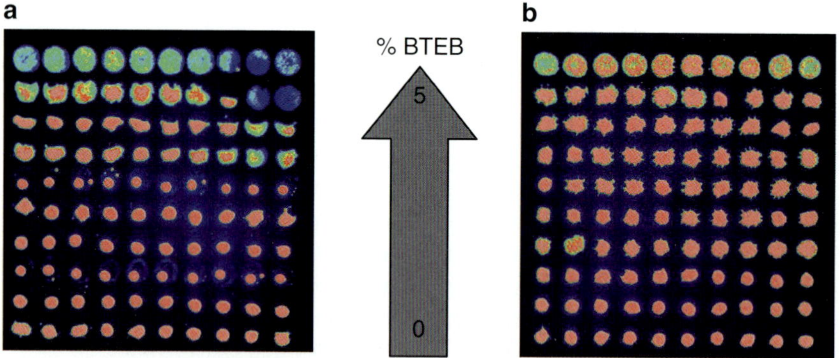

Fig. 16.18 False color CCD images from a $[Ru(dpp)_3]^{2+}$-doped array of xerogel-based sensors composed of different % BTEB: (**a**) Manually prepared; and (**b**) Robotically prepared. The center to center spot spacing is 350 μm in these images

Feature size (F) is an important factor in the fabrication of sensor arrays because it plays a key role in the final density of sensor elements within an array. For example, a hypothetical sensor array fabricated with a sensor element center-to-center (CTC) spacing of 200 $\propto$m ($F = 143–170\,\mu$m) has a density of 2,500 elements/cm^2 whereas an array with a CTC spacing of 100 μm ($F = 71–83\,\mu$m) has a density of 10,000 elements/cm^2. Figure 16.18 presents the false color CCD images associated with this series of pin-printed chemical sensor arrays that were produced manually using micropipettes (A) and using the ALHS (B). There are three interesting aspects of these results. First, formulations prepared using the ALHS showed better spot to spot uniformity. Second, as the concentration of BTEB increased the spot size increased. This behavior is more well-defined with the ALHS in comparison to the hand-made formulations. Third, by controlling the xerogel chemistry, one can produce feature sizes that are approximately twofold smaller than the actual pin diameter ($F \sim 80\,\mu$m; pin diameter = 200 μm).

6   Conclusions

These results demonstrate the remarkable capabilities of an automated production and screening system as a tool to produce and screen a wide array of materials in a short period of time. In the current embodiment, the system can readily produce and fully screen more than 100 samples per day. In a more restrictive mode (e.g., w/o TCSPC), one can increase the throughput to over 300 samples. In an imaging only mode, throughput approaches 1,000 samples per day. These results also demonstrate that automated systems can provide results with minimal user input, while providing better precision and accuracy in comparison to traditional manual methods. The experimental data also shows how the physicochemical properties of BP and xerogel-based materials can be tuned. Finally, the ability to rapidly prepare and characterize large numbers of formulations provides researchers with the ability to quickly select materials that best match specific applications.

Acknowledgments The work from our laboratories was generously supported by the National Science Foundation, the National Institute of Health, the Gerald A. Sterbutzel fund at UB, and the John R. Oishei Foundation.

References

1. Bastioli, C., *Handbook of Biodegradable Polymers*. Rapra Technology Limited: Shawbury, **2005**
2. De Jong, S. J.; Arias, E. R.; Rijkers, D. T. S.; Van Nostrum, C. F.; Kettenes-Van den Bosch, J. J.; Hennink, W. E., New insights into the hydrolytic degradation of poly(lactic acid): participation of the alcohol terminus. *Polymer* **2000**, 42, 2795–2802
3. Middleton, J. C.; Tipton, A. J., Synthetic biodegradable polymers as medical devices. *Medical Plastics and Biomaterials* **1998**, 31–38

4. Shikanov, A.; Kumar, N.; Domb, A. J., Biodegradable polymers: An update. *Israel Journal of Chemistry* **2005**, 45, 393–399
5. Albertsson, A. C.; Varma, I., Aliphatic Polyesters: Synthesis, Properties and Applications. In *Degradable Aliphatic Polyesters*, Albertsson, A. C., Ed. Springer: New York, **2002**; Vol. 157, pp 1–40
6. Amass, W.; Amass, A.; Tighe, B., A review of biodegradable polymers: Uses, current developments in the synthesis and characterization of biodegradable polyesters, blends of biodegradable polymers and recent advances in biodegradation studies. *Polym. Int.* **1998**, 47, 89–144
7. Park, J. H.; Ye, M.; Park, K., Biodegradable polymers for microencapsulation of drugs. *Molecules* **2005**, 10, 146–161
8. Ikada, Y.; Tsuji, H., Biodegradable polyesters for medical and ecological applications. *Macromol. Rapid Commun.* **1999**, 21(3), 117–132
9. Middleton, J. C.; Tipton, A. J., Synthetic biodegradable polymers as orthopedic devices. *Biomaterials* **2000**, 21, 2335–2346
10. Milella, E.; Barra, G.; Ramires, P. A.; Leo, G.; Aversa, P.; Romito, A., Poly(L-lactide)acid/ alginate composite membranes for guided tissue regeneration. *J. Biomed. Mater. Res.* **2001**, 57, 248–257
11. Chen, C.-C.; Chueh, J.-Y.; Tseng, H.; Huang, H.-M.; Lee, S.-Y., Preparation and characterization of biodegradable PLA polymeric blends. *Biomaterials* **2003**, 24, 1167–1173
12. Gunatillake, P.; Adhikari, R., Biodegradable synthetic polymers for tissue engineering. *Euro. Cells Mater.* **2003,** 5, 1–16
13. Daniels, A. U.; Chang, M. K. O.; Andriano, K. P., Mechanical properties of biodegradable polymers and composites proposed for internal fixation of bone. *J. Appl. Biomat.* **1990**, 1, 57–78
14. Utracki, L. A., *Commercial Polymer Blends*. 1st ed.; Springer,: New York, **1998**
15. Ljungberg, N.; Wesslen, B., Preparation and properties of plasticized poly(lactic acid) films. *Biomacromolecules* **2005**, 6(3), 1789–1796
16. Brinker, C. J.; Scherer, G. W., *Sol–Gel Science: The Physics and Chemistry of Sol–Gel Processing*, Academic: Boston, **1990**; p xiv, p. 908
17. Kato, M.; Sakai-Kato, K.; Toyo'oka, T., Silica sol–gel monolithic materials and their use in a variety of applications. *Journal of separation science* **2005**, 28, 1893–1908
18. Tang, Y.; Tehan, E. C.; Tao, Z.; Bright, F. V., Sol–gel-derived sensor materials that yield linear calibration plots, high sensitivity, and long-term stability. *Analytical Chemistry* **2003**, 75(10), 2407–2413
19. McEvoy, A. K.; McDonagh, C.; MacCraith, B. D., *Analyst* **1996**, 121, 785–788
20. Dunbar, R. A.; Jordan, J. D.; Bright, F. V., Development of chemical sensing platforms based on sol–gel-derived thin films: origin of film age vs. performance trade-offs. *Analytical Chemistry* **1996**, 68, 604–610
21. Rupcich, N.; Goldstein, A.; Brennan, J. D., Optimization of sol–gel formulations and surface treatments for the development of pin-printed protein microarrays. *Chemistry of Materials* **2003**, 15(9), 1803–1811
22. Ingersoll, C. M.; Bright, F. V., Using sol–gel-based platforms for chemical sensors. *CHEMTECH* **1997**, 27, 26–31
23. McDonagh, C.; MacCraith, B. D.; McEvoy, A. K., Tailoring of sol–gel films for optical sensing of oxygen in gas and aqueous phase. *Analytical Chemistry* **1998**, 70(1), 45–50
24. Lev, O.; Tsionsky, M.; Rabinovich, L.; Glezer, V.; Sampath, S.; Rankrator, I.; Gun, J., *Analytical Chemistry* **1995**, 67, 22A–30A
25. Collinson, M. M.; Howells, A. R., Sol gel and electrochemistry: Research at the intersection. *Analytical Chemistry* **2000**, 72(21), 702A–709A
26. Schottner, G., Hybrid sol–gel-derived polymers: applications of multifunctional materials. *Chemistry of Materials* **2001**, 13, 3422–3435
27. Maruszewski, K.; W. Strek, M. J.; Ucyk, A., Technology and applications of sol gel materials. *Radiation Effects and Defects in Solids* **2003**, 158, 439–450
28. Ciriminna, R.; Pagliaro, M., Catalysis by sol–gels: An advanced technology for organic chemistry. *Current Organic Chemistry* **2004**, 8, 1851–1862

29. Mahltig, B.; Haufe, H.; Bottcher, H., Functionalisation of textiles by inorganic sol–gel coatings. *Journal of Materials Chemistry* **2005**, 15, 4385–4398

30. Sanchez, C.; Julian, B.; Belleville, P.; Popall, M., Applications of hybrid organic–inorganic nanocomposites. *Journal of Materials Chemistry* **2005**, 15, 3559–3592

31. Innocenzi, P.; Lebeau, B., Organic-inorganic hybrid materials for non-linear optics. *Journal of Materials Chemistry* **2005**, 15, 3821–3831

32. Ogoshi, T.; Chujo, Y., Organic–inorganic polymer hybrids prepared by the sol gel method. *Composite Interfaces* **2005**, 11, 539–566

33. Avnir, D.; Coradin, T.; Lev, O.; Livage, J., Recent bio-applications of sol–gel materials. *Journal of Materials Chemistry* **2006**, 16, 1013–1030

34. Pagliaro, M.; Ciriminna, R.; Man, M. W. C.; Campestrini, S., Better chemistry through ceramics: the physical basis of the outstanding chemistry of ORMOSIL. *Journal of physical chemistry B* **2006**, 110, 1976–1988

35. Park, T. G.; Cohen, S.; Langer, R., Poly (L-lactic acid)/Pluronic blends: Characterization of phase separation behavior, degradation, and morphology and use as protein-releasing matrices. *Macromolecules* **1992**, 25, 116–122

36. Langer, R.; Chasin, M., *Biodegradable Polymers as Drug Delivery Systems*. Marcel Dekker: New York, NY, **1990**

37. Gombotz, W. R.; Pettit, D. K., Biodegradable polymers for protein and peptide drug delivery. *Bioconjugate Chem.* **1995**, 6, 332–351

38. Li, J. K.; Wang, N.; Wu, X. S., A novel biodegradable system based on gelatin nanoparticles and poly(lactic-co-glycolic acid) microspheres for protein and peptide drug delivery. *Journal of Pharmaceutical Sciences* **1997**, 86(8), 891–895

39. Chen, S., Piper, R., Webster, D., Singh, J., Triblock copolymers: synthesis, characterization, and delivery of a model protein. *International Journal of Pharmaceutics* **2005**, 288, 207–218

40. Collinson, M. M., Recent trends in analytical applications of organically modified silicate materials. *TrAC, Trends in Analytical Chemistry* **2002**, 21(1), 30–38

41. Livage, J., Biological applications of sol–gel glasses. In *Sol–Gel Technologies for Glass Producers and Users*, Aegerter, M. A.; Mennig, M., Eds. Springer: New York, **2004**; pp 399–402

42. MacCraith, B. D.; McDonagh, C., Optical chemical sensors. In *Sol–Gel Technologies for Glass Producers and Users*, Aegerter, M. A.; Mennig, M., Eds. Springer: New York, **2004**; pp 313–320

43. Rickus, J. L.; Dunn, B.; Zink, J. I., Optically based sol–gel biosensor materials. *Opt. Biosens.* **2002**, 427–456

44. Pandey, S.; Baker, G. A.; Kane, M. A.; Bonzagni, N. J.; Bright, F. V., On the microenvironments surrounding dansyl sequestered within class I and II xerogels. *Chemistry of Materials* **2000**, 12, (12), 3547–3551

45. Narang, U.; Jordan, J. D.; Bright, F. V.; Prasad, P. N., Probing the cybotactic region of PRODAN in tetramethylorthosilicate-derived sol–gels. *Journal of Physical Chemistry* **1994**, 98(33), 8101–8107

46. Narang, U.; Wang, R.; Prasad, P. N.; Bright, F. V., Effects of aging on the dynamics of rhodamine 6G in tetramethyl orthosilicate-derived sol–gels. *Journal of Physical Chemistry* **1994**, 98(1), 17–22

47. Jordan, J. D.; Dunbar, R. A.; Bright, F. V., Aerosol-generated sol–gel-derived thin films as biosensing platforms. *Analytica Chimica Acta* **1996**, 332(1), 83–91

48. Jordan, J. D.; Dunbar, R. A.; Hook, D. J.; Zhuang, H.; Gardella, J. A., Jr.; Colon, L. A.; Bright, F. V., Production, characterization and utilization of aerosol-deposited sol–gel-derived films. *Chemistry of Materials* **1998**, 10(4), 1041–1051

49. Bonzagni, N. J.; Baker, G. A.; Pandey, S.; Niemeyer, E. D.; Bright, F. V., On the origin of the heterogeneous emission from pyrene sequestered within tetramethylorthosilicate-based xerogels: a decay-associated spectra and O_2 quenching study. *Journal of Sol–Gel Science and Technology* **2000**, 17(1), 83–90

50. Bukowski, R. M.; Ciriminna, R.; Pagliaro, M.; Bright, F. V., High-performance quenchometric oxygen sensors based on fluorinated xerogels doped with [Ru(dpp)3]2+. *Analytical Chemistry* **2005**, 77(8), 2670–2672

51. Bukowski, R. M.; Davenport, M. D.; Titus, A. H.; Bright, F. V., O_2-responsive chemical sensors based on hybrid xerogels that contain fluorinated precursors. *Applied spectroscopy* **2006**, 60(9), 951–957

52. Tao, Z.; Tehan, E. C.; Tang, Y.; Bright, F. V., Stable sensors with tunable sensitivities based on class II xerogels. *Analytical Chemistry* **2006**, 78(6), 1939–1945

53. Shughart, E. L.; Ahsan, K.; Detty, M. R.; Bright, F. V., Site selectively templated and tagged xerogels for chemical sensors. *Analytical Chemistry* **2006**, 78(9), 3165–3170

54. Lochmuller, C. H.; Wenzel, T. J., Spectroscopic studies of pyrene at silica interfaces. *Journal of Physical Chemistry* **1990**, 94(10), 4230–4235

55. Chambers, R.; Haruvy, Y.; Fox, M. A., Excited-state dynamics in the structural characterization of solid alkyltrimethoxysilane-derived Sol–Gel films and glasses containing bound or unbound chromophores. *Chemistry of Materials* **1994**, 6(8), 1351–1357

56. Brennan, J. D.; Hartman, J. S.; Ilnicki, E. I.; Rakic, M., Fluorescence and NMR characterization and biomolecule entrapment studies of sol–gel-derived organic-inorganic composite materials formed by sonication of precursors. *Chemistry of Materials* **1999**, 11(7), 1853–1864

57. Goring, G. L. G.; Brennan, J. D., Fluorescence and physical characterization of sol–gel-derived nanocomposite films suitable for the entrapment of biomolecules. *Journal of Materials Chemistry* **2002**, 12(12), 3400–3406

58. Nishida, F.; McKiernan, J. M.; Dunn, B.; Zink, J. I.; Brinker, C. J.; Hurd, A. J., In situ fluorescence probing of the chemical changes during sol–gel thin film formation. *Journal of the American Ceramic Society* **1995**, 78(6), 1640–1648

59. Brennan, J. D., Using intrinsic fluorescence to investigate proteins entrapped in sol–gel derived materials. *Applied Spectroscopy* **1999**, 53(3), 106A–121A

60. Flora, K. K.; Dabrowski, M. A.; Musson, S. P.; Brennan, J. D., The effect of preparation and aging conditions on the internal environment of sol–gel derived materials as probed by 7-azaindole and pyranine fluorescence. *Canadian Journal of Chemistry* **1999**, 77(10), 1617–1625

61. Huang, M. H.; Soyez, H. M.; Dunn, B. S.; Zink, J. I., In situ fluorescence probing of molecular mobility and chemical changes during formation of dip-coated Sol–gel silica thin films. *Chemistry of Materials* **2000**, 12(1), 231–235

62. Flora, K. K.; Brennan, J. D., Characterization of the microenvironments of PRODAN entrapped in tetraethyl orthosilicate derived glasses. *Journal of Physical Chemistry B* **2001**, 105(48), 12003–12010

63. Keeling-Tucker, T.; Brennan, J. D., Fluorescent probes as reporters on the local structure and dynamics in sol–gel-derived nanocomposite materials. *Chemistry of Materials* **2001**, 13(10), 3331–3350

64. Tleugabulova, D.; Czardybon, W.; Brennan, J. D., Time-resolved fluorescence anisotropy in assessing side-chain and segmental motions in polyamines entrapped in sol–gel derived silica. *Journal of Physical Chemistry B* **2004**, 108(30), 10692–10699

65. Sui, X.; Lin, T.-Y.; Tleugabulova, D.; Chen, Y.; Brook, M. A.; Brennan, J. D., Monitoring the distribution of covalently tethered sugar moieties in sol–gel-based silica monoliths with fluorescence anisotropy: Implications for entrapped enzyme activity. *Chemistry of Materials* **2006**, 18(4), 887–896

66. Baker, G. A.; Pandey, S.; Maziarz, E. P., III; Bright, F. V., Toward tailored xerogel composites: local dipolarity and nanosecond dynamics within binary composites derived from tetraethyl-orthosilane and ORMOSILs, oligomers or surfactants. *Journal of Sol–Gel Science and Technology* **1999**, 15(1), 37–48

67. Baker, G. A.; Wenner, B. R.; Watkins, A. N.; Bright, F. V., Effects of processing temperature on the oxygen quenching behavior of tris(4,7′-diphenyl-1,10′-phenanthroline) ruthenium (II) sequestered within sol–gel-derived xerogel films. *Journal of Sol–Gel Science and Technology* **2000**, 17(1), 71–82

68. Tang, Y.; Tao, Z.; Bright, F. V., Sol hydrolysis and condensation time affects the sensitivity of thin film xerogel-based sensing Materials. *Journal of Sol–Gel Science and Technology* **2007**, 42(2), 127–133

69. Cho, E. J.; Tao, Z.; Tang, Y.; Tehan, E. C.; Bright, F. V.; Hicks, W. L., Jr.; Gardella, J. A., Jr.; Hard, R., Tools to rapidly produce and screen biodegradable polymer and sol–gel-derived xerogel formulations. *Applied Spectroscopy* **2002**, 56(11), 1385–1389

70. Cho, E. J.; Tao, Z.; Tang, Y.; Tehan, E. C.; Bright, F. V.; Hicks, W. L., Jr.; Gardella, J. A., Jr.; Hard, R., Tailored delivery of active keratinocyte growth-factor from biodegradable polymer formulations. *Journal of Biomedical Materials Research, Part A* **2003**, 66A, 417–424

71. Cho, E. J.; Bright, F. V., Pin-printed chemical sensor arrays for simultaneous multianalyte quantification. *Analytical Chemistry* **2002**, 74(6), 1462–1466

72. Dong, D. C.; Winnik, M., *Canadian Journal of Chemistry* **1984**, 62, 2560–2565

73. Wong, A. L.; Hunnicutt, M. L.; Harris, J. M., *Analytical Chemistry* **1991**, 63, 1076

74. Weber, G.; Farris, F. J., Synthesis and spectral properties of a hydrophobic fluorescent probe: 6-propionyl-2-(dimethylamino)naphthalene. *Biochemistry* **1979**, 18(14), 3075–3078

75. Drake, J. M., Photophysics and *cis–trans* isomerization of DCM. *Chemical Physics Letters* **1985**, 113(6), 530–534

76. Tang, Y.; Tao, Z.; Bukowski, R. M.; Tehan, E. C.; Karri, S.; Titus, A. H.; Bright, F. V., Tailored xerogel-based sensor arrays and artifical neural networks yield improved O_2 detection accuracy and precision. *Analyst* **2006**, 131, 1129–1136

Chapter 17
Diversity-Oriented Fluorescence Library Approach for Novel Sensor Development

Shenliang Wang and Young-Tae Chang

Abstract Fluorescence-based sensors are of extreme importance and have found applications in various fields. Although the demand for useful fluorescence sensors is acute, the conventional target-oriented approach has limited the scope and speed of novel sensor discovery. To make a breakthrough in this field, we introduced a new diversity-oriented fluorescence library approach (DOFLA) using combinatorial fluorescence compound libraries with huge structural diversity. Surprisingly, specific and unique sensors for a broad range of analytes from macromolecules including DNA, RNA, and proteins to signaling small molecules such as GTP, glutathione could be discovered from the library pools, demonstrating the universal applicability of this approach.

1 Fluorescence Sensor

Fluorescence sensors are dye molecules whose fluorescence property changes in response to the surrounding medium or through specific molecular recognition events.[1–3] Fluorescence sensors generally can be viewed as a signal transducer, which is a conjugate of receptor and fluorophore, as shown in Fig. 17.1. The receptor part changes its conformation, electron density distribution, or chemical bond break/formation upon receiving the inputs, and then leads to the change of the fluorescent property of the fluorophore part, such as wavelength shifts, increase or drop of quantum yields/fluorescence intensity, or lifetime difference, which can be used as the output for detection. Presence of specific analytes such as anions, cations, or some bio-relevant substances, enzymatic/chemical reactions, or environmental differences such as pH, polarity, viscosity can all be sensed by the receptor, and the binding event can be transformed into fluorescence readouts.

Y.-T. Chang (✉)
Department of Chemistry, National University of Singapore, Singapore 117543
chmcyt@nus.edu.sg

R.A. Potyrailo and V.M. Mirsky (eds.), *Combinatorial Methods for Chemical and Biological Sensors*, DOI: 10.1007/978-0-387-73713-3_17,

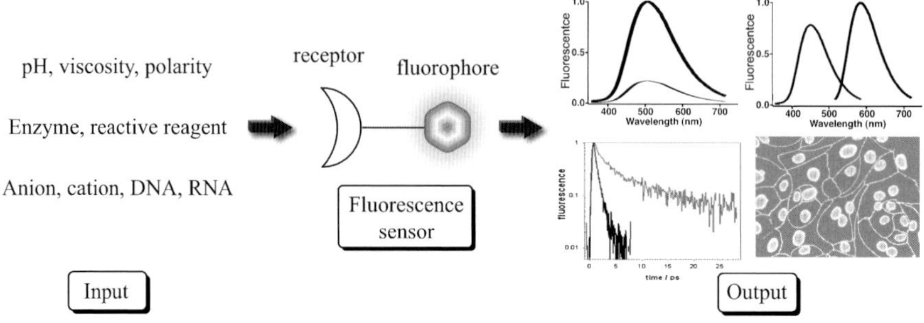

Fig. 17.1 General structure and mechanism of a conventional fluorescence sensor

Compared with fluorescent proteins and Q-dots, small fluorescence molecules have advantages such as less steric bulkiness, faster labeling, and easier handling,[4] and thus have been widely utilized as popular tools for chemical, biological, and medical applications. Although the demand for useful fluorescence sensors is acute, for decades, the conventional *target-oriented approach* remains as the dominating approach for new fluorescence sensor development, if not the only one, which has limited the scope and speed of novel sensor discovery.

1.1 Target-Oriented Approach

The normal strategy for fluorescence sensor design is to combine fluorescence dye molecules with designed receptors for specific analytes, in the wish that the recognition event between receptor and analyte will lead to a photophysical property change of the dye moiety. This approach has played the dominant role throughout the decades of sensor discovery. Successful examples include the evolution of zinc sensors based on the merging of dipyridineamine receptor with various fluorophores and calcium sensors by linking EGTA-based calcium acceptors with various kinds of fluorophores[5,6], as shown in Fig. 17.2.

Although many fluorescence sensors have been successfully developed through this approach, the discovery process might suffer from its inherent limitation. Knowledge about molecular interactions, such as electrostatic interactions, hydrogen bond and hydrophobic interactions are the major considerations in the receptor design.[7,8] Metal chelation effect is applied to cation receptor design, and derived metal-ligand complex can also be used for sensing anions. For communication between receptor and fluorophore, charge-transfer (CT), photoinduced electron transfer (PET), electronic energy transfer (EET), and monomer–excimer conversion are the main mechanism to direct the design of target-oriented fluorescence sensors.[9] Nevertheless, very limited structures such as cyclopolyamine, cyclopolyimidazole, crown ether, guanidine, boric acid, EDTA are repeatedly used as the receptor elements, which greatly limited the scope of new sensor discovery.

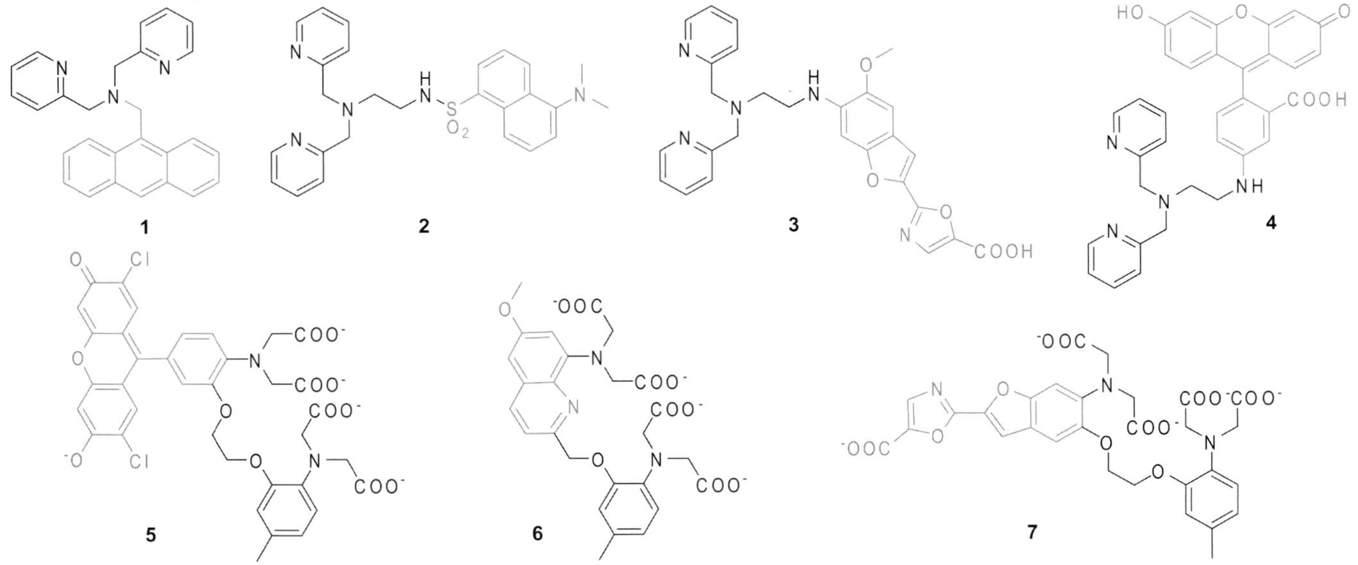

Fig. 17.2 Examples of typical designed fluorescent zinc (1–4) and calcium (5–7) sensors, with designed receptors in *black* and fluorophores in *gray*

The complexity of designed receptor structures usually requires cumbersome synthetic work for each individual candidate compound. Also, the sensor's scope of application is intrinsically limited to the preselected analytes that the sensor was designed for. These entire drawbacks make the target-oriented discovery approach slow, cumbersome, and of low chance for novel sensor structure discovery, thus complimentary approaches are highly demanded.

1.2 Diversity-Oriented Fluorescence Library Approach

Combinatorial chemistry is now being widely used in the chemical biology and medicinal/pharmaceutical field for the discovery of biologically active molecules or drug candidates. Not only has the combinatorial chemistry technical process proved to be convenient and efficient, but also more important is its general strategy of addressing problems with a diversity point of view.[10-12] Broader chemical space could be explored with collections of chemicals thus higher likelihood to the discovery of new small molecule–target interactions.[13] The development of fluorescence sensor is also probing the intermolecular interaction, if considered from a broader point of view, no matter the recognition event is between two small molecules, small molecule and macromolecule (proteins, DNA, RNA, etc.), or even response to environmental differences. Thus discovery and development of sensors with a diversity-oriented strategy is a natural extension sparked by mirroring the current application of combinatorial chemistry in other fields.[14]

The direct comparison between the target-oriented approach and diversity-oriented approach is schematically summarized in Fig. 17.3. The central goal of target-oriented approach is to find a feasible receptor for the target of interest and transform the receptor into a sensor by connecting it to a fluorophore. In contrast, diversity-oriented approach focuses on exploring the diverse chemical space directly around a fluorophore, which serves for both recognition and fluorescence property change. In the former approach, the prepared compounds would be evaluated with the target for fluorescence property change. Otherwise, the structure of receptor or fluorophore and the manner of their linkage should be further tuned and checked. In the diversity-oriented fluorescence library approach (DOFLA), the library of fluorescent compounds is screened against a broad range of analytes without a limitation. Once a fluorescence change is observed, the selected dye sensor is a novel lead structure for the given analyte for further development. Furthermore, it is totally possible to find multiple sensors for different analytes from one library based on their structural diversity.

The DOFLA has advantage in the following aspects (Table 17.1). This approach benefits from the combinatorial chemistry techniques, once an efficient synthetic route can be developed for a diverse set of dyes. Thus broader chemical space could be explored and unknown/unexpected molecular interactions might be discovered. The design and preparation of the fluorescent dye library is unbiased to any specific target analyte, and the library would be evaluated with quite distinct analytes to maximize the chance of application in different fields. Taken all these together,

Target Oriented Approach

1) design a receptor for the target of interest

2) connect fluorophores to the receptor

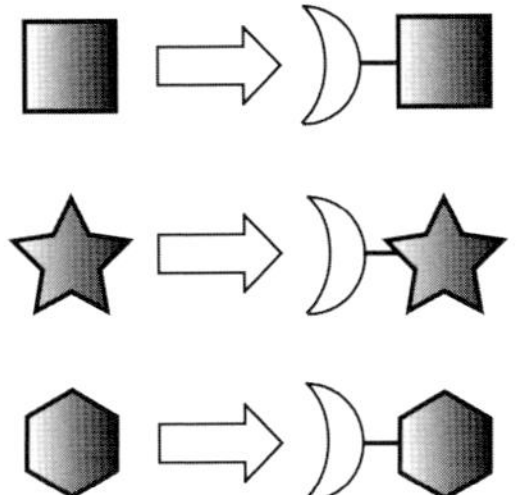

3) evaluate the fluorescent properties

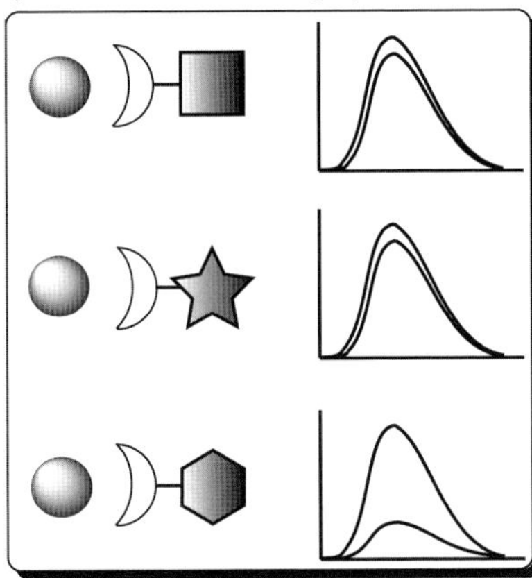

Diversity Oriented Fluorescence Library Approach

1) study the structures of fluorophores

2) prepare Diversity Oriented Fluoresence Library

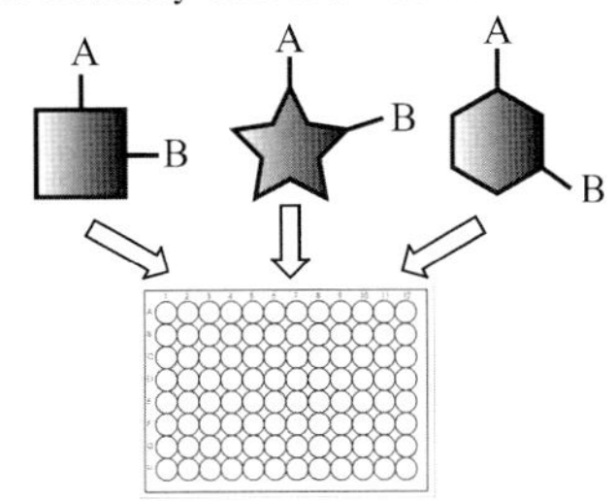

3) evaluate compounds with targets of interest to discover fluorescent sensor and target pair

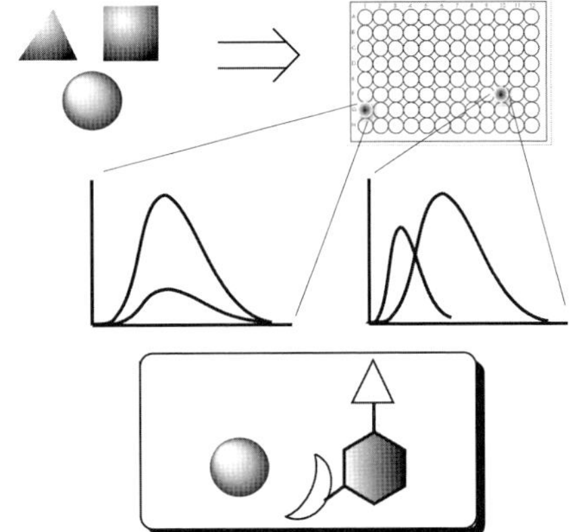

Fig. 17.3 Target-oriented approach and diversity-oriented fluorescence library approach

DOFLA would lead to higher chance for new discoveries and greatly speed up the discovery process and spur the whole filed to move forward.

On one side, application of combinatorial chemistry to sensor discovery could be traced back to very early time, as the receptor discovery has been recognized as an important applicable field for combinatorial chemistry as in medicinal chemistry.[15] In this approach, combinatorial techniques are utilized to compensate receptor design, and the fluorophore can be either incorporated into the receptor synthesis or remain independent for displacement assays. Diversity of the receptor part helps for probing the interactions between the receptor and target analyte, thus this approach can greatly facilitate the target-oriented approach in the situations that the receptor is unknown or not easily designed.[16] One recent example of mercury sensor development clearly demonstrated the power of this approach (Fig. 17.4).[17]

Table 17.1 Advantages of diversity-oriented fluorescence library approach

- Benefited from combinatorial chemistry techniques such as solid phase synthesis, preparation of a set of compounds with satisfying purity could be achieved with less tedious work.
- Fluorescence properties of final dye compounds such as emission wavelengths and quantum yields are diverse, providing large range of pools for multipurpose screenings.
- Broader chemical space directly around a fluorophore could be explored and unkown/unexpected molecular interactions might be discovered.
- The design and preparation of the fluorescent dye library is unbiased to any specific target analyte and the library would be evaluated with quite distinct analytes to maximize the chance of application in different fields.
- Together with the development of high-throughput screening technique, such as fluorescence microplate reader, small molecule microarray, and even high contents imager, a fast screening of the fluorescent dye library is practically feasible.

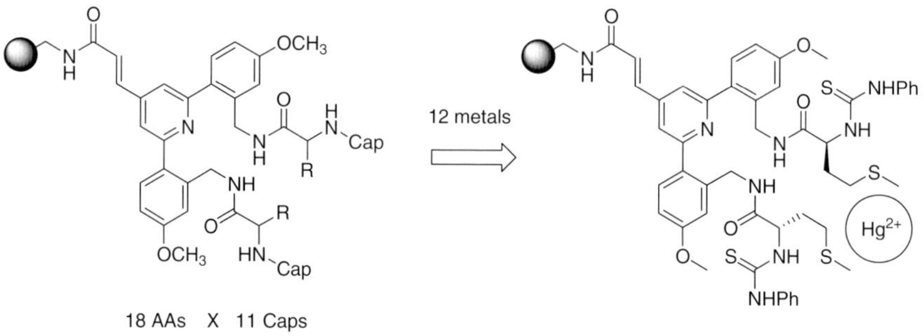

Fig. 17.4 Combinatorial discovery of a mercury (II) fluorescence sensor

The biarylpyridine fluorophore is known to be sensitive to conformation restriction and was chosen for this study to cooperate with a two-armed receptor. The receptor was designed and synthesized following the combinatorial chemistry methods: 18 amino acids and 11 acylating reagents as the amino group caps were sequentially assembled following the parallel synthesis method. The library was screened with a panel of 12 metal cations, and the final hit compound and mercury cation pair was selected. In contrast, DOFLA introduces the structural diversity directly on to the fluorophore itself by generating fluorescence dye library. The synthesis of corresponding fluorophore scaffold is designed where the diversity is introduced in the process of assembling the fluorophore. Thus the fluorescence properties of the final products are affected by the diversity of substituents and a broad chemical diversity is obtained. Examples of DOFLA scaffolds include coumarin,[18,19] dapoxyl,[20] styryl,[21] hemicyanine,[22] and rosamine[23] (Fig. 17.5). Each of the dye scaffolds could be derivatized at one or more positions to generate the diversity to form a library.

Benefited from solid phase synthesis and related techniques, preparation of a set of compounds with satisfying purity could be achieved with less tedious work. One possible problem is that these techniques usually limit the amount of the final products from large quantities. However, due to the sensitivity of fluorescence technique, usually large quantities of dyes are not required for the florescence-based

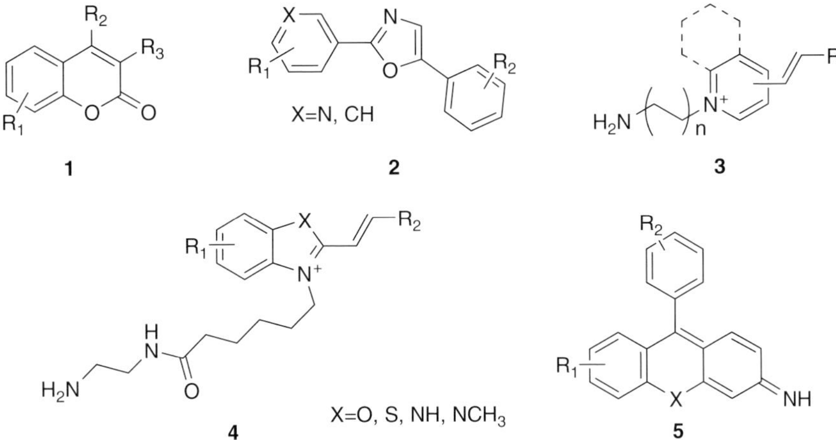

Fig. 17.5 Library of coumarin, Dapoxyl, styryl, hemicyanine, and rosamine scaffolds

primary screening for sensing. Thus the quantity will not be an issue for the application of solid phase synthesis in DOFLA.

Together with the development of high-throughput screening technique, such as fluorescence microplate reader, small molecule microarray, and even high contents imager, a fast screening of the fluorescent dye library is practically feasible. Although DOFLA to fluorescence sensor discovery is only in its infancy, the power of this approach has already been demonstrated with impressive discoveries of novel sensors. Herein, we will summarize several specific examples to further illustrate the approach.

2 Coumarin Dye Library and Applications

The investigation of fluorescence sensors via diversity-oriented fluorescence library approach was marked by the development of a coumarin library. For the first time, a library of compounds based on coumarin scaffold was synthesized at a time and compared with each other for the fluorescent properties.[18] Utilizing Suzuki (ten building blocks), Sonogashira–Hagihara (ten building blocks), and Heck (ten building blocks) type C–C bond formation reactions, 3-bromocoumarin (eight building blocks) were derived into a 151 (out of possible 240) member library. The emission maxima range from blue to yellow color (400–569 nm) and the quantum yields range from zero to near unity (Fig. 17.6). These compounds can possibly be explored as organic light-emitting diodes.

More recently, Wang et al. has published another coumarin library using azide-alkyne ligation (click chemistry).[19] The purpose for this research is to search for triazolylcoumarin with strong fluorescence, while the parent azidocoumarin has

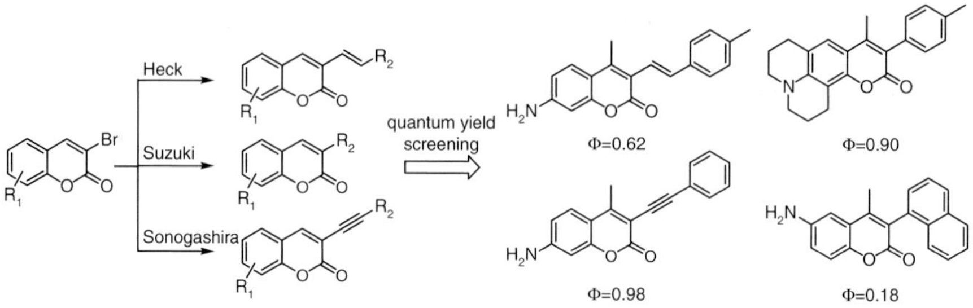

Fig. 17.6 The first coumarin library and their quantum yields

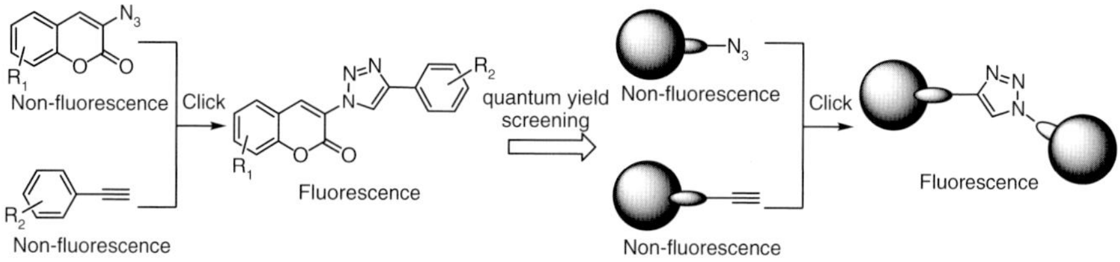

Fig. 17.7 Synthesis of a coumarin library by click chemistry

low or no fluorescence and apply these compounds in bioconjugation technology. 3-azidocoumarins (nine building blocks) were coupled with synthesized or commercial alkynes (24 building blocks) by 1,3-dipolar cycloaddition conditions to screen for bright triazolylcoumarin compounds. The final products cover an emission maxima range between 388 and 521 nm. This work confirmed the conclusion that electron-donating groups at 7-position of coumarins showed strong enhancement of the fluorescence and dominated the emission color. The quantum yields of the hit compounds are around 0.6–0.7, highly fluorescent compared to the parent coumarin compounds. The selected compounds could be applied in derivatization and attachment to other entities such as biomolecules, polymers, nanoparticles, and surfaces.

However, both of these libraries were screened only for their inherent fluorescence property, but not tested for their potential as sensors. As envisioned, fluorescence properties such as emission wavelengths and quantium yields of coumarin compounds (Fig. 17.7) in the same library span large diversity. This demonstrated that even a simple substitution on the fluorophore scaffold can dramatically affect the fluorescence properties, which cannot be delicately controlled through a designed approach. A large pool of fluorescent compounds with different properties provided potentially a large platform for screenings on different purposes.

3   Dapoxyl Dye Library and Human Serum Protein Sensor

Dapoxyl dyes are highly environment-sensitive fluorescent compounds. Their fluorescence in water is very low, but induces a large change of fluorescence intensity, Stokes' shifts, and extinction coefficients depending on pH and polarity of solvents.

A two-step solution phase synthesis was performed and two-dimensional diversities were introduced by starting materials of nicotinic acid (27 building blocks) and 2-amino acetophenone (nine building blocks),[20] as shown in Fig. 17.8. One hundred forty compounds were selected based on final product's purity, and these compounds were then subjected to aqueous environments at different pH or organic solvents and fluorescent spectra were collected and compared. Unique three hit compounds (**A2B1**, **A2B7**, and **A18B7**) showed large Stokes' shifts and strong fluorescence in aqueous buffers under different pH conditions, as well as in the presence of organic

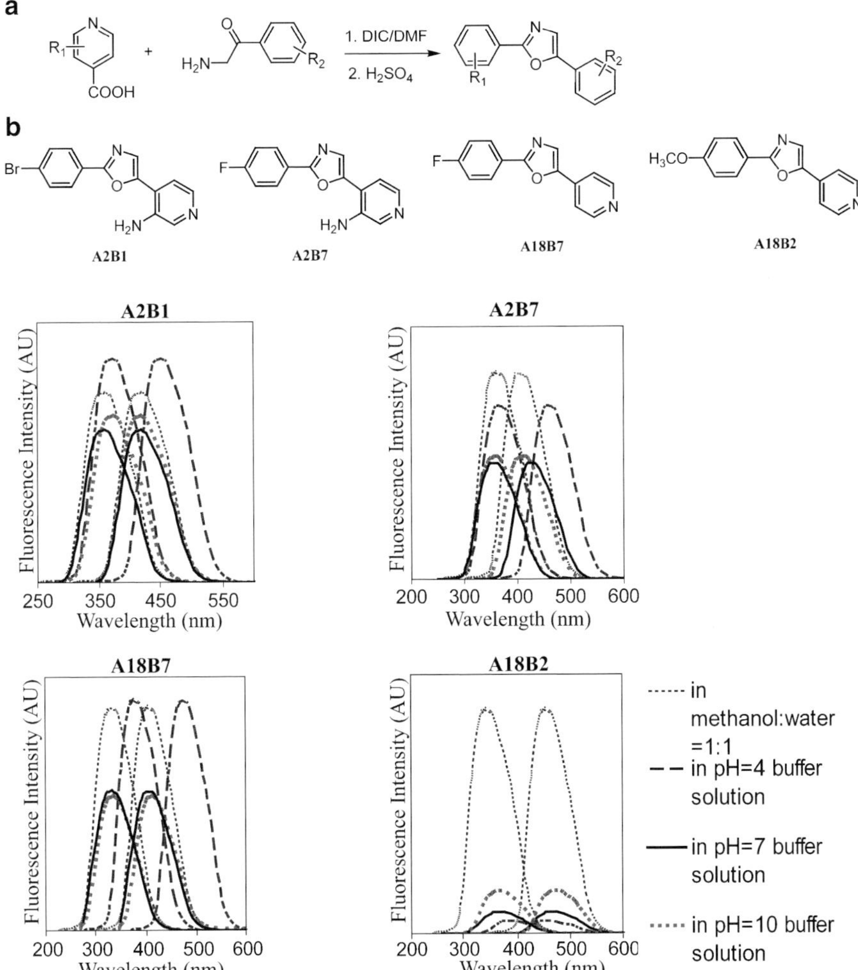

Fig. 17.8 (**a**) Synthetic scheme of a dapoxyl dye library; (**b**) structure of hit compounds from dapoxyl library and excitation (*left*) and emission (*right*) spectra in different environments

solvents, where most of other compounds have little or no fluorescence, such as the negative control compound **A18B2**, which shows strong pH as well as solvent-dependent fluorescence profiles. To facilitate the synthesis and purification, a solid phase synthesis route for dapoxyl library was also developed (Fig. 17.9).[24] To summarize, a phenol nicotinic acid was loaded onto 2-clorotrityl resin and 2-amino acetophenone building blocks were incorporated by HATU amide coupling. Burgess reagent and microwave conditions were applied for the cyclodehydration step, and finally the products were cleaved from the resin under mild acidic condition. Combined with the purified solution phase synthesized compounds, a library of 80 compounds was constructed. This library was screened with diverse analytes, and one notable result is the highly sensitive fluorescence response of compound **A41-S** to human serum protein (HSA). In a physiological condition, **A41-S** displayed a marked dose-dependent increase upon addition of HSA by ca. 55-fold (Fig. 17.10). To verify

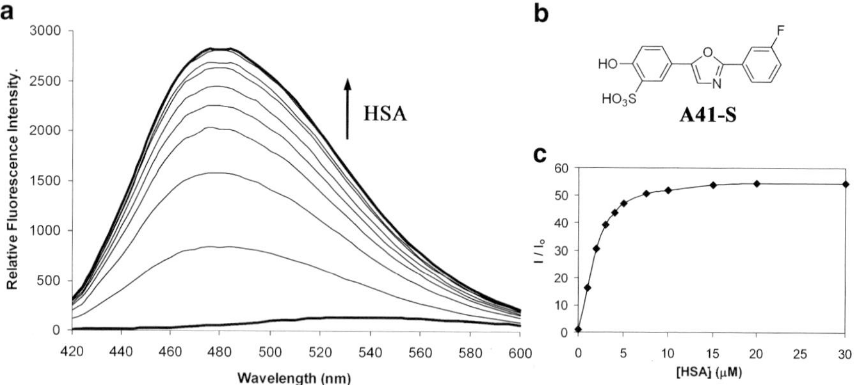

Fig. 17.9 Solid-phase synthesis of a dapoxyl dye library: (**a**) 2-chlorotrityl chloride resin, DIEA, THF; (**b**) 10% piperidine, DMF; (**c**) R_2–C_6H_5–COOH, HATU, DIEA, CH_2Cl_2; (**d**) Ph_3PCl_2, TEA, CH_2Cl_2; (**e**) 1%TFA, CH_2Cl_2

Fig. 17.10 Fluorescence response of A41-S to HSA: (**a**) Fluorescence emission spectra (excitation = 360 nm) of A41-S (5 µM) with 0, 1, 2, 3, 4, 5, 7.5, 10, 15, 20, 30 µM of HSA in 50 µM HEPES buffer (pH = 7.4); (**b**) structure of A41-S; (**c**) binding isotherm from the corresponding fluorescence intensity fold changes with/without HSA

binding sites of **A41-S** on HSA, the fluorescence displacement of **A41-S** from HSA was measured with site specific binding drugs: warfarin (site I)[25,26], ibuprofen (site II), digitoxin (sit III)[27] and triiodobenzoate (site I and II)[28]. The experimental data exhibited the selective binding of **A41-S** on site I of HSA.

In conclusion, two combinatorial dapoxyl dye libraries were synthesized in both solution and solid phase methods and a fluorescence-based screening identified the highly sensitive and selective binding dye (**A41-S**) toward site I of HSA, demonstrating that this dapoxyl dye library has a great potential to sense macromolecules with fluorescence intensity changes.

4 Styryl Dye Library and Applications

Styryl dyes have been used as mitochondrial labeling agents and membrane voltage-sensitive probes of cellular structure and function. From a structural point of view, styryl series of compounds can be viewed as a positive-charged pyridine connected with an aromatic moiety through a double bond. The synthesis is fulfilled by performing a secondary amine catalyzed Knoevenagel type condensation of an activated pyridinium and an aromatic aldehyde (Fig. 17.11). This is a perfect platform for combinatorial chemistry since both series of building blocks are easily synthesized or commercially available, and the reaction conditions are mild and yield is generally high. A library was prepared by the condensation of 41 aldehydes and 14 pyridinium salts in 96-well microplate.[29] The library was subjected to a cell-based microscope screening directly without further purification to test for the organelle localization property. Only 8 out of 855 compounds showed a strong nuclear localization (Fig. 17.12), and they were then subjected to DNA-based fluorescent screening to check for DNA response. Out of eight nuclear localizing compounds, only compound **1** showed a strong fluorescence increase to DNA.[30]

Compound **1(A41B2)** showed a linear increase of fluorescence intensity upon addition of double stranded DNA with 13.3-fold increase. A blue shift of 17 nm in the emission wavelength upon DNA addition was observed, without a significant excitation wavelength shift. The structure of compound **1** includes a 2,4,5-trimethoxy group from the benzaldehyde moiety and a unique adamantyl pyridinium functionality. Different trimethoxy isomers, **2** (3,4,5-trimethoxy) and **3** (2,3,4-trimethoxy) were synthesized to compare the positional effects of the methoxy groups in compound **1**. Although the responses of compound **2** and **3** to DNA treatment was similar to that

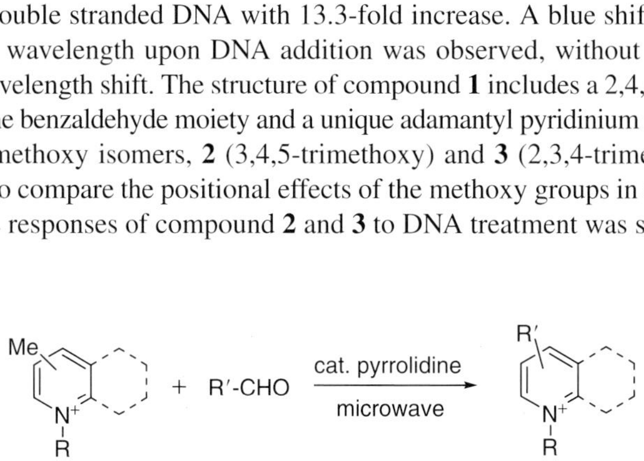

Fig. 17.11 Synthesis of a styryl compound library

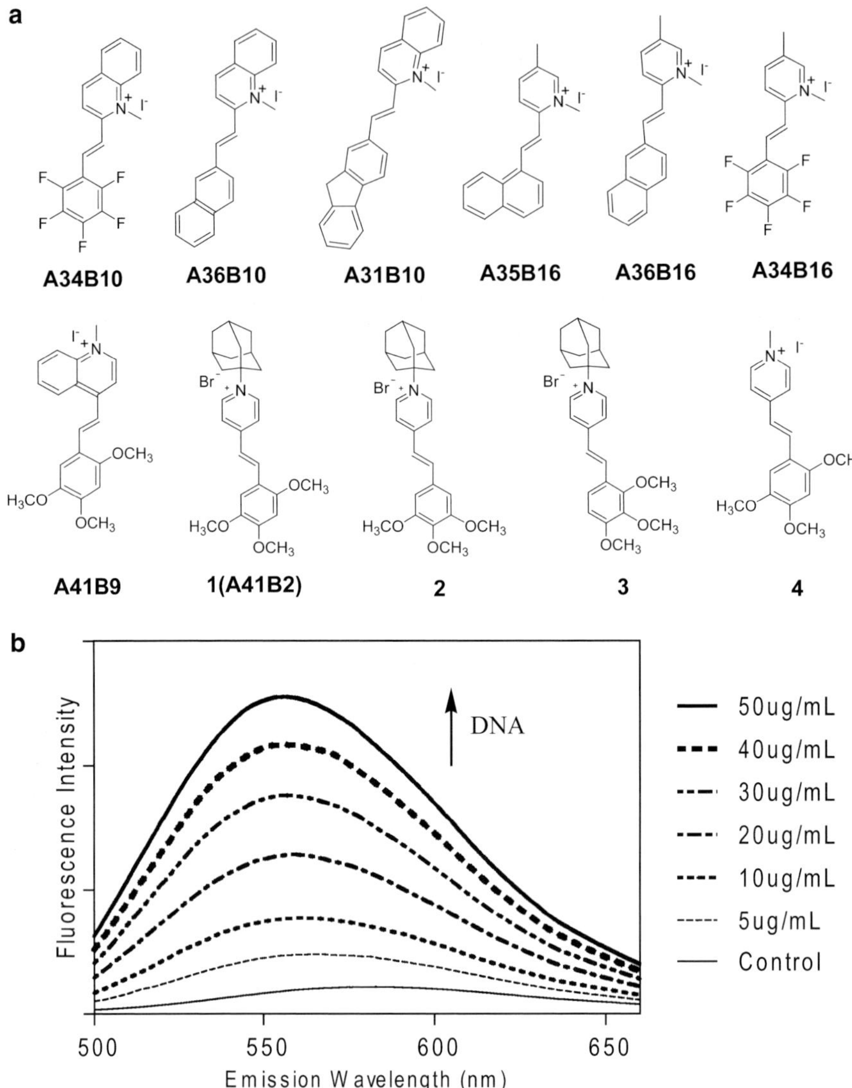

Fig. 17.12 (**a**) Eight nuclear localizing compounds from styryl library and three derivatives of compound 1; (**b**) fluorescence emission spectra (excitation = 394 nm) of compound 1 with indicated concentrations of double strand DNA in a buffer solution

of compound **1**, the fluorescence emission increase was much smaller in compound **2** (4.3-fold) and compound **3** (1.5-fold). Compound **4** was also resynthesized and tested to study the structural importance of the adamantyl group in compound **1**. Interestingly, the simple exchange of the adamantyl with a methyl group significantly reduced the DNA response in compound **4**. Therefore, both 2,4,5-trimethoxy groups and the adamantyl group are important in the specific interaction of compound **1** and DNA.

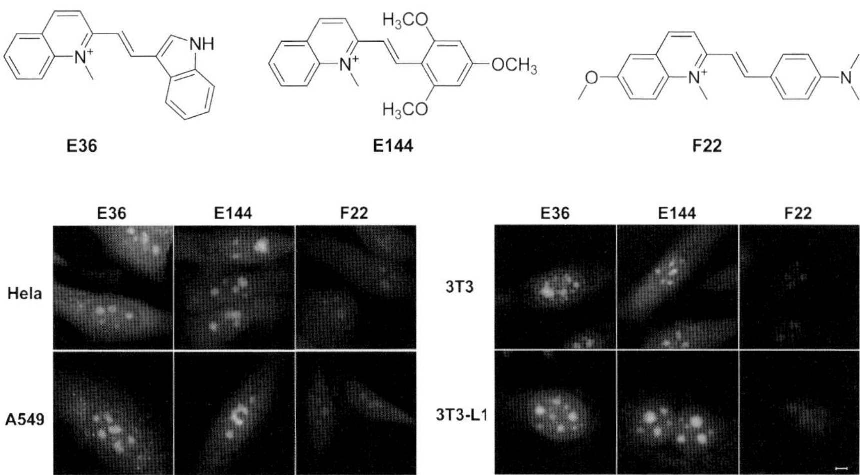

Fig. 17.13 Three hit compounds from styryl library and their live Cell RNA staining

To extend the application of styryl dyes on RNA selective sensors, an expanded library of 1,336 members was prepared in a similar manner.[31] Two primary screening of in vitro RNA response and nucleolar localization in live cells were performed. Three best compounds (**E36**, **E144**, and **F22**) were finally selected based on their strong RNA selective response, good nucleolar targeting, their high fluorescence intensity, low photobleaching, and the smallest effect on cell survival (Fig. 17.13). These three compounds were compared with commercially available SYTO RNASelect on the aspects of in solution response to RNA and DNA, cell tolerability, DNase and RNase digest experiments for DNA/RNA selectivity, and counterstain compatibility with Hoechst and DAPI. These dyes have shown low cell cytotoxicity, low phototoxicity, and high photostability. Their fluorescence excitation/emission is orthogonal to commercially available DNA-binding dyes, and they are found to be suitable for DNA-RNA colocalization experiments.

To develop a method of obtaining comparatively pure styryl compounds in a less cumbersome way, a solid phase synthesis route was also designed and performed.[21] Briefly, 2-chlorotrityl resin was functionalized with two different aminoalcohols, and the alcohol groups were sequentially mesylated and treatment with picoline or quinoline moieties (seven diversities). Subsequently, 64 aldehydes were condensed with the solid-supported pyridinium salts, and finally compounds were cleaved off from the resin. On the basis of the purity, 320 compounds were selected for further screening (Fig. 17.14).

Primary screening of amyloid sensing was carried out with insulin amyloid fibrils generated from fresh insulin, which were induced to aggregate at low pH conditions to form a cross-β-sheet secondary structure. Thirteen compounds from the 320 compounds in the library were selected based on the observed increase or enhancement in their fluorescence intensities upon addition of insulin amyloid fibril. These 13 selected compounds were further tested in synthetic Aβ40 and Aβ42 aggregates. The two Aβ variants, Aβ40 and Aβ42, which differ by truncation at the carboxyl

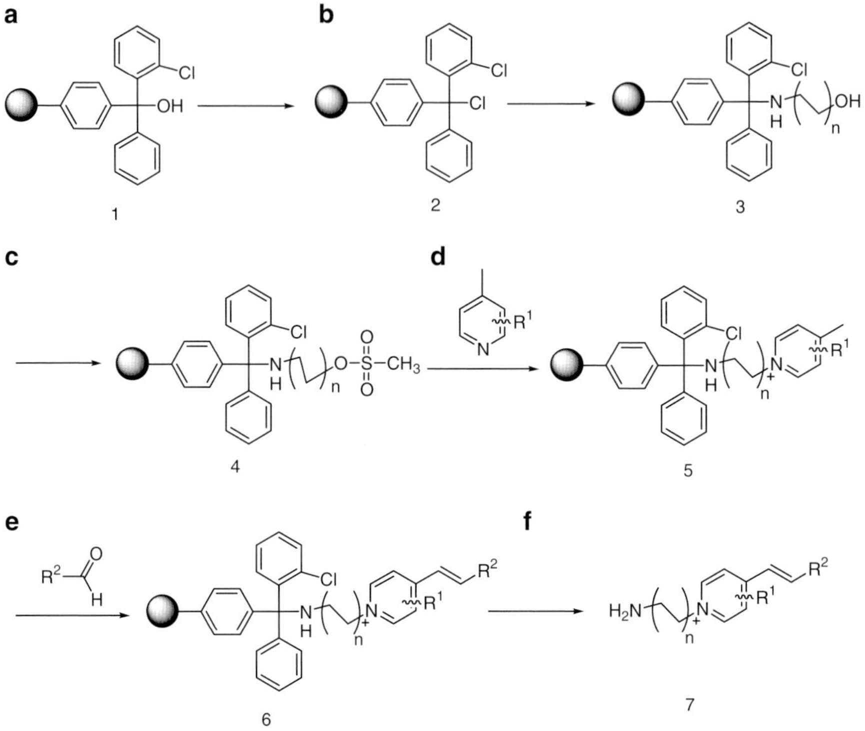

Fig. 17.14 Solid-phase synthesis of the styryl dyes. (**a**) thionyl chloride, CH_2Cl_2, r.t., 2 h; (**b**) ethanol amine ($n = 1$) or 6-amino 1-hexanol ($n = 3$), CH_2Cl_2, r.t., 3 h; (**c**) methanesulfonyl chloride, TEA, CH_2Cl_2, r.t., overnight; (**d**) pyridine derivatives (R^1), NMP, 80–90°C, overnight; (**e**) aldehydes (R^2), pyrrolidine, NMP, 80W microwave 6 min; (**f**) 1% TFA/CH_2Cl_2, 10 min

terminus, are the predominant plaque proteins in Alzheimer disease. All of the dyes showed comparable increase in their fluorescence emissions. Interestingly, all the styryl compounds showed higher sensitivities to fibril Aβ40 than to fibril Aβ42 (Aβ40:Aβ42 = 0.8–3.5), in contrast to thioflavin S, previously known by fluorescent Aβ sensor (Aβ40:Aβ42 = 0.4). When applied to Alzheimer disease mouse brain section, two selected styryl dyes, **2E10** and **2C40**, stained the plaque in green and red respectively, in contrast to the blue color of thioflavin S (Fig. 17.15).

5 Benzimidazolium Dye Library and GTP Sensor

Benzimidazolium scaffold is one of the typical scaffolds of the hemicyanine group. Condensation of benzimidazolium ring with 96 aromatic aldehydes provides extended conjugation and structural diversity.[22] To achieve longer wavelengths of the final fluorophore, which may be more useful for possible biological application, we introduced two Cl groups to the benzimidazolium ring (green–red range of

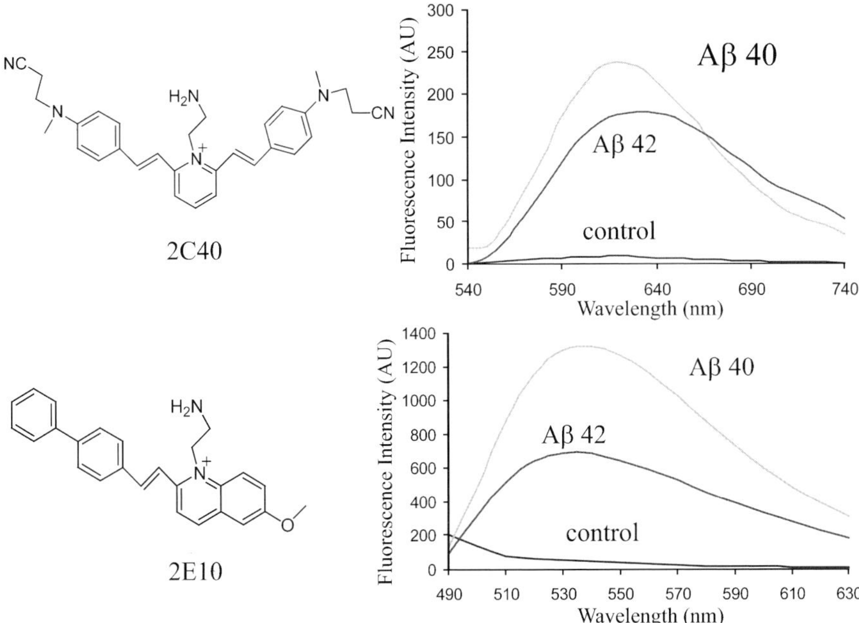

Fig. 17.15 Structure and fluorescence spectrum of 2C40 and 2E10 with 0.5 mg/mL Aβ40 fibril or Aβ42 fibril in 1 mM EDTA pH = 7.4 PBS buffer solution

emission) rather than using an unsubstituted benzimidazolium ring (UV-blue range of emission). It is noteworthy that the diversity elements (from aldehydes) constitute part of the conjugation system of the dye products and will also serve as recognition motifs for analyte binding. To facilitate the synthetic procedure, securing high purity compounds without further purification, we developed a novel solid-phase synthesis pathway for the benzimidazolium library. The optimized synthetic procedure is described in Fig. 17.16. The benzimidazolium scaffold with linker was prepared in solution phase and loaded onto ethylenediamine derivatized 2-chlorotrityl polystyrene solid support. Various lengths of the linker were tested and optimized for best loading of the benzimidazolium compound onto the resin. Aromatic aldehyde building blocks were then coupled to the benzimidazolium ring on solid support, and final products were collected by acidic cleavage. Because of the structural diversity, various excitation/emission wavelengths were observed.

Nucleotide anion detection has long intrigued researchers and witnessed continuous growth.[32–35] Assuming the cationic hemicyanine dye is a potential receptor of nucleotides due to electrostatic interactions, we chose to test the benzimidazolium library with nucleosides and nucleotides for novel fluorescence sensors. Two structurally related compounds (**G32** and **G49**, Fig. 17.17) showed dramatically increased fluorescence upon addition of GTP, while not responding to other nucleotides. Although GTP plays an important role in biological processes,[36,37] very little work has been done on the development of fluorescence sensors for it. Thus far, the best reported GTP sensor, which was designed rationally, showed around 90% quenching response at

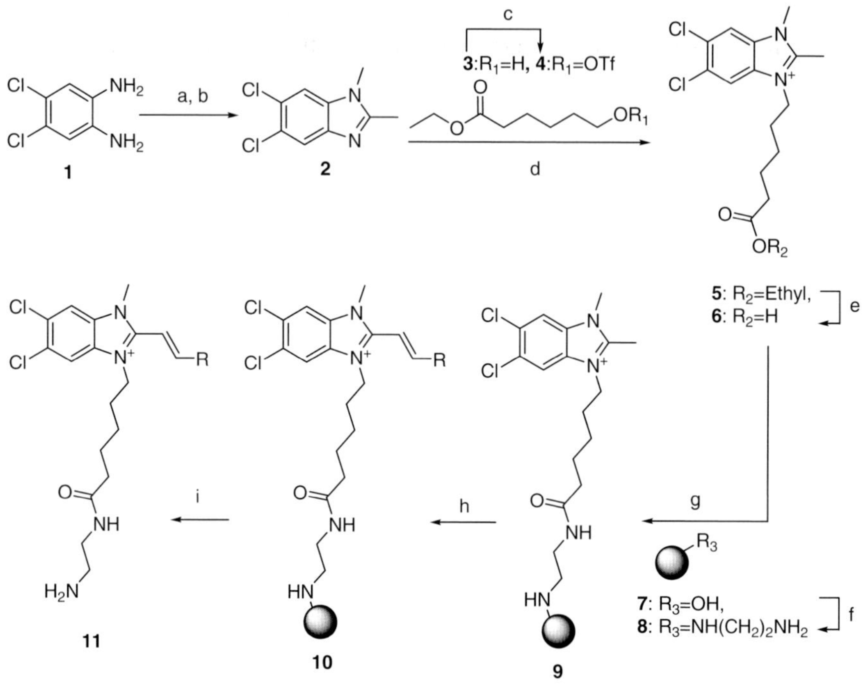

Fig. 17.16 Synthesis of benzimidazolium hemicyanine dyes. Reagents and conditions (**a**) triethyl orthoacetate, H$^+$, toluene, reflux; (**b**) KOH, MeI, acetone; (**c**) Tf$_2$O, poly(4-vinylpyridine), DCM; (**d**) 4, DCM; (**e**) 48% HBr, 65°C; (**f**) 2-chlorotrityl alcohol resin sequentially treated with thionyl chloride in DCM and ethylene diamine in DCM; (**g**) 8, HATU, DIPEA, 30% DMF/DCM; (**h**) R-CHO (96-aromatic aldehydes), pyrrolidine, NMP; (**i**) 5% TFA/DCM

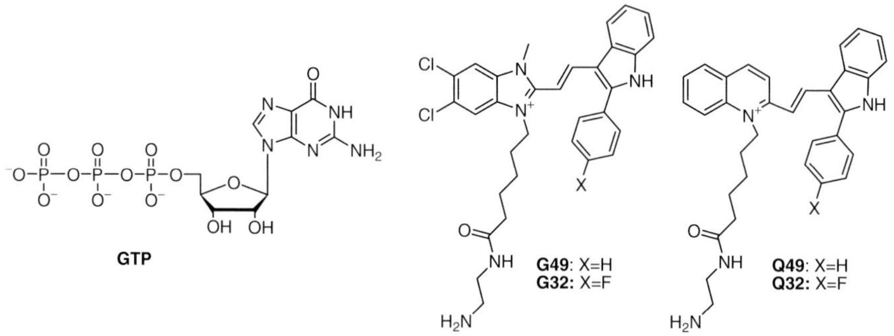

Fig. 17.17 Structure of GTP, G32 and G49, and their quinolinium analogues

around mM concentration of GTP,[34] and most of the known GTP sensors compete with ATP to some extent. To our knowledge, no turn-on fluorescence sensors for GTP have been reported yet. Thus these two compounds showed high potential to be developed as GTP fluorescence sensors and were further studied.

To fully check the selectivity of the two hit compounds, all the nucleosides (adenosine, uridine, cytidine, guanosine) and nucleotides (XNP, where X = A, U, C, G, and N = Mono, Di, Tri) were tested systematically. High selectivity of both **G49** and **G32** only to GTP was clearly exhibited without any obvious cross response to any of other nucleotides or nucleosides (Fig. 17.18). As we observed,

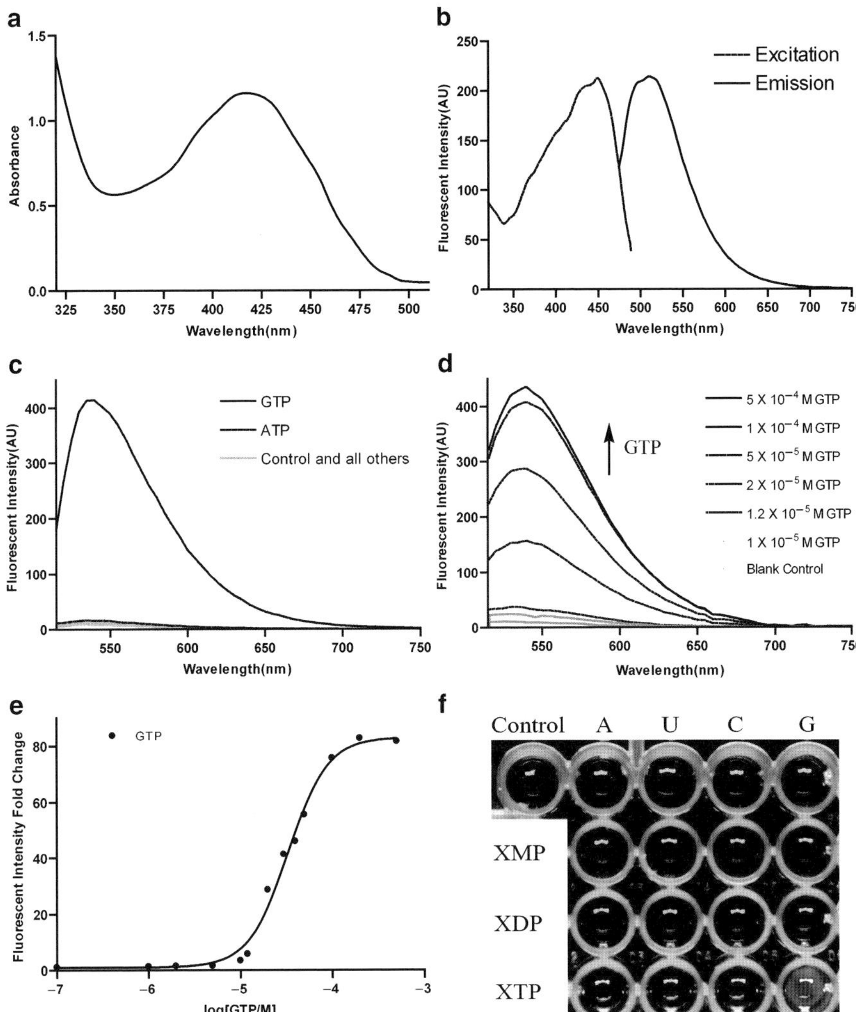

Fig. 17.18 Characterization of GTP Green: (**a**) absorption spectrum in MeOH; (**b**)excitation and emission spectra in MeOH; (**c**) fluorescence emission spectra (excitation = 480 nm)of G49 with 100 μM of GTP, ATP, all other 14 analytes and blank control in 10 mM HEPES buffer (pH = 7.4); (**d**) fluorescence emission spectra (excitation: 480 nm, cutoff: 495 nm) of G49 with 500, 100, 50, 20, 12, 10 μM GTP and blank control in 10 mM HEPES buffer (pH = 7.4); (**e**) binding isotherm from the fluorescence titration experiment with emission at 540 nm; (**f**) picture of G49 with analytes under 365 nm UV lamp light

G32 suffered from significant photo bleaching under strong irradiation light, we decided to focus on **G49** for further analysis. A maximum 80-fold fluorescence intensity increase at 540 nm was observed for GTP, while only two (ATP) or fewer fold changes were observed for all other analytes. The detection range is around 10–100 μM, with a binding constant of ~30,030 M^{-1}. Red shifts of both **G49** excitation and emission wavelength were observed after addition of GTP. A visual distinction was also possible when 5 μM of **G49** was used.

The quinolinium analogs of **G32** and **G49** were also prepared as control compounds, and neither of them showed any fluorescence change in the presence of GTP. Combined with the fact that none of the other 94 library members that originated from different aldehyde groups showed strong fluorescent response to GTP, we postulate that both the imidazonium and the 2-phenyl-indole moieties are important for selective GTP recognition.

Based on this unprecedented high selectivity of **G49** to GTP and its visual green fluorescence increase, we dubbed this compound "**GTP Green**." This is a clear example that unexpected molecular interactions could happen and be discovered through a diversity-oriented fluorescence library approach, even within two small molecules.

6 Rosamine Dye Library and Glutathione Sensor

Rhodamine is a highly favorable fluorescence scaffold due to its excellent photophysical properties such as high extinction coefficient, high quantum yield (often near 1), high photostability, pH-insensitivity (unlike fluorescein), and relatively long emission wavelength (>500 nm). These superior and constant optical properties of rhodamine partially originate from its rigid core structure. In addition to the fused xanthene structure, rhodamine carries 2′-carboxylic acid, which constrains the rotation of the 9-phenyl ring. Therefore, we envisioned that removal of the 2′-carboxylic acid group from rhodamine (called rosamine) will introduce flexibility onto the rhodamine scaffold, and thus would generate a possible sensor candidate whose fluorescence properties can be changed by an analyte environment.[23]

Initially three different 3-amino-6-nitro-9H-xanthone derivatives (S3, Y = O, NH, S) were synthesized. The first diversity was introduced by modifying the 3-amino group. Utilizing the 6-nitro group as a linker after reduction to an amino group, each intermediate was loaded on the 2-chlorotrityl chloride resin and reacted with 33 different Grignard reagents for the second diversity. An acidic cleavage from the resin resulted in the dehydration, giving the fully conjugated rosamine derivatives (Fig. 17.19).

For primary screening, the library compounds were screened for fluorescence intensity change toward a variety of bio-relevant analytes. Notably, **H22** exhibited a highly selective response toward reduced glutathione (GSH) compared with other analytes in physiological condition. **H22** showed a marked fluorescence increase upon addition of GSH (5 mM) by ca. 11-fold in 30 min (Fig. 17.20) and the quantum yield increased from 0.033 to 0.28. It is noteworthy that **H22** did not show any fluorescence response to GSSG of same concentration level, while several thiol-containing analytes such as DTT, β-mercaptoethanol and cysteine showed modest

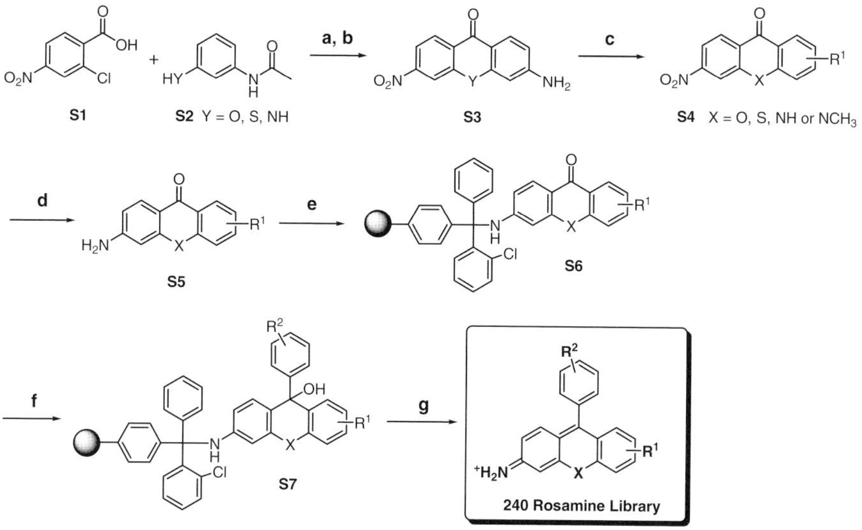

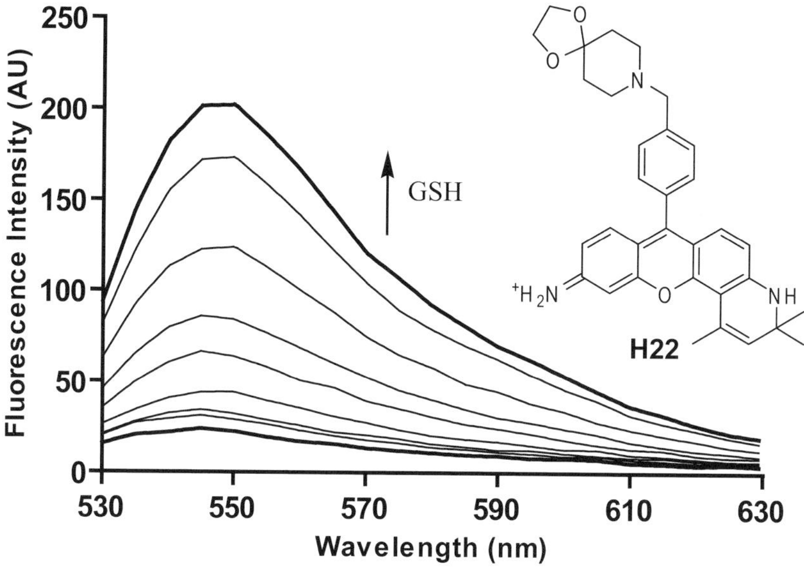

Fig. 17.19 Synthesis of rosamine library on solid support. (**a**) K$_2$CO$_3$, Cu, DMF, 130°C; (**b**) H$_2$SO$_4$, 80°C; (**c**) modification for R^1; (**d**) SnCl$_2$, EtOH, 90°C; (**e**) 2-chloro-trityl chloride resin, pyr, CH$_2$Cl$_2$/DMF; (**f**) R^2 Grignard reagent, THF, 62°C; (**g**) 1%TFA, CH$_2$Cl$_2$

Fig. 17.20 Fluorescence emission spectra (excitation = 500 nm) of H22 (3 μM) with GSH at 0, 0.01, 0.05, 0.1, 0.25, 0.5, 1.0, 2.5, 5 mM in 50 mM HEPES buffer (pH = 7.4) after 30 min incubation

responses to **H22**. To test its dynamic response toward GSH, the preincubated **H22** with GSH for 30 min was treated either with a thiol reactive reagent (*N*-methylmaleimide, NMM) or a thiol-oxidant (diamide) for another 30 min. The fluorescence intensity was decreased upon addition of either NMM or diamide in a dose-dependent manner.

The capability of **H22** to monitor GSH in a living cell was also proved and images were taken at different time. 3T3 cells were supplemented with R-lipoic acid, known to enhance the reduced GSH level in a variety of cells,[38] and subsequent staining of cells with **H22** showed a clear increase in the intracellular fluorescence intensity compared with nontreated cells. When NMM or diamide was supplemented to R-lipoic acid-treated cells stained with **H22**, a distinct decrease of fluorescence intensity was observed. In a similar manner, incubation of the cells with BSO (buthioninesulfoximine; GSH synthesis inhibitor[39]) also showed a similar intensity decrease, while untreated cells did not show such a decrease in the same experiment duration. Taken together, these experiments clearly demonstrate that **H22** is able to sense changes in the cellular GSH level in living cells.

7 Conclusion and Perspectives

Taken all these examples together, we have demonstrated the surprising power of *diversity-oriented fluorescence library approach (DOFLA)* in the discovery of new sensors. Specific and unique sensors for a broad range of analytes from macromolecules including DNA, RNA, and proteins to signaling small molecules such as GTP, glutathione could be discovered from the library pools, demonstrating the universal applicability of this approach. These discoveries went beyond the designed structures of fluorescence sensor, thus serves as a perfect complimentary to the traditional target-oriented approach.

Thus far, only several fluorescent scaffolds have been studied by the diversity-oriented approach and broader chemical space could be achieved for introducing higher diversity. Even more, there are other fluorescent dye scaffolds remain untouched yet. As more fluorescent compounds are synthesized systematically, richer information for structure–photophysical property relationships should be available, which in turn could help for the designing of target-specific fluorescence sensors.

Acknowledgments The research and discoveries summarized herein were mainly based on the continuous contributions from the previous and present Chang group members. We are very grateful to all the individuals involved with these projects, with particular appreciations to: Dr. Gustavo R. Rosania, Dr. Shao Q. Yao, Dr. Jaeki Min, Dr. Jaewook Lee, Dr. Qian Li, Dr. Young-Hoon Ahn, Mr. Jun-Seok Lee, and Ms. Yunkyung Kim.

References

1. Valeur, B. Molecular fluorescence: Principles and applications; Wiley: Weinheim, New York, 2002
2. Lakowicz, J. R. Principles of fluorescence spectroscopy; 2nd ed.; Kluwer/Plenum: New York, 1999
3. Czarnik, A. W. Fluorescent chemosensors for ion and molecule recognition; American Chemical Society: Washington, DC, 1993

4. Zhang, J.; Campbell, R. E.; Ting, A. Y.; Tsien, R. Y., Creating new fluorescent probes for cell biology, *Nat Rev Mol Cell Biol* 2002, 3, 906–918

5. Czarnik, A. W., Desperately seeking sensors, *Chem Biol* 1995, 2, 423–428

6. Jiang, P. J.; Guo, Z. J., Fluorescent detection of zinc in biological systems: Recent development on the design of chemosensors and biosensors, *Coord Chem Rev* 2004, 248, 205–229

7. Beer, P. D.; Gale, P. A., Anion recognition and sensing: The state of the art and future perspectives, *Angew Chem Int Ed Engl* 2001, 40, 486–516

8. Valeur, B.; Leray, I., Design principles of fluorescent molecular sensors for cation recognition, *Coord Chem Rev* 2000, 205, 3–40

9. deSilva, A.; Gunaratne, H.; Gunnlaugsson, T.; Huxley, A.; McCoy, C.; Rademacher, J.; Rice, T., Signaling recognition events with fluorescent sensors and switches, *Chem Rev* 1997, 97, 1515–1566

10. Tan, D. S., Diversity-oriented synthesis: Exploring the intersections between chemistry and biology, *Nat Chem Biol* 2005, 1, 74–84

11. Schreiber, S. L., Target-oriented and diversity-oriented organic synthesis in drug discovery, *Science* 2000, 287, 1964–1969

12. Fergus, S.; Bender, A.; Spring, D. R., Assessment of structural diversity in combinatorial synthesis, *Curr Opin Chem Biol* 2005, 9, 304–309

13. Dobson, C. M., Chemical space and biology, *Nature* 2004, 432, 824–828

14. Finney, N. S., Combinatorial discovery of fluorophores and fluorescent probes, *Curr Opin Chem Biol* 2006, 10, 238–245

15. De Miguel, Y. R., Sanders, J. K. M., Generation and screening of synthetic receptor libraries, *Curr Opin ChemBiol* 1998, 2, 417–421

16. Srinivasan, N.; Kilburn, J. D., Combinatorial approaches to synthetic receptors, Curr Opin *Chem Biol* 2004, 8, 305–310

17. Mello, J.; Finney, N., Reversing the discovery paradigm: A new approach to the combinatorial discovery of fluorescent chemosensors, *J Am Chem Soc* 2005, 127, 10124–10125

18. Schiedel, M. S.; Briehn, C. A.; Bauerle, P., Single-compound libraries of organic materials: Parallel synthesis and screening of fluorescent dyes, *Angew Chem Int Ed Engl* 2001, 40, 4677–4680

19. Sivakumar, K.; Xie, F.; Cash, B. M.; Long, S.; Barnhill, H. N.; Wang, Q., A fluorogenic 1,3-dipolar cycloaddition reaction of 3-azidocoumarins and acetylenes, *Org Lett* 2004, 6, 4603–4606

20. Zhu, Q.; Yoon, H. S.; Parikh, P. B.; Chang, Y. T.; Yao, S. Q., Combinatorial discovery of novel fluorescent dyes based on dapoxyl, *Tetrahedron Lett* 2002, 43, 5083–5086

21. Li, Q.; Lee, J.; Ha, C.; Park, C.; Yang, G.; Gan, W.; Chang, Y., Solid-phase synthesis of styryl dyes and their application as amyloid sensors, *Angew Chem Int Ed Engl* 2004, 43, 6331–6335

22. Wang, S.; Chang, Y. T., Combinatorial synthesis of benzimidazolium dyes and its diversity directed application toward GTP-selective fluorescent chemosensors, *J Am Chem Soc* 2006, 128, 10380–10381

23. Ahn, Y. H.; Lee, J. S.; Chang, Y. T., Combinatorial rosamine library and application to in vivo glutathione probe, *J Am Chem Soc* 2007, 129, 4510–4511

24. Min, J., Lee, J. W., Ahn, Y. H., Chang, Y. T., Combinatorial dapoxyl dye library and its application to site selective probe for human serum albumin, *J Comb Chem* 2007, 9, 1079–1083

25. Sudlow, G.; Birkett, D. J.; Wade, D. N., Characterization of 2 specific drug binding-sites on human-serum albumin, *Mol Pharmacol* 1975, 11, 824–832

26. Sudlow, G.; Birkett, D. J.; Wade, D. N., Further characterization of specific drug binding-sites on human-serum albumin, *Mol Pharmacol* 1976, 12, 1052–1061

27. Irikura, M.; Takadate, A.; Goya, S.; Otagiri, M., 7-alkylaminocoumarin-4-acetic acids as fluorescent-probe for studies of drug-binding sites on human serum-albumin, *Chem Pharm Bull* 1991, 39, 724–728

28. He, X. M.; Carter, D. C., Atomic-structure and chemistry of human serum-albumin, *Nature* 1992, 358, 209–215

29. Rosania, G. R.; Lee, J. W.; Ding, L.; Yoon, H. S.; Chang, Y. T., Combinatorial approach to organelle-targeted fluorescent library based on the styryl scaffold, *J Am Chem Soc* 2003, 125, 1130–1131

30. Lee, J. W.; Jung, M.; Rosania, G. R.; Chang, Y. T., Development of novel cell-permeable DNA sensitive dyes using combinatorial synthesis and cell-based screening, *Chem Comm* 2003, 1852–1853

31. Li, Q.; Kim, Y.; Namm, J.; Kulkarni, A.; Rosania, G. R.; Ahn, Y. H.; Chang, Y. T., RNA-selective, live cell imaging probes for studying nuclear structure and function, *Chem Biol* 2006, 13, 615–623

32. Mizukami, S.; Nagano, T.; Urano, Y.; Odani, A.; Kikuchi, K., A fluorescent anion sensor that works in neutral aqueous solution for bioanalytical application, *J Am Chem Soc* 2002, 124, 3920–3925

33. Li, C.; Numata, M.; Takeuchi, M.; Shinkai, S., A sensitive colorimetric and fluorescent probe based on a polythiophene derivative for the detection of ATP, *Angew Chem Int Ed Engl* 2005, 44, 6371–6374

34. Kwon, J.; Singh, N.; Kim, H.; Kim, S.; Kim, K.; Yoon, J., Fluorescent GTP-sensing in aqueous solution of physiological pH, *J Am Chem Soc* 2004, 126, 8892–8893

35. McCleskey, S.; Griffin, M.; Schneider, S.; McDevitt, J.; Anslyn, E., Differential receptors create patterns diagnostic for TP and GTP, *J Am Chem Soc* 2003, 125, 1114–1115

36. Burma, D. P., GTP acts as the master key in regulating diverse physiological processes, *J Sci Ind Res* 1988, 47, 65–80

37. Pogson, C. I., Guanine nucleotides and their significance in biochemical processes, *Am J Clin Nutr* 1974, 27, 380–402

38. Moini, H.; Tirosh, O.; Park, Y. C.; Cho, K. J.; Packer, L., R-alpha-lipoic acid action on cell redox status, the insulin receptor, and glucose uptake in 3T3-L1 adipocytes, *Arch Biochem Biophys* 2002, 397, 384–391

39. Deneke, S. M.; Fanburg, B. L., Regulation of cellular glutathione, *Am J Physiol* 1989, 257, L163–L173

Chapter 18
Construction of a Coumarin Library
for Development of Fluorescent Sensors

Tomoya Hirano and Hiroyuki Kagechika

Abstract Construction of chemical libraries is a useful approach to the discovery of better fluorescent materials, and several types, such as styryl dyes and cyanine dyes, have been reported. In this chapter, we focus on construction of a library of chemicals having a coumarin skeleton as the core structure. Coumarin and its derivatives are key structures in various bioactive or fluorescent molecules, and their fluorescence properties are dependent on the precise structure, including the positions of substituents.

1 Method to Develop Novel Fluorescent Sensors

Fluorescent sensors are functional fluorescent molecules, whose fluorescence properties (e.g., intensity, excitation wavelength, emission wavelength, etc.) can be changed by binding or reacting with target molecular species. They are useful in many fields of scientific research, especially analytical chemistry and cell biology.[1,2] Fluorescent sensors with appropriate specificity allow the visualization of cations, anions, small molecules or enzyme activity in living cells, tissues, or organisms by means of fluorescence microscopy.

There have been various attempts to develop novel and useful fluorescent sensors. One of the successes was the development of fluorescent sensors for calcium ion (Ca^{2+}), which has many physiologically important roles in biological systems. Fura-2 is a fluorescent Ca^{2+} sensor bearing a O,O'-bis(2-aminophenyl) ethyleneglycoyl-N,N,N',N'-tetraacetic acid (BAPTA) moiety as the Ca^{2+} recognition site, whose excitation wavelength is blue-shifted by binding to Ca^{2+} (Fig. 18.1).[3] Fluo-3 is another type of fluorescent sensor whose fluorescence intensity is increased by binding to Ca^{2+}.[4]

T. Hirano (✉) and H. Kagechika
School of Biomedical Sciences, Tokyo Medical and Dental University,
2-3-10 Kanda-Surugadai, Chiyoda-ku, Tokyo 101-0062, Japan
hiraomc@tmd.ac.jp

R.A. Potyrailo and V.M. Mirsky (eds.), *Combinatorial Methods for Chemical and Biological Sensors*, DOI: 10.1007/978-0-387-73713-3_18,

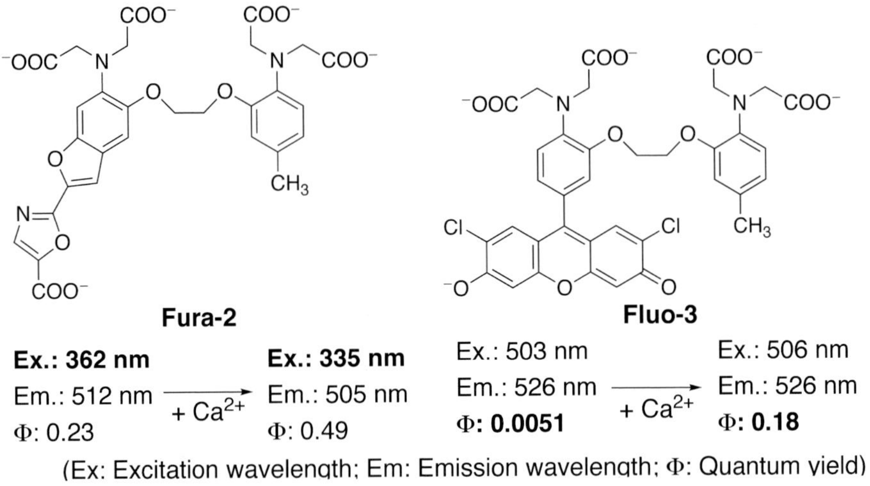

Fura-2

Ex.: 362 nm   **Ex.: 335 nm**   Ex.: 503 nm   Ex.: 506 nm

Em.: 512 nm $\xrightarrow{+\ Ca^{2+}}$ Em.: 505 nm   Em.: 526 nm $\xrightarrow{+\ Ca^{2+}}$ Em.: 526 nm

Φ: 0.23 Φ: 0.49   **Φ: 0.0051**   **Φ: 0.18**

Fluo-3

(Ex: Excitation wavelength; Em: Emission wavelength; Φ: Quantum yield)

Fig. 18.1 Fluorescent sensors for calcium ion

The structure of the fluorescent sensor Fluo-3 consists of BAPTA attached to a xanthene derivative as the fluorophore (Fig. 18.2). Many xanthene-based (sometimes referred to fluorescein-based) fluorescent sensors such as Fluo-3 have been reported. These compounds have an aromatic moiety that recognizes or reacts with the target at the 9-position of the xanthene ring.

Diaminofluoresceins (DAFs) were developed as fluorescent sensors for nitric oxide.[5] The structures of DAFs had been designed "empirically," based on reported substituent effects of fluorescein, i.e., that fluorescein substituted with an amino group at the upper benzoic acid moiety of fluorescein was weakly fluorescent, whereas amido-substituted fluorescein was strongly fluorescent, which indicated that *N*-acylation of aminofluorescein causes strong enhancement of the fluorescence intensity.[6] The reaction of the *o*-diamino group of DAFs with nitric oxide to generate a triazole ring was expected to have a similar effect on the fluorescence. As expected, the fluorescence intensity of DAFs was increased after the formation of triazole ring by reaction with nitric oxide (an oxygen molecule is also required), as reported in 1998.[5] Since then, other xanthene-based fluorescent sensors have been developed, such as DPAXs[7] and DMAXs,[8] whose fluorescence intensities are increased by reaction with singlet oxygen to form endoperoxide and ZnAFs, whose fluorescence intensities are increased by binding with zinc ion.[9,10]

The mechanism of the change of fluorescence intensity of xanthene-based fluorescent sensors was also studied, and it was hypothesized that photo-induced electron transfer between the xanthene and 9-aromatic moiety might be important. Nagano and Urano et al. systematically synthesized xanthene derivatives bearing various substituted phenyl groups at the 9-position, and confirmed that electron transfer has a significant role in their fluorescence properties; they also showed that the determinants of electron transfer efficiency are physical parameters of the xanthene or

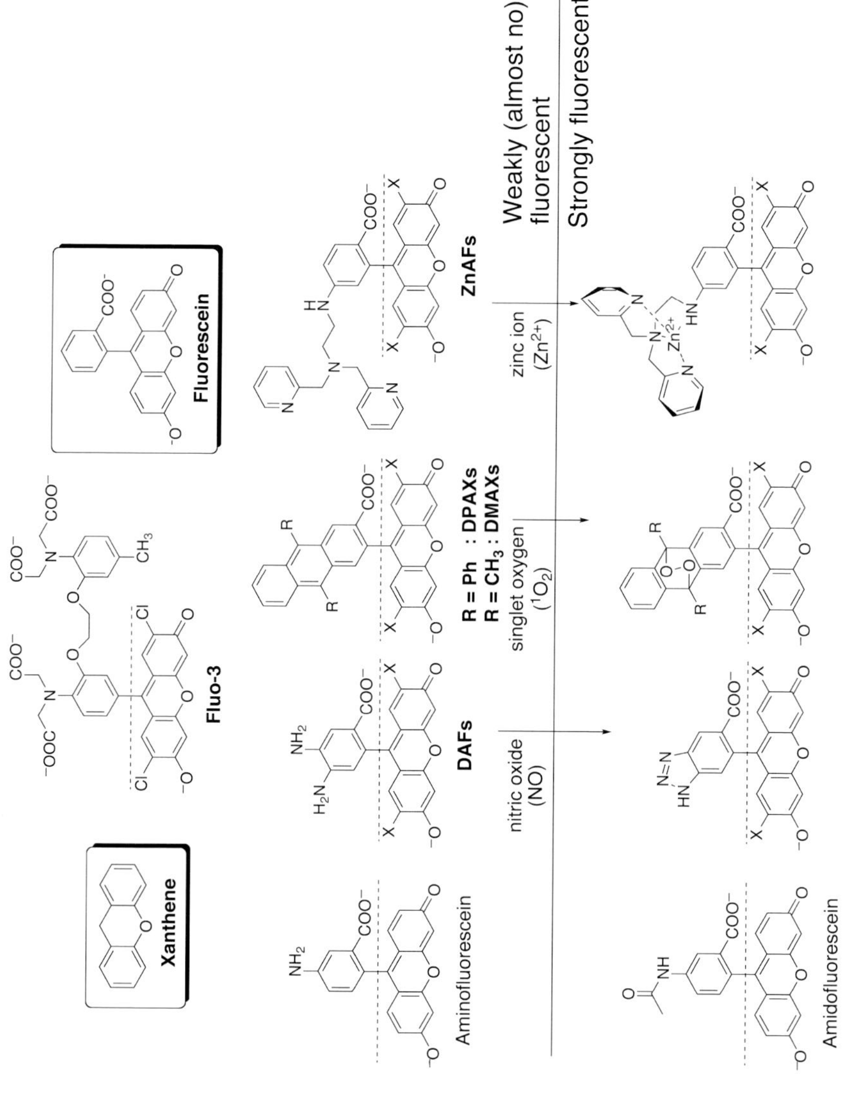

Fig. 18.2 Xanthene-based fluorescent sensors

benzoic moiety, such as the oxidation or reduction potentials.[11] This theoretical understanding enabled us to develop novel fluorescent sensors "rationally," not empirically, as exemplified by the development of glycosidase sensors.[12]

Both empirical and rational methods have been successful in developing novel fluorescent sensors. However, on the one hand, empirical design and synthesis may require considerable trial and error. On the other hand, the rational design approach described above is limited to analytes that can sufficiently change the oxidation or reduction potential. Further, even in the case of theoretically designed molecules, the fluorescence properties may be unexpectedly influenced by environmental factors. The construction of libraries of fluorescent molecules is one way to overcome some of these problems in the development of novel fluorescent sensors.

2 Construction of a Coumarin Library

Coumarin and its derivatives are found in many plants, and exhibit various biological activities.[13] For example, warfarin decreases blood coagulation by interfering with vitamin K metabolism, and coumafuryl (fumarin) is used as a rodenticide (Fig. 18.3).

In addition, they are widely used as fluorescent molecules for laser dyes,[14] in the emission layer in organic light-emitting diodes,[15] and as the fluorophore in fluorescent sensors.[16,17] Their fluorescence properties are known to depend upon the nature and position of substituents, and therefore several groups have constructed coumarin libraries to discover better coumarin-based materials.

2.1 Coumarin Library Constructed by Means of Palladium-Catalyzed Coupling Reactions

Bäuerle's group constructed a coumarin library with diversity at the 3-position via three well-known coupling reactions using palladium catalysts.[18,19] 3-bromo-coumarin

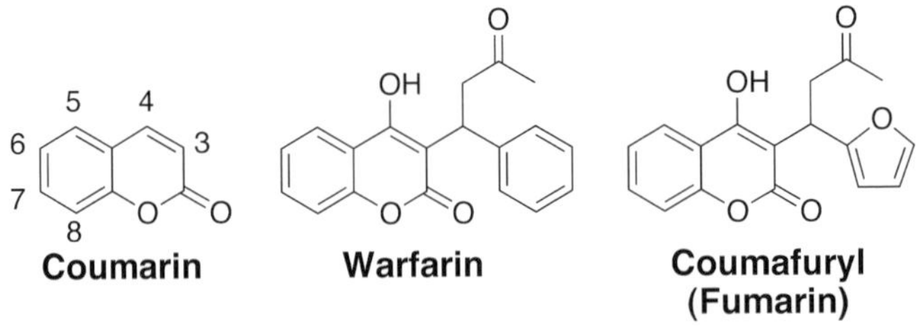

Fig. 18.3 Structures of coumarin and some biologically active derivatives

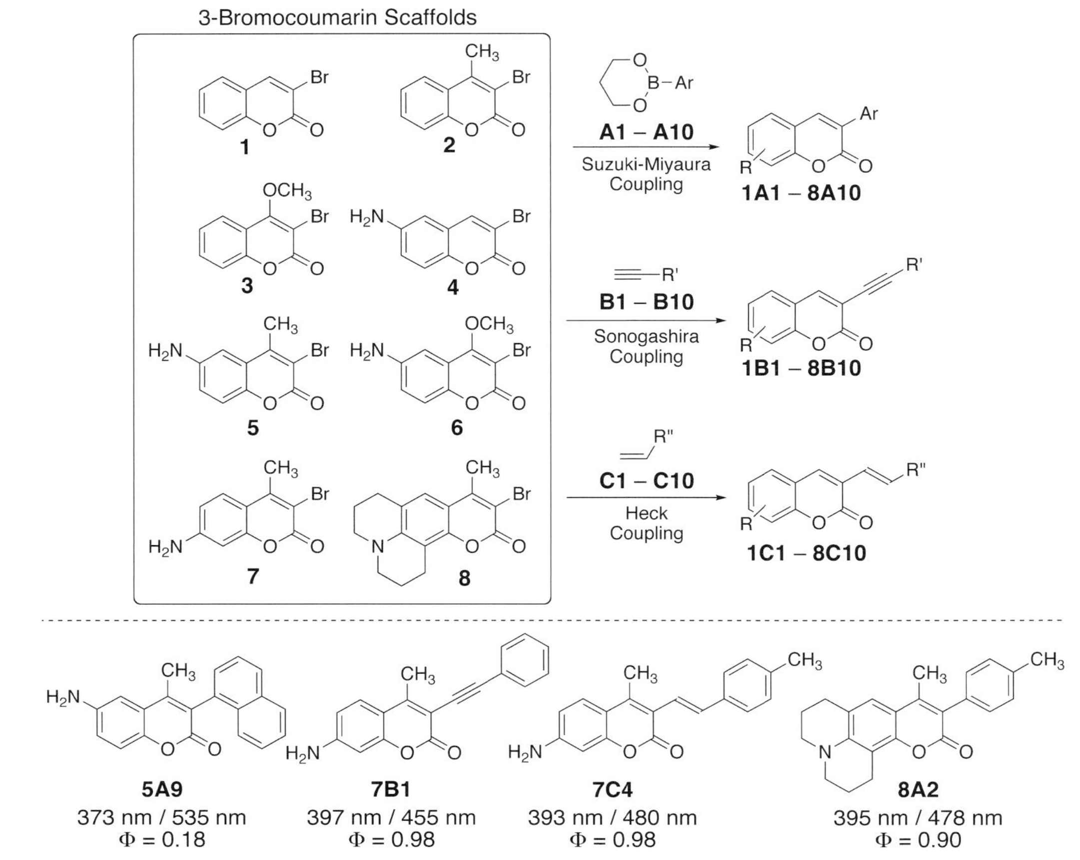

Fig. 18.4 Some members of a coumarin library constructed by means of palladium-catalyzed coupling reactions

was employed as a scaffold, and was reacted with various boronic acids, acetylenes, or olefins in solution-phase parallel synthesis (Fig. 18.4).

For such synthesis, optimization of reaction conditions is important. In the reaction of p-tolylboronic acid ester (**A2**) and scaffold **1** as coupling components, conditions with $Pd(PPh_3)_4$ as a catalyst, CsF as a base and dioxane as a solvent were found to be most effective, affording the coupling compound **1A2** in 94% yield. For Sonogashira-Hagihawa coupling reactions, $Pd(PPh_3)_2Cl_2/PPh_3/CuI$ as a catalyst and $NEt(i\text{-}Pr)_2$ as a base were found to be most appropriate. In the case of Heck reaction, the electronic nature of the alkene component is known to influence the effectiveness of coupling. Therefore, the $Pd_2(dba)_3/AgOAc/P(o\text{-}tol)_3/NEt(i\text{-}Pr)_2$ system was applied routinely, while in the reaction of electron-deficient alkenes, $Pd_2(dba)_3/AgOAc/P(t\text{-}Bu)_3/Cs_2CO_3$ afforded high yields. Under these conditions, parallel syntheses of **1** with ten boronic acid esters **A1–A10**, ten acetylenes **B1–B10** or ten olefins **C1–C10** were performed. The reaction mixtures were first passed through a reversed-phase chromatography column and then subjected to HPLC-MS. From scaffold **1**, ten 3-arylcoumarins **1A1–1A10** in yields of 67–97%, ten alkynylcoumarins **1B1–1B10** in yields 75–97% and ten vinylcoumarins **1C1–1C10** in yields of 30–94% were obtained. These reactions were also applied to seven other 3-bromo-coumarin scaffolds, **2–8**. Out of a possible 240 coumarin derivatives, 151 were isolated in >99% purity.

The fluorescence properties of the purified library compounds were evaluated by using a microplate reader, and selected compounds were studied further. Among them, **5A9**, **7B1**, **7C4**, and **8A2** were selected as "hits." In particular, **7B1** showed a higher fluorescence quantum yield than the comparable, commercially available coumarin 120.

2.2   *Coumarin Library Constructed by Click Chemistry*

Wang's group constructed a coumarin library by using so-called click chemistry, which involves Cu(I)-catalyzed Husigen 1,3-dipolar cycloaddition of azides with alkynes to afford quantitative formation of a triazole ring under mild conditions.[20] They prepared seven 3-azidocoumarin derivatives and one 4-azidocoumarin, each of which was reacted with 24 acetylenes (Fig. 18.5).

Because all of the azidocoumarin scaffold compounds show no or very weak fluorescence and most of the reactions would proceed quantitatively, they directly

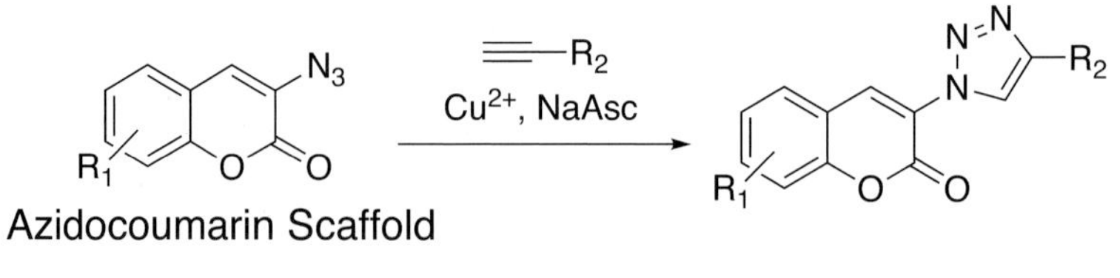

Fig. 18.5 Coumarin library synthesized by Wang's group

measured the fluorescence intensity of the diluted reaction solution as a preliminary screening. Based on this screening, several coumarin compounds were selected, synthesized in large quantity, and evaluated in detail. Although the purpose of this work is the development of the labeling reagent for biomolecules, the mild reaction conditions and easy work-up, including purification, made this approach to a coumarin library very effective.

3   Development of a 6-Arylcoumarin Library of Candidate Fluorescent Sensors

Our group recently reported the development of a coumarin library with diversity at the 6-position via Suzuki–Miyaura coupling.[21] The reaction conditions were optimized for the reaction of the 6-bromocoumarin scaffold **9** with the unsubstituted phenyl boronic acid pinacol ester **10a** (Figs. 18.6 and 18.7).

Under some reaction conditions, competitive debromination and ester-hydrolysis were expected to occur, reducing the yield of the coupling reaction, or complicating purification. In a protic solvent, THF-H_2O, with Pd(PPh$_3$)$_4$ as a catalyst and sodium carbonate as a base, the reaction yielded undesired by-products, and the coupling efficiency was relatively low (Entry 1). In an aprotic solvent, such as DMF or DME, the reaction did not afford the coupling product. The addition of one-third volume of ethanol and the use of a weaker base, CsF, allowed the coupling reaction to pro-

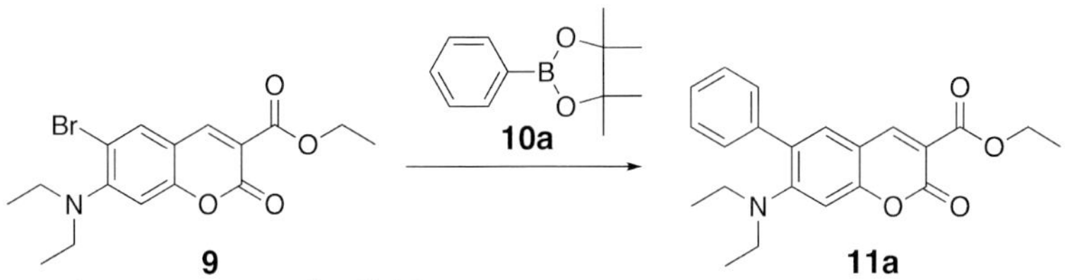

Entry	solvent	catalyst	base	yield
1	THF-H_2O (3:1, v/v)	Pd(PPh$_3$)$_4$	Na$_2$CO$_3$	trace
2	DMF-EtOH (3:1, v/v)	Pd(PPh$_3$)$_4$	CsF	53%
3	DME-EtOH (3:1, v/v)	Pd(PPh$_3$)$_4$	CsF	60%
4	toluene-EtOH (3:1, v/v)	Pd(PPh$_3$)$_4$	Cs$_2$CO$_3$	21%
5	DMF	PdCl$_2$(dppf)	Cs$_2$CO$_3$	32%
6	DMF	PdCl$_2$(dppf) (0.1eq)	CsF (2.5eq)	55%
7	DMF	PdCl$_2$(dppf) (0.2eq)	CsF (5.0eq)	72%

Fig. 18.6 Optimization of coupling reaction conditions

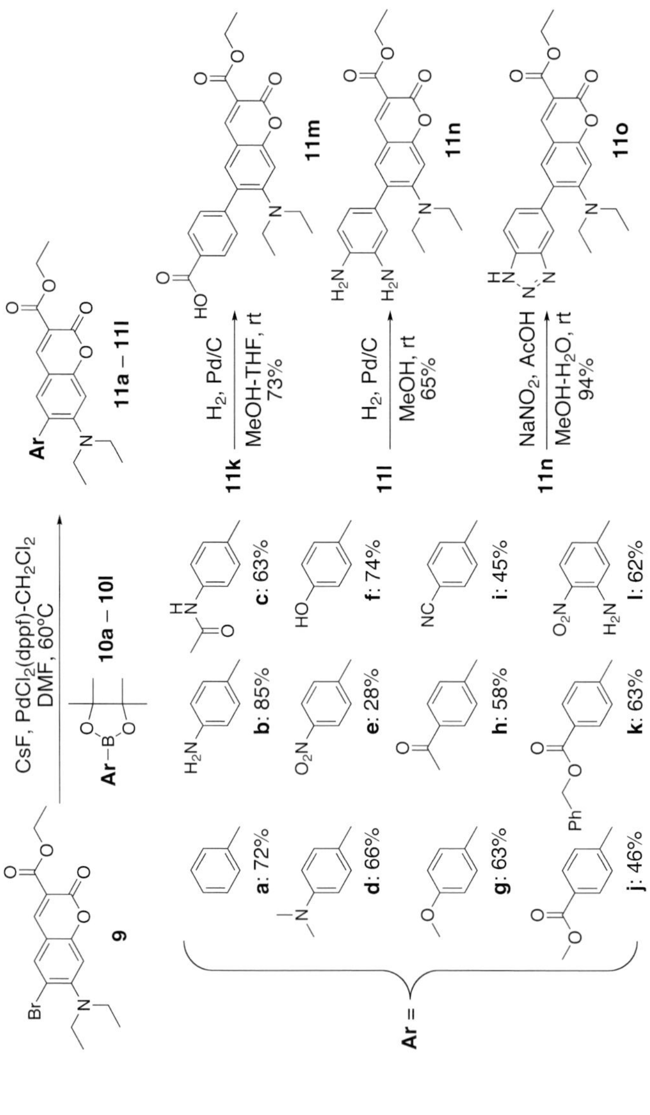

Fig. 18.7 Construction of a 6-arylcoumarin library

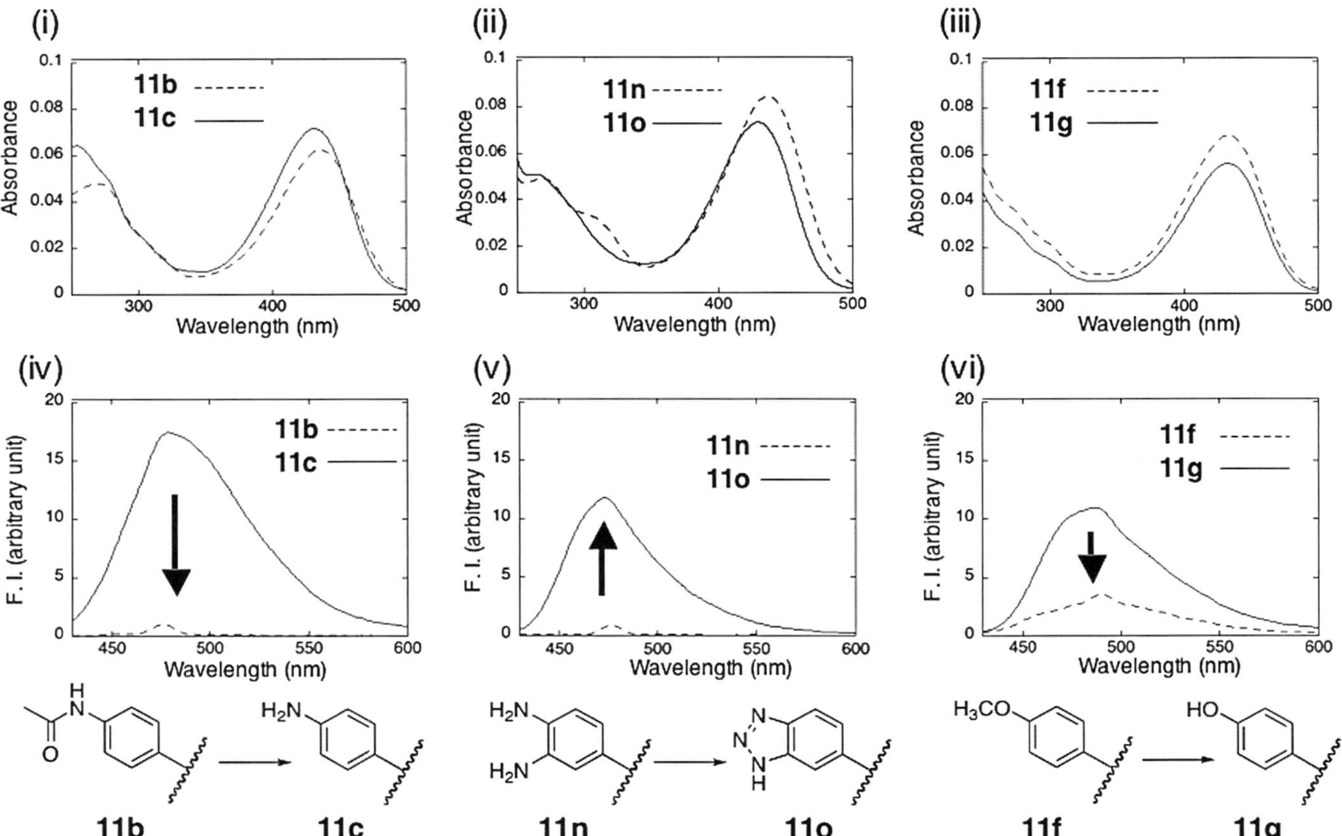

Fig. 18.8 Absorption and emission spectra of (i) (iv): **11b, 11c**; (ii) (v): **11n, 11o**; (iii) (vi) **11f, 11g** in Sodium Phosphate Buffer (pH 7.4)

ceed, affording **11a** in 50–60% yield, although the debrominated product was also obtained (Entries 2 and 3). Among several Pd catalysts examined, PdCl$_2$ (dppf) was most effective. Thus, a mixture of **9**, **10a** (1.5 eq), PdCl$_2$ (dppf) (0.1 eq) and CsF (2.5 eq) in DMF at 60°C afforded **11a** in 55% yield without debromination (Entry 6). In this case, the addition of ethanol was unnecessary. By increasing the amounts of **9a** to 3.0 eq, PdCl$_2$ (dppf) to 0.2 eq and CsF to 5.0 eq, the yield of **11a** was improved to 72% (Entry 7), and these conditions were adopted for the coupling reaction with other boronic acid derivatives.

Then, the scaffold **9** was reacted with 12 boronic acid pinacol ester derivatives (**10a–10l**), yielding the corresponding 6-arylcoumarins, **11a–11l**. Although the compounds with an electron-withdrawing group, such as nitro (**11e**), cyano (**11i**), or methoxycarbonyl (**11j**), were obtained in relatively low yield (28%, 45%, and 46%, respectively), most of the reactions proceeded in moderate yield. Some functional groups on the phenyl group were rather easily transformed, as observed in the syntheses of **11m**, **11n**, and **11o**.

Absorption and emission spectra of library compounds in acetonitrile, methanol, and water (1 mM sodium phosphate buffer pH 7.4) were measured, and selected spectra are shown in Fig. 18.8. Amino-substituted **11b** or *o*-diamino-substituted **11n** had almost no fluorescence, while the acetamido **11c** and triazole **11o** compounds had moderate fluorescence. These results suggested that 6-phenylcoumarin would be a good fluorophore for sensors of peptidase activity, which transforms an amido group to an amino group, and of nitric oxide, which reacts with an *o*-diamino group in the presence of oxygen to form triazole (Fig. 18.8 (iv) and (v)). The transformation of the methoxy group (**11g**) to a hydroxyl group (**11f**) also induced a change of fluorescence intensity, which suggested that this molecule might be useful as a fluorescent sensor for dealkylating enzymes, such as glycosidase (Fig. 18.8 (vi)). These characteristics of fluorescence intensity are similar to those seen with xanthene-based fluorescent sensors (Fig. 18.2), whose fluorescence properties are controlled by photo-induced electron transfer (PeT).

4 Conclusions

In this chapter, we described three examples of coumarin library synthesis, and showed that synthesis of libraries of fluorescent molecules is potentially useful for finding lead compounds in the development of novel functional fluorescent molecules. It should be possible to construct larger libraries by combinations of these methods with the introduction of diversity at various positions of the coumarin structure. Such libraries might well contain promising candidates for novel fluorescent sensors.

References

1. Lackowitz, J. R. *Principles of Fluorescence Spectroscopy, Third Edition*; Springer: New York, NY, 2006

2. de Silva, A. P.; Gunaratne, H. Q. N.; Gunnlaugsson, T.; Huxley, A. J. M.; McCoy, C. P.; Rademacher, J. T.; Rice, T. E., Signaling recognition events with fluorescent sensors and switches, *Chem. Rev.* **1997**, *97*, 1515–1566

3. Grynkiewicz, G.; Poenie, M.; Tsien, R. Y., A new generation of Ca^{2+} indicators with greatly improved fluorescence properties, *J. Biol. Chem.* **1985**, *260*, 3440–3450

4. Minta, A.; Kao, J. P. Y.; Tsien, R. Y., Fluorescent indicators for cytosolic calcium based on rhodamine and fluorescein chromophore, *J. Biol. Chem.* **1989**, *264*, 8171–8178

5. Kojima, H.; Nakatsubo, N.; Kikuchi, K.; Kawahara, S.; Kirino, Y.; Nagoshi, H.; Hirata, Y.; Nagano, T., Detection and imaging of nitric oxide with novel fluorescent indicators: diaminofluoresceins, *Anal. Chem.* **1998**, *70*, 2446–2453

6. Munkholm, C.; Parkinson, D. R.; Walt, D. R., Intramolecular fluorescence self-quenching of fluoresceinamine, *J. Am. Chem. Soc.* **1990**, *112*, 2608–2612

7. Umezawa, N.; Tanaka, K.; Urano, Y.; Kikuchi, K.; Higuchi, T.; Nagano, T., Novel fluorescent probes for singlet oxygen, *Angew. Chem. Int. Ed.* **1999**, *38*, 2899–2901

8. Tanaka, K.; Miura, T.; Umezawa, N.; Urano, Y.; Kikuchi, K.; Higuchi, T.; Nagano, T., Rational design of fluorescein-based fluorescence probes -mechanism-based design of a maximum fluorescence probe for singlet oxygen-, *J. Am. Chem. Soc.* **2001**, *123*, 2530–2536

9. Hirano, T.; Kikuchi, K.; Urano, Y.; Higuchi, T.; Nagano, T., Highly zinc-selective fluorescent sensor molecules suitable for biological applications, *J. Am. Chem. Soc.* **2000**, *122*, 12399–12400

10. Hirano, T.; Kikuchi, K.; Urano, Y.; Nagano, T., Improvement and biological applications of fluorescent probes for zinc, ZnAFs, *J. Am. Chem. Soc.* **2002**, *124*, 6555–6562

11. Miura, T.; Urano, Y.; Tanaka, K.; Nagano, T.; Ohkubo, K.; Fukuzumi, S., Rational design principle for modulating fluorescence properties of fluorescein-based probes by photoinduced electron transfer, *J. Am. Chem. Soc.* **2003**, *125*, 8666–8671

12. Urano, Y.; Kamiya, M.; Kanda, K.; Ueno, T.; Hirose, K.; Nagano, T., Evolution of fluorescein as a platform for finely tunable fluorescence probes, *J. Am. Chem. Soc.* **2005**, *127*, 4888–4894

13. O'Kennedy, R.; Thornes, R. D. *Coumarins: Biology, Applications, and Mode of Actions*; Wiley: England, 1997

14. Koefod, R. S.; Mann, K. R., Preparation, photochemistry, and electronic-structures of coumarin laser-dye complex of cyclopentadienylruthenium(II), *Inorg. Chem.* **1989**, *28*, 2285–2290.

15. Kido, J.; Iizumi, Y., Fabrication of highly efficient organic electroluminescent devices, *Appl. Phys. Lett.* **1998**, *73*, 2721–2723

16. Komatsu, H.; Miki, T.; Citterio, D.; Kubota, T.; Shindo, Y.; Kitamura, Y.; Oka, K.; Suzuki, K., Single molecular multianalyte (Ca^{2+}, Mg^{2+}) fluorescent probe and applications to bioimaging, *J. Am. Chem. Soc.* **2005**, *127*, 10798–10799

17. Setsukinai, K.; Urano, Y.; Kikuchi, K.; Higuchi, T.; Nagano, T., Fluorescence switching by *o*-dearylation of 7-aryloxycoumarins. Development of novel fluorescence probes to detect reactive oxygen species with high selectivity, *J. Chem. Soc., Perkin Trans. 2* **2000**, 2453–2457

18. Schiedel, M.-S.; Briehen, C. A.; Bäuerle, P., Single-compound libraries of organic materials: Parallel synthesis and screening of fluorescent dyes, *Angew. Chem. Int. Ed.* **2001**, *40*, 4677–4680

19. Schiedel, M.-S.; Briehen, C. A.; Bäuerle, P., C-C cross-coupling reactions for the combinatorial synthesis of novel organic materials, *J. Organomet. Chem.* **2002**, *653*, 200–208

20. Silvakumar, K.; Xie, F.; Cash, B. M.; Long, S.; Barnhill, H. N.; Wang, Q., A fluorogenic 1,3-dipolar cycloaddition reaction of 3-azidocoumarins and acetylenes, *Org. Lett.* **2004**, *6*, 4603–4606

21. Hirano, T.; Hiromoto, K.; Kagechika, H., Development of a library of 6-arylcoumarins as candidate fluorescent sensors, *Org. Lett.* **2007**, *9*, 1315–1318

Section 8
Mining of New Knowledge on Sensing Materials

Chapter 19
Determination of Quantitative Structure–Property Relationships of Solvent Resistance of Polycarbonate Copolymers Using a Resonant Multisensor System

Radislav A. Potyrailo, Ronald J. Wroczynski, Patrick J. McCloskey, and William G. Morris

Abstract In sensor and microfluidic applications, the need is to have an adequate solvent resistance of polymers to prevent degradation of the substrate surface upon deposition of sensor formilations, to prevent contamination of the solvent-containing sensor formulations or contamination of organic liquid reactions in microfluidic channels. Unfortunately, no comprehensive quantitative reference solubility data of unstressed copolymers is available to date. In this study, we evaluate solvent-resistance of several polycarbonate copolymers prepared from the reaction of hydroquinone (HQ), resorcinol (RS), and bisphenol A (BPA). Our high-throughput polymer evaluation approach permitted the construction of detailed solvent-resistance maps, the development of quantitative structure–property relationships for BPA-HQ-RS copolymers and provided new knowledge for the further development of the polymeric sensor and microfluidic components.

1   Introduction

Solvent-resistant polymers are attractive for a variety of microanalytical applications. For chemical sensing, solvent-resistant polymers are important as supports for deposition of solvent-based polymeric sensing formulations.[1] Otherwise, a solvent that is used for the preparation of the sensor formulation can attack a plastic substrate of choice forcing the use of either less attractive substrate materials or the use of a complicated sensor-assembly process.[2,3] Solvent-resistant polymers also attract interest for microfluidic applications as an alternative to glass and silicon.[4,5] Examples of solvent-resistant polymeric microfluidic systems include those for organic-phase synthesis,[6] polymer synthesis,[7] studies of polymeric and colloidal

R.A. Potyrailo (✉)
Chemical and Biological Sensing Laboratory, Chemistry Technologies and Material Characterization, General Electric Global Research, Niskayuna, New York, NY 12309, USA
potyrailo@crd.ge.com

R.A. Potyrailo and V.M. Mirsky (eds.), *Combinatorial Methods for Chemical and Biological Sensors*, DOI: 10.1007/978-0-387-73713-3_19,
© Springer Science+Business Media, LLC 2009

formulations,[5] high-throughput measurements of immiscible fluids,[7] and microfluidic combinatorial polymer research.[8]

A variety of recently reported parallel combinatorial polymerization reactors[9–12] can be applied to accelerate synthesis of polymers. However, analytical tools for rapid evaluation of combinatorially produced polymers are much less advanced and are highly desired.[11] For example, conventional evaluations of solvent effects on engineering polymers require exposure of a relatively large amount of a sample to a solvent and rely on measurements of solvent-induced cracks under stress.[13–16] It is not possible to predict the solvent crazing resistance of a polymer across the whole spectrum of organic liquids.[15] Thus, environmental stress cracking resistance is required in applications where engineering polymers are in contact with solvents because of the possibility of premature cracking and embrittlement of stressed or strained polymers.[13–16] It has been reported that solubility of a variety of polymers in different solvents correlated[13] or did not correlate[14,16] with the environmental stress cracking resistance. Polymer-solvent evaluations also rely on measurements of solvent uptake[4,17,18] and weight loss.[19] Since all these determinations are manually performed, they cannot handle multiple samples to match the throughput of combinatorial polymerization reactor arrays.

We have applied our sensor-based high-throughput screening infrastructure[20–22] to better understand solvent-resistance of polycarbonate copolymers for sensing and microfluidic applications.[23] Polycarbonates, their copolymers, and blends are used as materials for fabrication of microfluidic components[24–28] and gas sensor materials.[29,30] Einhorn demonstrated reactions of hydroquinone (HQ) or resorcinol (RS) with phosgene to prepare polycarbonates.[31] HQ polycarbonates were studied by Kricheldorf and Lubbers.[32] Several binary and ternary copolycarbonates were prepared using monomers such as hydroquinone, biphenol, and substituted hydroquinones. Solvent-resistance of HQ polycarbonates has been reported.[33] Brunelle reviewed in detail blends and copolymers of polycarbonate with dramatically improved solvent resistance.[34] It was pointed out that polymerization of the HQ and bisphenol A (BPA) co-cyclics via anionically initiated, ring-opening polymerization leads to high molecular weight semicrystalline polymers with the capability of incorporating of HQ into the polycarbonate.[34] These polycarbonates showed dramatically increased solvent resistance.

In this study, we evaluate solvent-resistance of several polycarbonate copolymers prepared from the reaction of HQ, RS, and BPA.[22,23] In sensor and microfluidic applications, our need is to have good solvent resistance of polymers to prevent degradation of the substrate surface upon deposition of a sensor formulation, to prevent contamination of the solvent-containing sensor formulations or contamination of organic liquid reactions in microfluidic channels. Unfortunately, no comprehensive quantitative reference solubility data of unstressed copolymers is available to date.[13–16] Our high-throughput polymer evaluation approach permitted the construction of detailed solvent-resistance maps, the development of quantitative structure–property relationships for BPA-HQ-RS copolymers and provided new knowledge for the further development of the polymeric sensor and microfluidic components.

2   Concept of Combinatorial Screening of Copolymer–Solvent Interactions

Synthesis of BPA-HQ-RS copolymers has been described by Brunelle.[34] The composition space of these copolymers is presented in Fig. 19.1. The amounts of monomers employed for copolymer synthesis are presented in Table 19.1. Pure HQ and RS polymers have poor mechanical properties[34] and are of little practical interest. Thus, only a composition range shown in Fig. 19.1 was studied. High-throughput evaluation of solvent-resistance of BPA copolymers was performed in different solvents of practical importance such as chloroform, tetrahydrofuran (THF), and methyl ethyl ketone (MEK). Measurements were performed using our acoustic-wave sensor array system designed for high-throughput materials characterization and operated in the thickness shear mode (TSM)[20–22]. TSM sensors are widely used for materials characterization.[35]

Our screening concept is depicted in Fig. 19.2. For simplicity, operation of only one of the sensors from the sensor array is illustrated. An initially clean sensor crystal is exposed to a solvent containing a small amount of polymer followed by sensor withdrawal from the solution (Fig. 19.2a). Quantification of the residual dissolved polymer deposited onto the sensor is done by the measurements of the mass increase of the crystal, which is proportional to the amount of polymer film deposited onto the sensor from a polymer solution. Depending on the solvent-resistance of the polymers, the dissolution rate of polymers in solvents will be different as determined from multiple measurements of deposited polymers during the experiment (Fig. 19.2b). The measurements of frequency changes are done when the sensors are periodically withdrawn from the solvents and the solvents are evaporated. In this way, the measured signal change is indicative of the amount of the deposited material from the solution.

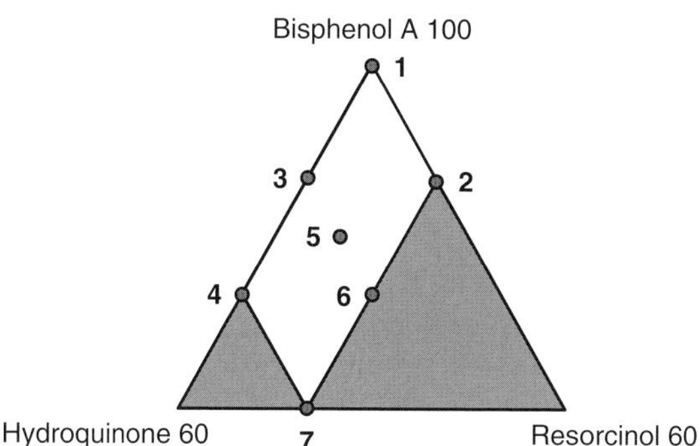

Fig. 19.1 Composition space of BPA-HQ-RS copolymers evaluated using high-throughput sensor-based system. Data points signify materials that were employed to build quantitative structure–property relationships. *Numbers* are percent of monomers

Table 19.1 Composition of copolymers

Sample no.	Bisphenol A monomer (mol%)	Hydroquinone monomer (mol%)	Resorcinol monomer (mol%)
1	100	0	0
2	80	0	20
3	80	20	0
4	60	40	0
5	70	20	10
6	60	20	20
7	40	40	20

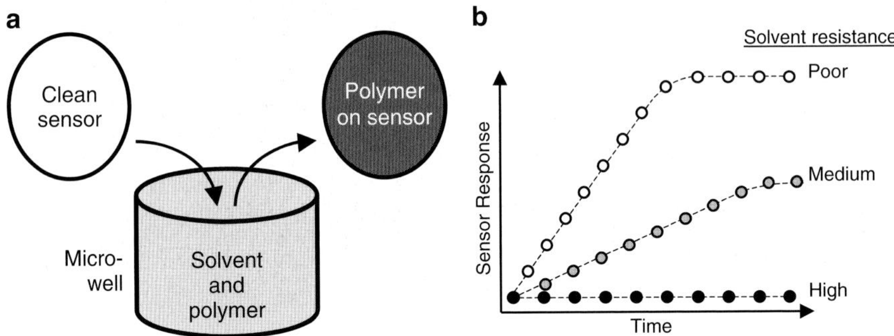

Fig. 19.2 Concept of determination of polymer solvent-resistance using TSM acoustic-wave sensors: (**a**) Periodic exposure and withdrawal of a sensor resonator into a polymer solution. (**b**) Determination of the dissolution rate of polymers in solvents from repetitive measurements of deposited polymers during the experiment

3 Sensor Array for High-Throughput Screening of Polymer–Solvent Interactions

Measurements were performed using a 24-channel acoustic-wave sensor array system shown in Fig. 19.3. The sensor system was arranged on two printed circuit boards (Fig. 19.3a). One circuit board contained sensor resonators arranged as a 6 × 4 array (Fig. 19.3b), compatible with available 24-well plates with 16-mm well diameter and positioned onto a Z-stage. Sensor resonators were 8-mm diameter polished AT-cut quartz crystals and operated at ~20 MHz. The crystals had 3-mm diameter Au electrode plated in the center of each face of the crystal. The mass resolution for a 20-MHz crystal was 10 ng.

Connections were made from this printed circuit board to a second printed circuit board, which contained 24 integrated circuit oscillators. For parallel determinations of polymer solvent-resistance, minute amounts (three pellets, ~15–30 mg total mass) of polymers were arranged in individual wells. The initially

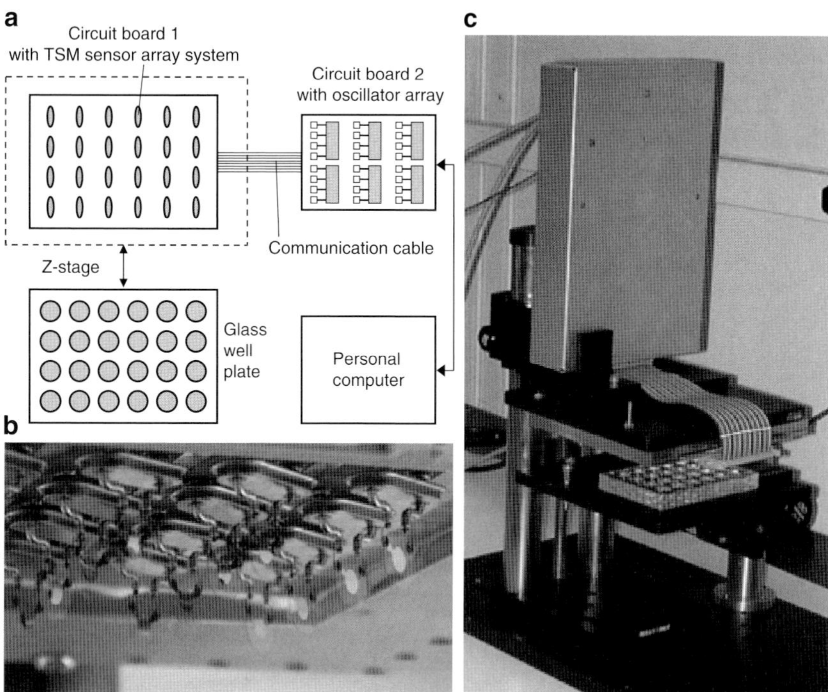

Fig. 19.3 Acoustic-wave TSM sensor array system for mapping of solvent-resistance of polymers. (**a**) System layout. (**b**) Sensor resonators (8-mm diameter) arranged as a 6 × 4 array, compatible with available 24-well plates. (**c**) General view of the screening system. (**a**, **b**) Reprinted with permission from Potyrailo et al.[23] Copyright 2006 American Chemical Society. (**c**) Reprinted with permission from Potyrailo et al.[22] Copyright 2004 American Chemical Society

clean sensor crystals were periodically immersed into the wells with these polymer/ solvent systems. Upon dissolution of a given polymer, dissolved polymer was deposited onto the sensor. In a typical experiment, 2–4 replicates of the same materials were used. A meticulous sensor cleaning was achieved by washing off polymer deposits from the sensors with a pure solvent immediately after the experiment. Monitoring the frequency of the sensors ensured the quality of the cleaning procedure.[22] Upon complete cleaning, the frequency reached the value approaching that before polymer dissolution experiment to within several Hertz. A general view of the sensor system in depicted in Fig. 19.3c. Wettability of TSM resonator surfaces by different solvents was studied using a network analyzer.[36,37]

Typical molecular weights of all synthesized copolymers were in the range of 22,000–33,000 g/mol with polycarbonate standards used for calibration. These slight differences in molecular weights did not produce detectable changes in solvent-resistance for the same copolymer compositions as determined by traditional solvent-resistance measurements.[16]

4   Variability of System Performance

Because of the 6 × 4 configuration of the array, we can test a variety of solvents and polymers or other materials in a single experiment. Such studies not only are advantageous because they are done in parallel but also because they reduce the variability between the measurements by reducing possible uncontrolled environmental variations that may affect individual kinetic experiments (lab temperature, atmospheric pressure, etc.). To study the solubility of the polymers, measurements can be performed of the dissolved amount and/or dissolution rate.

With the goal to use this sensor system for detailed evaluation of polymer solubility, we determined the variation sources in the response of the sensors. Some of these sources included the initial condition of the sensor crystals, position of polymer samples in each well, and differences in the initial mass of polymer samples. To reduce the variability in analysis as much as possible, we attempted to minimize identified variation sources in sensor response. First, a good initial condition of the sensor crystals was insured by checking for mechanical integrity of crystals and their cleanness before the measurements. The cleanness of crystals was ensured by observing no changes in measured frequency upon crystal washing with a solvent. Second, to minimize the variation in dissolving rate of polymers in each well and to ensure the homogeneous distribution of dissolved polymer in each well, solutions in the wells were stirred after withdrawal of sensors from solutions. Third, to compensate for the differences in the initial mass of polymer samples in each replicate well, the frequency response of each sensor was normalized by the initial total mass of polymer pellets in individual wells of the 24-well plate. This total mass of polymer in each well was available from measurements before the solubility experiment.

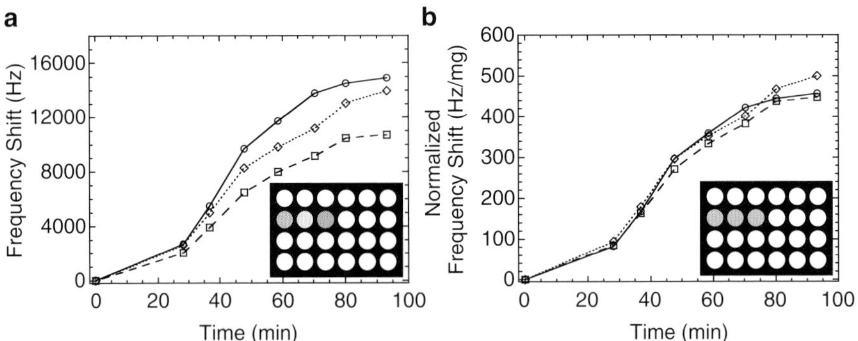

Fig. 19.4 Significant reduction of variability in sensor response due to differences in initial polymer amounts in each well of the 24-well plate. Response of three representative sensors (**a**) before and (**b**) after the normalization by the initial mass of the polymer in each well. *Gray levels* of three replicate wells in insets in (**a**) and (**b**) illustrate the normalization approach. Reprinted with permission from Potyrailo et al.[22] Copyright 2004 American Chemical Society

Typical achieved improvements are illustrated in Fig. 19.4. When measured at 60 min after beginning of the experiment, the relative standard deviation of frequency measurements was improved from ~20% to less than 4%.

In addition, other potential sources of variability were considered. For example, the sensor surface may potentially differently adsorb deposited polymers. However, the inert nature of quartz crystal and gold sensor electrode,[36] with respect to the employed solvents and polymers, resulted in no detectable irreversible adsorption of polymers to the sensor surface as measured by the recovery of the sensor baseline after washing-off deposited polymers. Thus, the sensors were completely reusable because the solvents and dissolved polymers did not chemically interact with the quartz crystal and gold electrodes of the sensor.

5   Wettability of Sensor Resonators

Our initial experiments were focused on understanding of the wettability of sensor resonators by different solvents of interest. Sensor response was determined by scanning the frequency through the resonance region and measuring the magnitude and phase angle of the impedance[36] in different environments that included air, chloroform, THF, and MEK. The phase angle of the resonator in air had a sharp transition approaching 90°. The transitions of the resonator in chloroform, THF, and MEK all had negative phase angles and much smaller slopes than response in air (Fig. 19.5). This demonstrated that wettability of the resonator with these solvents was very similar, attractive for our application. The employed solvents had

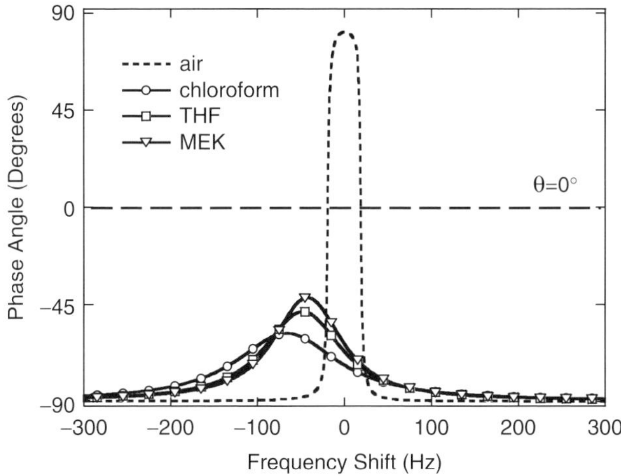

Fig. 19.5 Phase angle of a ~20-MHz polished quartz sensor resonator measured in different environments such as air, chloroform, THF, and MEK. Reprinted with permission from ref. 23. Copyright 2006 American Chemical Society

very close surface tension values of 24.6, 26.4, and 27.5 mN/m for MEK, THF, and chloroform, respectively, compared with 72.8 mN/m for water, all at 20°C.[38,39]

Thus, the employed pure solvents showed similar spreading behavior on the gold electrodes of the sensors. Similar spreading behavior on gold electrodes was also observed for dissolved copolymers that were from the same chemical family of polycarbonates. For more diverse solvents and polymers, one of possible solutions to reduce the differences in spreading of dissolved polymers may be to employ unpolished sensor crystals.[35,40,41]

6   High-Throughput Screening of Copolymers

The sensor system was applied to the quantification of solvent-resistance of BPA-HQ-RS polycarbonate copolymers in chloroform, THF, and MEK. In a typical experiment, the 6 × 4 sensor array was periodically immersed into the wells containing copolymers in respective solvents. The analytically useful frequency changes were measured upon sensor removal from the solvent and solvent evaporation. Comparison of solvent-resistance of copolymers in the solvents is presented in Fig. 19.6 where data were obtained from measurements of sensor signal change after 46–48 min from the beginning of the experiment. Copolymer solvent-resistance is represented as the absolute sensor frequency shift per mass of polymer in each well

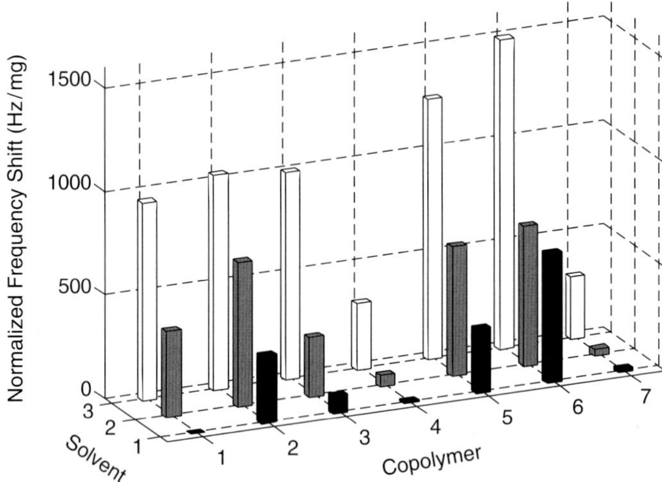

Fig. 19.6 Comparison of solvent-resistance of copolymers in solvents of different nature. Copolymer solvent-resistance is represented as a sensor frequency shift per mass of polymer in each well of a 24-well plate. Copolymers are listed in Table 19.1. Solvents: (1) MEK; (2), chloroform; (3) THF. Reprinted with permission from Potyrailo et al.[23] Copyright 2006 American Chemical Society

of a 24-well plate. Sensor frequency shifts induced by a deposited material are commonly reported in sensor applications.[42–44] The mass m_F of the deposited polymer film onto the sensor crystal can be calculated from the frequency shift Δf_F induced by the polymer film as:[35]

$$\Delta f_F \equiv -2.26 \times 10^{-6} f_0^2 m_F / A. \tag{1}$$

For sensor crystals operating at a frequency $f_0 = 2 \times 10^7$ Hz and with the active surface area of one face of the crystal $A = 0.07\,cm^2$, a 1,000-Hz frequency shift was produced by 150 ng of deposited polymer film. The calculated film thickness was 9 nm per 1,000 Hz of the frequency change induced by the deposited polymer film. The maximum film thickness was 0.4 μm.

7 Property/Composition Mapping and Structure–Property Relationships

To construct property-composition maps and to determine structure–property relationships in BPA-HQ-RS copolymers, we applied a D-optimal mixture design and selected copolymers on the ternary diagram shown in Fig. 19.1. Mixture designs are statistically based techniques, which permit an investigation of linear, nonlinear, and interactive effects of independent variables (known as factors) on a response. Mixture designs are used when the response is a function of the proportions of the ingredients in a composition, which normally corresponds to a total of 100 mol%.[45,46] D-optimal mixture design was selected because such optimization minimizes the general variance of the coefficients in the polynomial model and maximizes the determinant of the Fisher information matrix.[47]

D-optimal selection chooses points from the candidate point set, which are spread throughout the design region. We used 21 data points to construct the solvent-resistance/composition maps. These data points were seven copolymer compositions with three replicate solvent-resistance tests for each of them. These seven BPA-HQ-RS compositions covered the corner points, the centroid, and two longest edges (Fig. 19.1) and were sufficient for modeling because this data set provided 14 degrees of freedom for pure error.[48] For solvent-resistance/composition mapping, we applied a special cubic model that took into the account the primary mixture terms such as BPA, HQ, and RS, binary interaction terms such as BPA-HQ, BPA-RS, and HQ-RS, and a ternary interaction term BPA-HQ-RS. The special cubic model considers interactions between all factors and is based on a canonical mixture polynomial[48]:

$$Y = \sum_{i=1}^{q} \beta_i x_i + \sum_{i<j}^{q} \beta_{ij} x_i x_j + \sum_{i<j<m=q}^{k=q} \beta_{ijk} x_i x_j x_k, \tag{2}$$

where Y is the response variable (sensor response), x_i are the factors (the equivalent mole fractions of each monomer), q is the number of factors, β_i are the linear regression coefficients, β_{ij} are the binary regression coefficients, and β_{ijk} is the ternary regression coefficient.

To verify the adequacy of the developed models of solvent-resistance in THF, chloroform, and MEK, normal probability plots were evaluated. Typical normal probability plots of residuals should be close to a straight line as shown in Fig. 19.7 because the underlying error distribution is expected to be normal.[47] This means that the normality assumption is valid for the proposed model. Residuals that were intensified in the middle of straight line indicated that data were normally distributed. Also, there were no outliers in the model as indicated by absence of significant deviations from the straight line. A combination of the normal distribution of the model residuals (Fig. 19.7) and the very high values of adjusted R^2 demonstrated a good quality of the model.

Response surfaces were constructed to study the simultaneous effect of BPA, HQ, and RS monomers on the solvent-resistance of resultant copolymers. Response surface methodology is a collection of statistical methods that are useful for the modeling and analysis of problems in which a response of interest is influenced by several factors. The objective is to find a desirable location in the design space, which could be a maximum, a minimum, or an area where the response is stable over a range of the factors.[48] Response surfaces for resistance of ternary BPA-HQ-RS compositions to three solvents are presented in Fig. 19.8. The largest absolute solvent-resistance for all copolymers was achieved in MEK, followed by chloroform and THF. The solvent-resistance/composition maps demonstrated that the HQ component has the best effect on solvent resistance for all studied solvents. Also, at a constant HQ level, increasing BPA and decreasing RS improved solvent resistance slightly. Other polymers of polycarbonate family also demonstrated an improvement of solvent-resistance with addition of HQ monomer[33,34] while RS monomer is responsible for the good melt-flow during polymer molding.[49]

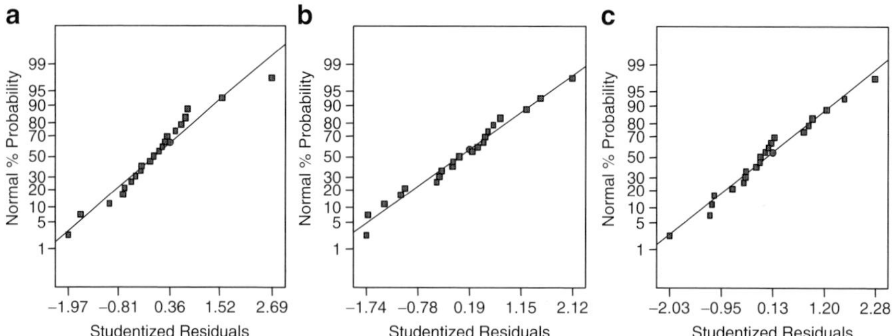

Fig. 19.7 Plots of normal probability vs. studentized residuals for models describing solvent resistance of BPA-HQ-RS copolymer compositions in: (**a**) THF, (**b**) chloroform, (**c**) MEK. Reprinted with permission from Potyrailo et al.[23] Copyright 2006 American Chemical Society

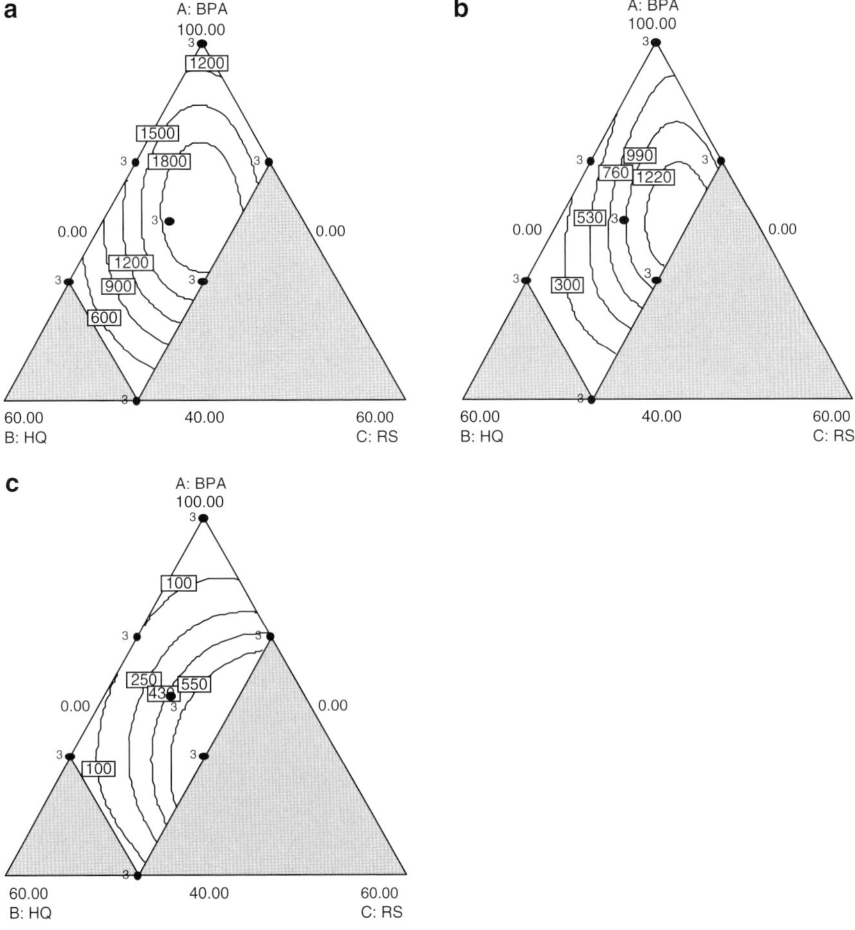

Fig. 19.8 Property/composition mapping of solvent-resistance of BPA-HQ-RS copolymers in different solvents: (**a**) THF, (**b**) chloroform, (**c**) MEK. *Numbers* in the contour lines are normalized sensor frequency shift values (Hz per mg of polymer in a well). Reprinted with permission from Potyrailo et al.[23] Copyright 2006 American Chemical Society

Effects of BPA, HQ, and RS on solvent-resistance were further established by relating resistance of BPA-HQ-RS compositions to THF, chloroform, and MEK solvents. The structure–property relationships in BPA-HQ-RS copolymers established using the developed high-throughput screening approach are summarized in Fig. 19.9. The solvent-resistance of BPA-HQ-RS copolymers was plotted as a function of HQ monomer concentration because HQ had the strongest effect on solvent-resistance (Fig. 19.8). At maximum tested HQ concentration of 40%, the solvent resistance was the best for all tested solvents. Also, different concentrations of BPA (40 and 60%) and RS (0 and 20%) did not affect the solvent resistance, while without

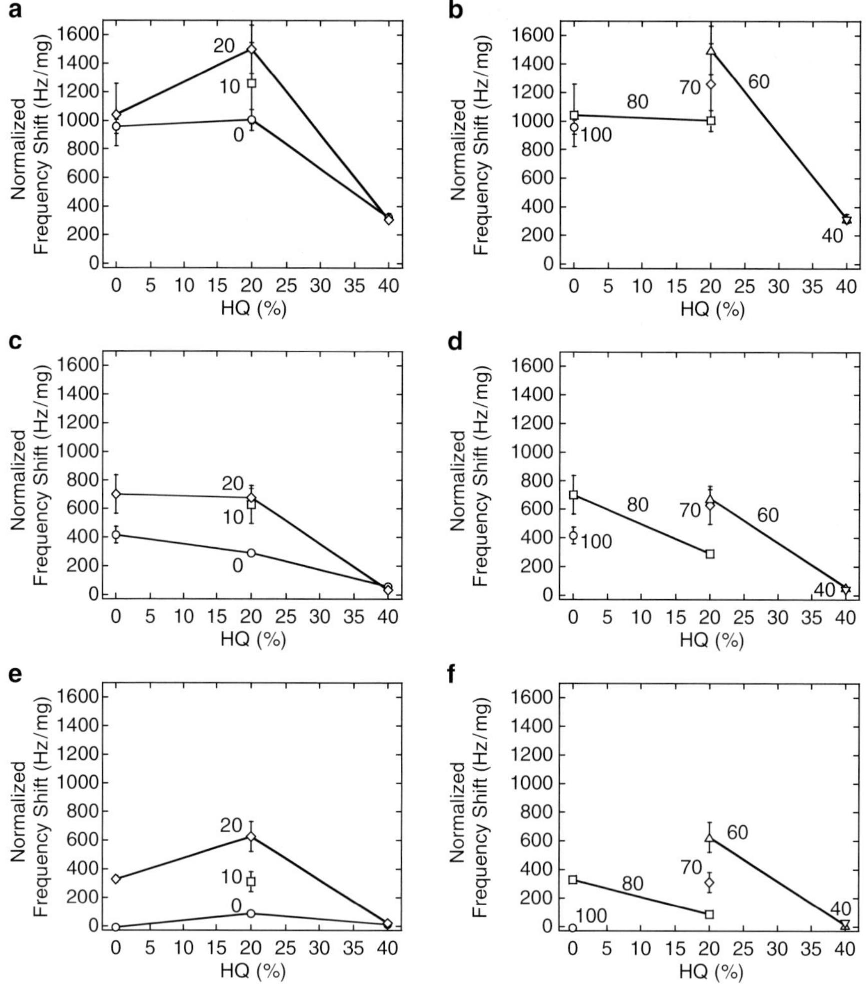

Fig. 19.9 Structure–property relationships in BPA-HQ-RS copolymers measured using developed high-throughput screening approach. Relation between HQ monomer concentration and BPA-HQ-RS copolymer solvent-resistance measured as the normalized frequency shift of sensor response for different levels of RS and BPA. Solvents: (**a, b**) THF, (**c, d**) chloroform, (**e, f**) MEK. RS monomer: **a, c**, and **e**. BPA monomer: **b, d**, and **f**. Error bars, 1 SD from three replicate solvent-resistance testes. Reprinted with permission from Potyrailo et al.[23] Copyright 2006 American Chemical Society

HQ, solvent resistance was decreasing with an increase of RS and decrease of BPA. Overall, with an increase of HQ concentration from 0 to 40%, the solvent resistance of BPA-HQ-RS copolymers was improved by up to three times in THF, by 21 times in chloroform, and by 32 times in MEK.

8 Applications of Polycarbonate Copolymers as Sensor Substrates

Polycarbonate copolymers were further applied as substrates for solvent-based polymeric sensing formulations for optical detection of ionic species in water.[1] In one application, a sensing film for pH measurements was prepared by dissolving cellulose acetate and bromothymol blue in MEK and depositing this formulation onto a polycarbonate copolymer substrate. As illustrated in Fig. 19.10, upon an exposure of the dried sensing film to solutions of low and high pH, the sensing film that was deposited using an organic solvent did not degrade the sensor substrate and

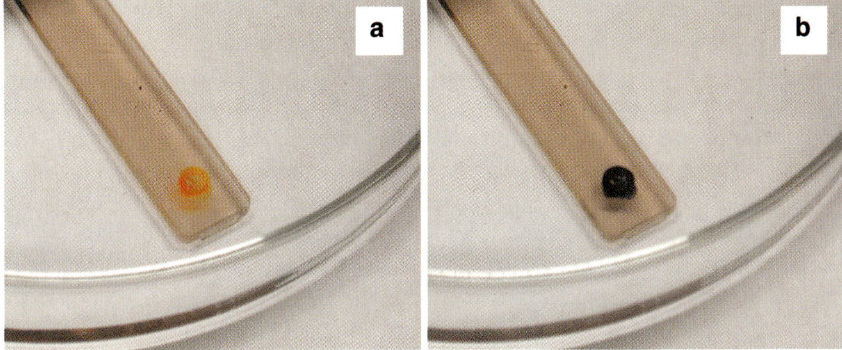

Fig. 19.10 Inspection of the pH reagent immobilized in a cellulose acetate film and deposited onto a polycarbonate copolymer material upon immersion in solutions of low (**a**) and high (**b**) pH

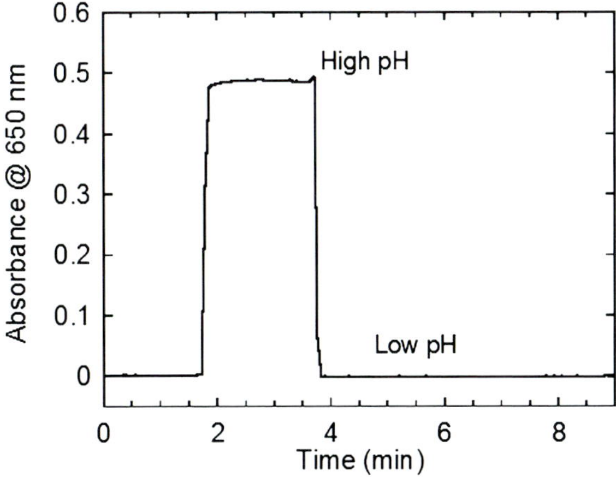

Fig. 19.11 Stability of the immobilized reagent over the relevant time of measurement in solutions of low and high pH

the immobilized pH reagent exhibited preserved analyte-sensitive properties. Stability of the immobilized reagent in solutions of low and high pH over the relevant time of measurement is demonstrated in Fig. 19.11.

9 Conclusions

Application of the developed sensor system provided previously unavailable capabilities of parallel evaluation of polymer solvent-resistance making possible analysis of nanogram quantities of polymers in small amounts of solvent and permitted the simultaneous analysis of multiple samples to build solvent-resistance maps of polycarbonate copolymers. A 2-mL volume per well employed in our screening experiments is an intermediate volume between the high-throughput biological analyses performed using 96, 384, 1,564-well arrays that have assay volumes down to $0.5–10\,\mu L^{50}$ and a traditional scale testing that involves hundreds of milliliters of solvents per test. While the reagent cost is much less in polymer testing compared to biological testing, the attractiveness of our small-scale approach is in at least 100 times less generated organic solvent and solid waste.

The described sensor-based polymer-screening approach provided new knowledge on quantitative relations between the composition and solvent-resistance of copolymers in several industrially important solvents. Such knowledge was difficult to obtain using conventional one-sample-at-a-time approach. In the future, high mass resolution of the resonant TSM sensors, use of only minute volume of a solvent, and parallel operation would make this system a good fit with available polymer combinatorial synthesis equipment. We further employed some of these polycarbonate copolymers as solvent-resistant supports for direct deposition of solvent-containing sensing formulations that provided improved optical quality and stability of sensor layers.

Acknowledgment Authors are grateful to William D. Richards for helpful discussions on conventional test methods of environmental stress cracking resistance.

References

1. Potyrailo, R. A.; McCloskey, P. J.; Ramesh, N.; Surman, C. M., *Sensor devices containing co-polymer substrates for analysis of chemical and biological species in water and air*; US Patent Application 2005133697: 2005.
2. Johnson, R. D.; Badr, I. H. A.; Barrett, G.; Lai, S.; Lu, Y.; Madou, M. J.; Bachas, L. G., Development of a fully integrated analysis system for ions based on ion-selective optodes and centrifugal microfluidics, *Anal. Chem.* **2001**, *73*, 3940–3946.
3. Badr, I. H. A.; Johnson, R. D.; Madou, M. J.; Bachas, L. G., Fluorescent ion-selective optode membranes incorporated onto a centrifugal microfluidics platform, *Anal. Chem.* **2002**, *74*, 5569–5575.
4. Rolland, J. P.; Van Dam, R. M.; Schorzman, D. A.; Quake, S. R.; DeSimone, J. M., Solvent-resistant photocurable liquid fluoropolymers for microfluidic device fabrication, *J. Am. Chem. Soc.* **2004**, *126*, 2322–2323.

5. Harrison, C.; Cabral, J. T.; Stafford, C. M.; Karim, A.; Amis, E. J., A rapid prototyping technique for the fabrication of solvent-resistant structures, *J. Micromech. Microeng.* **2004**, *14*, 153–158.

6. Cygan, Z. T.; Cabral, J. T.; Beers, K. L.; Amis, E. J., Microfluidic platform for the generation of organic-phase microreactors, *Langmuir* **2005**, *21*, 3629–3634.

7. Cabral, J. T.; Hudson, S. D.; Wu, T.; Beers, K. L.; Douglas, J. F.; Karim, A.; Amis, E. J., Microfluidic combinatorial polymer research, *Polym. Mater.: Sci. Eng.* **2004**, *90*, 337–338.

8. Amis, E. J., Combinatorial materials science reaching beyond discovery, *Nat. Mater.* **2004**, *3*, 83–85.

9. Komon, Z. J. A.; Diamond, G. M.; Leclerc, M. K.; Murphy, V.; Okazaki, M.; Bazan, G. C., Triple tandem catalyst mixtures for the synthesis of polyethylenes with varying structures, *J. Am. Chem. Soc.* **2002**, *124*, 15280–15285.

10. Boussie, T. R.; Diamond, G. M.; Goh, C.; Hall, K. A.; LaPointe, A. M.; Leclerc, M.; Lund, C.; Murphy, V.; Shoemaker, J. A. W.; Tracht, U.; Turner, H.; Zhang, J.; Uno, T.; Rosen, R. K.; Stevens, J. C., A fully integrated high-throughput screening methodology for the discovery of new polyolefin catalysts: Discovery of a new class of high temperature single-site group (iv) copolymerization catalysts, *J. Am. Chem. Soc.* **2003**, *125*, 4306–4317.

11. Hoogenboom, R.; Meier, M. A. R.; Schubert, U. S., Combinatorial methods, automated synthesis and high-throughput screening in polymer research: Past and present, *Macromol. Rapid Commun.* **2003**, *24*, 15–32.

12. Potyrailo, R. A.; Wroczynski, R. J.; Lemmon, J. P.; Flanagan, W. P.; Siclovan, O. P., Fluorescence spectroscopy and multivariate spectral descriptor analysis for high-throughput multiparameter optimization of polymerization conditions of combinatorial 96-microreactor arrays, *J. Comb. Chem.* **2003**, *5*, 8–17.

13. Bernier, G. A.; Kambour, R. P., The role of organic agents in the stress crazing and cracking of poly(2,6-dimethyl-1,4-phenylene oxide), *Macromolecules* **1968**, *1*, 393–400.

14. Kambour, R. P.; Romagosa, E. E.; Gruner, C. L., Swelling, crazing, and cracking of an aromatic copolyether-sulfone in organic media, *Macromolecules* **1972**, *5*, 335–340.

15. Kambour, R. P.; Gruner, C. L.; Romagosa, E. E., Biphenol-A polycarbonate immersed in organic media. Swelling and response to stress, *Macromolecules* **1974**, *7*, 248–253.

16. Li, X., Environmental stress cracking resistance of a new copolymer of bisphenol-A, *Polym. Degrad. Stab.* **2005**, *90*, 44–52.

17. Huang, J.-C.; Zhu, Z.-k.; Yin, J.; Qian, X.-F.; Sun, Y.-Y., Poly(etherimide)/montmorillonite nanocomposites prepared by melt intercalation: Morphology, solvent resistance properties and thermal properties, *Polymer* **2001**, *42*, 873–877.

18. Takeichi, T.; Ujiie, K.; Inoue, K., High performance poly(urethane-imide) prepared by introducing imide blocks into the polyurethane backbone, *Polymer* **2005**, *46*, 11225–11231.

19. Qi, Y.; Ding, J.; Day, M.; Jiang, J.; Callender, C. L., Cross-linkable highly fluorinated poly(arylene ether ketones/sulfones) for optical waveguiding applications, *Chem. Mater.* **2005**, *17*, 676–682.

20. Potyrailo, R. A.; Morris, W. G.; Wroczynski, R. J., Acoustic-wave sensors for high-throughput screening of materials, In *High Throughput Analysis: A Tool for Combinatorial Materials Science*; R. A. Potyrailo and E. J. Amis, Eds.; Kluwer/Plenum: New York, NY, 2003; ch. 11.

21. Potyrailo, R. A.; Morris, W. G.; Wroczynski, R. J., Multifunctional sensor system for high-throughput primary, secondary, and tertiary screening of combinatorially developed materials, *Rev. Sci. Instrum.* **2004**, *75*, 2177–2186.

22. Potyrailo, R. A.; Morris, W. G.; Wroczynski, R. J.; McCloskey, P. J., Resonant multisensor system for high-throughput determinations of solvent-polymer interactions, *J. Comb. Chem.* **2004**, *6*, 869–873.

23. Potyrailo, R. A.; McCloskey, P. J.; Wroczynski, R. J.; Morris, W. G., High-throughput determination of quantitative structure-property relationships using resonant multisensor system: Solvent-resistance of bisphenol a polycarbonate copolymers, *Anal. Chem.* **2006**, *78*, 3090–3096.

24. Martin, P. M.; Matson, D. W.; Bennett, W. D.; Lin, Y.; Hammerstrom, D. J., Laminated plastic microfluidic components for biological and chemical systems, *J. Vac. Sci. Technol.* **1999**, *A 17*, 2264–2269.

25. Becker, H.; Locascio, L. E., Polymer microfluidic devices, *Talanta* **2002**, *56*, 267–287.
26. Soper, S. A.; Henry, A. C.; Vaidya, B.; Galloway, M.; Wabuyele, M.; McCarley, R. L., Surface modification of polymer-based microfluidic devices, *Anal. Chim. Acta* **2002**, *470*, 87–99.
27. Erickson, D.; Li, D., Integrated microfluidic devices, *Anal. Chim. Acta* **2004**, *507*, 11–26.
28. Madou, M. J., *Fundamentals of Microfabrication. The Science of Miniaturization*; CRC Press: Boca Raton, FL, 2002.
29. Freud, M. S.; Lewis, N. S., A chemically diverse conducting polymer-based "electronic nose", *Proc. Natl. Acad. Sci. U S A* **1995**, *92*, 2652–2656.
30. Sivavec, T. M.; Potyrailo, R. A., *Polymer coatings for chemical sensors*; US Patent 6,357,278 B1: 2002.
31. Einhorn, A., Ueber die carbonate der dioxybenzole, *Liebigs Ann. Chem.* **1898**, *300*, 135–155.
32. Kricheldorf, H. R.; Lübbers, D., Polymers of carbonic acid, 1. Synthesis of thermotropic aromatic polycarbonates by means of bis(trichloromethyl) carbonate, *Makromol. Chem. Rapid Commun.* **1989**, *10*, 383–386.
33. Schnell, H., Polycarbonate, eine gruppe neuartiger thermoplastischer kunststoffe. Herstellung und eigenschaften aromatischer polyester der kohlensäure, *Angew. Chem.* **1956**, *68*, 633–640.
34. Brunelle, D. J., Solvent-resistant polycarbonates, *Trends Polym. Sci.* **1995**, *3*, 154–158.
35. Thompson, M.; Stone, D. C., *Surface-Launched Acoustic Wave Sensors: Chemical Sensing and Thin-Film Characterization*; Wiley: New York, NY, 1997, pp 196.
36. Duncan-Hewitt, W. C.; Thompson, M., Four-layer theory for the acoustic shear wave sensor in liquids incorporating interfacial slip and liquid structure, *Anal. Chem.* **1992**, *64*, 94–105.
37. Kanazawa, K. K., Mechanical behaviour of films on the quartz microbalance, *Faraday Discuss.* **1997**, *107*, 77–90.
38. Jasper, J. J., The surface tension of pure liquid compounds, *J. Phys. Chem. Ref. Data* **1972**, *1*, 841–1009.
39. Riddle, F. L., Jr.; Fowkes, F. M., Spectral shifts in acid-base chemistry. 1. Van der Waals contributions to acceptor numbers, *J. Am. Chem. Soc.* **1990**, *112*, 3259–3264.
40. Ballantine, D. S., Jr.; White, R. M.; Martin, S. J.; Ricco, A. J.; Frye, G. C.; Zellers, E. T.; Wohltjen, H. *Acoustic Wave Sensors: Theory, Design, and Physico-Chemical Applications*; Academic Press: San Diego, CA, 1997, pp. 436.
41. Daikhin, L.; Urbakh, M., Influence of surface roughness on the quartz crystal microbalance response in a solution new configuration for qcm studies, *Faraday Discuss.* **1997**, *107*, 27–38.
42. Finklea, H. O.; Phillippi, M. A.; Lompert, E.; Grate, J. W., Highly sorbent films derived from $Ni(SCN)_2(4\text{-picoline})_4$ for the detection of chlorinated and aromatic hydrocarbons with quartz crystal microbalance sensors, *Anal. Chem.* **1998**, *70*, 1268–1276.
43. Grate, J. W.; Patrash, S. J.; Kaganove, S. N.; Wise, B. M., Hydrogen bond acidic polymers for surface acoustic wave vapor sensors and arrays, *Anal. Chem.* **1999**, *71*, 1033–1040.
44. Park, J.; Groves, W. A.; Zellers, E. T., Vapor recognition with small arrays of polymer-coated microsensors. A comprehensive analysis, *Anal. Chem.* **1999**, *71*, 3877–3886.
45. Khuri, A. I.; Cornell, J. A., *Response Surfaces: Designs and Analyses*; Marcel Dekker: New York, NY, 1996.
46. Santafé-Moros, A.; Gozálvez-Zafrilla, J. M.; Lora-García, J.; García-Díaz, J. C., Mixture design applied to describe the influence of ionic composition on the removal of nitrate ions using nanofiltration, *Desalination* **2005**, *185*, 289–296.
47. Segurola, J.; Allen, N. S.; Edge, M.; Mc Mahon, A., Design of eutectic photoinitiator blends for UV/visible curable acrylated printing inks and coatings, *Prog. Org. Coat.* **1999**, *37*, 23–37.
48. Cornell, J. A., *Experiments with mixtures. Design, Models, and the Analysis of Mixture Data*; Wiley: New York, NY, 1981.
49. LeGrand, D. G.; Bendler, J. T., (Eds.), *Handbook of Polycarbonate Science and Technology*; Marcel Dekker: New York, NY, 2000.
50. Potyrailo, R. A., Combinatorial screening, In *Encyclopedia of materials: Science and technology*; K. H. J. Buschow; R. W. Cahn; M. C. Flemings; B. Ilschner; E. J. Kramer and S. Mahajan, Eds.; Elsevier: Amsterdam, The Netherlands, 2001; Vol. 2; pp. 1329–1343.

Chapter 20
Computational Approaches to Design and Evaluation of Chemical Sensing Materials

Margaret A. Ryan and Abhijit V. Shevade

Abstract Materials for use as chemical sensors may be evaluated and selected with computational and experimental approaches. Computational methods have focused on developing fundamental electronic and atomic level descriptions of materials to insight into chemical interactions between targeted analytes and sensing materials. Computational methods also include use of statistical and computational approaches to characterize measured and experimentally observed analyte-sensing material interactions and sensing material responses to the presence of analyte. In the following chapter, we have provided an overview of various approaches that have been used to investigate and select chemical sensing materials.

1 Introduction

Materials for chemical and biological sensing, catalysis, drug design and various industrial applications have been investigated using computational, experimental and computational-experimental approaches. Methods developed in these approaches have been used to evaluate and select candidate materials as well as to provide inputs to design, synthesis and evaluation of new materials. This chapter discusses examples of computational approaches that have been used successfully in designing and selecting materials for chemical sensing devices.

Understanding the sensing properties of materials depends on understanding the chemical interactions between sensing material and other chemical species which may be present in the sensed environment, including both targeted analyte and background species such as oxygen, carbon dioxide, water, etc. These chemical interactions are, for the most part, van der Waals and Coulombic interactions. For chemical sensing, we are generally interested in reversible binding of the target

M.A. Ryan (✉) and A.V. Shevade
Jet Propulsion Laboratory, California Institute of Technology
4800 Oak Grove Drive, Pasadena, CA 91109, USA

R.A. Potyrailo and V.M. Mirsky (eds.), *Combinatorial Methods for Chemical and Biological Sensors*, DOI: 10.1007/978-0-387-73713-3_20,
© Springer Science + Business Media, LLC 2009

molecule with the sensing film and minimal interaction of both target molecule and sensing material with the background. Quantum mechanical (QM) calculations of binding energies between sensing materials and target analytes provide a theoretical approach to screening sensor materials. This, followed by experimental confirmation of the sensing materials, is a rapid method for selecting sensing materials for an array when exhaustive materials development and testing is not an option.

Evaluation of chemical sensing materials using purely computational methods has generally focused on developing a fundamental electronic and atomic level description of materials to provide insight into chemical interactions of materials. Computational methods used in materials simulations have made tremendous strides in the last two decades, including development of approaches that use first principles QM techniques, molecular dynamics (MD) or atomistic techniques, statistical mechanical and multiscale approaches.[1] In addition to purely computational methods, statistical and multivariate methods have been widely used; these approaches include semiempirical and combinatorial methods that use molecular descriptors, as in quantitative structural activity relationships/quantitative structure–property relationships (QSAR/QSPR)[2] and linear solvation energy relationships.[3]

On a fundamental level, describing or characterizing the properties of materials depends on understanding chemical interactions, which may take place between the material and other chemical species with which it may come into contact; this understanding, in turn, involves understanding the electronic and atomic level descriptions of materials. The length-time scale of these descriptions is the nanometer–picosecond level. The hierarchy of materials simulations used to obtain engineered materials of desired properties starts with first principles quantum mechanics (electrons), followed by MD (atoms), meso-scale (segments/grain), and finally, finite element analysis (grids) level descriptions for predicting properties of new materials.[1] QM calculations are the most accurate, and the state-of-the-art calculations can handle system sizes up to 50 atoms. The time needed for QM simulation generally scales as $N^3 - N^5$ for characteristic methods, where N is the system size, but new approaches have achieved N^2 scaling. Information derived from QM is used to develop a force field, an empirical functional form, for atomistic level simulations for use in MD techniques. Although MD simulations can be used to study system sizes as much as five orders of magnitude larger than systems studied in QM, at times material design requires simulations on larger time-length scales (μs/s-μm/mm) than MD can handle. This need leads to the next steps, considering grains (meso-scale) and grids (finite elements) for design and evaluation of materials. All these methods provide an opportunity to design, characterize, and optimize materials prior to expensive experimentation. Reliable simulation methodology for real materials can be of significant benefit to several areas of materials research; moreover, such methods can provide new material candidates.

2 Application of Computational Techniques

Array-based sensing, in which an array of several semiselective chemical sensors is used to characterize selected analytes, has given rise to electronic noses, which focus on detecting and characterizing vapor phase chemical species in air. In electronic nose research, both computational and computational-experimental methods for sensor design and array element selection can be applied effectively. High-throughput screening and associated techniques are also used, as are knowledge-based systems and chemical reactivities. The process of selecting the components of an array considers both type and identity of sensing materials, the optimum number of sensors to be used in an array for a particular set of analytes, and how the responses of sensors will be treated in data analysis. Selection of elements for a chemical sensing array may involve purely computational approaches, semiempirical approaches, or strictly experimental approaches to determine sensor response. Following collection of data on sensor response statistical or other methods may be used to select the various elements for an array. Prior to recent advances in computing techniques, many researchers have used experimental methods and experimental-computational methods to select sensing materials.

The computational techniques discussed here have been applied to design and selection of materials for sensing arrays. Statistical methods have been used to analyze experimental data from sensing arrays[4] and to assign array response to specific stimulus or target analytes; these methods include principal component analysis (PCA) and cluster analysis methods (e.g., Hierarchical cluster analysis). PCA aims at representing large amounts of multidimensional data as a more intuitive, low-dimensional representation. Cluster analysis methods are aimed toward partitioning a data set into classes or categories consisting of elements of comparable similarity. Regression methods include simple and multiple linear regression methods (MLR), stepwise MLR using genetic function approximation (GFA) methods, and partial least square methods (PLS). In addition, there are more recent neural network-based approaches,[5] particularly multilayer neural networks with back propagation. Although these methods are not specifically designed to enable sensor material design and evaluation, they can be used in selecting which materials to include in an array. These statistical methods will point out which sensing materials contribute to distinguishing among different stimuli, and which do not.

In the following sections, we provide an overview of various computational modeling approaches that researchers have used to investigate chemical sensing problems with respect to sensing arrays, including design of new sensing materials, selection of materials, and finally modeling the sensing material response when incorporated in a device.

3　Materials Modeling and Evaluation of Sensing Materials

Many of the QM and atomistic modeling efforts on sensing materials have been done on organic sensing materials, primarily polymers. The following sections describe efforts by researchers to design new polymer-based sensors by understanding binding energies with target analytes.

Screening and selection of chemical sensing materials is a multistep process. New polymer materials can be designed using a QM atomistic screening approach. The process starts with identifying and evaluating sets of potential sensing material candidates, followed by narrowing the search to a set of promising candidates based on experimental and statistical data. The choice of the final set of materials is made after performing reliability studies on the most promising materials and making a performance comparison. Although it may be preferable to do exhaustive studies to determine the best set of materials for an array, when time and resources are limited a rapid initial screening to identify potential sensing material candidates is needed.

An approach based on first principles QM rapid screening of organic sensor materials has been reported by Ryan and coworkers.[6,7] This method was applied to development of sensors for detection of sulfur dioxide (SO_2) and elemental mercury ($Hg°$) in air at parts per-million (ppm) and parts per-billion (ppb) concentration levels using polymer-based sensors. The QM screening methodology involves calculating binding energies for organic sensing materials with SO_2 and $Hg°$. Organic sensing materials considered were common classes of organic groups, which could be a part of a polymer chain, either in the backbone or as side groups. These organic groups include alkanes, alkenes, aromatics, primary and secondary amines, aldehydes, and carboxylic acids. Interaction energies are calculated using quantum mechanics (QM) using the B3LYP and X3LYP flavors of density functional theory (DFT) with basis superposition error (BSSE) corrections.[8] These QM results were used to develop a first principles force field for use in the calculation of interaction energies of SO_2 molecules and $Hg°$ atoms with sensing materials during MD simulations.

The calculated binding energies for organic-SO_2 and organic-$Hg°$ systems indicate that a polymer candidate for both SO_2 and $Hg°$ detection would be one containing primary or secondary amines. Other chemical functionalities in the polymer that have strong binding with SO_2 are amides, aldehydes, and carboxylic acids. To validate the QM findings, polymers containing recommended chemical functionalities were synthesized and tested to detect SO_2 and $Hg°$. The experimental results validate the modeling predictions. Experimental results show that this approach is a good method for ranking the performance of various sensing materials.[6,7] The binding energy curves were also fitted by empirical functional form to get a force field to perform large scale polymer simulations.

A de novo structure-based design approach for the unbiased construction of complementary host architecture consisting of selective organic hosts for the development of many types of sensors has recently been reported.[9] Hay and coworkers developed a computational methodology specifically tailored to discover host architectures for small guest molecules. This approach is capable of generating and evaluating millions of candidate structures in minutes on a personal computer, allowing users to rapidly identify three-dimensional architectures for binding a targeted guest species. This approach has been shown to be an effective way to search for cation hosts containing aliphatic ether oxygen groups and anion hosts containing urea groups. The computational method applies fundamental information about structure and bonding as a basis to search for host architectures that are highly organized for guest complexation. Complex host architectures are tackled by considering two or three simple host components; for example, an 18-crown-6 macrocycle can be broken down into two triglyme components, three diglyme components, or six dimethylether components and only the component is considered in defining a guest structure.

In addition to computational methodologies discussed above for polymer-based sensing films, nanostructured metal-oxide films and nanowires, and carbon nanotubes-based sensors have received considerable interest and attention.[10,11] These sensor devices show changes in electrical properties of the active sensing element (metal oxide or CNT) when subjected to an externally applied perturbation such as exposure to chemical or biomolecules or stress. A fundamental study of such a sensor system by quantum chemical modeling based on DFT to compute bond-rearrangements, binding energy, charge transfer, and changes to the electronic structure has been made by Maiti and coworkers; these authors have also studied electronic transport in metal oxide nanowire systems using Non-Equilibrium Green's Function.[10,11]

4 Selection of a Sensor Set from Evaluated Materials: Modeling Sensor Response

Once a decision of the chemical functionality or host structure is made and a sensing film is included in a sensor device, the next goal would be to model the sensor response of the film in the device. Sensor response to an analyte is a complex function of the partitioning of the target analytes based on the interactions within the film as well as the transport properties of the analyte in the sensor. The sensor responses for polymer-based sensors have been modeled by various approaches using (1) first principles techniques such as Hansen solubilities, (2) multivariate techniques such as QSAR to correlate sensor response with molecular descriptors, and (3) simulations and empirical formulations used to calculate the partition coefficient, such as linear solvation energy relationships, to provide a measure of selectivity and sensitivity of the material under consideration.

Belmares et al.[12] have reported a first principle MD model of sensor-analyte response that measures resistance change in polymer-carbon composite films. The relative change in resistance is assumed to be directly proportional to the target analyte permeability. The resistive responses of the sensors are correlated with the Hansen components of the cohesive energy of the polymer and solvent as well as the molar volume of the solvent.

$$\Delta R / R = R_0 \exp(-\gamma V_s) \exp[\sum_{i=1}^{3} \beta_i (\delta_i^s - \delta_i^p)], \tag{1}$$

where $\Delta R/R$ is the normalized sensor response of resistive polymer–carbon composite film when exposed to a target analyte; δ_i^s ($i = 1,2,3$) are the cohesive energy density components of the target analyte, where $i = 1$, 2, and 3 refer to the electrostatic, dispersion, and hydrogen bond components respectively; δ_i^p is the i-th cohesive energy component of the polymer sensing film; γV_s is the activation energy of diffusion of the solute in the polymer, proportional to the molar volume of the target molecule, V_s; and the exponential factor γ is a best-fit parameter. A multisample MD method was used, providing a feasible tool for estimating Hildebrand and Hansen solubility parameters without the need for experimental data.

A QSAR study using genetic function approximations (GFA) has been reported by Shevade et al.[2] QSAR was used to describe the activities of a polymer–carbon composite chemical vapor sensor using a novel approach to selecting a molecular descriptor set. The measured sensor responses are conductivity changes in polymer–carbon composite films upon exposure to target vapors (analytes) at parts-per-million (ppm) concentrations. The descriptor set combines the default analyte descriptor set commonly used in QSAR studies with descriptors for sensing film-analyte interactions, which describes the sensor response. The default analyte descriptors are obtained using a combination of empirical and semiempirical QSPR methods. The descriptors for the sensing film-analyte interactions are calculated using molecular modeling and simulation tools. A statistically validated QSAR model was developed for training data for a sensing array of 16 sensors and 17 analyte molecules. The applicability of this model was tested for its effectiveness in predicting sensor activities for three test analytes not considered in the training set. The functional form of a sensor activity correlated to molecular descriptors for one such polymer–carbon composite sensor, polyethylene oxide-carbon is

$$\text{Sensor activity} = 0.15207\ \text{Epa} + 0.116727\ \text{HB}_D^2 + 0.000241\ \text{MR}^2 \qquad (2)$$

In the above QSAR equation (2), the analyte descriptors that appear along with a term describing the energy of polymer–analyte interaction term (E_{pa}) are hydrogen bond donor site (HB_D) and molar refractivity (MR). The E_{pa} term is calculated using atomistic simulations, and the HB_D and MR descriptors are calculated using empirical functional forms. This functional form may also be used to provide an insight into the mechanism of sensor response. On the basis of the PEO monomer chemical structure, it is known that there is one hydrogen bond acceptor site, so it would be expected that a descriptor that represents the hydrogen bond donor nature of the analyte should appear in the equation as seen. The analyte descriptor MR is a combined measure of molecule size and polarizability, and is calculated from the refractive index, molecular weight, and density of the analyte. As swelling in the polymer–carbon composite film is one mechanism for change in resistance of the sensor, or sensor response,[13] it is logical that molecular size of the analyte will appear in the equation describing sensor response.

Another thermodynamic property used to describe the selectivity or measure of polymer sensing film response is the partition coefficient, K, which is the equilibrium distribution coefficient of the target analyte concentrations between the gas phase and the sorbent phase (or sensing film). Grate and coworkers have modeled vapor sorption by sorbents such as polymers using linear solvation energy relationships (LSER).[14] In this method, $log\ K$ is modeled as a linear combination of terms related to particular types of interactions terms, dipolarity/polarizability ($s\pi_2^H$), polarizability ($rR2$), hydrogen-bonding ($\Sigma\alpha_2^H$), hydrogen-bonding ($b\ \Sigma\beta_2^H$), and combination of dispersion interaction and cavity term ($l\ log\ L^{16}$). The LSER equation for vapor sorption is given as

$$\log K = c + r\,R_2 + s\,\pi_2^{H} + a\,\Sigma\alpha_2^{H} + b\,\Sigma\beta_2^{H} + l\,\log L^{16}. \tag{3}$$

Each term other than the constant contains a solvation parameter related to the vapor's solubility properties as a solute (R_2, π_2^{H}, $\Sigma\alpha_2^{H}$, $\Sigma\beta_2^{H}$, or $\log L^{16}$) and a coefficient (r, s, a, b, or the letter l). All of these parameters except R_2 are free energy related. The parameter scales were derived from measurements of complexation or partitioning equilibria. Grate et al. have applied this method successfully to evaluate polymer sensors on acoustic sensor arrays to study the sorption of target analytes in polymer films.[3,14]

The partition coefficient can also be predicted using atomistic simulations using Grand Canonical Monte Carlo Simulations (GCMC) as investigated by Nakamoto et al.[15] Good agreement was found with their acoustic sensor experiments. When compared with the previous methods, no experimental inputs or empirical parameters are needed in GCMC approaches.

The ability to predict sensor responses or activities accurately using the above approaches will be of great help in characterizing sensing materials. Training an array in an electronic nose for a given set of analytes and a given set of environmental conditions (temperature, pressure, and humidity) is time consuming; in addition, developing training sets and calibration information may impinge on the useful lifetime of the sensors. Approaches described by Belmares et al.[12] and Shevade et al.[2] aim at developing a representative equation for each sensor material in an array; by populating the equation with analyte properties, the response of any material can be estimated. Similarly, computation of the partition coefficient of an analyte in a sensing material allows estimation of the response of that material to the analyte. In the future, the development of "n" equations to describe a sensing array will facilitate the generation of virtual training sets for any given sensor array for analytes that may not easily tested, such as highly toxic or explosive compounds. Subsequently, fewer experimental tests will need to be run on any given sensor array.

5 Conclusions

There are several computational approaches that have been used to design and evaluate materials for chemical sensors. With the increasing use of sensing arrays, such approaches offer complementary information to that developed through experimental approaches. Experimental approaches for sensor material design, evaluation, and selection include methods to screen chemical materials using high-throughput screening, designing novel experiments, and combining experimentation and data-analysis inputs. Experimental techniques such as high-throughput screening, which apply combinatorial strategies to screen large sets (tens and hundreds) of sensing materials, are very popular. Numerous functional parameters of sensing materials can be tailored to achieve their desirable capabilities. HT experiments use sensing arrays coated with large number of materials, which are tested at various process conditions. These procedures are very elaborate and intensive. The process starts with identifying and evaluating sets of potential sensing material candidates, followed by narrowing the search to a set of promising candidates based on the experimental and statistical data. The choice of the final set of materials is made after performing reliability studies on the most promising materials and making a performance comparison. Potyrailo has used combinatorial and high-throughput experimentation for the development of new sensor materials.[16]

Computational methods cannot generally produce the level of confidence in sensor performance that can be found with experimental data; however, initial screening with computational methods could be used to reduce the time and effort necessary to optimize the materials selected for a sensing array. Once a set of possible materials has been selected, optimizing the array with the best set must be done using computational-experimental methods.

There have been reports of electronic noses with as few as four and as many as 100 sensors in the array[17,18]; statistical methods to determine the optimum number of sensors in an array will greatly assist in sensor selection. Statistical methods based on experimental data have been used successfully to optimize an array; statistical methods may also be used with data simulated using purely computational approaches. Recent work in using neural networks for array selection in addition to analyte identification shows promise as an approach.[19]

Use of computational methods in designing, selecting, and ultimately in optimizing chemical sensing materials and sensor sets for arrays is a growing field, which will assist in development of sensing devices for various applications.

Acknowledgment This work was carried out at the Jet Propulsion Laboratory, California Institute of Technology, under a contract with the National Aeronautics and Space Administration.

References

1. Goddard, W. A.; Cagin, T.; Blanco, M.; Vaidehi, N.; Dasgupta, S.; Floriano, W.; Belmares, M.; Kua, J.; Zamanakos, G.; Kashihara, S.; Iotov, M.; Gao, G.H., Strategies for multiscale modeling and simulation of organic materials: polymers and biopolymers, *Computational and Theoretical Polymer Science* **2001**, *11*, 329–343.
2. Shevade, A.V.; Homer, M.L.; Taylor, C.J.; Zhou, H.; Manatt, K.; Jewell, A. D.; Kisor, A.; Yen, S.P.S.; Ryan, M.A., Correlating polymer-carbon composite sensor response with molecular descriptors, *Journal of Electrochemistry* **2006**, *153*, H209-H21.
3. Grate, J.W.; Patrash, S.J.; Abraham, M.H., Method for estimating polymer-coated acoustic-wave vapor sensor responses, *Analytical Chemistry* **1995**, *67*, 2162–2169.
4. Livingstone, D., *Data Analysis for Chemists*, Oxford University Press&rlenis;, New York, 1995.
5. Bartlett, P. N.; Gardner, J. W., *Electronic Noses: Principles and Applications*, Oxford University Press, Oxford, 1999; pp. 140 –183, and references therein.
6. Ryan, M. A.; Homer, M. L.; Zhou H.; Manatt, K.; Jewell, A. D.; Kisor, A.; Shevade, A.; Yen, S. P. S.; Blanco, M.; Goddard, W. A., Expanding the capabilities of the JPL electronic nose

for an international space station technology demonstration, *Proceedings of the 36th International Conference on Environmental Systems*, SAE ICES-2179, 2006.

7. Shevade, A. V; Homer, M. L.; Zhou, H.; Jewell, A. D.; Kisor, A. K.; Manatt, K. S.; Torres, J.; Soler, J.; Yen, S. P. S; Ryan, M. A., Development of the third generation JPL electronic nose for international space station technology demonstration, *Proceedings of the 37th International Conference on Environmental Systems*, SAE ICES-3149, 2007.

8. Lee, C.; Yang, W.; Parr, R. G., Development of the colle-salvetti correlation-energy formula into a functional of the electron density, *Physical Review B* **1988**, *37*, 785–789.

9. Hay, B. P.; Firman, T. K.; Bryantsev, V. S., *HostDesigner User's Manual*, PNNL 13850, Pacific Northwest National Laboratory, Richland, WA, 2006.

10. Maiti, A.; Rodriguez, J.A.; Law M; Kung, P.; McKinney, J.R.; Yang, P.D., SnO_2 nanoribbons as NO_2 sensors: Insights from first principles calculations, *Nano Letters* **2003**, *3*, 1025–1028.

11. Maiti, A., Electromechanical and chemical sensing at the nanoscale: Molecular modeling applications, *Molecular Simulations* **2004**, *30*, 191–198.

12. Belmares, M.; Blanco, M.; Goddard, W. A.; Ross, R. B.; Caldwell, G.; Chou, S.-H.; Pham, J.; Olofson, P. M.; Thomas, C. J., Hildebrand and Hansen solubility parameters from molecular dynamics with applications to electronic nose polymer sensors, *Journal of Computational Chemistry* **2004**, *25*, 1814–1826.

13. Severin, E. J.; Lewis, N. S., Relationships among resonant frequency changes on a coated quartz crystal microbalance, thickness changes, and resistance responses of polymer-carbon black composite chemiresistors, *Analytical Chemistry* **2000**, *72*, 2008–2015.

14. Grate, J. W.; Abraham, M. H.; McGill, R. A., Sorbent polymer coatings for chemical sensors and arrays, In *Handbook of Biosensors: Medicine, Food, and the Environment*, E. Kress-Rogers and S. Nicklin, eds., CRC Press, Boca Raton, FL, 1996; pp. 593–612.

15. Nakamura, K.; Nakamoto, T.; Moriizumi, T., Prediction of QCM gas sensor responses and calculation of electrostatic contribution to sensor responses using a computational chemistry method, *Materials Science and Engineering C* **2000**, *12*, 3–7.

16. Potyrailo, R. A., Polymeric sensor materials: Toward an alliance of combinatiorial and rational design tools? *Angewandte Chemie-International Edition* **2006**, *45*, 702–723.

17. Kramer K. E.; Rose-Pehrsson, S. L.; Hammond, M. H.; Tillett, D.; Streckert, H. H., Detection and classification of gaseous sulfur compounds by solid electrolyte cyclic voltammetry of cermet sensor array, *Analytica Chemica Acta* **2007**, *584*, 78–88.

18. Lewis, N. S., Comparisons between mammalian and artificial olfaction based on arrays of carbon black-polymer composite vapor detectors, *Accounts of Chemical Research* **2004**, *37*, 663–672.

19. Tchoupo, G. N.; Guiseppi-Elie, A., On pattern recognition dependency of desorption heat, activation energy, and temperature of polymer-based VOC sensors for the electronic NOSE, *Sensors and Actuators B-Chemical* **2005**, *110*, 81–88.

Section 9
Outlook

Chapter 21
Combinatorial Methods for Chemical and Biological Sensors: Outlook

Radislav A. Potyrailo and Vladimir M. Mirsky

Abstract This chapter provides a summary of status of combinatorial development of materials for chemical and biological sensors and an outlook for the future developments.

A theoretical dimensionality of the hyperspace of independent chemical sensor features has been estimated to be ~10^{21} (Fig. 21.1) and includes the permutations of varying sensing materials, transducer principles, and modes of operation for each sensor/transducer combination.[1–3] As shown in the present book, all types of sensing materials have been explored with combinatorial technologies, which demonstrates the desire of the sensing community for the accelerated development of sensing materials using newly introduced research tools.

Combinatorial and high-throughput technologies in materials science have been successfully accepted by the research groups in the academia and governmental laboratories that have overcome the entry barrier of dealing with new emerging aspects in materials research such as automation and robotics, computer programming, informatics and materials data mining. The main driving forces for combinatorial materials science in industry include broader and more detailed explored materials and process parameters space and faster time to market. Industrial research laboratories working on new catalysts and inorganic luminescent materials were among the first adopters of combinatorial methodologies in industry. The classical example of an effort by Mittasch who has spent 10 years (over 1900–1909) to conduct 6,500 screening experiments with 2,500 catalyst candidates to find a catalyst

R.A. Potyrailo (✉)
Chemical and Biological Sensing Laboratory, Chemistry Technologies and Material Characterization, General Electric Global Research, Niskayuna, New York, NY 12309, USA
potyrailo@crd.ge.com

V.M. Mirsky
Lausitz University of Applied Sciences, Department of Nanobiotechnology, Senftenberg, Germany
vmirsky@fhlausitz.de

R.A. Potyrailo and V.M. Mirsky (eds.), *Combinatorial Methods for Chemical and Biological Sensors*, DOI: 10.1007/978-0-387-73713-3_21,
© Springer Science + Business Media, LLC 2009

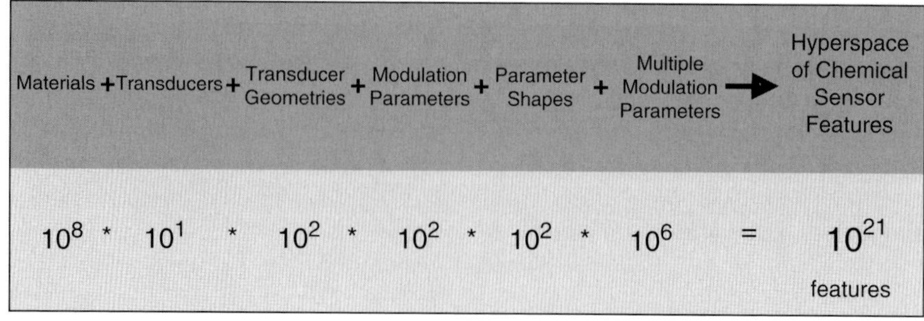

Fig. 21.1 Hyperspace of chemical sensor features with about 10^{21} independent features. Reprinted with permission from Göpel et al.[1] Copyright 1998 Elsevier

for industrial ammonia synthesis[4] will never happen again because of the availability and affordability of modern tools for high-throughput synthesis and characterization.

In the area of sensing materials, reported examples of significant screening efforts are less dramatic, yet also breathtaking. For example, a decade ago, Cammann, Shulga, and coworkers[5] reported an "extensive systematic study" of more that 500 compositions to optimize vapor sensing polymeric materials. Walt and coworkers[6] reported screening of over 100 polymer candidates in search for "their ability to serve as sensing matrices" for solvatochromic reagents. Seitz and coworkers[7] have investigated the influence of multicomponent compositions on the properties of pH-swellable polymers by designing $3 \times 3 \times 3 \times 2$ factorial experiments. Clearly, combinatorial technologies have been introduced at the right time to make the search for new materials more *intellectually rewarding*. Naturally, numerous academic groups that were involved in the development of new sensing materials turned to combinatorial methodologies to speed up knowledge discovery.[8–14]

From numerous results achieved using combinatorial and high-throughput methods, the most successful have been in the areas of molecular imprinting, polymeric compositions, catalytic metals for field-effect devices, and metal oxides for conductometric sensors. In those materials, the desired selectivity and sensitivity have been achieved by the exploration of multidimensional chemical composition and process parameters space at a previously unavailable level of detail at a fraction of time required for conventional one-at-a-time experiments. These new tools provided the opportunity for the more challenging, yet more rewarding explorations that previously were too time consuming to pursue.

The future advances in combinatorial development of sensing materials will be related to several key remaining unmet needs that prevent researchers of having a complete combinatorial workflow and to "analyze in a day what is made in a day."[15]

First, new fabrication methods of combinatorial libraries of sensing materials will be implemented ranging from those adapted from other materials synthesis and fabrication approaches[16,17] to those developed specifically for sensing applications.[18]

Second, while evaluation of performance properties of sensing material have been automated and numerous sensing systems have been developed to collect reliable response data from sensing materials, the remaining need is to develop screening tools for high-throughput characterization of intrinsic materials properties in order to keep up with the rates of performance screening of sensing materials candidates. For example, in the area of conductometric metal oxide sensors, a variety of employed techniques (e.g., Hall, catalytic conversion, and work function measurements, DRIFT spectroscopy, etc.) are at different stages on their high-throughput screening capabilities.

Third, certain portions of the data management aspects of the combinatorial workflow are still under development as summarized in Table 21.1.[19] However, over the last several years, there have been a growing number of reports on data mining in sensing materials.[20–23] "Searching for a needle in the haystack" has been popular in the early days of combinatorial materials science.[24–26] At present, it has been realized that screening of the whole materials and process parameters space is still too costly and time prohibitive even with the availability of existing tools. Instead, designing the high-throughput experiments to discover relevant descriptors will become more attractive.[27]

Fourth, predictive models of behavior of sensing materials under realistic conditions over long periods of time are needed. These modeling efforts will require inputs not only from screening of the performance and intrinsic properties of sensing materials but also screening of the effects of interfaces between sensing materials and transducers.

Fifth, the very different approaches developed in many laboratories for screening of sensing properties of materials and data analysis should be unified. This will require a significant work on the definition of common test protocols, data format, and data analysis. Such effort could result in a development of a comprehensive data bank describing sensing properties of known materials.

Chapters in this book critically analyzed the benefits of combinatorial technologies from the standpoint of practitioners of these tools. Perhaps, the best response to one's possible skeptical arguments that "this is not intellectually satisfying," "this is not science," and "this is too Edisonian" are two observations. The first observation is a quote from a book chapter by Göpel and Reinhardt[28] published in 1996 before the broad acceptance of combinatorial technologies into materials science. Göpel and Reinhardt mentioned "…it is surprising that no sensor group has so far screened systematically the many well-established metal oxide based catalysts for their potential use as sensor materials. On the other hand, it is surprising that only a few catalysis groups make use of the possibility of characterizing their catalysts by complementary monitoring their sensor properties." The second observation is that 10 years later, the multidisciplinary essence of combinatorial technologies has brought together sensor and catalysis groups[13,20,29,30] and many other diverse research groups and has impacted researchers as well. At present, an effective combinatorial scientist is acquiring skills as diverse as experimental planning, automated synthesis, basics of high-throughput materials characterization, chemometrics, and data mining. These new skills can be now obtained through the growing network of practitioners and through the new generation of scientists educated across the world in combinatorial methodologies. Combinatorial and high-throughput experimentation was able to bring together several previously disjoined disciplines and to combine valuable complementary attributes from each of them into a new scientific approach.

Table 21.1 Functions of data management system[19]

Function	Current capabilities	Remaining needs
Experimental planning	Composition parameters	Unification of functional tests of sensing materials
	Process parameters	Iterative planning based on results from virtual or experimental libraries
	Library design	
Data base	Entry/search of composition/process variables	Unification of formats for description of sensing properties
	Operation with heterogeneous data	Storage and manipulation (search) of large amounts of data
	Unification of data between different instruments	Development of common data bases for sensing materials
Instrument control	Operation of diverse instruments	Development of software for simple design and control of high-throughput measurements systems
		Inter-instrument calibration
		Full instrument diagnostics
		Plug'n'play multiple instrument configurations
Data analysis	Visualization of composition/process conditions and measured parameters of library elements	Unification of data analysis, development of a common approach
	Univariate/multivariate processing of steady-state and dynamic data	Advanced data compression
	Quantitative analysis	Processing of large amounts of data, cloud computing when required
	Outlier detection	
Data mining	Prediction of only certain properties of a limited number of new materials	Identification of appropriate descriptors
	Virtual libraries	Unification of data mining, development of a common approach
	Cluster analysis	
	Molecular modeling	
	QSAR	

References

1. Göpel, W., Chemical imaging: I. Concepts and visions for electronic and bioelectronic noses, *Sens. Actuators B* **1998**, *52*, 125–142.
2. Weimar, U.; Göpel, W., Chemical imaging: II. Trends in practical multiparameter sensor systems, *Sens. Actuators B* **1998**, *52*, 143–161.

3. Mitrovics, J.; Ulmer, H.; Weimar, U.; Göpel, W., Modular sensor systems for gas sensing and odor monitoring: The moses concept, *Acc. Chem. Res.* **1998**, *31*, 307–315.

4. Ertl, G., Elementary steps in heterogeneous catalysis, *Angew. Chem. Int. Ed.* **1990**, *29*, 1219–1227.

5. Buhlmann, K.; Schlatt, B.; Cammann, K.; Shulga, A., Plasticised polymeric electrolytes: New extremely versatile receptor materials for gas sensors (VOC monitoring) and electronic noses (odour identification:discrimination), *Sens. Actuators B* **1998**, *49*, 156–165.

6. Walt, D. R.; Dickinson, T.; White, J.; Kauer, J.; Johnson, S.; Engelhardt, H.; Sutter, J.; Jurs, P., Optical sensor arrays for odor recognition, *Biosens. Bioelectron.* **1998**, *13*, 697–699.

7. Conway, V. L.; Hassen, K. P.; Zhang, L.; Seitz, W. R.; Gross, T. S., The influence of composition on the properties of pH-swellable polymers for chemical sensors, *Sens. Actuators B* **1997**, *45*, 1–9.

8. Lundström, I.; Sundgren, H.; Winquist, F.; Eriksson, M.; Krantz-Rülcker, C.; Lloyd-Spetz, A., Twenty-five years of field effect gas sensor research in Linköping, *Sens. Actuators B* **2007**, *121*, 247–262.

9. Dickinson, T. A.; Walt, D. R.; White, J.; Kauer, J. S., Generating sensor diversity through combinatorial polymer synthesis, *Anal. Chem.* **1997**, *69*, 3413–3418.

10. Cho, E. J.; Tao, Z.; Tang, Y.; Tehan, E. C.; Bright, F. V.; Hicks, W. L., Jr.; Gardella, J. A., Jr.; Hard, R., Tools to rapidly produce and screen biodegradable polymcr and sol-gel-derived xerogel formulations, *Appl. Spectrosc.* **2002**, *56*, 1385–1389.

11. Apostolidis, A.; Klimant, I.; Andrzejewski, D.; Wolfbeis, O. S., A combinatorial approach for development of materials for optical sensing of gases, *J. Comb. Chem.* **2004**, *6*, 325–331.

12. Simon, U.; Sanders, D.; Jockel, J.; Heppel, C.; Brinz, T., Design strategies for multielectrode arrays applicable for high-throughput impedance spectroscopy on novel gas sensor materials, *J. Comb. Chem.* **2002**, *4*, 511–515.

13. Frantzen, A.; Scheidtmann, J.; Frenzer, G.; Maier, W. F.; Jockel, J.; Brinz, T.; Sanders, D.; Simon, U., High-throughput method for the impedance spectroscopic characterization of resistive gas sensors, *Angew. Chem. Int. Ed.* **2004**, *43*, 752–754.

14. Mirsky, V. M.; Kulikov, V.; Hao, Q.; Wolfbeis, O. S., Multiparameter high throughput characterization of combinatorial chemical microarrays of chemosensitive polymers, *Macromol. Rapid Commun.* **2004**, *25*, 253–258.

15. Cohan, P. E., Combinatorial materials science applied – mini case studies, lessons and strategies, In *2002 Combi – The 4th Annual International Symposium on Combinatorial Approaches for New Materials Discovery*, Knowledge Foundation, Arlington, VA, 2002.

16. de Gans, B.-J.; Wijnans, S.; Woutes, D.; Schubert, U. S., Sector spin coating for fast preparation of polymer libraries, *J. Comb. Chem.* **2005**, *7*, 952–957.

17. Egger, S.; Higuchi, S.; Nakayama, T., A method for combinatorial fabrication and characterization of organic/inorganic thin film devices in uhv, *J. Comb. Chem.* **2006**, *8*, 275–279.

18. Potyrailo, R. A.; Morris, W. G.; Leach, A. M.; Hassib, L.; Krishnan, K.; Surman, C.; Wroczynski, R.; Boyette, S.; Xiao, C.; Shrikhande, P.; Agree, A.; Cecconie, T., Theory and practice of ubiquitous quantitative chemical analysis using conventional computer optical disk drives, *Appl. Opt.* **2007**, *46*, 7007–7017.

19. Potyrailo, R. A.; Maier, W. F., Combinatorial materials and catalysts development: Where are we and how far can we go?, In *Combinatorial and High-Throughput Discovery and Optimization Of Catalysts and Materials*; R. A. Potyrailo and W. F. Maier, Eds.; CRC Press: Boca Raton, FL, 2006; 3–16.

20. Frenzer, G.; Frantzen, A.; Sanders, D.; Simon, U.; Maier, W. F., Wet chemical synthesis and screening of thick porous oxide films for resistive gas sensing applications, *Sensors* **2006**, *6*, 1568–1586.

21. Villoslada, F. N.; Takeuchi, T., Multivariate analysis and experimental design in the screening of combinatorial libraries of molecular imprinted polymers, *Bull. Chem. Soc. Jpn.* **2005**, *78*, 1354–1361.

22. Mijangos, I.; Navarro-Villoslada, F.; Guerreiro, A.; Piletska, E.; Chianella, I.; Karim, K.; Turner, A.; Piletsky, S., Influence of initiator and different polymerisation conditions on performance of molecularly imprinted polymers, *Biosens. Bioelectron.* **2006**, *22*, 381–387.
23. Potyrailo, R. A.; McCloskey, P. J.; Wroczynski, R. J.; Morris, W. G., High-throughput determination of quantitative structure-property relationships using resonant multisensor system: Solvent-resistance of bisphenol A polycarbonate copolymers, *Anal. Chem.* **2006**, *78*, 3090–3096.
24. Jandeleit, B.; Schaefer, D. J.; Powers, T. S.; Turner, H. W.; Weinberg, W. H., Combinatorial materials science and catalysis, *Angew. Chem. Int. Ed.* **1999**, *38*, 2494–2532.
25. Potyrailo, R. A.; Olson, D. R.; Chisholm, B. J.; Brennan, M. J.; Lemmon, J. P.; Cawse, J. N.; Flanagan, W. P.; Shaffer, R. E.; Leib, T. K. High throughput analysis of polymer materials and coatings, In: *Invited Symposium "Analytical Tools For High Throughput Chemical Analysis And Combinatorial Materials Science", Pittsburgh Conference on Analytical Chemistry and Applied Spectroscopy, March 4–9*, New Orleans, Louisiana, 2001.
26. Jansen, M., A concept for synthesis planning in solid-state chemistry, *Angew. Chem. Int. Ed.* **2002**, *41*, 3746–3766.
27. Potyrailo, R. A., High-throughput experimentation in early 21st century: Searching for materials descriptors, not for a needle in the haystack, *6th DPI Workshop on Combinatorial and High-Throughput Approaches in Polymer Science, September 10–11*, Darmstadt, Germany, 2007.
28. Göpel, W.; Reinhardt, G., Metal oxide sensors: New devices through tailoring interfaces on the atomic scale, In *Sensors Update, vol. 1*; H. Baltes; W. Göpel and J. Hesse, Eds.; VCH: Weinheim, 1996; 47–120.
29. Frantzen, A.; Sanders, D.; Scheidtmann, J.; Simon, U.; Maier, W. F., A flexible database for combinatorial and high-throughput materials science, *QSAR Comb. Sci.* **2005**, *24*, 22–28.
30. Sieg, S. C.; Suh, C.; Schmidt, T.; Stukowski, M.; Rajan, K.; Maier, W. F., Principal component analysis of catalytic functions in the composition space of heterogeneous catalysts, *QSAR Comb. Sci.* **2007**, *26*, 528–535.

Index

Printed in the United States of America